# Ordinary Differential Equations in the Complex Domain

EINAR HILLE

DOVER PUBLICATIONS, INC.
Mineola, New York

### Bibliographical Note

This Dover edition, first published in 1997, is an unabridged and unaltered republication of the work first published by John Wiley & Sons, Inc., New York, in 1976 in the *Wiley-Interscience Series in Pure and Applied Mathematics.*

### Library of Congress Cataloging-in-Publication Data

Hille, Einar, 1894–
  Ordinary differential equations in the complex domain / Einar Hille.
      p.    cm.
  "An unabridged and unaltered republication of the work first published by John Wiley & Sons, Inc., New York, in 1976 in the Wiley-Interscience series in pure and applied mathematics"—T.p. verso.
  Includes bibliographical references (p.     –     ) and index.
  ISBN 0-486-69620-0 (pbk.)
  1. Differential equations.   2. Functions of complex variables.   I. Title.
QA372.H56   1997
515'.35—dc21                                                      97-70
                                                                  CIP

Manufactured in the United States of America
Dover Publications, Inc., 31 East 2nd Street, Mineola, N.Y. 11501

*To the Memory of My Parents
and
the Welfare of My Sons*

# PREFACE

A friend recently wrote: "Mathematics is the creation of flesh and blood, not just novelty-curious and wise automatons. We ought to devote some part of our efforts to increasing understanding of the observable universe." This book is a contribution to these efforts. We praise famous men, men who created beautiful structures and directed the course of mathematics. A quotation from the *Edda* may be appropriate: "One thing I know that never dies, the call after a dead man."

The structures that the masters built are not just beautiful to the eye; they are also eminently useful. In these days when the need is felt for "applicable mathematics" and "utilitas mathematica," it is fitting to recall that few domains of mathematics are so widely applicable as the theory of ordinary differential equations. This range of ideas is dear to my heart: for close to 60 years much of my time has been given to the cultivation of differential equations.

The book deals with ordinary differential equations in the complex domain. It covers the usual ground, more or less. Here and there features are introduced that are less canonical. There is a general emphasis on growth questions: the dominants and minorants of Section 2.7 constitute a variation of the majorant theme. The Nevanlinna theory of value distribution plays an important role: it is applied to the Malmquist-Wittich-Yosida theorem (Sections 4.5 and 4.6) and to Boutroux's investigations (Sections 11.2 and 12.3). The Papperitz-Wirtinger account of Riemann's lectures on hypergeometric functions and their uniformization by elliptic modular functions has been rescued from oblivion (Section 10.5). Finally, the second half of Chapter 12 presents the Emden-Fowler and the Thomas-Fermi equations, quadratic systems, and Russell Smith's recent work on polynomial autonomous systems, all matters of some novelty.

The reader is expected to have some knowledge of complex variables, a subject in which our students are frequently weak: they comprehend little, and often their knowledge is too abstract and is of the wrong kind. Elementary manipulative skill is too often atrophied. Hence the second half of Chapter 1 of this book is devoted to complex analysis. Chapter 11 has an appendix on elliptic functions, and modular and theta functions are discussed at some length in Sections 7.3 and 10.5. These sections should help the reader.

Each chapter has a list of references to the literature, and there is a bibliography at the end of the book. The exercises at the ends of sections comprise some 675 items.

The book was written at the behest of Harry Hochstadt, who scoffed at my misgivings and attempts to escape; he has aided and abetted my efforts, and I owe him hearty thanks. May the book live up to his expectations. Thanks are also due to numerous friends who have helped with advice, bibliographical and biographical information, and constructive criticism. Specific mention should be made of L. V. Ahlfors, O. Borůvka, W. N. Everitt, C. Frymann, Ih-Ching Hsu, S. Kakutani, Z. Nehari, D. Rosenthal, I. Schoenberg, R. Smith, H. Wittich, C. C. Yang and K. Yosida. J. A. Donaldson and H. Hochstadt have kindly helped with the proofreading. Further, I am grateful to Addison-Wesley Publishing Co. and to the R. Society of Edinburgh for permission to use copyrighted material. I am also indebted to the Department of Mathematics of the University of California at San Diego for Xerox copying and to the National Science Foundation for support (Grant GP 41127). Finally, I owe much to my family, wife and sons, for encouragement, help, interest, and patience.

EINAR HILLE

*La Jolla, California*
*February 1976*

# CONTENTS

# 1
# INTRODUCTION

In this chapter we shall list, with or without proof, various facts which will be used in the following. They will fall under two general headings: (I) algebraic and geometric structures, and (II) analytic structures. Under I we shall remind the reader of abstract spaces, metrics, linear vector spaces, norms, fixed point theorems, functional inequalities, partial ordering, linear transformations, matrices, algebras, etc. Under II we discuss analytic functions: analyticity, Cauchy's integral, Taylor and Maclaurin series, entire and meromorphic functions, power series, growth, analytic continuation, and permanency of functional equations. This is quite an ambitious program, and the reader may find the density of ideas per page somewhat overwhelming. He is advised to skim over the pages in the first reading and to return to the relevant material as, if, and when needed.

## I. Algebraic and Geometric Structures

### 1.1. VECTOR SPACES

The term *abstract space* is often used as a synonym for *set* or *point set*, but the term usually indicates that the author intends to endow the set with an algebraic or geometric structure or both. If a Euclidean space $R^n$ serves as a prototype of a space, we obtain an abstract space by abstracting (= withdrawing) some of its properties while keeping others. Incidentally, *property* is an undefined term (we can obviously not use the definition ascribed to Jean Jacques Rousseau: La propriété c'est le vol!). We denote our space by $X$ and its elements by $x, y, z, \ldots$. We say that the space has an algebraic structure if one or more algebraic operations can be performed on the elements, or if a notion of order is meaningful at least for some elements.

1

The set is a *linear vector space* if the operations of *addition* and *scalar multiplication* can be performed. It is required that the set be an Abelian group under addition, that is, $x + y$ is defined as an element of $X$, addition is associative and commutative, there is a unique *neutral element* $0$ such that $x + 0 = x$ for all $x$, and every element $x$ has a unique *negative*, $-x$, with $x + (-x) = 0$.

To define scalar multiplication we need a field of scalars, which is almost always taken to be the real field $R$ or the complex field $C$. For any scalar $\alpha$ and element $x$ there is a unique element $\alpha x$; scalar multiplication is associative, it is distributive with respect to addition, and $1 \cdot x = x$, where 1 is the unit element of the scalars.

We speak of a *real* or a *complex vector space* according as the scalar field is $R$ or $C$. The elements of $X$ are now called *vectors*. A linear vector space which also contains the product of any two of its elements is called an *algebra*. The set of all polynomials in a variable $t$ is obviously an algebra, and so is the set of all functions $t \mapsto f(t)$ which are continuous at a point $t_0$.

Consider a set of $n$ vectors $x_i$ in $X$, and let the underlying scalar field be denoted by $F$. Then the vectors $x_j$ are *linearly independent* over $F$ if

$$\alpha_1 x_1 + \alpha_2 x_2 + \cdots + \alpha_n x_n = 0 \qquad (1.1.1)$$

implies that all the $\alpha$'s are zero. They are *linearly dependent* over $F$ if multipliers $\alpha_j$ can be found so that (1.1.1) holds with $|\alpha_1| + |\alpha_2| + \cdots + |\alpha_n| > 0$. Emphasis should be placed on "over $F$," for restricting $F$ to a subfield $F^0$ or extending it to a larger field $F^*$ affects the independence relations. Thus 1 and $2^{1/2}$ are linearly independent over $Q$, the field of rational numbers, but not over $A$, the field of algebraic numbers. The space $X$ is said to be of *dimension $n$* if it contains a set of $n$ linearly independent vectors while any $n + 1$ vectors are linearly dependent. It is of infinite dimension if $n$ linearly independent vectors can be found for any $n$.

The notion of *partial ordering* is another form of algebraic structure. We say that $X$ is *partially ordered* if for some pairs $x$, $y$ of $X$ there is an ordering relation $x \leq y$ (equivalently, $y \geq x$) which is *reflexive, proper,* and *transitive*, that is, (i) $x \leq x$ for all $x$, (ii) $x \leq y$ and $y \leq x$ imply $x = y$, (iii) $x \leq y$, $y \leq z$ imply $x \leq z$. If $X$ is linear as well as partially ordered, we should have

$$x \leq y \qquad \text{implies} \qquad x + a \leq y + a \quad \text{for all } a, \qquad (1.1.2)$$

$$x \leq y \qquad \text{implies} \qquad \alpha x \leq \alpha y \qquad \text{for } \alpha > 0. \qquad (1.1.3)$$

In this case $X$ has a *positive cone* $X^+$, defined as the set of all elements $x \in X$ such that $0 \leq x$. This positive cone is invariant under addition and multiplication by positive scalars. It contains **0**, the *neutral element*, usually referred to as the *zero element*. We now have $\mathbf{x} \leqslant \mathbf{y}$ if $\mathbf{y} - \mathbf{x} \in \mathbf{X}^+$.

The set of real valued continuous functions on the closed interval [0, 1], say $C[0, 1]$, is an algebra. We define its positive cone $\mathbf{X}^+$ as the set of functions $t \mapsto f(t)$ whose values on [0, 1] are nonnegative. We have $f \leqslant g$ if $g(t) - f(t)$ is nonnegative in [0, 1].

A more prosaic example may be helpful: the fowl in a hen-yard are partially ordered under the pecking order.

### EXERCISE 1.1

1.  Consider the space of all polynomials $P(t)$ in a real variable which take on real values. Show that **X** is an algebra.

2.  An order relation $P < Q$ is established in **X** by defining $P$ as positive if its values are positive for all large positive values of $t$. Show that this ordering is a *trichotomy* in the sense that for a given $P$ there are only three possibilities: (i) $P$ is positive, (ii) $-P$ is positive, or (iii) $P = 0$.

3.  An order relation is said to be *Archimedean* if $\mathbf{x} < \mathbf{y}$ implies the existence of an integer $n$ such that $\mathbf{y} < n\mathbf{x}$. (The natural ordering of the reals is Archimedean.) Show that the order defined in Problem 2 is non-Archimedean inasmuch as the elements fall into rank classes $R_k$, where $R_k$ consists of all polynomials of exact degree $k$, each $R_k$ is Archimedean, but if $f$ is a positive element of $R_j$ and $g$ a positive element of $R_k$ with $j < k$, then $f < g$ and $nf < g$ for all $n$. Verify.

4.  Prove that $1, t, t^2, \ldots, t^n$ are linearly independent over **R**. What is the dimension of the space formed by these elements?

### 1.2. METRIC SPACES

A *metric space* is one in which there is defined a notion of *distance* subject to the following conditions:

$D_1$.   For any pair of points $P$ and $Q$ of **X** a number $d(P, Q) \geqslant 0$ is defined, called the distance from $P$ to $Q$ such that $d(P, Q) = 0$ iff $P = Q$.
$D_2$.   $d(P, Q) = d(Q, P)$.
$D_3$.   For any $R$ we have $d(P, Q) \leqslant d(P, R) + d(R, Q)$.

These notions go back to the work in the 1890's of Hermann Minkowski (1864–1909) on what he called the "geometry of numbers." He was chiefly concerned with the extremal properties of linear and of quadratic forms, for which he found alternative definitions of distance, adjusted to the problem in hand. Minkowski did not always require $D_2$. Condition $D_3$ is the *triangle inequality*.

We say that a linear vector space is *normed* if the following conditions hold:

$N_1$.   For each $x \in X$ there is assigned a number $\|x\| \geqslant 0$ such that $\|x\| = 0$ iff $x = 0$.

$N_2$.   $\|\alpha x\| = |\alpha| \|x\|$ for each $\alpha$ in the scalar field.

$N_3$.   $\|x + y\| \leqslant \|x\| + \|y\|$.

A normed linear vector space becomes a metric space by setting

$$d(x, y) = \|x - y\|. \tag{1.2.1}$$

In a metric space we can do analysis since the fundamental operation of analysis, that of finding limits of a sequence, becomes meaningful. If $\{x_n\}$ is a sequence in the metric space $X$, we say that $x_n$ converges to $x_0$ and

$$x_0 = \lim x_n \qquad \text{if} \qquad \lim_{n \to \infty} d(x_0, x_n) = 0. \tag{1.2.2}$$

We say that $\{x_n\}$ is a *Cauchy sequence* if, given any $\epsilon > 0$, there exists an $N$ such that

$$d(x_m, x_n) < \epsilon \qquad \text{for } m, n > N. \tag{1.2.3}$$

If (1.2.2) holds, it follows that $\{x_n\}$ is a Cauchy sequence, but the converse is not necessarily true, for there may be gaps in the space. A metric space $X$ is said to be *complete* if all Cauchy sequences converge to elements of the space. Euclidian spaces are complete, and so are various function spaces that will be encountered in the following. The space $Q$ of rational numbers is not complete.

Various notions of real analysis are meaningful in complete metric spaces, such as the concepts of *closure, open set, closed set,* and $\epsilon$-neighborhood. The Bolzano-Weierstrass theorem need not be valid in a complete metric space, that is, there may be bounded infinite point sets without a limit point. Incidentally, "bounded" means that the set can be enclosed in a "sphere" $d(x, 0) < R$. The *topological diameter* $d(S)$ of a subset of $X$ is the least upper bound of the distances $d(x, y)$ for $x$ and $y$ in $S$.

## EXERCISE 1.2

1.   The Euclidean norm $\|x\|_2$ of $x$ in $C^n$ is $[\Sigma_{j=1}^n |x_j|^2]^{1/2}$, where $x = (x_1, x_2, \ldots, x_n)$ and the $x$'s are complex numbers. Alternative norms are $\|x\|_1 = \Sigma_{j=1}^n |x_j|$ and $\|x\|_\infty = \sup |x_j|$. Show that they are indeed acceptable norms. Between what limits do they lie if $\|x\|_2 = 1$?

2.   How do you define an open set in these three normed topologies? Show that a set open in one of them is also open with respect to the others. Verify that $C^n$ is complete in all three metrics.

3. Let $X = C[0, 1]$ be the set of all functions, $t \mapsto f(t)$, continuous in the closed interval $[0, 1]$. Define a Cauchy sequence if $\|f\| = \sup_{0 \leqslant t \leqslant 1} |f(t)|$, and show that the space is complete.

## 1.3. MAPPINGS

We shall study mappings from a metric space $X$ into a metric space $Y$, both being complete. We shall often have $Y = X$.

The mapping $T$ is a pairing of points $x$ of $X$ with points $y$ of $Y$, say $\langle x, y \rangle$. Here to every $x$ of $X$ is ordered a unique $y$ of $Y$. To $x_1 \neq x_2$ correspond the two values $y_1$ and $y_2$, which may or may not be distinct, in fact every $x \in X$ may be mapped on the same point $y_0 \in Y$. The mapping is *onto* (a *surjection* in the Bourbaki language) if every point of $Y$ is the image $y$ of at least one $x$ in $X$. It is $(1, 1)$ (read "one to one") if

$$x_1 \neq x_2 \quad \text{implies} \quad y_1 = T(x_1) \neq T(x_2) = y_2. \qquad (1.3.1)$$

The mapping is *bounded* if there exists a finite $M$ such that

$$d[T(x_1), T(x_2)] \leqslant M \, d(x_1, x_2). \qquad (1.3.2)$$

This is a generalized *Lipschitz condition* and implies continuity of $T(x)$ with respect to $x$.

If $X$ and $Y$ are linear vector spaces over the same scalar field, and if

$$T(\alpha_1 x_1 + \alpha_2 x_2) = \alpha_1 T(x_1) + \alpha_2 T(x_2), \qquad (1.3.3)$$

then $T$ is called a *linear transformation*. It is bounded iff

$$d[T(x), 0] \leqslant M \, d(x, 0). \qquad (1.3.4)$$

The most important case is that in which $X$ and $Y$ are complete normed linear vector spaces, in which case the spaces are called *Banach spaces* after the Polish mathematician Stefan Banach (1892–1945), who termed them B-spaces. In this case (1.3.4) takes the form

$$\|T[x]\| \leqslant M \|x\|. \qquad (1.3.5)$$

If $T$ is a linear transformation, then $T(0) = 0$ and the transformation is $(1, 1)$ if

$$T(x) = 0 \quad \text{implies} \quad x = 0.$$

If $X$ and $Y$ are B-spaces, the set $E(X, Y)$ of linear bounded transformations on $X$ to $Y$ is also a B-space under the norm

$$\|T\| = \sup \|T[x]\|, \qquad (1.3.6)$$

the supremum being taken with respect to all elements $x$ of $X$ of norm 1. The algebraic operations in $E(X, Y)$ are defined in the obvious manner by

$$[T_1 + T_2][x] = T_1[x] + T_2[x], \qquad (\alpha T)[x] = \alpha T[x]. \tag{1.3.7}$$

If $Y = X$, we write $E[X]$ for $E[X, X]$ and note that products are definable in the obvious manner by

$$(T_1 T_2)[x] = T_1(T_2[x]). \tag{1.3.8}$$

This gives

$$\|T_1 T_2\| \leq \|T_1\| \|T_2\|. \tag{1.3.9}$$

It may be shown that $E(X, Y)$ and $E(X)$ are complete in the normad metric, so they are B-spaces. Also, $E(X)$, which is a normed algebra, actually is a B-algebra since it is a B-space and satisfies (1.3.9).

If $T \in E(X)$ and is $(1, 1)$, there is an *inverse transformation* $T^{-1}$ such that

$$T^{-1}[T[x]] = x, \quad \forall x, \qquad T[T^{-1}[y]] = y \quad \text{if } y = T[x]. \tag{1.3.10}$$

### EXERCISE 1.3

1. Show that $E(C^n)$ is complete. Use any of the metrics for $C^n$ listed in Problem 1.2: 1.
2. If $T$ is a linear transformation, verify that $T(0) = 0$. Here on the right stands the zero element of $Y$, while on the left we operate on the zero element of $X$. [*Hint*: $T(0 + 0) = T(0)$.]
3. If $T$ is $(1, 1)$, why does $T(x) = 0$ imply $x = 0$ and vice versa?
4. Why is (1.3.6) a norm? Show that it is the least value that $M$ can have in (1.3.5).
5. Prove (1.3.9).

### 1.4.  LINEAR TRANSFORMATIONS ON $C^n$
### INTO ITSELF; MATRICES

The simplest of all linear transformations are those which map $C^n$ into itself. If $T$ is such a transformation, then $T$ is uniquely determined by linearity and its effect on the *basis* of $C^n$. Any set of $n$ linearly independent vectors would serve as a basis, but we may just as well use the unit vectors

$$e_j = (\delta_{jk}), \tag{1.4.1}$$

where $\delta_{jk}$ is the Kronecker delta, that is, the vector whose $j$th component is one, all others being zero. This gives

$$x = x_1 e_1 + x_2 e_2 + \cdots + x_n e_n, \tag{1.4.2}$$

if $x = (x_1, x_2, \ldots, x_n)$ in the coordinate system defined by the vectors $e_j$. Now

$T$ takes vectors into vectors, so there are $n^2$ complex numbers $a_{jk}$ such that

$$T[\mathbf{e}_k] = a_{1k}\mathbf{e}_1 + a_{2k}\mathbf{e}_2 + \cdots + a_{nk}\mathbf{e}_n, \quad k = 1, 2, \ldots, n. \qquad (1.4.3)$$

The linearity of $T$ then gives

$$T[\mathbf{x}] = T\left(\sum_{k=1}^{n} x_k \mathbf{e}_k\right) = \sum_{k=1}^{n} x_k T[\mathbf{e}_k]$$

or

$$T[\mathbf{x}] = \sum_{j=1}^{n} \left(\sum_{k=1}^{n} a_{jk}x_k\right)\mathbf{e}_j \equiv \mathbf{y}, \qquad (1.4.4)$$

from which we can read off the components of the vector $\mathbf{y}$.

The quadratic array

$$\mathscr{A} = \begin{bmatrix} a_{11} & a_{12} & \cdots & a_{1n} \\ a_{21} & a_{22} & \cdots & a_{2n} \\ \cdots\cdots\cdots\cdots\cdots\cdots \\ a_{n1} & a_{n2} & \cdots & a_{nn} \end{bmatrix} \qquad (1.4.5)$$

is known as a *matrix*—more precisely, the matrix of the transformation $T$ with respect to the chosen basis. We can now write $T$ symbolically as

$$\mathbf{y} = T[\mathbf{x}] = \mathscr{A} \cdot \mathbf{x}, \qquad (1.4.6)$$

where the last member may be considered as the product of the matrix $\mathscr{A}$ with the column vector $\mathbf{x}$, the result being the column vector $\mathbf{y}$.

We have to decide when the mapping defined by $T$ is $(1, 1)$. Here the condition $T[\mathbf{x}] = \mathbf{0}$ implies that $\mathbf{x} = \mathbf{0}$ now takes the form that the homogeneous system

$$\sum a_{jk}x_k = 0, \qquad j = 1, 2, \ldots, n, \qquad (1.4.7)$$

must have the unique solution

$$x_1 = x_2 = \cdots = x_n = 0.$$

This will happen as long as

$$\det(\mathscr{A}) \neq 0. \qquad (1.4.8)$$

In this case the mapping is also onto, since for a given vector $\mathbf{y}$ we can solve the system

$$\sum_{k=1}^{n} a_{jk}x_k = y_j, \qquad j = 1, 2, \ldots, n, \qquad (1.4.9)$$

uniquely for $\mathbf{x} = (x_1, x_2, \ldots, x_n)$. It follows that $T$ has a unique inverse, also an element of $E[C^n]$, i.e., a linear bounded transformation of $C^n$ into itself. With this transformation goes a matrix $\mathscr{A}^{-1}$, which we refer to as the *inverse* of $\mathscr{A}$. The fact that its elements may be computed from (1.4.9) shows that the element in the place $(j, k)$ is $A_{kj}/\Delta$, where $A_{jk}$ is the cofactor of $a_{jk}$ in the determinant $\Delta = \det(\mathscr{A})$.

We can define algebraic operations and a norm in the set $\mathfrak{M}_n$ of *n*-by-*n* matrices in terms of which the set becomes a Banach algebra. This follows from the fact that there is a (1, 1) correspondence between the linear transformations $T$ in $E(C^n)$ and their matrices. Then to $T_1 + T_2$, $\alpha T$, and $T_1 T_2$ correspond

$$(a_{jk} + b_{jk}) \equiv \mathscr{A} + \mathscr{B}, \tag{1.4.10}$$

$$(\alpha a_{jk}) \equiv \alpha \mathscr{A}, \tag{1.4.11}$$

$$\left( \sum_{m=1}^{n} a_{jm} b_{mk} \right) \equiv \mathscr{A}\mathscr{B}. \tag{1.4.12}$$

A number of different but equivalent norms may be defined for $\mathfrak{M}_n$. A suitable one for analysis is

$$\|\mathscr{A}\| = \max_j \sum_{k=1}^{n} |a_{jk}|. \tag{1.4.13}$$

We have then

$$\|\mathscr{A}\mathscr{B}\| \leq \|\mathscr{A}\| \, \|\mathscr{B}\|.$$

Since $\mathfrak{M}_n$ is complete in the normed metric (why?), it follows that $\mathfrak{M}_n$ is a B-algebra.

We have seen that $\mathscr{A}$ has an inverse iff $\det(\mathscr{A}) \neq 0$. If this is the case, $\mathscr{A}$ is said to be *regular*; otherwise, *singular*. Together with the given matrix $\mathscr{A}$ we consider the family of matrices

$$\lambda \mathscr{E} - \mathscr{A},$$

where $\lambda$ runs through the complex field $C$ and $\mathscr{E} = (\delta_{jk})$ is the *n*-by-*n* unit matrix. These matrices are normally regular, but there exist $n$ values of $\lambda$ for which $\lambda \mathscr{E} - \mathscr{A}$ is singular: the $n$ roots of the *characteristic equation* of $\mathscr{A}$,

$$\det(\lambda \mathscr{E} - \mathscr{A}) = 0. \tag{1.4.14}$$

The roots $\lambda_1, \lambda_2, \ldots, \lambda_n$ form the *spectrum* $\sigma(\mathscr{A})$ of $\mathscr{A}$. They are known as *characteristic values*, *latent roots*, or *eigenvalues*. For these values of $\lambda$ one can find vectors $\mathbf{x}_k$ in $C^n$ of norm 1 such that

$$\mathscr{A} \cdot \mathbf{x}_k = \lambda_k \mathbf{x}_k. \tag{1.4.15}$$

The *characteristic vectors* $\mathbf{x}_k$ are linearly independent and may be chosen so that they form an *orthogonal system*; in this case the *inner product*

$$(\mathbf{x}, \mathbf{y}) = \sum_{j=1}^{n} x_j \bar{y}_j = 0 \tag{1.4.16}$$

for $\mathbf{x} = \mathbf{x}_k$, $\mathbf{y} = \mathbf{x}_m$, $k \neq m$. This holds even if (1.4.14) has multiple roots. A matrix $\mathcal{A}$ is singular iff zero belongs to the spectrum.

### EXERCISE 1.4

1. Find the elements of $\mathcal{A}^{-1}$ when $\mathcal{A}$ is regular.
2. Verify the inequality for the norm of the matrix product.
3. Prove the Hamilton-Cayley theorem, which asserts that the matrix $\mathcal{A}$ satisfies its own characteristic equation.

## 1.5. FIXED POINT THEOREMS

The Dutch mathematician L. E. J. Brouwer proved in 1912 that a continuous map of the unit ball in $\mathbf{R}^n$ into itself must necessarily leave at least one point invariant. Such a point is known as a *fixed point*, and an assertion about the existence of fixed points is known as a *fixed point theorem*. We shall prove some theorems of this nature. We start with a theorem proved by S. Banach in his Krakow dissertation of 1922. It refers to mappings of a complete metric space by a *contraction*, i.e., a bounded transformation of the space $X$ into itself such that

$$d[T(x), T(y)] \leqslant k d(x, y), \tag{1.5.1}$$

where $k$ is a fixed constant, $0 < k < 1$. Such a mapping evidently tries to shrink the object. Banach's theorem states that there is a point which does not move.

### THEOREM 1.5.1

*If* T *is a contraction defined on a complete metric space* X, *then there is one and only one fixed point.*

*Proof.* The triangle inequality plays a basic role here. We start with an arbitrary point $x_1 \in X$ and form its successive transforms under $T$:

$$x_{n+1} = T(x_n), \qquad n = 1, 2, \ldots. \tag{1.5.2}$$

These elements form a Cauchy sequence, and $X$ being a complete metric space, $x_0 = \lim x_n$ exists and is to be proved to be a fixed point—in fact, the only such point. Now it is sufficient to prove that, given any $\epsilon > 0$, there is an $N$ such that

$$d(x_n, x_{n+p}) < \epsilon, \qquad n > N, \quad p = 1, 2, \ldots.$$

To this end we note that by the triangle inequality the left member does not exceed

$$d(x_n, x_{n+1}) + d(x_{n+1}, x_{n+2}) + \cdots + d(x_{n+p-1}, x_{n+p}).$$

Now, using the contraction hypothesis, we see that

$$d(x_m, x_{m+1}) \le kd(x_{m-1}, x_m) \le \cdots \le k^{m-1}d(x_1, x_2)$$

and thus

$$d(x_n, x_{n+p}) \le (k^{n-1} + k^n + \cdots + k^{n+p-1})d(x_1, x_2)$$
$$< \frac{k^{n-1}}{1-k} d(x_1, x_2).$$

This expression can be made as small as we please by choosing $n$ large enough. Hence $\{x_n\}$ is a Cauchy sequence regardless of the choice of $x_1$, and the limit $x_0$ exists. Since

$$x_{n+1} = T(x_n),$$

we conclude that

$$x_0 = \lim x_{n+1} = \lim T(x_n) = T[\lim x_n] = T(x_0),$$

where we have used the continuity of $T$. It is seen that $x_0$ is indeed a fixed point.

Suppose that $y_0$ is a fixed point. Then

$$d(x_0, y_0) = d[T(x_0), T(y_0)] \le kd(x_0, y_0).$$

Since $k < 1$, this implies that $d(x_0, y_0) = 0$ or $y_0 = x_0$, so there is one and only one fixed point. ■

The restriction to contraction operators is a drawback, but it can occasionally be avoided by observing the following.

## COROLLARY

*There is a unique fixed point if some power of* T *is a contraction.*

*Proof.* If $T^m$ is a contraction, then there exists a fixed point $x_0$ such that $T^m(x_0) = x_0$. We have also

$$\lim (T^m)^n (x_1) = x_0$$

for any choice of $x_1$. Here we set $x_1 = T(x_0)$ and find that

$$(T^m)^n [T(x_0)] = T[(T^m)^n (x_0)] = T(x_0).$$

When $n$ becomes infinite, the first member tends to $x_0$, so we have $T(x_0) = x_0$ or $T$ admits $x_0$ as a fixed point. Since any fixed point of $T$ is a fixed point of $T^m$ and the latter has a unique fixed point, it follows that $x_0$ is the unique fixed point of $T$. ∎

Vito Volterra (1860–1940), in his discussion of integral equations with variable upper limits of integration, proved the uniqueness of the solutions. From his work in the 1890's we can distill a fixed point theorem which is rather useful.

### THEOREM 1.5.2

Let **X** be a B-space. Let $z_0$ be a given element of **X**, and let S belong to E(**X**) and be such that

$$\sum_0^\infty \|S^n\| < \infty. \tag{1.5.3}$$

Then the transformation

$$T(x) = z_0 + S[x] \tag{1.5.4}$$

has a unique fixed point $x_0$ given by

$$x_0 = z_0 + \sum_{n=1}^\infty S^n[z_0]. \tag{1.5.5}$$

It is enough to observe that the series converges in norm by (1.5.3) and $S$ can be applied termwise to the series and shows that $T(x_0) = x_0$. The uniqueness proof is left to the reader.

Consider, in particular, Volterra's equation

$$f(t) = g(t) + \int_0^t K(s, t)f(s)\, ds. \tag{1.5.6}$$

Here the kernel $K(s, t)$ and $g(t)$ are known and $f(t)$ is to be found. We consider the particular case in which the kernel is a function of $s$ alone.

### THEOREM 1.5.3

Suppose that $g(t) \in C[0, a]$ and $K(s) \in L(0, a)$. Then the equation

$$f(t) = g(t) + \int_0^t K(s)f(s)\, ds \tag{1.5.7}$$

has a unique solution in $C[0, a]$, namely,

$$f(t) = g(t) + \int_0^t K(s)\, exp\left[\int_s^t K(u)\, du\right] g(s)\, ds. \tag{1.5.8}$$

The verification is left to the reader.

**COROLLARY**

*For* $K(t) \equiv K$ *the equation*

$$f(t) = g(t) + K \int_0^t f(s) \, ds \qquad (1.5.9)$$

*has the unique solution*

$$f(t) = g(t) + K \int_0^t exp \, [K(t-s)]g(s) \, ds. \qquad (1.5.10)$$

**EXERCISE 1.5**

1. Verify Theorem 1.5.2.
2. Verify Theorem 1.5.3.
3. Verify the corollary.
4. If $T[f](t) = K \int_0^t f(s) \, ds$, find $T^n$. Is $T$ a contraction? What are the conditions? For a given interval $[0, a]$ find $m$ so that $T^m$ is a contraction.
5. Let $C^+[0, a]$ denote the set of nonnegative functions continuous in $[0, a]$. The mapping

$$T[f](t) = \int_0^t (t - s)f(s) \, ds$$

   is clearly a mapping of **X** into itself. Is it a contraction? Find $T^m$, and study its contractive properties. Find a fixed point by inspection.

## 1.6. FUNCTIONAL INEQUALITIES

This is a field of increasing importance. We shall consider inequalities of the form

$$f(t) \leq T[f](t). \qquad (1.6.1)$$

We consider a complete metric space **X**, the elements of which are mappings from some interval $[a, b]$ in $R^1$. Here $T$ is a mapping of **X** into itself, and the problem is to discuss the inequality. Can it be satisfied by elements $f$ of **X**? Is it trivial in the sense that it is satisfied by all $f$'s in **X**? If neither of the above is true, characterize the elements of **X** for which (1.6.1) holds. Is it so restrictive that it holds for one and only one $f$? It can be seen that there are a number of pertinent questions.

The discussion in Section 1.5 leads to several functional inequalities which are *categorical* or *determinative* in the sense that there is a single element of the space under consideration which satisfies the inequality. If in (1.5.7) and (1.5.9) we assume $g$ to be identically zero, then $f$ is identically zero. This suggests

## THEOREM 1.6.1

*Let* **X** *be the positive cone of* C[0, a], $0 < a < \infty$. *Let* K(t) $\in$ L(0, a), *continuous and nonnegative on the half-open interval* (0, a]. *If* f $\in$ **X** *and if for* $0 \leqslant t \leqslant a$

$$f(t) \leqslant \int_0^t K(s)f(s) \, ds, \tag{1.6.2}$$

*then* f *is identically zero.*

*Proof.* This functional inequality is very important for the theory of differential equations (DE's) since it underlies uniqueness proofs based on a Lipschitz or, more generally, a Carathéodory condition. We shall obtain the theorem as a consequence of more general theorems, but in view of its importance it is desirable to give a direct short proof. Set

$$F(t) = \int_0^t K(s)f(s) \, ds. \tag{1.6.3}$$

This is an element of **X** and $F(0) = 0$. Furthermore, for $0 < t$

$$F'(t) = K(t)f(t) \leqslant K(t)F(t),$$

so that (1.6.2) implies that

$$F'(t) - K(t)F(t) \leqslant 0. \tag{1.6.4}$$

This we multiply by the positive function $\exp[-\int_0^t K(u) \, du]$, and the result is an exact derivative so that

$$\frac{d}{dt} \left\{ F(t) \exp\left[ -\int_0^t K(u) \, du \right] \right\} \leqslant 0.$$

Since $F(0) = 0$, this shows that $F(t) \leqslant 0$ for $0 < t \leqslant a$. But we already know that $F(t) \geqslant 0$. To satisfy both inequalities we must have $F(t) \equiv 0$. This implies that $f(t) \equiv 0$, as asserted. ∎

A uniqueness theorem due to Mitio Nagumo (1926) goes back to the following functional inequality:

## THEOREM 1.6.2

*Let* **X** *be the subspace of* $C^+[0, a]$, *the elements of which satisfy* f(0) = 0, $lim_{h \downarrow 0} f(h)/h = 0$. *Then, if* f $\in$ **X**, *and if*

$$f(t) \leqslant \int_0^t f(s) \frac{ds}{s},$$

then

$$f(t) \equiv 0. \tag{1.6.5}$$

A proof may be given using (1.6.3) with $K(s)$ replaced by $s^{-1}$, which is not integrable down to the origin. The proof is left to the reader.

There are several other uniqueness theorems which also go back to functional inequalities. We shall not pursue these cases any further but instead proceed to the use of fixed point theorems in the discussion of functional inequalities. We have

**THEOREM 1.6.3**

*Let* **X** *be a complete metric space which is partially ordered in such a manner that if* $\{x_n\}$ *is an increasing sequence in* **X**, *so that* $x_n \leq x_{n+1}$ *for all* n, *and if* $lim_{n \to \infty} x_n \equiv x_0$ *exists in the sense of the metric, then* $x_n \leq x_0$ *for all* n. *Let* T *be an order-preserving mapping of* **X** *into* **X** *such that* $T^m$ *is a contraction for some* m. *Let* $f_0$ *be the unique fixed point of* T. *Then*

$$f \leq T[f] \quad implies \quad f \leq f_0. \quad (1.6.6)$$

*Proof.* We say that $T$ is order-preserving if for $f_1, f_2 \in$ **X**

$$f_1 \leq f_2 \quad implies \quad T[f_1] \leq T[f_2]. \quad (1.6.7)$$

Suppose that $f \in$ **X**$_0$, the subset of **X** for which the inequality is meaningful. Since **X**$_0$ contains $f_0$ at least, it is not void. Then

$$f \leq T[f] \leq T^2[f] \leq \cdots \leq T^n[f] \leq \cdots.$$

Now $T^n[f]$ tends to the limit $f_0$ as $n \to \infty$ for

$$\lim_{k \to \infty} T^{km}[T^j(f)] = f_0, \quad j = 0, 1, \ldots, m-1,$$

since $T^m$ is a contraction and the limit is the same for all elements of **X**. It follows that the increasing sequence $\{T^n[f]\}$ converges to $f_0$. Since order is preserved under the limit operation, we have $f \leq f_0$ and the theorem is proved. ■

If $T[f]$ is defined by (1.6.3), $T$ is order-preserving since the kernel $K(s)$ is nonnegative and the space **X** is linear. Here $T$ usually does not define a contraction, but all powers $T^m$ with a sufficiently large $m$ are contractions for

$$T^n[f](t) = \frac{1}{(n-1)!} \int_0^t K(s) \left[ \int_s^t K(u) \, du \right]^{n-1} f(s) \, ds, \quad (1.6.8)$$

the norm of which goes rapidly to zero as $n$ becomes infinite. This provides another proof for Theorem 1.6.1.

We can also apply the Volterra fixed point theorem to functional inequalities.

**THEOREM 1.6.4**

Let $\mathbf{X}$ be a partially ordered B-space such that the positive cone $\mathbf{X}^+$ is a closed point set. Let S be a linear bounded positive transformation on $\mathbf{X}$ to $\mathbf{X}$ and such that

$$\sum_{n=1}^{\infty} \|S^n\| < \infty. \tag{1.6.9}$$

Let $\mathbf{g}$ be a given element of $\mathbf{X}^+$, and $\mathbf{f}_0$ the unique fixed point of

$$T[f] = g + S[f]. \tag{1.6.10}$$

Then

$$f \leqslant g + S[f] \qquad implies \qquad f \leqslant f_0. \tag{1.6.11}$$

*Proof.* That $S$ is positive means that it maps $\mathbf{X}^+$ into itself. All the powers of $S$ are then also positive and $S$ is order-preserving. The existence of a unique fixed point follows from Theorem 1.5.2. We have $\lim_{n \to \infty} T^n[f] = f_0$ for any $f$—in particular, for an $f$ satisfying the first inequality under (1.6.11). Now we have

$$T^n[f] \leqslant g + S[g] + S^2[g] + \cdots + S^{n-1}[g] + S^n[f]. \tag{1.6.12}$$

This is an increasing sequence which goes to the limit $f_0$ as $n$ becomes infinite. This, combined with $f \leqslant T[f]$ and the order-preserving properties of limits, leads to the desired result. ∎

We state a couple of applications of this theorem which are of special importance to the theory of DE's.

**THEOREM 1.6.5**

Let $K \in C^+(a, b) \cap L(a, b)$, and let $\mathbf{g}$ and $\mathbf{f}$ belong to $C^+[a, b]$. Suppose that for all $t$ in $[a, b]$

$$f(t) \leqslant g(t) + \int_a^t K(s)f(s)\,ds. \tag{1.6.13}$$

Then

$$f(t) \leqslant g(t) + \int_a^t K(s)\,exp\left[\int_s^t K(u)\,du\right]g(s)\,ds. \tag{1.6.14}$$

The proof is left to the reader. We see that if $g(t) \equiv 0$; then $f(t) \equiv 0$ and Theorem 1.6.1 is again obtained. It is worth while stating the case $K(t) \equiv K$ as a separate result.

**THEOREM 1.6.6**

*If* $K(t) \equiv K$ *and* f *and* g *are in* $C^+[a, b]$, *then*

$$f(t) \leqslant g(t) + K \int_a^t f(s)\, ds \qquad (1.6.15)$$

*implies that*

$$f(t) \leqslant g(t) + K \int_a^t exp\ [K(t-s)]g(s)\, ds. \qquad (1.6.16)$$

*If* g(t) *is also a constant, then*

$$f(t) \leqslant g + K \int_a^t f(s)\, ds \qquad (1.6.17)$$

*implies that*

$$f(t) \leqslant g\ exp\ [K(t-a)]. \qquad (1.6.18)$$

The inequalities listed under Theorems 1.6.5 and 1.6.6 are known as *Gronwall's lemma* after the Swedish-American mathematician Thomas Hakon Gronwall (1877–1932), who found a special case in 1918 when investigating the dependence of a system of DE's with respect to a parameter. Gronwall was a pupil of Gösta Mittag-Leffler (1846–1927), who also taught Ivar Bendixson (1861–1935), Ivar Fredholm (1866–1927), Helge von Koch (1870–1924), and Johannes Malmquist (1888–1952). All of them will figure somewhere in this treatise; the first three were among the teachers of the present author. A theorem proved by Bendixson in 1896 (see Theorem 2.8.2) may be regarded as a forerunner of Gronwall's lemma.

**EXERCISE 1.6**

1. Prove Theorem 1.6.2.
2. Prove the following analogue of Theorem 1.6.5 for Nagumo's kernel. Let $X$ be the space of functions $f$ defined on $[0, a]$ so that (i) $f \in C^+[0, a]$, (ii) $\lim_{t \downarrow 0} f(t) = 0$, (iii) $\lim_{t \downarrow 0} f(t)/t = 0$, (iv) $f(t)/t^2 \in L(0, a)$. Define a norm in terms of which $X$ becomes a complete metric space. If $f$ and $g$ belong to $X$ ($g$ fixed), then

$$f(t) \leqslant g(t) + \int_0^t f(s)s^{-1}\, ds \qquad \text{implies} \qquad f(t) \leqslant g(t) + t \int_0^t g(s)s^{-2}\, ds.$$

3. Verify (1.6.8).
4. Prove Theorem 1.6.5.
5. Prove Theorem 1.6.6.
6. Prove the following theorem. If $f$ and $g \in C^+[a, b]$, and if

$$f(t) \leqslant g(t) + K^2 \int_a^t (t-s)f(s)\, ds,$$

then

$$f(t) \le g(t) + K \int_a^t \sinh \left[ K(t - s) \right] g(s) \, ds.$$

7. Show that the inequality

$$f(t) \le -1 - [f(t)]^2$$

is absurd for real-valued elements of $C[a, b]$.

8. Show that the inequality $4f(t) \le 3 + [f(t)]^4$ is trivial for real-valued elements of $C[a, b]$.

9. Show that $6f(6t) \le 3f(3t) + 2f(2t)$ is determinative for $f \in C^+[0, a]$, $0 < a \le +\infty$. Show that there are solutions unbounded for approach to zero of the form $t^{-p}$ if $p > p_0$, the positive root of a certain transcendental equation. There are also solutions of the form $-t^{-q}$ for $q < p_0$. Can such solutions be combined? If so, how?

# II. Analytical Structures

## 1.7. HOLOMORPHIC FUNCTIONS

There are three essentially different approaches to analytic function theory, associated with the names of Bernhard Riemann (1826–1866), Augustin Louis (Baron) Cauchy (1789–1857), and Karl Weierstrass (1815–1897), respectively. Riemann's approach was largely geometric and physical (potential theory). His conformal mapping theorem will figure later, particularly in Chapter 10, but in the main our approach will be a mixture of that of Cauchy and that of Weierstrass.

The complex variable $z = x + iy$ is represented geometrically by the points of $\mathbf{R}^2$, the euclidean plane, so that to $z_0 = x_0 + iy_0$ in $\mathbf{C}$ corresponds the point $(x_0, y_0)$ of $\mathbf{R}^2$. We have

$$z = r(\cos \theta + i \sin \theta) \equiv r e^{i\theta}, \tag{1.7.1}$$

where

$$r = +(x^2 + y^2)^{1/2}, \qquad \tan \theta = \frac{y}{x}. \tag{1.7.2}$$

We write $x = \mathrm{Re}\,(z)$ (read "$x$ is the *real part* of $z$"), $y = \mathrm{Im}\,(z)$ (read "$y$ is the *imaginary part* of $z$"). Furthermore, $r = |z|$ is the *absolute value of $z$*, and $\theta = \arg z$ is the *argument* of $z$, which is determined only up to multiples of $2\pi$. We have

$$|z_1 + z_2| \le |z_1| + |z_2|, \qquad |az| = |a|\,|z|, \qquad |z_1 z_2| = |z_1|\,|z_2|, \tag{1.7.3}$$

so that (assuming a higher standpoint) $\mathbf{C}$ is a B-algebra, the simplest of the species.

The equation $|z - z_0| = r$ represents a circle in the complex plane with center at $z_0$ and radius $r$. The interior and the exterior of this circle are given by the inequalities $|z - z_0| < r$ and $|z - z_0| > r$, respectively. The former will often be called a *circular disk* in the following discussion and is a *circular neighborhood* of $z_0$. The upper half-plane is given by Im $(z) > 0$; the left half-plane, $x < a$, by Re $(z) < a$. If $\lim |z_0 - z_n| = 0$, we say that $\lim z_n = z_0$ and $\{z_n\}$ is a *Cauchy sequence*. The complex plane is a complete metric space in terms of the norm defined by the absolute value. Every bounded infinite point set $S$ has at least one limit point (theorem of Bolzano-Weierstrass). A set which contains its limit points is said to be *closed*. A set $G$ is *open* if, whenever $z_0 \in G$, there is an $\epsilon$-neighborhood, $|z - z_0| < \epsilon$, the points of which belong to $G$. We shall call a set $S$ in the complex plane *connected* if for any two of its points, $z_1$, $z_2$ say, there is a polygonal line joining $z_1$ with $z_2$ all the points of which are in $S$. An open connected set is called a *domain* and will be denoted by $D$. The term *region* is also used, but we reserve this term for sets which have a nonempty *interior* Int $(S) \neq \emptyset$ and contain some of their boundary points.

So far we have dealt with the finite plane. We extend this by adjoining an ideal point: *the point at infinity. Stereographic projection* makes it possible to visualize this addition. Take a sphere of radius $\frac{1}{2}$ tangent to the $z$-plane at the origin, the "south pole" on the sphere. Join the point $z$ in the plane with the "north pole" by a straight line. This line has a second intersection $P$ with the surface of the sphere, and $P$ is taken as the spherical representative of $z$. The finite plane is mapped on the sphere omitting the north pole, and the latter becomes the representative of the point at infinity. The *chordal distance* between two points, $z_1$ and $z_2$, in the plane is the length of the chord joining their representatives, $P_1$ and $P_2$, on the sphere. Chordal distances are bounded, being equal at most to one.

We have now a sufficient set of notions in terms of which we can define *holomorphic* functions. Let $D$ be a domain in the $z$-plane and consider a $(1, 1)$ mapping of $D$ into C. This mapping $T$ defines a function $z \mapsto f(z)$, where $f(z)$ denotes the image of $z$. The mapping is *continuous* at $z = z_0 \in D$ if for every $\epsilon$-neighborhood of $f(z_0)$, there is a $\delta$-neighborhood of $z_0$ whose image is restricted to an $\epsilon$-neighborhood of $f(z_0)$. We can say that $\lim f(z_n)$ exists and equals $f(z_0)$ if $\lim z_n = z_0$. Or, again, Cauchy sequences in the $z$-plane correspond to Cauchy sequences in the $w$-plane, $w = f(z)$.

We now go one step further. Suppose that $f(z)$ is defined and continuous in $D$, and that

$$\lim_{h \to 0} \frac{1}{h} [f(z + h) - f(z)] \tag{1.7.4}$$

exists and is independent of the manner in which $h \to 0$. Then $f(z)$ has a *unique derivative* everywhere in $D$. Such a function is said to be *holomorphic* in $D$.

The usual rules of the calculus apply. Thus sums, constant multiples, and products of differentiable functions are differentiable. A quotient is differentiable at all points where the denominator is not zero. This enables us to assert that polynomials in $z$ are holomorphic in the finite plane, while rational functions are holomorphic except at the zeros of the denominator. In particular, a *fractional linear function*,

$$z \mapsto \frac{az+b}{cz+d} = w, \qquad ad - bc \neq 0, \tag{1.7.5}$$

is holomorphic except at $z = -d/c$. The mapping of the extended $z$-plane into the extended $w$-plane is $(1, 1)$, and the family of circles and straight lines in the $z$-plane goes into the family of circles and straight lines in the $w$-plane. The formula contains three essential constants and is completely determined if the images of three points are given.

Suppose that $z = x + iy$ and $w = u + iv$. For the existence of a unique derivative of $w = f(z)$ it is necessary (but not sufficient) that the *Cauchy-Riemann differential equations* hold:

$$\frac{\partial u}{\partial x} = \frac{\partial v}{\partial y}, \qquad \frac{\partial u}{\partial y} = -\frac{\partial v}{\partial x}. \tag{1.7.6}$$

This says that the real and the imaginary parts of $f(z)$ have partial derivatives with respect to the real and imaginary parts of $z$. If in addition the partials are continuous, $f(z)$ is differentiable at $z = z_0$ or in $D$, as the case may be. As will be seen later, a holomorphic function has derivatives of all orders. In particular, $u(x, y) = \operatorname{Re}[f(z)]$, $v(x, y) = \operatorname{Im}[f(z)]$ have second partials with respect to $x$ and $y$, and these functions are (logarithmic) *potential functions* so that Laplace's equation holds:

$$\Delta u \equiv \frac{\partial^2 u}{\partial x^2} + \frac{\partial^2 u}{\partial y^2} = 0, \qquad \Delta v \equiv \frac{\partial^2 v}{\partial x^2} + \frac{\partial^2 v}{\partial y^2} = 0. \tag{1.7.7}$$

We can define the exponential function $\exp(z)$ or $e^z$ for complex arguments $z$ in various ways, say by the exponential series or by the property that it equals its derivative and is one for $z = 0$. We can also set

$$e^z = e^x e^{iy} = e^x(\cos y + i \sin y). \tag{1.7.8}$$

The last member has continuous partial derivatives of all orders, and the first order partials satisfy the Cauchy-Riemann equations. Hence this convention defines a function holomorphic in the finite plane which agrees with $e^x$ on the real axis ($y = 0$). We shall see later that such an extension by a holomorphic function is unique. We see that $e^z$ is *periodic with period* $2\pi i$.

Formula (1.7.8) shows that we can define a *logarithm* by

$$\log z = \log(r e^{i\theta}) = \log r + i\theta, \qquad (1.7.9)$$

and this is a holomorphic function in any simply connected domain (roughly without holes) which does not contain the origin. The point $z = 0$ is obviously singular since the real part (i.e., $\log r$) goes to $-\infty$ when $r \to 0$. Moreover, the function is not single valued in any domain containing the origin since the argument is then free to take on infinitely many values for a fixed $z$. There are two ways to cope with this difficulty. One method is to restrict the variability of $z$ by introducing a cut, not to be crossed, say along the negative real axis, and define $\log z$ so that $-\pi < \text{Im}(\log z) < \pi$. Instead, we can extend the domain of definition to be a surface with infinitely many sheets, with passage from one sheet to the next along the negative real axis: *up* (increasing values of $\theta$) by crossing from positive values of $y$ to negative, *down* by going in the opposite direction.

We have of course the same difficulties with roots. The $n$th root of $z$ has $n$ determinations. If $z = r e^{i\theta}$, then

$$z^{1/n} = r^{1/n} \exp\left[\frac{1}{n}(\theta + 2k\pi)i\right], \qquad k = 0, 1, \ldots, n-1. \quad (1.7.10)$$

Here the extended domain of definition is a Riemann surface with $n$ sheets, and passage from one to the next is across, say, the negative real axis. Here we run into visual difficulties. By going around the origin $n$ times in the positive sense, we are in the top sheet and one more turn takes us back to the first sheet.

### EXERCISE 1.7

1. What is $\arg(z_1 z_2)$?
2. Construct the sum of two vectors, $z_1$ and $z_2$, in the complex plane.
3. The vector $\bar{z} = x - iy$ is known as the *complex conjugate* of $z$. Show that $|z|^2 = z\bar{z}$. Find $\arg \bar{z}$.
4. Mark the four points $0$, $z_1$, $z_1 + z_2$, and $z_2$, and draw the parallelogram with these points as vertices. Prove the *parallelogram law* (the sum of the squares of the lengths of the diagonals equals the sum of the squares of the lengths of the sides).
5. From the definition of an ellipse derive its equation in complex variables.
6. Show that chordal distances define a metric for $\mathbf{C}$.
7. Derive the Cauchy-Riemann equations and Laplace's equation.

8. Verify that $\log r$ is *harmonic*, i.e., satisfies (1.7.7). Is any power of $r$ harmonic?
9. Verify that (1.7.5) maps circles and lines into circles and lines.

## 1.8. POWER SERIES

Power series served as the foundation of the function theory developed by Weierstrass. He did not invent them, but he perfected their theory. Let there be given a sequence $\{a_n\}$ of complex numbers; form the series

$$\sum_{n=0}^{\infty} a_n z^n. \qquad (1.8.1)$$

Suppose that the series does not diverge for all $z \neq 0$. The basic observation of Weierstrass is

### THEOREM 1.8.1

*Suppose that there is a $z_0 \neq 0$ such that the sequence $\{|a_n z_0^n|\}$ is bounded. Then the series (1.8.1) is absolutely convergent for $|z| < |z_0|$, uniformly for $|z| \leq |z_0| - \delta$, $\delta > 0$.*

This means that the positive reals fall into two classes:

1. $K_1$ contains those numbers $r$ for which $\{|a_n| r^n\}$ is a bounded sequence.
2. $K_2$ contains the remaining positive numbers.

There is a number $R$ which is the supremum of the numbers in class $K_1$ (equivalently, the infimum of the numbers in $K_2$), and now it is seen that the series converges absolutely for $|z| < R$ and diverges for $|z| > R$. For $R$ we have the expression

$$\frac{1}{R} = \limsup_{n \to \infty} |a_n|^{1/n}. \qquad (1.8.2)$$

The sequence $\{|a_n|^{1/n}\}$ may have a single limit, in which case we have $1/R$. This was the case considered by Cauchy. The sequence may, however, have more than one limit point, even infinitely many. In any case there is a largest limit point, and this is the superior limit. The general formula is due to Jacques Hadamard (1865–1963). This $R$ is the *radius of convergence* of the power series. It can take any value in $[0, \infty]$.

If $R > 0$, the sum of the series is evidently a continuous function of $z$ in the circle of convergence $|z| < R$, say $f(z)$. Moreover, we can differentiate term by term to obtain the power series

$$f_1(z) \equiv \sum_{n=1}^{\infty} n a_n z^{n-1} \qquad (1.8.3)$$

with the same circle of convergence. Furthermore, for $|z| \le R_0 < R$, $|z + h| \le R_0$ we find that

$$\left| \frac{1}{h} [f(z+h) - f(z)] - f_1(z) \right| \le \tfrac{1}{2} |h| \sum_{n=2}^{\infty} n(n-1) |a_n| R_0^{n-2}, \qquad (1.8.4)$$

which goes to zero with $|h|$. It follows that $f(z)$ is differentiable and $f'(z) = f_1(z)$. It follows also that the sum of the power series is a holomorphic function of $z$. Using the same technique, one sees that $f(z)$ *has derivatives of all orders for* $|z| < R$.

The function $z \mapsto f(z)$ can be expanded in a Taylor series about any point $a$ in the circle of convergence, and the resulting series, a power series in $z - a$, has a radius of convergence $R_a$:

$$R - |a| \le R_a \le R + |a|. \qquad (1.8.5)$$

Weierstrass obtained this by elementary expansions and rearrangements. We have

$$z^n = (z - a + a)^n = \sum_{k=0}^{n} \binom{n}{k} (z-a)^k a^{n-k}. \qquad (1.8.6)$$

Substitution of this in (1.8.1) gives a double series,

$$\sum_{n=0}^{\infty} a_n \sum_{k=0}^{n} \binom{n}{k} (z-a)^k a^{n-k}. \qquad (1.8.7)$$

This series is absolutely convergent for $|z - a| + |a| < R$. Now, an absolutely convergent double series can be summed by columns as well as by rows and, in fact, by any process that is exhaustive. Here (1.8.7) is the sum by rows. Summing by columns, we get a power series in $z - a$, where the coefficient of $(z - a)^k$ is

$$\frac{1}{k!} \sum_{n=k}^{\infty} n(n-1) \ldots (n-k+1) a_n a^{n-k} = \frac{f^{(k)}(a)}{k!}, \qquad (1.8.8)$$

so that

$$f(z) = \sum_{k=0}^{\infty} \frac{f^{(k)}(a)}{k!} (z-a)^k, \qquad (1.8.9)$$

and this representation is guaranteed to hold for $|z - a| < R - |a|$.

Now this is a power series with a radius of convergence $R_a$ which may very well exceed $R - |a|$. In the circle of convergence, $|z - a| < R_a$, (1.8.6) defines a holomorphic function of $z$; this function coincides with $f(z)$ in the lens-shaped region of intersection of the two disks, where they are defined, and the union of the two disks is the domain of definition of a single holomorphic function which is represented by (1.8.1) in one disk and by (1.8.9) in the other. We can repeat this process for all points $a$ with $|a| < R$. We obtain a family of power series (1.8.9) which represent the same holomorphic function in their domains of convergence. The union of all the disks is the domain of definition that can be reached by direct rearrangements. There is at least one point on the circle of convergence $|z| = R$ which stays outside of all disks $|z - a| < R_a$ with $|a| < R$. This is a *singular point*, and *every power series admits of at least one singular point on its circle of convergence*.

It is possible for all points on the boundary to be singular. This will happen iff $R_a = R - |a|$ for all $a$ with $|a| < R$. We then say that $z = R$ is the *natural boundary* of $f(z)$. An extreme example of this phenomenon is furnished by the series

$$f(z) = \sum_{n=0}^{\infty} \frac{1}{n!} z^{2^n} \qquad (1.8.10)$$

with $R = 1$. Here the series, as well as all the derived series, converge absolutely on $|z| = 1$. The unit circle nevertheless is the natural boundary. Such natural boundaries are typical for so-called *lacunary series*, where there are long and widening gaps in the expansion.

The German mathematician Alfred Pringsheim (1850–1941) proved that, if the coefficients $a_n$ are nonnegative and infinitely many are positive, then $z = R$ is a singular point of the function defined by the series. Since this theorem has important applications to the theory of DE's, we shall sketch a proof. If the theorem were false, then for an $a < R$ but close to $R$ the series (1.8.9) would be convergent for a value of $z$ on the real axis beyond $z = R$; i.e., the series

$$\sum_{k=0}^{\infty} (z - \alpha R)^k \sum_{n=k}^{\infty} \binom{n}{k} a_n (\alpha R)^{n-k}, \qquad 0 < \alpha < 1,$$

for a suitable choice of $\alpha$ near to one would converge for a $z = R(1 + \delta)$ for some $\delta > 0$. Thus

$$\sum_{k=0}^{\infty} (1 - \alpha + \delta)^k R^k \sum_{n=k}^{\infty} \binom{n}{k} a_n (\alpha R)^{n-k}$$

would converge. But this double series has positive terms, so we can interchange the order of summation, leading to the absurd conclusion that the series

$$\sum_{n=0}^{\infty} a_n [(1 + \delta) R]^n$$

would converge. It follows that for $0 < a < R$ we have always $R_a = R$, so $z = R$ is a singular point.

**EXERCISE 1.8**

1. Verify that $f(z)$ and its derived series $f_1(z)$ have the same radius of convergence.
2. How is (1.8.2) obtained?
3. Give examples of power series where $R = 0$, 1, or $\infty$.
4. Verify (1.8.4).
5. Fill in missing details in the proof of (1.8.9).
6. Why does (1.8.8) hold?
7. For the function $z \mapsto f(z)$ defined by (1.8.10) the point $z = 1$ is singular by Pringsheim's theorem. Show that $z = -1$ is singular and that $z = i$ and $z = -i$ are singular. Extend to $2^n$th roots of unity, the union of which is dense on the unit circle.
8. Fill in missing details in the proof of Pringsheim's theorem.
9. In a power series (1.8.1) all the coefficients $a_n$, where $n$ is not a multiple of 3, are zero while $a_{3k} \geq 0$ with infinitely many larger than zero. What can be said about singularities on $|z| = R$?
10. Why is $R_a \leq R + |a|$?

**1.9.  CAUCHY INTEGRALS**

Our three founders of analytic function theory differed in almost all respects and not least in their attitude toward publication. Weierstrass, who was a perfectionist and late in gaining acclaim, published sparingly; his fame rests mainly on his lectures at the University of Berlin, 1864–1892. Riemann was a man of genius but was shy and plagued by poor health; a considerable part of his work was published after his death. Cauchy, on the other hand, overwhelmed the periodicals with his notes.

His first publication on integration between imaginary limits dates from 1825, but he seems to have had the basic ideas as early as 1814. Consider a function $z \mapsto f(z)$, holomorphic in a simply connected domain $D$. Set

$$f(z) = U(z) + iV(z), \tag{1.9.1}$$

and define the integral

$$\int_{z_1}^{z_2} F(z)\, dz = \int_{z_1}^{z_2} [U\, dx - V\, dy] + i \int_{z_1}^{z_2} [U\, dy + V\, dx], \tag{1.9.2}$$

where the path of integration is a curve joining $z_1$ with $z_2$ in $D$ and the integrals are line integrals in the sense of the calculus. Since $U$ and $V$

satisfy the Cauchy-Riemann equations (1.7.6), Cauchy claimed that the integral is independent of the path joining $z_1$ and $z_2$, or, in other words, the integral along a closed contour $C$ is 0:

$$\int_C f(z)\, dz = 0. \tag{1.9.3}$$

There are many questionable points in this argument, and objections go back to the 1880's; desirable precision and generality were reached around 1900. It is required that the curve $C$ have an arc length (= be rectifiable) which requires a representation

$$z = z(t), \qquad 0 \le t \le L, \tag{1.9.4}$$

where $L$ is the length of $C$ and $t \mapsto z(t)$ is a continuous function of *bounded variation*. The integral then becomes a so-called *Riemann-Stieltjes integral*:

$$\int_0^L f[z(t)]\, dz(t), \tag{1.9.5}$$

which is the limit of *Riemann-Stieltjes sums*,

$$\sum_{j=0}^n f[z(t_j)][z(t_j) - z(t_{j-1})]. \tag{1.9.6}$$

The limit exists for any continuous *integrand* $f$ and *integrator* $z$ of bounded variation.

In this setting one proves the theorem for $f = 1$ and $f = z$. Then one observes that (i) $C$ may be approximated arbitrarily closely by a closed polygon, (ii) a polygon may be triangulated, (iii) the theorem is proved for a small triangle, and (iv) hence is true for a polygon and an arbitrary rectifiable curve.

The integral is *additive* with respect to the path and *linear* with respect to the integrand. If $D$ is not simply connected, the integral along $C$ equals a sum of integrals around the holes that are inside $C$.

The Cauchy integral is an exceedingly powerful tool. If $z \mapsto f(z)$ is holomorphic in a simply connected domain $D$ and on its rectifiable boundary $C$, then

$$\frac{1}{2\pi i} \int_C \frac{f(t)\, dt}{t - z} = \begin{cases} 0 & \text{if } z \text{ is outside } C, \\ f(z) & \text{if } z \text{ is inside } C. \end{cases} \tag{1.9.7}$$

We can differentiate under the sign of integration as often as we please, and the formal $n$th derivative represents $f^{(n)}(z)$:

$$\frac{n!}{2\pi i} \int_C \frac{f(t)\, dt}{(t - z)^{n+1}} = f^{(n)}(z). \tag{1.9.8}$$

Thus *a holomorphic function has derivatives of all orders,* a property proved for power series in Section 1.8. It has been observed that the property of defining a holomorphic function of $z$ inside $C$ resides, not in the factor $f(t)$ in the integral, but in the *Cauchy kernel* $1/(t - z)$, for we can replace $f(t)$ by any continuous function $F(t)$ without losing analyticity of the integral, which, however, may not be zero outside of $C$ in this case. It is the Cauchy kernel which is a holomorphic function of $z$ as long as $z$ is kept away from the contour of integration $C$. We can expand the kernel in powers of $z$ or of $1/z$, multiply by $f(t)$, and integrate term by term, as is usually permitted by uniform convergence of the series. The linearity in $f$ often implies continuity with respect to $f$. Among the many results obtainable by such considerations we list

**THEOREM 1.9.1**

*If* $z \mapsto f(z)$ *is holomorphic in* D, *if the disk* $|z - a| < R$ *lies in* D, *then* $f(z)$ *can be expanded in the Taylor series*

$$f(z) = \sum_{k=0}^{\infty} \frac{f^{(k)}(a)}{k!} (z - a)^k, \qquad (1.9.9)$$

*the series being absolutely convergent in the disk.*

*Proof.* We have

$$\frac{1}{t - z} = \frac{1}{(t - a) - (z - a)} = \sum_{k=0}^{\infty} \frac{(z - a)^k}{(t - a)^{k+1}}, \qquad (1.9.10)$$

which converges uniformly in $z$ and $t$ if $|z - a| \le R - \delta, |t - a| = R$. Multiplication by $f(t)$ and termwise integration yields (1.9.9) in view of (1.9.8), where $z$ is replaced by $a$. ∎

In a similar manner one obtains the *Laurent series,* discovered by Pierre Alphonse Laurent (1813–1854) in 1843 and known to Weierstrass in 1841.

**THEOREM 1.9.2**

*If* $z \mapsto f(z)$ *is holomorphic in an annulus.*

$$0 \le R_1 < |z - a| < R_2 \le \infty, \qquad (1.9.11)$$

*then*

$$f(z) = \sum_{-\infty}^{\infty} a_n (z - a)^n, \qquad a_n = \frac{1}{2\pi i} \int_C \frac{f(t)\, dt}{(t - a)^{n+1}}, \qquad (1.9.12)$$

*where* C: $|t - a| = r, R_1 < r < R_2$.

We shall not give the proof but mention that in addition to (1.9.10) there is needed

$$\frac{1}{t-z} = -\sum_{n=0}^{\infty} \frac{(t-a)^n}{(z-a)^{n+1}}, \tag{1.9.13}$$

which is valid for $|z-a| > |t-a|$.

The case in which $z = a$ is an isolated singularity is particularly important. Here $R_1 = 0$, and the negative powers in (1.9.12) constitute the *principal part* of the singularity. There are three different possibilities.

1. No negative powers. We define $f(a) = a_0$; the singularity is *removable*.

2. A finite number of negative powers, $a_n = 0$ for $n < -m$ but $a_{-m} \neq 0$. This is a *pole of order m*, and $(z-a)^m f(z)$ is holomorphic at $z = a$.

3. Infinitely many negative powers. Here $z = a$ is an *essential singularity*. In any neighborhood of $z = a$ the function $z \mapsto f(z)$ assumes any preassigned value $c$ infinitely often with at most two exceptions (theorem of Émile Picard).

The property of being holomorphic may be said to be hard to acquire, but once acquired it persists. It can be expected to survive a passage to the limit. The simplest case is

**THEOREM 1.9.3**

*Suppose that $\{f_n(z)\}$ is a sequence of functions, holomorphic in a domain D, which converges uniformly to a function f(z) in D; then f(z) is holomorphic.*

The functions holomorphic in $D$ form a normed algebra under the sup norm $\|f\| = \sup_{z \in D} |f(z)|$. Convergence in the norm is uniform convergence in the ordinary sense and the algebra is complete, so the theorem follows. This case is almost trivial, but we can greatly weaken the assumptions by using *induced convergence*. Here is an example. Instead of assuming convergence in a domain, we can assume it in a subset from which it spreads to the whole domain.

**THEOREM 1.9.4**

*Let $\{f_n(z)\}$ be a sequence of functions holomorphic in a domain D. Let C be a simple, closed, rectifiable oriented curve which, together with its interior, lies in D. Suppose that the sequence $\{f_n(t)\}$ converges uniformly with respect to t on C. Then there exists a function $z \mapsto f(z)$ holomorphic in the interior of C such that $f_n(z)$ converges to $f(z)$ uniformly in the interior of C. Moreover, if S is any subset of the interior of C having a positive distance from C, and if p is any positive integer, then the sequence $\{f_n^{(p)}(z)\}$ converges uniformly to $f^{(p)}(z)$ in S.*

We shall not prove this theorem, but we call attention to the fact that the *principle of the maximum* (see below) implies that a Cauchy sequence $\{f_n(t)\}$ on $C$ is also a Cauchy sequence inside and on $C$.

A *zero* of $f(z)$ is by definition a point where the function is zero. It is of order $m$ if the Taylor expansion starts with the term $a_m(z-a)^m$, $a_m \neq 0$. The zeros of a holomorphic function can not have a cluster point in the interior of a domain where $f(z)$ is holomorphic except when the function is identically zero. If $z = a$ is a limit point of zeros of $f(z)$, then in the Taylor expansion

$$f(z) = a_0 + a_1(z-a) + a_2(z-a)^2 + \cdots$$

the constant term $a_0 = f(a)$ is zero by the continuity of the function. But then

$$f_1(z) = a_1 + a_2(z-a) + a_3(z-a)^2 + \cdots$$

also has infinitely many zeros with $z = a$ as a limit point. This forces $a_1$ to be zero and so on; all coefficients are zero, and $f(z)$ is identically zero.

This implies that, if two functions $f(z)$ and $g(z)$ holomorphic in a domain $D$ coincide for infinitely many values of $z$ with a limit point in $D$, they are identical in all of $D$, for $h(z) = f(z) - g(z)$ has infinitely many zeros and is thus identically zero. This is known as the *identity theorem*. Instead of zeros we may of course consider any other fixed value of the function. We see that limit points of zeros or of a value $c$ are singular points of the function.

The *calculus of residues* occupied a central position in Cauchy's work. Suppose that $z = a$ is an isolated singular point of a function $z \mapsto f(z)$ in the neighborhood of which $f(z)$ is single valued. There is then an associated Laurent expansion (1.9.12). The coefficient $a_{-1}$ is the *residue* of $f(z)$ at $z = a$. The reason for the name is that $a_{-1}$ is all that is left when we form

$$\frac{1}{2\pi i} \int_C f(t) \, dt,$$

where $C$ is a small circle, $|t - a| = r$, for we can substitute the Laurent series and integrate termwise since the series is uniformly convergent on the circle. We have

$$\int_C (t-a)^{-n-1} \, dt = r^{-n} i \int_0^{2\pi} \exp(-ni\theta) \, d\theta \qquad (1.9.14)$$

and the integral is zero unless $n = 0$, when it equals $2\pi i$. This gives the residue theorem.

**THEOREM 1.9.5**

*If* f(z) *is holomorphic in a simply connected domain* D *except for isolated singularities at* $z = s, s_2, \ldots, s_n$, *then*

$$\int_C f(t)\, dt = 2\pi i \sum_{j=1}^{n} r_j, \tag{1.9.15}$$

*where* $r_j$ *is the residue of* f(z) *at* $z = s_j$ *and* C *is the boundary of* D *supposed to be rectifiable.*

This is so because the integral along $C$ equals the sum of the integrals around the small circles surrounding the singularities $s_j$. In all these formulas the integrals are taken in the positive sense.

An important consequence is

**THEOREM 1.9.6**

*If* f(z) *is holomorphic inside and on* C *except for poles, then*

$$\frac{1}{2\pi i} \int_C \frac{f'(t)}{f(t)}\, dt = Z_f - P_f, \tag{1.9.16}$$

*where* $Z_f$ *is the number of zeros, and* $P_f$ *the number of poles inside* C.

*Proof.* It is assumed that neither zeros nor poles are located on $C$. The integrand is then holomorphic inside and on $C$ except for simple poles at the zeros and poles of $f(z)$. At a zero the residue equals the multiplicity of the zero, whereas at a pole the residue is the negative of the multiplicity of the pole. The conclusion then follows from Theorem 1.9.5. ∎

The integral in (1.9.16) can be evaluated directly since the integrand is the derivative of

$$\log f(t) = \log |f(t)| + i \arg f(t).$$

Here the real part returns to its original value when $z$ returns to the starting point after having described $C$ once in the positive sense. The imaginary part, however, does not necessarily return to its initial value but will differ from it by a multiple of $2\pi$. From Theorem 1.9.6 we then get the so-called *principle of the argument*:

**THEOREM 1.9.7**

*Under the assumptions of Theorem* 1.9.6 *the increase in the argument of* f(z) *after* C *has been described once in the positive sense is* $2\pi(Z_f - P_f)$.

A useful addition to Theorem 1.9.6 is given by

**THEOREM 1.9.8**

*Under the assumptions of Theorems 1.9.6 and 1.9.7, suppose in addition that* g(z) *is holomorphic inside and on* C. *Then*

$$\frac{1}{2\pi i} \int_C g(t) \frac{f'(t)}{f(t)} \, dt = \sum g(a_j) - \sum g(b_k), \qquad (1.9.17)$$

*where the summation is extended over the zeros* $a_j$ *and poles* $b_k$ *of* f(z), *and each summand is repeated as often as the multiplicity of the zero or pole requires.*

The zeros and poles of $f(z)$ are still simple poles of the integrand, and at a zero of $f$ of multiplicity $\mu_j$ the residue is $\mu_j g(a_j)$, and similarly at the poles.

Cauchy's formula (1.9.7) invites some comments. Replace $z$ by $z_0$, and let the path of integration be the circle $t = z_0 + r e^{i\theta}$, where $\theta$ goes from zero to $2\pi$. The result is

$$f(z_0) = \frac{1}{2\pi} \int_0^{2\pi} f(z_0 + r e^{i\theta}) \, d\theta. \qquad (1.9.18)$$

The right-hand side is the mean value in the sense of the integral calculus over the interval $(0, 2\pi)$ of the integrand. This is a basic property of holomorphic functions but is shared with harmonic (logarithmic) potential functions. From (1.9.18) we also get

$$|f(z_0)| \leq \max |f(z)| \quad \text{for } |z - z_0| = r. \qquad (1.9.19)$$

In this relation the inequality normally holds, equality can hold iff $|f(z)|$ equals its maximum for all $z$—to start with, all $z$ on the circle and ultimately all $z$ in the plane, and $f(z) = M e^{i\alpha}$, where $M$ is the maximum and $\alpha$ is real, fixed. This is a form of the *principle of the maximum*. The principle asserts that the absolute value of a holomorphic function $f(z)$ cannot have a local maximum unless it is a constant. If in 3-space we plot the surface

$$u = |f(x + iy)|^2, \qquad (1.9.20)$$

where $f$ is not a constant, if $f(x_0 + iy_0) = f(z_0) \neq 0$, there are paths on the surface leading from $z = z_0$, $u = u_0$ along which $u$ is strictly increasing, and also paths along which $u$ is strictly decreasing. The latter type of path naturally is missing if $u_0 = 0$. We formulate a form of the principle which is sufficient for our purposes.

**THEOREM 1.9.9**

*If* f(z) *is holomorphic inside and on the rectifiable curve* C, *and if* M(f, C) *is the maximum of* |f(z)| *on* C, *then for all* z *inside* C

$$|f(z)| \leq M(f, C). \tag{1.9.21}$$

*Proof.* The use of Cauchy's integral below is due to Edmund Landau (1877–1938), a German mathematician who made profound contributions to function theory and analytic number theory. We have

$$[f(z)]^k = \frac{1}{2\pi i} \int_C \frac{[f(t)]^k}{t - z} \, dt, \tag{1.9.22}$$

whence

$$|f(z)|^k \leq [2\pi d(z, C)]^{-1} L [M(f, C)]^k.$$

Here $d(z, C)$ is the distance of $C$ from the point $z$ in its interior, and $L$ is the length of $C$. We extract the $k$th root and pass to the limit with $k$ to obtain (1.9.21). ∎

**EXERCISE 1.9**

1.  If $C$ is rectifiable, then $\Sigma_{j=1}^n |z(t_j) - z(t_{j-1})| \leq L$ and the sums in (1.9.6) are dominated by $M(f, C)L$. Verify.

2.  Verify (1.9.7) and (1.9.8). For $n = 1$ verify that the difference quotient $(1/h)[f(z + h) - f(z)]$ tends to the formal derivative uniformly if $z$ and $z + h$ have a distance from $C$ which exceeds a $\delta > 0$.

3.  Fill in details in the proofs of Theorems 1.9.1 and 1.9.2.

4.  Use the principle of the maximum to prove that a Cauchy sequence $\{f_n(t)\}$ on $C$ generates a Cauchy sequence $\{f_n(z)\}$ in the interior.

5.  Verify Theorems 1.9.5 and 1.9.6.

6.  Show that a harmonic function, not a constant, cannot have a local maximum.

**1.10. ESTIMATES OF GROWTH**

Given a power series

$$f(z) = \sum_{n=0}^{\infty} a_n z^n \tag{1.10.1}$$

with radius of convergence $R > 0$, let $M(r, f)$ be its *maximum modulus*:

$$M(r, f) = \max_{0 \leq \theta < 2\pi} |f(r e^{i\theta})|. \tag{1.10.2}$$

Formula (1.9.8) then gives the Cauchy estimates,

$$|a_n| \leq M(r, f) r^{-n}, \qquad r < R. \tag{1.10.3}$$

Here $r$ is arbitrary, so the question of optimizing the estimate arises. If $R < \infty$, the choice $r = [1 - (1/n)]R$ is often a good one.

The maximum principle shows that for a nonconstant holomorphic function $M(r, f)$ is an increasing function of $r$. For $R < \infty$ it may very well be bounded. This is not the case for $R = \infty$, however, as shown by Joseph Liouville (1809–1882). A sharper form of his theorem is

**THEOREM 1.10.1**

*Suppose that* $R = \infty$ *and that there are positive constants* A, B, *and* c *such that*

$$M(r, f) \leq A + Br^c \qquad (1.10.4)$$

*for all* r. *Then* $z \mapsto f(z)$ *is a polynomial in* z *of degree not exceeding* c.

*Proof.* By (1.10.3) we have

$$|a_n| \leq (A + Br^c)r^{-n},$$

and this goes to zero as $r \to \infty$ if $n > c$. This means that all coefficients $a_n$ are zero for $n > c$, so $z \mapsto f(z)$ is a polynomial of degree not exceeding $c$ as asserted. ∎

A power series with $R = \infty$ is called an *entire function* (*integral function* in Great Britain). Such a function may be *algebraic* or *transcendental*, according as it is a polynomial or not. In the transcendental case $M(r, f)$ grows faster than any power of $r$ by Liouville's theorem, but there is neither a slowest nor a fastest possible growth of the maximum modulus. Given an increasing function $r \mapsto G(r)$ such that $\log G(r)/\log r$ becomes infinite with $r$, one can find an entire function whose maximum modulus grows faster than $G(r)$ and an entire function whose maximum modulus grows slower than $G(r)$, so that

$$\lim_{r \to \infty} \frac{M(r, f_1)}{G(r)} = +\infty, \qquad \lim_{r \to \infty} \frac{M(r, f_2)}{G(r)} = 0,$$

respectively.

More properties of entire functions will be encountered in Chapters 4 and 5. In addition to entire functions, we shall also ultimately have to consider *meromorphic functions*, i.e., functions having no other singularities than poles in the finite domain. Again we have two types, algebraic and transcendental meromorphic functions. The first is the class of rational functions, functions which are meromorphic in the extended plane. A function of the second class either has a finite number of poles plus an essential singularity at $z = \infty$ or infinitely many poles which have no limit point in the finite part of the plane.

In Chapter 4 we shall encounter various generalizations of the maximum modulus which are suitable for a study of meromorphic functions. Here we shall consider just a generalization of Liouville's theorem for an isolated singular point.

**THEOREM 1.10.2**

*Suppose that* $z = a$ *is an isolated singularity of* $z \mapsto f(z)$, *where* $f(z)$ *is holomorphic in a punctured disk,* $0 < |z - a| < R$. *Set*

$$M_a(r, f) = max |f(a + r e^{i\theta})|, \qquad 0 < r < R. \qquad (1.10.5)$$

*Suppose there exist positive numbers* A, B, *and* c *such that*

$$M_a(r, f) \le A + Br^{-c}. \qquad (1.10.6)$$

*Then* $z = a$ *is either a removable singularity or a pole of order* $\le c$.

*Proof.* The assumptions imply that there is a Laurent expansion (1.9.12), and we have

$$|a_{-k}| \le M_a(r, f)r^k < (A + Br^{-c})r^k ;$$

this goes to zero with $r$ if $k > c$. It follows that $z = a$ is either a removable singularity or a pole of order not exceeding $c$, as asserted. ∎

Exercises bearing on this section will be found after Section 1.11.

### 1.11. ANALYTIC CONTINUATION; PERMANENCY OF FUNCTIONAL EQUATIONS

Analytic continuation is a concept introduced by Weierstrass and basic for his attack on function theory. A function $z \mapsto f(z)$ is defined originally by a power series, say,

$$f(z) = \sum_{n=0}^{\infty} a_n z^n \qquad (1.11.1)$$

with a radius of convergence $R$. If $R = \infty$ and the function is entire, there is no continuation problem. Also, $R = 0$ is out (we disregard the possibility that the series may be summable by some method or other). If $0 < R < \infty$, there is a continuation problem. We saw in Section 1.8 that $f(z)$ admits of expansions in a Taylor series,

$$f(z; a) = \sum_{k=0}^{\infty} \frac{f^{(k)}(a)}{k!} (z - a)^k, \qquad (1.11.2)$$

obtained by direct rearrangement of (1.11.1) after setting $z = a + (z - a)$ and expanding. The series $f(z; a)$ converges for $|z - a| < R_a$, where

$R - |a| \leq R_a \leq R + |a|$. If $R_a = R - |a|$, the point of contact of $|z - a| = R_a$ with $|z| = R$ is a singular point of $f(z)$ and analytic continuation in the direction $\arg z = \arg a$ is not possible. On the other hand, if $R - |a| < R_a$, the disk $|z - a| < R_a$ is partly outside $|z| < R$, and in the lens-shaped overhang $f(z; a)$ defines an analytic continuation of $f(z)$. This process is repeated for all $a$ with $|a| < R$. If for all such $a$'s we get $R_a = R - |a|$, no analytic continuation is possible and $|z| = R$ is the natural boundary of $f(z)$.

If, on the other hand, $R_a > R - |a|$ for some values of $a$, the union of the disks $|z - a| < R_a$ is a simply connected domain $D_1$ in which our function is defined by one of the series $f(z; a)$ with $|a| < R$. Moreover, if a point $z_0 \in D_1$, it belongs to infinitely many disks $|z - a| < R_a$ and the corresponding series $f(z; a)$ all assign the same value to $f(z)$ at $z = z_0$. We now repeat the process for points $a_1$ in $D_1$ with $|a_1| \geq R$. This gives a set of power series $f(z; a, a_1)$ obtained by double rearrangements: of the original series at $z = a$ with $|a| < R$, and of the series $f(z; a)$ at $z = a_1$. This gives a definition of $f$ in the union of all the disks. Since this set may be self-overlapping, we are no longer assured of the consistency of the definition but have to keep account of the steps involved.

Suppose that we can find a sequence of points $a_0, a_1, \ldots, a_n$ such that power series in terms of $z - a_j$ are obtained by repeated rearrangements of (1.11.1). Suppose that

$$|a_0| < R, \qquad |a_1 - a_0| < R_{a_0}, \qquad \ldots, \qquad |a_n - a_{n-1}| < R_{a_{n-1}}. \qquad (1.11.3)$$

Here

$$f_j(z) = \sum a_{j,k}(z - a_j)^k, \qquad |z - a_j| < R_j.$$

There is then defined a branch of $f(z)$ at $z = a_n$ obtained by analytic continuation of $f(z)$ using the intermediary points $a_0, a_1, \ldots, a_n$. According to Weierstrass, *the totality of such series constitutes an analytic function $f(z)$*. To these *regular elements* of $f$ are further adjoined the following:

1. *Polar elements* $(z - a)^{-m} \sum_{j=0}^{\infty} b_j(z - a)^j$, one for each pole.
2. *Algebraic elements* $(z - a)^{-k/p} \sum_{j=0}^{\infty} c_j(z - a)^{jk/p}$ for algebraic branch points.
3. *Elements at infinity*. These may be regular: $\sum_{j=0}^{\infty} d_j z^{-j}$, polar: $z^m \sum_0^{\infty} d_j z^{-j}$, or algebraic: $z^{k/p} \sum_0^{\infty} d_j z^{-jk/p}$.

Essential singular points are not considered as belonging to the domain of definition of the function and contribute no functional elements. Logarithmic singularities also are noncontributing.

At a given point $z = a$ there are normally infinitely many elements, regular and/or singular. Henri Poincaré (1854–1912) was a famous French mathematician whose name will be encountered again and again in this book. At this stage we need a result of his according to which the distinct elements of $f(z)$ at $z = a$ form a countable set. This we see as follows. An element at $z = a$ is the end product of a chain of rearrangements. We can order these first according to the number of steps involved, say $n$. For each $n$ we have rearrangements at points $a_0, a_1, \ldots, a_n$. Without loss of generality we may assume that the $a$'s used have rational coordinates (these numbers are dense in $C$). Now the points with rational coordinates form a denumerable set. Thus we are dealing with a countable number of countable sets, and such a set is itself countable. Hence the different determinations at $z = a$ form either a finite set or a countable one.

Leaving these general considerations, we turn to the principle of permanence of functional equations, which is a very important fact in the theory of DE's. It is hard to find a definition of the term *functional equation*, and the mathematicians who deal with the subject normally exclude differential and integral equations from consideration, so we cannot expect much help from that quarter. Before going any further we have to prove the double series theorem of Weierstrass, which will be needed here and later.

**THEOREM 1.11.1**

*Suppose that the functions* $z \mapsto f_n(z)$, $n = 0, 1, 2, \ldots$, *are holomorphic in a disk* $|z - a| < R$ *and that the series*

$$\sum_{n=0}^{\infty} f_n(z) \equiv f(z) \qquad (1.11.4)$$

*converges uniformly in* $|z - a| \leq \rho R$ *for each* $\rho$ *with* $0 < \rho < 1$. *Suppose also that*

$$f_n(z) = \sum_{k=0}^{\infty} a_{k,n}(z - a)^k. \qquad (1.11.5)$$

*Then each of the series*

$$\sum_{n=0}^{\infty} a_{k,n} \equiv A_k, \qquad k = 0, 1, 2, \ldots, \qquad (1.11.6)$$

*converges and, for* $|z - a| < R$,

$$f(z) = \sum_{k=0}^{\infty} A_k(z - a)^k. \qquad (1.11.7)$$

*Proof.* This follows from Theorem 1.9.4, which implies that the sum of the series (1.11.4) is a holomorphic function $f(z)$, that the series may be differentiated term by term $p$ times, and that the resulting sum of $p$th derivatives is the $p$th derivative of the sum $f(z)$, where $p$ is any integer. In particular, we have convergence for $z = a$ *and this implies the assertions.* ■

The name *double series theorem* refers to the fact that under the stated conditions the order of summation in the series

$$\sum_{n=0}^{\infty} \left[ \sum_{k=0}^{\infty} a_{k,n} (z - a)^k \right] \tag{1.11.8}$$

may be interchanged.

We shall prove a special form of the principle of permanence of functional equations, which is sufficient for most of our needs and is connected with our work on functional inequalities in Section 1.6. There it was found that, if $X$ is a partially ordered metric space with elements $f$, if $T$ is a mapping of $X$ into itself, and if $T$ or one of its powers is a contraction with fixed point $g$, then the elements of $X$ which satisfy the inequality

$$f \leq T[f] \tag{1.11.9}$$

also satisfy

$$f \leq g, \quad \text{where } g = T[g]. \tag{1.11.10}$$

Now this is a functional equation and obeys the principle of permanence if the data are given an analytical form. We consider a function $T(z, w)$ of two complex variables given by the series

$$T(z, w) = \sum_{j=0}^{\infty} \sum_{k=0}^{\infty} a_{jk} z^j w^k, \qquad a_{00} = 0. \tag{1.11.11}$$

It is assumed that there exist values of $z$ and $w$, different from $(0, 0)$, for which the series is absolutely convergent, say $z = a$, $w = b$. These numbers $a$ and $b$ may be assumed to be positive. The reader should observe that a double power series does not have a unique radius of convergence, though there exist pairs of associated radii: if you lower one, you can increase the other. Our assumption implies that the terms of the series are bounded for $z = a$, $w = b$,

$$|a_{jk}| a^j b^k \leq M. \tag{1.11.12}$$

By an obvious generalization of Theorem 1.8.1 this implies that the series (1.11.11) is absolutely convergent for $|z| < a$, $|w| < b$. Next, suppose that we have found a function

$$z \mapsto f(z) \equiv \sum_{m=1}^{\infty} c_m z^m \tag{1.11.13}$$

such that (i) there exist a pair of numbers $s < a$, $t < b$ and (1.11.13) is absolutely convergent for $|z| \leqslant s$ and $|f(z)| \leqslant t$ for such values, and (ii) if $T[z, f(z)] \equiv F(z)$ for $|z| \leqslant s$, then $f(z) = F(z)$ or

$$f(z) = T[z, f(z)]. \qquad (1.11.14)$$

We have then the following principle of permanence of functional equations:

**THEOREM 1.11.2**

*If* T *satisfies the conditions just stated, and if the solution* f(z) *as well as the composite function* F(z) = T[z, f(z)] *can be continued along the same path from the origin, then all along the path* (1.11.14) *holds; i.e., the continuation of the solution is the solution of the continuation of the equation.*

*Proof.* We assume that the continuation involves a chain of disks $D_0$, $D_1, \ldots, D_n$, where $D_0 = [z; |z| < s]$ and the center $a_j$ of $D_j$ lies in $D_{j-1}$. In $D_j$ we have functional elements $f_j(z)$ and $F_j(z) = T[z, f_j(z)]$ represented by convergent power series in $z - a_j$. Now, in $D_0$ we have $f(z) \equiv F(z)$, and this equality holds at $z = a_1$ and in some neighborhood thereof. But the identity theorem then requires that $f_1(z) \equiv F_1(z)$ in $D_1$—in particular, at $z = a_2$ so that $f_2(z) \equiv F_2(z)$, and so on. ∎

The reader should notice that this is not an existence proof. It is by no means clear that our assumptions are strong enough to guarantee the existence of a solution. All that is claimed is that, *if* we have found an analytic solution and we can continue it analytically together with the right member of the equation, then it remains a solution.

It should also be observed that (1.11.10) is a rather special case of a functional equation and that the law of permanence holds in much more general situations.

**EXERCISES 1.10–1.11**

1.  For a power series (1.10.1) $R = 1$ and $M(r, f) < (1 - r)^{-c}$, where $c \geqslant 0$, find a realistic estimate for $|a_n|$.

2.  If $z \mapsto (1 - z)^{-c}$, where $c \geqslant 0$, find how fast the binomial coefficients grow.

3.  If $z \mapsto f(z)$ is a transcendental entire function and $M(r, f) < \exp(Ar^c)$, where $c$ and $A$ are positive, find a realistic estimate of the Maclaurin coefficients, $a_n$.

4.  If $z \mapsto f(z)$ has a pole of order $m$ with leading coefficient $a_{-m}$, how fast does $M_a(r, f)$ grow?

5.  Is $z \mapsto \tan \pi z$ a transcendental meromorphic function? Where are the poles? Show that they are simple.

6. Show that $\cos z = \frac{1}{2}[\exp(iz) + \exp(-iz)]$. Show that for this function the maximum modulus is attained on the imaginary axis and equals $\cosh r$ (the hyperbolic cosine of $r$).

7. Show that $|\sin(x + iy)|^2 = \sin^2 x + \sinh^2 y$.

8. If $e^z = 1 + \Sigma_1^\infty z^n/n!$, prove by series multiplication or otherwise that $e^z = e^{z-a} e^a$.

9. If $P(t) = a + bt + ct^2$, express $\Sigma_{n=0}^\infty P(n)z^n/n!$ in terms of the exponential function.

10. Where in the complex plane is $\sin(x + iy)$ real, and where purely imaginary?

11. What is the limit, if any, of $\tan(x + iy)$ as $y \to +\infty$, $y \to -\infty$?

12. What is the series expansion for $f(z, a)$ if $f(z) = (1 - z)^{-1}$ and $|a| < 1$? How does $R_a$ vary with $a$?

13. Show that in this case the domain $D_1$ where the first rearrangements converge is bounded by a cardioid with the cusp at $z = 1$ and passing through $z = -3$.

14. Prove that the same curve gives the boundary of $D_1$ for all the binomial functions $(1 - z)^c$, where $c$ is not a positive integer.

15. Take $c = \frac{1}{2}$, and consider second rearrangements $f(z; a, a_1)$. Try to form some notion of what part of the plane is covered by $D_2$ and show that there are two distinct elements of the function with center at $z = 2$.

16. Consider the analytic function $z \mapsto [(1 - z)^{1/2} - (1 + z)^{1/2}]^{1/4}$. Where are the branch points? Describe the functional elements with centers at the branch points. What are the elements at infinity? The roots are given all possible determinations.

17. Two power series have radius of convergence $= 1$. Express the product of the two series as a simple series, and determine its radius of convergence. Can it be greater than one?

18. When are the following series absolutely convergent: (i) $\Sigma_1^\infty (zw)^n$, (ii) $\Sigma_0^\infty (z + w)^n$, (iii) $\Sigma_0^\infty \Sigma_0^\infty z^k z^n$?

19. Let $T(z, w) = \frac{1}{2}[z + \Sigma_2^\infty (z + w)^n]$. Let $f = T(z, f)$ be the solution which is 0 at $z = 0$. Show that the solution is holomorphic except for two branch points. Find their locations and natures.

20. State the conditions under which the principle of permanence would apply to the functional equation involved in Problem 1.6:6.


## LITERATURE

As collateral reading for Chapter 1 the author may perhaps be permitted to refer to three of his own books:

Hille, E. *Lectures on Ordinary Differential Equations*. Addison-Wesley, Reading, Mass. 1969.

——. *Methods in Classical and Functional Analysis*. Addison-Wesley, Reading, Mass. 1970.

_____. *Analytic Function Theory.* Vols. I, and II, 2nd ed. Chelsea, New York, 1973; 1st ed., Ginn, Boston, 1959.

These books will be referred to hereafter as LODE, MCFA, and AFT, respectively.

For Section 1.1 consult MCFA, Chapters 1 and 2, or LODE, Chapter 1. The latter can be used also for Sections 1.2–1.6. Matrices and mappings are discussed in both books.

For Sections 1.5 and 1.6 consult Chapters 5 and 12 of MCFA. For differential and integral inequalities the basic work is:

Walter, W. *Differential- und Integralungleichungen.* Springer Tracts on Natural Philosophy, Vol. 2. Springer-Verlag, Berlin, 1966.

The original Gronwall Lemma occurs in:

Gronwall, T. H. Note on the derivative with respect to a parameter of the solutions of a system of differential equations. *Ann. Math.*, (2), **20** (1918), 292–296.

See also:

Bendixson, I. Démonstration de l'existence de l'intégral d'une équation aux dérivées partielles linéaires. *Bull. Soc. Math. France*, **24** (1896), 220–225.

For this question see also Chapter 3 of LODE.

For Part II consult AFT. For Section 1.7 Chapters 1–4 will round out the presentation. Section 1.8 and Chapter 5 go together. For Section 1.9 consult Chapters 7–9. Functions of bounded variation and the Riemann-Stieltjes integral are treated in Appendix C. Analytic continuation is discussed in Chapters 5 and 10 of Vol. II, where also the principle of permanence of functional equations is treated under different assumptions.

The standard treatise on functional equations is:

Aczél, J. *Lectures on Functional Equations and Their Applications.* Academic Press, New York, 1964.

# 2
# EXISTENCE AND UNIQUENESS THEOREMS

This chapter is devoted to generalities about differential equations (DE's) and their solutions. The main subject matter will be existence and uniqueness theorems for analytic DE's: the use of fixed point methods, successive approximations, majorant methods, the Cauchy majorant, the majorant of Lindelöf, and dominants and minorants. We shall also consider variations of parameters, internal and external.

## 2.1. EQUATIONS AND SOLUTIONS

Our first question is, What is meant by a DE and by a solution? A crude definition of a DE would be a functional equation (also undefined!) involving derivatives of unknown functions. Such derivatives may be ordinary or partial, according to whether the number of independent variables is one or more. We restrict ourselves to functions of a single variable. This does not exclude the possibility that the equation involves one or more parameters which occur explicitly in it.

We have now excluded partial DE's from consideration. But not every functional equation involving derivatives of the unknown with respect to a single variable (which need not occur explicitly in the equation) is a bona fide ordinary DE. To clarify the ideas, let us consider the following equations from the point of view of their being acceptable for study in this book:

$$w'(z) = f_0(z) + f_1(z)w(z) + f_2(z)[w(z)]^2, \qquad (2.1.1)$$

$$w'(z) = z^{-2}w(z), \qquad (2.1.2)$$

$$zw''(z) = [w(z)]^2, \qquad (2.1.3)$$

$$w'(z) = w(z + 1), \qquad\qquad\qquad (2.1.4)$$

$$w'(z) = \int_0^z \{1 + [w(s)]^2\}\, ds, \qquad\qquad (2.1.5)$$

$$w(z) = \int_0^1 \{[w'(s)]^2 + [w(z)]^2\}^{1/2}\, ds. \qquad (2.1.6)$$

The notation is intended to suggest that the functions involved are analytic functions of a complex variable. The first three equations are acceptable; the others are not. The first is a nonlinear first order (Riccati) equation, discussed in Chapter 4, the second is a trivial first order equation (Chapter 5), and the third is a nonlinear second order equation, a special case of Emden's equation (Chapter 12). Equation (2.1.4) is rejected because the unknown is subjected to a translation of the independent variable besides the differentiation. The remaining equations are excluded because of the integral operators. Equation (2.1.5), however, leads to a bona fide second order DE:

$$w''(z) = 1 + [w(z)]^2 \qquad\qquad\qquad (2.1.7)$$

upon differentiation. However, the two equations are not equivalent: all solutions of (2.1.5) satisfy (2.1.7), but not conversely.

In the following we shall consider only ordinary DE's or systems of such equations. Up to and including Chapter 4 the equations will be of the first order and of the form

$$w'(z) = F[z, w(z)]. \qquad\qquad\qquad (2.1.8)$$

Here there are two alternatives: $F(z, w)$ is an analytic function of the two variables $z$ and $w$, holomorphic in a given dicylinder in $C^2$, or (second alternative) $w(z)$ is vector valued in $C^n$, $z$ is a complex variable, and $F$ is a holomorphic mapping of $C^{n+1}$ into $C^n$. The latter alternative enables us to treat $n$th order DE's as a special case of (2.1.8). Note that a DE is of *order n* if the order of the highest derivative entering into the equation is $n$.

The notion of what is meant by a *solution* of a DE has varied considerably over the three centuries that have elapsed since Isaac Newton made the first classification of what he called *fluxional equations* in 1671. He even invented the method of series expansions with indeterminate coefficients. However, since Newton's work was not published until 1736, long after his death, it had no influence on the development of the subject. The first published paper on DE's is due to Jacob Bernoulli in his study of the *isochrone* (1696); this was rapidly followed by a number of investigations by Johann and Daniel Bernoulli, Clairaut, Euler,

Leibniz, Riccati, and others during the eighteenth century. To these early founding fathers of calculus and analysis, the problem was to find a function of two variables $(z, w) \mapsto G(z, w)$ such that

$$G_z(z, w)\, dz + G_w(z, w)\, dw = 0 \qquad (2.1.9)$$

is implied by (2.1.8) and vice versa. This would require

$$F(z, w) = -\frac{G_w(z, w)}{G_z(z, w)}, \qquad (2.1.10)$$

and the solution would be

$$G(z, w) = C. \qquad (2.1.11)$$

In (2.1.9) and (2.1.10) the subscripts $z$ and $w$ indicate partial differentiation with respect to $z$ and $w$, respectively. Since in those days the only acceptable functions were those built up by composition of a finite number of what later became known as elementary functions, i.e., powers, exponentials, logarithms, trigonometric functions, and their inverses, the integrable types were soon exhausted. Just as most functions could not be integrated, most DE's were insolvable.

Here Cauchy showed the way out of the difficulty. The founding fathers had asked too much: they had demanded global solutions in terms of elementary functions. This was an algebraic straitjacket that was not suitable and prevented further progress. The typical operation of analysis is the limit process, which is very powerful but can require a local point of view.

Let us return to (2.1.8). Suppose that $(z, w) \mapsto F(z, w)$ is holomorphic in a neighborhood of $(z_0, w_0)$. Cauchy asked, Is there a function of $z$, say

$$w = w(z; z_0, w_0), \qquad (2.1.12)$$

such that it is holomorphic in some neighborhood of $z = z_0$ and for all $z$ in the neighborhood, with $w_0 = w(z_0; z_0, w_0)$,

$$w'(z; z_0, w_0) = F[z, w(z; z_0, w_0)]? \qquad (2.1.13)$$

Cauchy showed the existence and uniqueness of such a solution. To this end he developed several ingenious devices. For the real case he worked out a step-by-step method using linear approximations. For the complex domain he employed series expansions and the majorant method.

The global problem still remains. Here the method of analytic continuation used by Weierstrass provided at least a partial solution. A local solution given by a power series at $z = z_0$ is continued analytically along a path $C$ from $z_0$ to $z_1$. If $F[z, w(z)]$ can also be continued analytically along $C$, then by the law of permanence of functional equations (see Section 1.10) *the continuation of the solution is the solution of the continuation of the equation.*

This leaves out of consideration the singularities. They are either *fixed*, given by the equation, or *movable*, i.e., dependent on the initial values. Thus (2.1.2) has only fixed singularities, namely, 0 and $\infty$, whereas (2.1.1) has both kinds. There the singularities of $f_0$, $f_1$, $f_2$ and the zeros of $f_2$ are fixed. The movable singularities are simple poles. This is illustrated by the special example

$$w' = 1 + w^2, \qquad (2.1.14)$$

which is satisfied by $w = \tan(z - z_0)$ with poles at $z = z_0 + \frac{1}{2}(2n + 1)\pi$. Equation (2.1.3) has fixed singularities at $z = 0$ and $\infty$ and movable singularities, which may be placed anywhere in the plane and are of very complicated nature.

For a further study of (2.1.8) we must determine the fixed singularities and the nature of the solutions at such a point. We must also decide why, when, and where movable singularities occur and what the actual natures of these singularities are.

Some equations can be characterized by postulating the position and nature of the singularities together with the relations joining these elements. This problem was first posed by Riemann for the hypergeometric equation. There is a *Riemann problem* for linear DE's, one of the many Riemann problems. This involves global characterization of equations and solutions. Usually global information is hard to get; there are, however, equations for which we can assert that the only movable singularities are poles, and equations of which the solutions are single valued.

A solution of (2.1.8) which involves an arbitrary constant is called the *general solution*. A solution obtained by giving the constant a particular value is known as a *particular solution*. Thus (2.1.14) has the general solution $\tan(z - z_0)$, and $\tan z$ is a particular solution. This equation has also two *singular solutions*, $w(z) \equiv i$ and $w(z) \equiv -i$. They are not obtainable from the general solution by specifying the constant $z_0$.

### EXERCISE 2.1

Find the fixed singularities of the following DE's, and discuss how the nature of the singularity varies with the parameters. All the equations are elementary.

1. $zw' = aw$.
2. $(z - a)w' = (z - b)w$.

3.  $(z - a)^2 w'' + (z - a) w' - b^2 w = 0.$

4.  The functional equation (not a DE)

$$f'(z) = 3^{\frac{1}{2}} \int_0^1 \{1 + [f(s)]^2\}^{\frac{1}{2}} s \, ds$$

has solutions of the form $f(z) = Cz$ (why?). Determine the value of $C$ if $1^{1/2} = + 1$.

Determine the DE of the lowest order which is satisfied by the following conics:

5.  $b^2 x^2 + a^2 y^2 = a^2 b^2.$

6.  $(x - a^2)^2 + (y - a)^2 = a^2 + a^4.$

7.  $(x - a)^2 + (y - b)^2 = 1.$
    Obtain the envelope of the curve family when it exists and show that it is also a (singular) solution.

8.  Find the general solution of $w' = w^2 - 1$, and find the singular solutions. Discuss the movable singularities (nature, frequency, and dependence on the constant of integration).

## 2.2.  THE FIXED POINT METHOD

Historically this is the most recent method proposed for proving existence and uniqueness theorems for ordinary DE's. It goes back to a paper by G. D. Birkhoff and O. D. Kellogg in 1922. They used a fixed point theorem by L. E. J. Brouwer. That Banach's fixed point theorem (Theorem 1.5.1) could be applied for the same purpose was pointed out by R. Caccioppoli in 1930.

We are concerned with the equation

$$w' = F(z, w), \tag{2.2.1}$$

where $(z, w) \mapsto F(z, w)$ is holomorphic in the *dicylinder*

$$D : |z = z_0| \le a, \qquad |w - w_0| \le b. \tag{2.2.2}$$

It is required to find a function $z \mapsto w(z; z_0, w_0)$, holomorphic in some disk $|z - z_0| < r \le a$, such that

$$\begin{aligned} w'(z; z_0, w_0) &= F[z, w(z; z_0, w_0)], \\ w(z_0; z_0, w_0) &= w_0. \end{aligned} \tag{2.2.3}$$

To apply Theorem 1.5.1 we have to exhibit a complete metric space **X** consisting of functions $z \mapsto g(z)$ and define an operator $T$ which maps **X** into itself and is a strict contraction. Also, **X** has to be chosen so that the existing unique fixed point is the desired solution. As a first step we replace (2.2.1) by the integral equation

$$f(z) = w_0 + \int_{z_0}^{z} F[s, f(s)] \, ds. \tag{2.2.4}$$

If this equation has a unique solution $f$, then $f(z)$ also satisfies (2.2.1) including the initial condition. This suggests defining an operator $T$ by

$$T[g](z) = w_0 + \int_{z_0}^{z} F[s, g(s)] \, ds, \tag{2.2.5}$$

where $T$ operates on a space to be defined. We note first that $F$ satisfies two conditions used in the following, namely,

$$|F(z, w)| < M, \tag{2.2.6}$$
$$|F(z, u) - F(z, v)| < K|u - v|, \tag{2.2.7}$$

for suitably chosen constants $K$ and $M$ and $(z, w)$, $(z, u)$, and $(z, v)$ in $D$. Since $D$ is closed, $F$ is certainly bounded in $D$ and so is $F_w(z, w)$, the partial of $F$ with respect to $w$. The Lipschitz condition (2.2.7) is implied by the boundedness of the partial derivative. We can now state and prove

**THEOREM 2.2.1**

*Under the stated assumptions on* F, *in the disk* $D_0$: $|z - z_0| < r$, *where*

$$r < min \left( a, \frac{b}{M}, \frac{1}{K} \right), \tag{2.2.8}$$

*(2.2.1)* *has a unique holomorphic solution satisfying the initial-value condition* $w(z_0; z_0, w_0) = w_0$.

*Proof.* Let **X** be the set of all functions $z \mapsto g(z)$, holomorphic and bounded in $D_0$ so that

$$g(z_0) = w_0 \qquad \text{and} \qquad \|g - w_0\| \leq b, \tag{2.2.9}$$

where

$$\|g - w_0\| = \sup_{z \in D_0} |g(z) - w_0|. \tag{2.2.10}$$

Under these assumptions the composite function $z \mapsto F[z, g(z)]$ exists and is holomorphic in $D_0$, and its norm is at most $M$. The holomorphism follows from an applications of the double series theorem (Theorem 1.11.1) plus analytic continuation. This implies that $z \mapsto T[g](z)$ is also holomorphic (the integral from $z_0$ to $z$ of a holomorphic function is holomorphic); it takes the value $w_0$ at $z = z_0$ and

$$\|T[g] - w_0\| \leq Mr \leq b$$

by (2.2.8). Hence $T$ maps $\mathbf{X}$ into itself. The contraction property follows from the Lipschitz condition, for we have

$$|T[g](z) - T[h](z)| = \left| \int_{z_0}^{z} \{F[s, g(s)] - F[s, h(s)]\} \, ds \right|$$

$$\leq K \left| \int_{z_0}^{z} |g(s) - h(s)| \, |ds| \right| \leq Kr \|g - h\|.$$

Since $Kr < 1$ by (2.2.8), it is seen that

$$\|T[g] - T[h]\| \leq Kr\|g - h\| = k\|g - h\|, \tag{2.2.11}$$

where $k = Kr < 1$. Also, $\mathbf{X}$ is a metric space under the sup norm; it is complete, for if a sequence $\{g_n\}$ of $\mathbf{X}$ is Cauchy, then $\lim g_n(z) = g(z)$ exists uniformly in $D_0$. Furthermore, $g$ is holomorphic in $D_0$, $g(z_0) = w_0$, and $\|g - w_0\| \leq b$. Thus $g \in \mathbf{X}$ and $\mathbf{X}$ is complete. All the assumptions of Theorem 1.5.1 are satisfied, and we conclude that $\mathbf{X}$ has a unique fixed point under $T$. This function $z \mapsto f(z)$ satisfies (2.2.4) and hence is our required solution $w(z; z_0, w_0)$ of (2.2.1).

By the proof of Theorem 1.5.1 we could start with any element $g$ of $\mathbf{X}$ and obtain the invariant element $f$ as

$$f = \lim_{n \to \infty} T^n[g]. \tag{2.2.12}$$

The convergence of this sequence is comparatively slow. Convergence as the geometric series $\sum_1^{\infty} k^n$ is all that this method gives. The method of successive approximations of Picard gives convergence as an exponential series and also dispenses with the obnoxious condition $Kr < 1$. ■

The fixed point method also applies to the vector case referred to above in connection with (2.1.8). We state the result without proof.

**THEOREM 2.2.2**

*Let* $(z, \mathbf{w}) \mapsto \mathbf{F}(z, \mathbf{w})$ *be a mapping of* $\mathbf{C}^{n+1}$ *into* $\mathbf{C}^n$ *defined in an* $(n + 1)$-*cylinder*:

$$D: |z - z_0| \leq a, |w_1 - w_{10}| \leq b_1, \ldots, |w_n - w_{n0}| \leq b_n, \tag{2.2.13}$$

*where* $\mathbf{w} = (w_1, w_2, \ldots, w_n)$, *and we set* $(w_{10}, w_{20}, \ldots, w_{n0}) = \mathbf{w}_0$. *We set* $\mathbf{F} = (F_1, F_2, \ldots, F_n)$. *In* $\mathbf{C}^n$ *and* $\mathbf{C}^{n+1}$ *we use the maximum coordinate norm, i.e.,*

$$\|\mathbf{w}\| = max \, |w_j|, \|(z, \mathbf{w})\| = max \, (|z|, |w_1|, \ldots, |w_n|). \tag{2.2.14}$$

*Assume that* $\mathbf{F}(z, \mathbf{w})$ *is holomorphic in* $D$ *and also that*

$$max \, \|\mathbf{F}(z, \mathbf{w})\| = M, \|\mathbf{F}(z, \mathbf{u}) - \mathbf{F}(z, \mathbf{v})\| \leq K\|\mathbf{u} - \mathbf{v}\| \tag{2.2.15}$$

*for* $(z, \mathbf{w})$, $(z, \mathbf{u})$, *and* $(z, \mathbf{v})$ *in* D. *Define a disk* $D_0$: $|z - z_0| \leq r$, *where*

$$r < min \left\{ a, \frac{b_1}{M}, \frac{b_2}{M}, \ldots, \frac{b_n}{M}, \frac{1}{K} \right\}. \tag{2.2.16}$$

*Then there exists uniquely a vector function* $z \mapsto \mathbf{w}(z; z_0, \mathbf{w}_0)$, *defined and holomorphic in* $D_0$, *such that*

*and*

$$\mathbf{w}'(z; z_0, \mathbf{w}_0) = \mathbf{F}[z, \mathbf{w}(z, z_0, \mathbf{w}_0)] \tag{2.2.17}$$

$$\mathbf{w}(z_0; z_0, \mathbf{w}_0) = \mathbf{w}_0. \tag{2.2.18}$$

Note that the derivative of a vector function is the vector whose components are the derivatives of the components of the given vector. Similarly, by definition the integral of a vector function is the vector whose components are the integrals of the components.

### EXERCISE 2.2

For the following *autonomous* equations (i.e., DE's in which the independent variable does not occur explicitly), find a value of $r$ for which Theorem 2.2.1 guarantees a solution. Take $z_0 = 0$.

1. $w'(z) = w(z)$, $w_0 \neq 0$.
2. $w'(z) = 1 + [w(z)]^2$.
3. [Optimization] What are the best values of $r$ obtainable in Problems 1 and 2? Would removal of the condition $rK < 1$ lead to better values?
4. What is the justification for the Lipschitz condition in Theorem 2.2.2?
5. Write a complete proof of Theorem 2.2.2.
6. Reduce $w'' - zw' + z^2 w = 0$, $w(0) = 0$, $w'(0) = 1$ to vector form, and estimate $r$ by Theorem 2.2.2. Take $a = 1$, $b_1 = b_2 = b$.

## 2.3 THE METHOD OF SUCCESSIVE APPROXIMATIONS

Going back in time to 1890, we find an important memoir by Émile Picard (1856–1941), introducing the method of successive approximations. This rapidly became the basic tool for proving the existence of solutions of DE's as well as other functional equations including various forms of the *implicit function* theorem. Basically it is the old method of trial and error, which dates back to Sir Isaac Newton. There is a profound difference, however: Picard systematized the trial and estimated the error.

We again consider

$$w' = F(z, w), \qquad w(z_0) = w_0. \tag{2.3.1}$$

Here $(z, w) \mapsto F(z, w)$ is holomorphic in the closed dicylinder

$$D: |z - z| \leq a, |w - w_0| \leq b, \tag{2.3.2}$$

and, with $1 \leq K$,

$$|F(z, w)| \leq M, \tag{2.3.3}$$

$$|F(z, u) - F(z, v)| \leq K|u - v| \tag{2.3.4}$$

for $(z, w)$, $(z, u)$, and $(z, v)$ in $D$. We introduce a disk,

$$D_0: |z - z_0| \leq r \qquad \text{with } r < \min\left(a, \frac{b}{M}\right). \tag{2.3.5}$$

Note that the condition $rK < 1$ is no longer imposed.

### THEOREM 2.3.1

*Under the stated conditions* (2.3.1) *has a unique holomorphic solution in* $D_0$.

*Proof.*   The desired solution must satisfy

$$w(z) = w_0 + \int_{z_0}^{z} F[s, w(s)] \, ds. \tag{2.3.6}$$

We now define a sequence of approximations $w_n(z)$ recursively by

$$w_0(z) = w_0, \qquad w_n(z) = w_0 + \int_{z_0}^{z} F[s, w_{n-1}(s)] \, ds, \quad n > 0. \tag{2.3.7}$$

It should now be shown that the definitions make sense for $z$ in $D_0$. This involves showing that the approximations exist as holomorphic functions in $D_0$ and that $|w_n(z) - w_0| \leq b$. Suppose that this has been achieved for $n < m$. The implication is that $w_{m-1}(z)$ exists in $D_0$, where it is holomorphic and satisfies $|w_{m-1}(z) - w_0| \leq b$. This in turn implies that the composite function $z \mapsto F[z, w_{m-1}(z)]$ exists and is holomorphic in $D_0$. It follows that the integral exists and is a holomorphic function of $z$ in $D_0$. Furthermore,

$$\left| \int_{z_0}^{z} F[s, w_{m-1}(s)] \, ds \right| \leq M|z - z_0| \leq Mr < b, \tag{2.3.8}$$

so that $w_m(z)$ is holomorphic in $D_0$ and $|w_m(z) - w_0| \leq b$. Thus the approximations exist for all $n$, are holomorphic in $D_0$, and take the value $w_0$ at $z = z_0$.

Convergence must now be proved. This follows from the Lipschitz condition, which has not been used so far. Now we have

$$|w_1(z) - w_0| \leq M|z - z_0| \leq KM|z - z_0|,$$

so that

$$w_2(z) - w_1(z) = \int_{z_0}^{z} \{F[s, w_1(s)] - F[s, w_0]\} \, ds$$

and

$$\begin{aligned} |w_2(z) - w_1(z)| &\leq K \left| \int_{z_0}^{z} |w_1(s) - w_0| \, |ds| \right| \\ &\leq K^2 M \left| \int_{z_0}^{z} |s - z_0| \, |ds| \right| \\ &= K^2 M \tfrac{1}{2} |z - z_0|^2. \end{aligned}$$

[To evaluate the integral, set $s = z_0 + (z - z_0)t$, where $t$ goes from zero to one.] This suggests taking as induction hypothesis the assumption

$$|w_k(z) - w_{k-1}(z)| \leq M \frac{K^k}{k!} |z - z_0|^k, \qquad (2.3.9)$$

which holds for $k = 2$, giving

$$\begin{aligned} |w_{k+1}(z) - w_k(z)| &= \left| \int_{z_0}^{z} \{F[s, w_k(s)] - F[s, w_{k-1}(s)]\} \, ds \right| \\ &\leq K \left| \int_{z_0}^{z} |w_k(s) - w_{k-1}(s)| \, |ds| \right| \\ &\leq \frac{K^{k+1}}{k!} \left| \int_{z_0}^{z} |s - z_0|^k \, |ds| \right| \\ &= M \frac{K^{k+1}}{(k+1)!} |z - z_0|^{k+1}. \end{aligned}$$

Thus (2.3.9) holds for all $k$, and it follows that the series

$$w(z) \equiv w_0 + \sum_{n=1}^{\infty} [w_n(z) - w_{n-1}(z)] \qquad (2.3.10)$$

converges absolutely and uniformly in $D_0$. This implies also that $w(z)$ is holomorphic, at least in $\mathrm{Int}(D_0)$, and continuous on the boundary. Furthermore,

$$\lim w_n(z) = w(z), \qquad \lim F[z, w_{n-1}(z)] = F[z, w(z)],$$

and finally

$$w(z) = \omega_0 + \int_{z_0}^{z} F[s, w(s)] \, ds$$

as desired. This is clearly a solution of (2.3.1).

It remains to show that the solution is unique. This may be proved in a number of different ways by invoking the Lipschitz condition again. Suppose that $f(z)$ is a solution; we may assume that it is also holomorphic in $D_0$. Again we have

$$f(z) = z_0 + \int_{z_0}^{z} F[s, f(s)] \, ds.$$

From (2.3.7) we obtain

$$f(z) - w_n(z) = \int_{z_0}^{z} \{F[s, f(s)] - F[z, w_{n-1}(s)]\}\, ds, \qquad (2.3.11)$$

whence

$$|f(z) - w_n(z)| \leqslant K \left| \int_{z_0}^{z} |f(s) - w_{n-1}(s)|\, |ds| \right|$$

$$\leqslant K^2 \left| \int_{z_0}^{z} |ds| \left| \int_{z_0}^{s} |f(t) - w_{n-2}(t)|\, |dt| \right| \right|$$

$$= K^2 \left| \int_{z_0}^{z} |s - z_0|\, |f(s) - w_{n-2}(s)|\, |ds| \right|. \qquad (2.3.12)$$

Repeated use of the same device finally gives

$$|f(z) - w_n(z)| \leqslant \frac{K^n}{(n-1)!} \left| \int_{z_0}^{z} |s - z_0|^{n-1} |f(s) - w_0|\, |ds| \right|.$$

Here we may assume that $|f(s) - w_0| \leqslant b$, so that

$$|f(z) - w_n(z)| \leqslant \frac{(Kr)^n}{n!}\, b, \qquad (2.2.13)$$

and this goes to zero as $n$ becomes infinite. Hence $f(z) = w(z)$ and the solution is unique, thus proving the theorem. ∎

We can get another uniqueness proof from Theorem 1.6.1. If we have two solutions, $w(z)$ and $f(z)$, their difference satisfies

$$f(z) - w(z) = \int_{z_0}^{z} \{F[s, f(s)] - F[s, w(s)]\}\, ds, \qquad (2.3.14)$$

and if $|f(z) - w(z)| = g(z)$ then

$$g(z) \leqslant K \left| \int_{z_0}^{z} g(s)|ds| \right|.$$

Here we set, with $\theta = \arg(z - z_0)$,

$$s = z_0 + t\, e^{i\theta}, \qquad z = z_0 + u\, e^{i\theta}, \qquad g(s) = h(t), \qquad g(z) = h(u)$$

and obtain the inequality

$$h(u) \leqslant K \int_0^u h(t)\, dt. \qquad (2.3.15)$$

Here $h(t)$ is continuous and nonnegative. But by Theorem 1.6.1 the only solution of this inequality in $C^+[0, a]$ is $h(t) \equiv 0$. This gives $g(z) \equiv 0$ and $f(z) \equiv w(z)$.

Finally, the same type of argument applies to the vector case. We can drop the condition $rK < 1$ there also. The result will be cited as Theorem 2.3.2 in the following.

**EXERCISE 2.3**

1. Find a sequence of successive approximations for the case $w' = w$, $w(0) = 1$.
2. Consider $w' = 1 + w^2$ with $w(0) = 0$. The successive approximations are polynomials in $z$. Find the degree of the $n$th polynomial. Take $w_0(z) = z$.
3. Prove that the only continuous nonnegative solution of (2.3.15) is identically zero by the method of successive substitution:

$$0 \le h(u) \le K \int_0^u h(t)\, dt \le K^2 \int_0^u dt \int_0^t g(v)\, dv \le \cdots .$$

4. C. Carathéodory in 1918 proved existence and uniqueness theorems, replacing (2.3.3) by $|F(z, w)| \le MK(|z - z_0|)$ and (2.3.4) by $|F(s, t_1) - F(s, t_2)| \le K(|s - z_0|)|t_1 - t_2|$. Here $K(u) \ge 0$ and $K(u) \in C(0, a] \cup L(0, a)$. Carry through the proof, and use Theorem 1.6.1 for uniqueness and the method of successive approximations.

**2.4. MAJORANTS AND MAJORANT METHODS**

Given the DE

$$w'(z) = F[z, w(z)], \qquad w(z_0) = w_0, \tag{2.4.1}$$

where $F(z, w)$ is holomorphic in the dicylinder

$$D: |z - z_0| \le a, |w - w_0| \le b, \tag{2.4.2}$$

it is natural to try to satisfy the equation by a power series

$$w - w_0 = \sum_{n=1}^{\infty} c_n (z - z_0)^n. \tag{2.4.3}$$

Here the coefficients $c_n$ have to be chosen so that the series formally satisfies (2.4.1). "Formally" means that, if the series is substituted in (2.4.1) and the right-hand member is written as a power series in $z - z_0$, then for all $k$ the coefficient on the right of $(z - z_0)^k$ equals $(k + 1)c_{k+1}$, which is the coefficient of $(z - z_0)^k$ on the left. As we shall see, the $c_k$'s can be determined successively and uniquely.

This *method of indeterminate coefficients* was discovered by Isaac Newton in 1671 and published in 1736. The method was rediscovered on the continent and was commonly used by the analysts of the eighteenth century. The procedure involves a number of doubtful steps, all of which go back to the question of whether or not the resulting series is absolutely convergent in some disk $D_0$: $|z - z_0| < r$. Such finicky questions did not bother the founding fathers, who believed that a well-defined analytic expression always has a sense. That this is not necessarily true was discovered much later.

As a case in point we may take the equation

$$zw' = (1 + z)w - z \tag{2.4.4}$$

with the formal series

$$\sum_{n=0}^{\infty} (-1)^n n! z^{-n}, \tag{2.4.5}$$

which diverges for all finite values of $z$. The equation can be solved by elementary methods, and one solution is

$$z e^z \int_z^{\infty} e^{-t} \frac{dt}{t}, \tag{2.4.6}$$

which is actually represented asymptotically by the series.

Divergent series were banished from serious consideration by Abel and Cauchy in the 1820's (they were resurrected by Poincaré in the 1880's). Cauchy, who had laid solid foundations for analytic function theory as well as for the theory of DE's, also made the method of indeterminate coefficients rigorous by his *calcul des limites*. In this connection *limites* should be understood as "bounds" rather than "limits." It is a method of *majorants*, a term introduced by Poincaré, who also was the first to use the symbol $\ll$ for the relation between the given function and one of its majorants.

Let

$$f(z) = \sum_{n=0}^{\infty} c_n z^n \tag{2.4.7}$$

be a power series with a positive radius of convergence, and let

$$g(z) = \sum_{n=0}^{\infty} C_n z^n \tag{2.4.8}$$

be a power series with nonnegative coefficients and the radius of convergence $R$.

**DEFINITION 2.4.1**

g(z) *is a majorant of* f(z), *symbolized by*

$$f(z) \ll g(z), \tag{2.4.9}$$

*if*

$$|c_n| \le C_n, \qquad \forall n. \tag{2.4.10}$$

This clearly implies that the radius of convergence of (2.4.7) is at least $R$. The relation $\ll$ is *transitive*, i.e.,

$$f \ll g \quad \text{and} \quad g \ll h \quad \text{imply} \quad f \ll h. \tag{2.4.11}$$

**LEMMA 2.4.1**

*If* m *is a positive integer and if* f $\ll$ g, *then*

$$[f(z)]^m \ll [g(z)]^m. \tag{2.4.12}$$

*Furthermore,*

$$f^{(m)}(z) \ll g^{(m)}(z), \tag{2.4.13}$$

$$\int_0^z f(t)\,dt \ll \int_0^z g(t)\,dt. \tag{2.4.14}$$

*Repeated integration also preserves the majorant relation.*

The proof is left to the reader.

The majorant relation is also preserved by some composite mappings.

**LEMMA 2.4.2**

*Let*

$$h(w) = \sum_{n=0}^{\infty} h_n w^n, \qquad H(w) = \sum_{n=0}^{\infty} H_n w^n \quad \text{with } h(w) \ll H(w), \tag{2.4.15}$$

*the series being convergent for* $|w| < R_0$. *Suppose that* f $\ll$ g, f(0) = g(0) = 0, *and* $|g(z)| < R_0$ *for* $|z| < R$. *Then*

$$h[f(z)] \ll H[g(z)]. \tag{2.4.16}$$

*Proof.* Suppose that $f_j$ and $g_j$, $j = 1, 2, \ldots, m$, are power series, the $g_j$'s being absolutely convergent for $|z| < R$, and suppose that

$$f_j \ll g_j, \qquad j = 1, 2, \ldots, m. \tag{2.4.17}$$

Let $a_1, a_2, \ldots, a_m$ be arbitrary numbers, and let $|a_j| \le A_j$. Then

$$a_1 f_1 + a_2 f_2 + \cdots + a_m f_m \ll A_1 g_1 + A_2 g_2 + \cdots + A_m g_m. \tag{2.4.18}$$

Since the right member is a power series in $z$ convergent for $|z| < R$, the same holds for the left member.

We now apply this to the case

$$f_j(z) = [f(z)]^j, \qquad g_j(z) = [g(z)]^j, \qquad a_j = h_j, \qquad A_j = H_j. \tag{2.4.19}$$

Then for each $m$

$$h_1 f(z) + h_2 [f(z)]^2 + \cdots + h_m [f(z)]^m$$

$$\ll H_1 g(z) + H_2 [g(z)]^2 + \cdots + H_m [g(z)]^m. \tag{2.4.20}$$

Here the right-hand member is the $m$th partial sum of the series $\sum_{1}^{\infty} H_n w^n$ with $w$ replaced by $g(z)$. In the disk $|z| \leq R - \delta$ we can find an $\epsilon > 0$ such that $|g(z)| \leq R_0 - \epsilon$. Since the series $\sum_{1}^{\infty} H_n (R_0 - \epsilon)^n$ converges, the series $\sum_{1}^{\infty} H_n [g(z)]^n$ converges uniformly in $|z| \leq R_0 - \delta$ to a holomorphic function $L(z)$ with nonnegative coefficients in its Maclaurin expansion. The series

$$K(z) = \sum_{n=1}^{\infty} h_n [f(z)]^n \qquad (2.4.21)$$

is also convergent for $|z| < R_0$, and the majorant relation (2.4.20), which holds for the partial sums, holds also in the limit so that $K(z) \ll L(z)$, as asserted. ∎

The majorant concept extends to functions of several variables. Suppose, in particular, that

$$F(z, w) = \sum_{j=0}^{\infty} \sum_{k=0}^{\infty} c_{jk} z^j w^k \qquad (2.4.22)$$

is a holomorphic function of $(z, w)$ in the dicylinder

$$D: |z| \leq a, |w| \leq b.$$

Suppose that

$$G(z, w) = \sum_{j=0}^{\infty} \sum_{k=0}^{\infty} C_{jk} z^j w^k \qquad (2.4.23)$$

is also holomorphic in $D$. We say that $G$ is a majorant of $F$:

$$F(z, w) \ll G(z, w) \qquad \text{if } |c_{jk}| \leq C_{jk}, \qquad \forall j, k. \qquad (2.4.24)$$

We now consider (2.4.1), where for simplicity we take $z_0 = w_0 = 0$. Here the fundamental fact is that *a majorant relation for the right-hand sides implies the corresponding majorant relation for the solutions.* We state and prove

**THEOREM 2.4.1**

*Let* $F(z, w)$ *be defined by the series* (2.4.22), *and let* $G(z, w)$ *be a majorant of* $F(z, w)$ *defined by* (2.4.23) *and* (2.4.24). *Suppose that*

$$W'(z) = G[z, W(z)], \qquad W(0) = 0 \qquad (2.4.25)$$

*has a solution*

$$W(z) = \sum_{j=1}^{\infty} C_j z^j \qquad (2.4.26)$$

*convergent for* $|z| < r$. *Let*

$$w(z) = \sum_{j=1}^{\infty} c_j z^j \qquad (2.4.27)$$

*be a formal solution of*

$$w'(z) = F[z, w(z)], \qquad w(0) = 0. \tag{2.4.28}$$

*Then*

$$w(z) \ll W(z), \tag{2.4.29}$$

*and the series* (2.4.27) *is absolutely convergent for* $|z| < r$ *and is the unique solution of* (2.4.28).

*Proof.* The derivation of (2.4.27) involves a number of operations on power series which are legitimate iff the series are absolutely convergent. The series are "formal" if the required operations have been performed without regard to legitimacy, the justification being given *a posteriori* when it is found that the series are indeed absolutely convergent.

We have to form the composite series

$$\sum_{j=0}^{\infty} \sum_{k=0}^{\infty} c_{jk} z^j \left[ \sum_{p=1}^{\infty} c_p z^p \right]^k. \tag{2.4.30}$$

Here we start by forming the $k$th powers by using Cauchy's product formula (valid for absolutely convergent series and leading to absolutely convergent series). The $k$th series is multiplied by $c_{jk} z^j$, and the result is summed for $j$ and $k$. We then rearrange the result as a power series in $z$. This can be justified with the aid of the double series theorem of Weierstrass, provided that we have absolute convergence. In the result the coefficient of $z^n$ will be a multinomial in the $c_{jk}$'s and the $c_p$'s, say,

$$M_n[c_{jk}; c_p]. \tag{2.4.31}$$

Here the big question is, What $c_{jk}$'s occur, and what $c_p$? To the first question we can reply that a necessary condition for $c_{jk}$ to occur in $M_n$ is that $j + k \le n$. This is gratifying, for it means that $c_{n+1}$ depends only on a finite number of the known coefficients $c_{jk}$, at most $\frac{1}{2}(n+1)(n+2)$ in number. Furthermore, the only $c_p$'s that can enter are $c_1, c_2, \ldots, c_n$, all of which have been determined by the time the $(n+1)$th coefficient is considered. This means that the coefficients $c_p$ can be determined successively and uniquely. The numerical constants which enter when the $k$th powers are formed are positive integers, a fact which is also important. The series (2.4.27) is a formal solution if

$$(n+1)c_{n+1} = M_n(c_{jk}; c_p) \tag{2.4.32}$$

for all $n$. The first three coefficients are

$$
\begin{aligned}
c_1 &= c_{00}, \\
c_2 &= \tfrac{1}{2}(c_{10} + c_{01}c_{00}), \\
c_3 &= \tfrac{1}{2}[c_{20} + c_{11}c_{00} + c_{02}(c_{00})^2 + \tfrac{1}{2}c_{01}c_{10} + \tfrac{1}{2}(c_{01})^2 c_{00}].
\end{aligned} \tag{2.4.33}
$$

This means that all coefficients $c_p$ are ultimately determined in terms of the coefficients of $F(z, w)$.

If the same procedure is applied to the majorant equation (2.4.25), we get exactly the same formulas for determining the $C_n$'s, provided that we replace the lower case letters in $M_n$ by the corresponding capitals. It is then clear that

$$|c_n| \leq C_n, \qquad \forall n, \qquad (2.4.34)$$

so $W(z)$ is a majorant for $w(z)$ provided that the series (2.4.26) has a disk of convergence. If we know that this is the case, it follows that (2.4.27) has a radius of convergence at least as large as that of $W(z)$. And now all the operations performed to get the coefficients are justified, and the formal series is an actual solution.

The coefficients $c_n$ are uniquely determined so that (2.4.27) is the only solution which is holomorphic in some neighborhood of $z = 0$. There is still the possibility of the existence of a nonholomorphic solution. If this would be given by a series in terms of fractional powers of $z$, either the series itself or some derivative thereof would become infinite as $z \to 0$. This cannot be a bona fide solution of an equation of type (2.4.1), for a solution of such an equation must possess derivatives of all orders continuous at $z = 0$. Nonanalytic solutions can also be excluded by one of the uniqueness theorems of Section 2.3. ■

What remains to be done with the series is to find a suitable majorant equation having an absolutely convergent solution series. This problem will be taken up in the next two sections.

Before leaving the general majorant problem, let us consider another variant of the majorant method which leads, not necessarily to a majorant series, but rather to a *dominant function* which may give alternative convergence proofs.

It was seen above that positivity plays an important role in our problems. If the coefficients $c_{jk}$ are nonnegative, the solution coefficients are also nonnegative. This means in the first place that, for $z$ real positive, $z = x > 0$; then the sequence of partial sums

$$S_n(x) = \sum_{j=1}^{n} c_j x^j \qquad (2.4.35)$$

forms a nondecreasing sequence, so that it is sufficient to show that it is bounded for some $x$. The supremum of the $x$'s for which the partial sums are bounded gives the radius of convergence of the series.

In the second place we have the Pringsheim theorem (see the end of Section 1.8), which states that for a power series with positive coefficients the point $z = R$ is a singularity of the function defined by the series. If the coefficients are increased, the radius of convergence cannot increase and the singularity either stays put or moves toward the origin. Both observations show the importance of having a convenient dominant for $F(z, w)$ at one's disposal. This question will be examined in greater detail in Section 2.7.

### EXERCISE 2.4

1. Show that the function (2.4.6) equals

$$1 - z^{-1} + 2!z^{-2} - \cdots + (-1)^n n! z^{-n} - (-1)^n (n+1)! z e^z \int_z^\infty t^{-n-2} e^{-t} \, dt.$$

   Use integration by parts. Estimate the remainder.
2. Verify Lemma 2.4.1.
3. Take the system $w' = z^2 + w^2$, $w(0) = 0$, and verify that the formal solution at the origin is $z^3$ times a power series in $z^4$. What does this imply for the singularities of the solution?
4. In problem 3 $w'(x) > [w(x)]^2$ if $z = x > 0$. Deduce from this that $w(x)$ cannot stay bounded for all $x > 0$ and that there is a finite $\omega$ such that $w(x) \uparrow +\infty$ as $x \uparrow \omega$. Furthermore, $(\omega - x)w(x) < 1$ for $0 < x < \omega$.
5. We have $w'(x) < \omega^2 + [w(x)]^2$ for $0 < x < \omega$. Deduce from this that arc tan $[w(x)/\omega] < \omega x$ and $\omega > (\frac{1}{2}\pi)^{1/2}$. Thus the radius of convergence of the series giving the solution of Problem 3 is at least $(\frac{1}{2}\pi)^{1/2}$.

### 2.5. THE CAUCHY MAJORANT

We consider the system

$$w'(z) = F[z, w(z)], \qquad w(0) = 0, \tag{2.5.1}$$

with

$$F(z, w) = \sum_{j=0}^\infty \sum_{k=0}^\infty c_{jk} z^j w^k. \tag{2.5.2}$$

Here the double series is absolutely convergent in the dicylinder

$$D: |z| \le a, \qquad |w| \le b. \tag{2.5.3}$$

It is desired to prove the convergence of the formal solution

$$\sum_{n=1}^\infty c_n z^n \tag{2.5.4}$$

by exhibiting a suitable majorant $G(z, w)$ of $F(z, w)$ for which the system

$$W'(z) = G[z, W(z)], \qquad W(0) = 0. \tag{2.5.5}$$

has an absolutely convergent series solution,

$$W(z) = \sum_{n=1}^{\infty} C_n z^n.$$  (2.5.6)

To this end we note that the series

$$M(a, b) = \sum_{j=0}^{\infty} \sum_{k=0}^{\infty} |c_{jk}| a^j b^k$$  (2.5.7)

is convergent. In particular its terms are bounded; $M(a, b) \equiv M$ is trivially a bound so that

$$|c_{jk}| \leq M a^{-j} b^{-k}, \qquad \forall j, k.$$  (2.5.8)

Thus we can take

$$G(z, w) = M \sum_{j=0}^{\infty} \sum_{k=0}^{\infty} \left(\frac{z}{a}\right)^j \left(\frac{w}{b}\right)^k = \frac{Mab}{(a - z)(b - w)}.$$  (2.5.9)

Now the system

$$W'(z) = Mab(a - z)^{-1}[b - W(z)]^{-1}, \qquad W(0) = 0$$  (2.5.10)

can be solved explicitly. We have

$$W' - \frac{1}{b} WW' = \frac{Ma}{a - z}$$

and

$$W(z) - \frac{1}{2b} [W(z)]^2 = -Ma \log \left(1 - \frac{z}{a}\right),$$

whence

$$W(z) = b - \left[b^2 + 2abM \log \left(1 - \frac{z}{a}\right)\right]^{1/2}.$$  (2.5.11)

Here the logarithm has its principal value ($\log u$ real if $u > 0$) and $(b^2)^{1/2} = + b$. The majorant series is the Maclaurin series for $z \mapsto W(z)$. It has positive coefficients, and the series converges for $|z| < R$, where $R$ is the singularity on the positive real axis nearest to the origin. The singularities of $W(z)$ are (i) singularities of the logarithm and (ii) singularities of the square root. Here $z = a$ is the only finite singularity of the logarithm. The square root becomes singular at the point where the expression inside the square root is zero; this occurs for

$$z = a\left[1 - \exp \left(-\frac{b}{2aM}\right)\right] \equiv R$$  (2.5.12)

since it is $< a$. Thus we have proved

**THEOREM 2.5.1**

*The system* (2.5.1) *has a series solution* (2.5.4), *which is absolutely convergent in the disk* $|z| < R$ *with* R *defined by* (2.5.12). *The coefficients are uniquely defined by* (2.4.32).

The quantity $R$ is defined by (2.5.12) in terms of $a$, $b$, and $M$. These quantities are not uniquely determined by $F(z, w)$, as observed in Section 1.11 in the discussion of the transformation $T(z, w)$. Here we have two alternatives. We can regard the result as a pure existence proof: there exists a disk $D_0$: $|z| < r$ in which the formal series solution is absolutely convergent and defines an actual solution. Or we can try to find the optimal choice of $a$ and $b$ (which determine $M$), i.e., a choice that gives the largest possible value for $R$. This is usually a difficult problem, and one may have to be satisfied with something less than the best result.

As an illustration, let us take

$$F(z, w) = (1 - z - w)^{-1} = \sum_{j=0}^{\infty} \sum_{k=0}^{\infty} \frac{(j + k)!}{j!k!} z^j w^k. \qquad (2.5.13)$$

Here we take $a = p$, $b = 1 - p$, $0 < p < 1$. The series diverges for this choice, but the terms are bounded. In fact, we may take $M = 1$ since

$$1 = [p + (1 - p)]^n = \sum_{j=0}^{n} \frac{n!}{j!(n - j)!} p^j (1 - p)^{n-j}.$$

This choice gives

$$r(p) = p\left[1 - \exp\left(\frac{1}{2} - \frac{1}{2p}\right)\right],$$

and now the problem is to maximize $r(p)$ by a proper choice of $p$. Now $r(p)$ has a unique maximum, and

$$0.211 < \max r(p) < 0.213. \qquad (2.5.14)$$

How close is this value to the actual radius of convergence of the series? Since the coefficients of $F(z, w)$ are positive, those of the series solution will have the same property and $z = R$ is the singularity nearest to the origin. This means that the essential features of the solution are discernible in the real domain. Now the solution curve

$$y = f(x) \qquad (2.5.15)$$

goes from the origin to a point $P_0$ on the line $x + y = 1$. Its slope is 1 at the origin and $+\infty$ at $P_0(x_0, 1 - x_0)$. This means that the integral curve is confined to the triangle

$$\triangle: x < y < 1 - x, \qquad \text{and} \qquad R = x_0 < \tfrac{1}{2}. \qquad (2.5.16)$$

As we shall see later, the singularity at $z = x_0$ is a branch point of order 1:

$$y = 1 - x_0 - [2(x_0 - x)]^{1/2} + \cdots . \tag{2.5.17}$$

As a matter of fact, we can determine $R = x_0$ exactly, for $x$ as a function of $y$ satisfies the linear DE

$$\frac{dx}{dy} = 1 - x - y,$$

and the solution with the initial value $x = 0$ for $y = 0$ equals

$$x = 2 - y - 2e^{-y}. \tag{2.5.18}$$

For $y = y_0$ the slope of the curve (2.5.18), i.e., $x'(y_0)$, is 0 since $y'(x_0) = +\infty$, and this shows that $y_0 = \log 2$, $x_0 = 1 - \log 2 = 0.30685$. Since this is the value of $R$, the estimate (2.5.14) is rather far off the mark.

### EXERCISE 2.5

1. Verify (2.5.18).
2. Find a lower bound for $R$ in the system $w' = z^2 + w^2$, $w(0) = 0$, and compare your result with that obtained in Problem 2.4:5.
3. It is desired to extend Theorem 2.5.1 to systems (compare Theorem 2.2.2). Take $z_0 = w_{10} = w_{20} = \cdots = w_{n0} = 0$ and

$$F_m(z, w) = \sum c_{j,k_1,\ldots,k_n}^{(m)} z^j w_1^{k_1} \cdots w_n^{k_n}, \qquad 1 \leqslant m \leqslant n.$$

Suppose that the terms of these series are uniformly bounded by $M$ for $z = a$, $w_1 = \cdots = w_n = b$, where $a$ and $b$ are positive numbers. Show that each $F_m$ admits of

$$M\left(1 - \frac{z}{a}\right)^{-1}\left(1 - \frac{w}{b}\right)^{-n}$$

as a majorant. Use this fact to find a common majorant for the solution series and corresponding lower bounds for the radii of convergence.

### 2.6. THE LINDELÖF MAJORANT

Cauchy's *calcul des limites* goes back to 1839, and for almost 60 years it remained the only method of majorants. In 1896 Ernst Lindelöf (1870–1940) observed that the best majorant of $F(z, w)$ is obtained by replacing the coefficients $c_{jk}$ by their absolute values, thus taking

$$C_{jk} = |c_{jk}|, \qquad \forall j, k. \tag{2.6.1}$$

He also found that the whole discussion can then be carried out in the real domain and the question of convergence can be replaced by questions of boundedness, which are simpler to handle. This shows again the importance of positivity.

Lindelöf's attack on this problem was simple, direct, and based on first principles, an approach characteristic of his life work. He became the founder of the school of analysis in Finland and had a number of eminent pupils. Among them were L. V. Ahlfors and F. and R. Nevanlinna, whose work will be discussed in Chapter 4.

As usual we set

$$G(x, y) = \sum_{j=0}^{\infty} \sum_{k=0}^{\infty} C_{jk} x^j y^k \qquad (2.6.2)$$

for the majorant of $F(x, y)$. Here the variables are real positive. Consider the $n$th partial sum,

$$P_n(x) = \sum_{m=0}^{n} C_m x^m, \qquad (2.6.3)$$

of the formal series solution of the problem

$$y'(x) = G[x, y(x)], \qquad y(0) = 0. \qquad (2.6.4)$$

Suppose that the series (2.6.2) is convergent for $x = a$, $y = b$, and set

$$\sum_{j=0}^{\infty} \sum_{k=0}^{\infty} C_{jk} a^j b^k \equiv M(a, b) < \infty. \qquad (2.6.5)$$

Next we note that

$$P'_{n+1}(x) \le \sum_{j=0}^{n} \sum_{k=0}^{n} C_{jk} x^j [P_n(x)]^k, \qquad \forall n. \qquad (2.6.6)$$

This follows from the structure of the recursion formulas for the coefficients which define $C_p$. From the discussion in Section 2.4 we recall that $C_p$ depends on $C_1, C_2, \ldots, C_{p-1}$ and the $C_{jk}$ with $0 \le j + k \le p$. This means that all the terms on the left will cancel against terms on the right. The right member would normally involve additional positive terms not canceled by terms on the left.

Suppose now that

$$r = \min \left[ a, \frac{b}{M(a, b)} \right]. \qquad (2.6.7)$$

Suppose that it is known that for some $n$

$$P_n(x) \le b \qquad \text{for } 0 \le x \le r. \qquad (2.6.8)$$

We have then

$$P'_{n+1}(x) \le \sum_{j=0}^{n} \sum_{k=0}^{n} C_{jk} r^j [P_n(r)]^k$$

$$\le \sum_{j=0}^{n} \sum_{k=0}^{n} C_{jk} a^j b^k \le M(a, b).$$

For $0 \leq x \leq r$ this gives

$$P_{n+1}(x) \leq x M(a, b) \leq r M(a, b) \leq b. \tag{2.6.9}$$

Now

$$P_1(x) = C_{00} x \leq r M(a, b) \leq b,$$

so that the estimate for $P_n$ holds for all $n$.

Now the sequence $P_n(x)$ is nondecreasing for fixed $x$, and since it is bounded for $0 \leq x \leq r$ the partial sums $P_n(x)$ of the formal solution series converge to a finite limit, i.e., the formal solution is an actual solution and its radius of convergence is at least equal to $r$ defined by (2.6.7).

Again we have an optimization problem on our hands. How should $a$ and $b$ be chosen to get the best possible result? Unless other information is available, it may be advantageous to choose $a$ and $b$ so that

$$a M(a, b) = b \tag{2.6.10}$$

and then to maximize with this side condition.

We apply this technique to the system defined in (2.5.13):

$$w' = (1 - z - w)^{-1} w, \qquad w(0) = 0. \tag{2.6.11}$$

Here $M(a, b) = (1 - a - b)^{-1}$, and condition (2.6.10) gives

$$a = \frac{b - b^2}{1 + b}, \tag{2.6.12}$$

the maximum of which is attained for $b = 2^{1/2} - 1$. Then $a_{\max} = 3 - 2 \cdot 2^{1/2} = 0.1716 \ldots$. This is max $r(a, b)$ as obtained by this device, and it is considerably worse than (2.5.14).

As a second example consider the system

$$w' = 2z^3 + w^3, \qquad w(0) = 0. \tag{2.6.13}$$

Here

$$\frac{b}{M(a, b)} = \frac{b}{2a^3 + b^3}.$$

For a fixed $a > 0$ the maximum of this ratio is $\frac{1}{3} a^{-2}$. Hence the optimal choice is $a = \frac{1}{3} a^{-2}$ or $a = 3^{-1/3} = 0.693 \ldots$. This estimate is not too bad. The method of the next section (see Problem 2.6:5) gives as a lower bound for $R$

$$2^{1/9} 3^{-1/6} \pi^{1/3} = 1.316.$$

In 1899 Lindelöf extended his method to give a proof of the *implicit function theorem*, which we state without proof. Given the equation

$$w = F(z, w) = b_{10} z + \sum \sum' b_{jk} z^j w^k, \tag{2.6.14}$$

where the prime indicates that $j + k > 1$, it is desired to solve (2.6.14) for $w$. We set

$$G(z, w) = B_{10}z + \sum_j \sum_k{}' B_{jk}z^j w^k, \qquad B_{jk} = |b_{jk}|. \qquad (2.6.15)$$

**THEOREM 2.6.1**

*Let* a *and* b *be positive numbers such that the series* (2.6.15) *converges for* $z = a$, $w = b$. *Then the formal series solution of* $w = F(z, w)$, $w(0) = 0$, *converges absolutely for* $|z| < R_{a,b}$, *where*

$$R_{a,b} = min\left\{ a, \left[ \frac{1}{G(a, b)} \right]^2 \right\}. \qquad (2.6.16)$$

Lindelöf showed how to find the exact radius of convergence when $G(z, w)$ is an entire function of $z$ and $w$, either a polynomial of degree $\geq 2$ or transcendental in $w$. We shall not consider this question; Theorem 2.6.1 is good enough for our purposes. One further remark about Lindelöf: his monograph, *Le Calcul des Residues et ses Applications à la Théorie des Fonctions* (Gauthier-Villars, Paris, 1909; Chelsea reprint, New York, 1947), is still the best book in this field.

**EXERCISE 2.6**

1. Determine an admissible $r$ for $w' = z^2 + w^2$, $w(0) = 0$.
2. For the system of simultaneous equations in Problem 2.5:3 find Lindelöf majorants and corresponding lower bounds for the radii of convergence of the solution series.

Find admissible values of $r$ for each of the following systems:

3. $w_1' = -w_2$, $w_2' = w_1$, $w_1(0) = 1$, $w_2(0) = 0$.
4. $w_1' = w_2 w_3$, $w_2' = -w_3 w_1$, $w_3' = -k^2 w_1 w_2$, $w_1(0) = 0$, $w_2(0) = w_3(0) = 1$, $0 < k < 1$.
5. For the solutions of Problems 3 and 4 show that

$$[w_1(z)]^2 + [w_2(z)]^2 \equiv 1, \qquad k^2[w_1(z)]^2 + [w_3(z)]^2 \equiv 1.$$

[*Remark*: These three functions are the *elliptic functions of Jacobi* or, in Gudermann's notation, sn $(z; k)$, cn $(z; k)$, and dn $(z; k)$.]

6. Solve the system in Problem 3.
7. What is the simplest differential equation satisfied by $e^z$? Use a uniqueness theorem to show that $e^z \neq 0$ for all $z$.

## 2.7. THE USE OF DOMINANTS AND MINORANTS

A power series has a finite radius of convergence, $R$ say, iff the function defined by the series

$$z \mapsto f(z) = \sum_0^\infty f_n z^n \qquad (2.7.1)$$

has no singularity in $|z| < R$ but has a singular point on the rim of the disk $|z| \leqslant R$. Naturally this applies also to the solution of the DE

$$w' = F(z, w) \qquad \text{with } w(0) = 0, \qquad (2.7.2)$$

but neither the method of Cauchy nor that of Lindelöf gives much information about the distance to the nearest singularity, much less about its nature.

These defects may be partly remedied for the class of DE's

$$w' = P(z, w), \qquad w(0) = 0, \qquad (2.7.3)$$

where $P(z, w)$ is a polynomial in $z$ and $w$ with positive coefficients. In this case the solution

$$w(z) = \sum_{1}^{\infty} C_n z^n \qquad (2.7.4)$$

also has nonnegative coefficients, and the point $z = R$ is a singularity of $w(z)$. Suppose that

$$P(z, w) = \sum \sum C_{jk} z^j w^k, \qquad (2.7.5)$$

where only a finite number of the coefficients are different from zero. More precisely, we assume the existence of integers $m, n, p$ such that

$$C_{jk} \geqslant 0, \quad \forall j, k; \qquad C_{jk} = 0, \quad j > m > 1, \ k > n > 1; \qquad (2.7.6)$$

$$C_{j0} = 0, \quad 0 \leqslant j < p \leqslant m, \qquad C_{p0} > 0; \qquad C_{mn} > 0. \qquad (2.7.7)$$

The first result in this setting is now

**THEOREM 2.7.1**

*Under the stated assumptions on* $P(z, w)$ *the radius of convergence* R *of the solution (2.7.4) is finite. The solution tends to* $+\infty$ *as* $z \uparrow R$ *along the real axis. Moreover, there exists numbers* A *and* B *depending only on* m, n, p, $C_{p0}$, *and* $C_{mn}$ *such that*

$$0 < A < R < B < \infty. \qquad (2.7.8)$$

*Proof.* We take $z = x$, $w \equiv y$ as real positive; then with $w(z) = y(x)$ we have

$$y'(x) > C_{p0} x^p, \qquad 0 < x < R, \qquad (2.7.9)$$

so that

$$y(x) > \frac{C_{p0}}{p+1} x^{p+1}. \qquad (2.7.10)$$

Next we note that $y(x)$ increases with $x$ and tends to a limit as $x$ tends to $R$. If $R$ should be infinite, then obviously

$$\lim_{x \uparrow R} y(x) = +\infty. \tag{2.7.11}$$

But this is true also if $R$ is finite, for if $\lim_{x \uparrow R} y(x)$ is finite, equal to $y_0$ say, then $P(z, w)$ is holomorphic at $(R, y_0)$ and the initial-value problem

$$w'(z) = P[z, w(z)], \qquad w(R) = y_0$$

has a unique solution holomorphic in some disk $|z - R| < \delta$. This solution must coincide with the solution (2.7.2) on the interval $(R - \delta, R)$ and hence furnish the analytic continuation of $w(z)$ in the disk $|z - R| < \delta$. This contradicts the fact that $z = R$ is a singularity of $w(z)$.

Thus (2.7.11) holds whether or not $R$ is finite. We proceed to prove that $R$ is finite. Since

$$y'(x) > C_{mn} x^m [y(x)]^n, \tag{2.7.12}$$

we have, for $0 < x_1 < x_2 < R$,

$$[y(x_1)]^{1-n} - [y(x_2)]^{1-n} > \frac{n-1}{m+1} C_{mn} (x_2^{m+1} - x_1^{m+1}). \tag{2.7.13}$$

Here the left-hand side is less than the first term, while the right member becomes infinite with $x_2$. This shows that $R$ must be finite.

We can now let $x_2$ increase to $R$. Then for all $x_1$ in $(0, R)$

$$R^{m+1} \leq x_1^{m+1} + \frac{m+1}{n-1} (C_{mn})^{-1} [y(x_1)]^{1-n}.$$

For the last term we get an upper bound from (2.7.10). The result is of the form

$$R^{m+1} \leq x_1^{m+1} + C(x_1)^{-(n-1)(p+1)}, \tag{2.7.14}$$

where $C$ is independent of $x_1$. This holds for all $x_1$, $0 < x_1 < R$, and, in particular, for the value $x_1 = A$, for which the right member becomes a minimum. Here $0 < A < R$. Now define $B$ by

$$B^{m+1} = A^{m+1} + CA^{-(n-1)(p+1)}. \tag{2.7.15}$$

We have then clearly $0 < A < R < B$, and the theorem is proved. ∎

We have shown that $y(x) \to +\infty$ as $x$ increases to $R$. We can actually get an estimate of the rate of growth. To this end we consider (2.7.13) once more. We set $x_1 = x$, and let $x_2 \to R$. This gives

$$[y(x)]^{1-n} > \frac{n-1}{m+1} C_{mn} (R^{m+1} - x^{m+1})$$

$$= (n-1) C_{mn} x_0^m (R - x).$$

Here the mean value theorem has been used, and $x < x_0 < R$. Hence

$$y(x) < [(n-1) C_{mn} x_0^m]^{-1/(n-1)} (R-x)^{-1/(n-1)} \tag{2.7.16}$$

and

$$\limsup_{x \to R} [(R - x)^{1/(n-1)} y(x)] \leq [(n - 1)C_{mn}R^m]^{-1/(n-1)}. \qquad (2.7.17)$$

On the other hand, for a small $\delta > 0$ we can find an $\epsilon > 0$ such that, for $R - \delta < x < R$,

$$y'(x) < (1 + \epsilon)C_{mn}R^m [y(x)]^n. \qquad (2.7.18)$$

This gives

$$[y(x)]^{1-n} < (1 + \epsilon)(n - 1)C_{mn}R^m (R - x)$$

and

$$y(x) > [(1 + \epsilon)(n - 1)C_{mn}R^m]^{-1/(n-1)}(R - x)^{-1/(n-1)}, \qquad (2.7.19)$$

whence

$$\liminf_{x \to R} (R - x)^{1/(n-1)} y(x) \geq [(n - 1)C_{mn}R^m]^{-1/(n-1)}. \qquad (2.7.20)$$

Combining (2.7.17) and (2.7.20), we obtain

$$\lim_{x \to R} [(R - x)^{1/(n-1)} y(x)] = [(n - 1)C_{mn}R^m]^{-1/(n-1)}. \qquad (2.7.21)$$

Thus we have proved   ∎

### THEOREM 2.7.2

*Under the stated assumptions on* P(z, w) *the solution* w(z) *of (2.7.3) has a singularity at* z = R, *where*

$$\lim_{z \to R} [(R - z)^{1/(n-1)} w(z)] = [(n - 1)C_{mn}R^m]^{-1/(n-1)}, \qquad (2.7.22)$$

*at least for radial approach of* z *to* R.

The right-hand members of (2.7.9) and (2.7.12) may be referred to as *minorants* of $P(x, y)$, while the right member of (2.7.18) ranks as a *dominant*. We can obviously find other useful dominants and minorants for this class of equations. Thus $P(R, y)$ is evidently a dominant, and from

$$y'(x) \leq P[R, y(x)], \qquad 0 < x < R \qquad (2.7.23)$$

we get

$$\int_0^\infty \frac{dt}{P(R, t)} \leq R. \qquad (2.7.24)$$

Since the left side is a function of $R$ alone, say $Q(R)$, we have the inequality

$$Q(R) \leq R. \qquad (2.7.25)$$

Now $Q(R)$ is evidently a decreasing function of $R$, so the equation

$$Q(r) = r \qquad (2.7.26)$$

has one and only one positive root, $r_0$ say, and this implies

$$r_0 \leqslant R. \qquad (2.7.27)$$

This inequality is sometimes a useful one.

**EXERCISE 2.7**

1.  Is it essential for the discussion given above that $P(0, 0) = 0$ and that $P(z, 0)$ is not identically zero? What modifications in the argument are needed if these assumptions are dropped?
2.  Instead of $w(0) = 0$, would $w(0) = y_0 > 0$ be manageable?
3.  The restriction of $P(z, w)$ to be a polynomial in $z$ and $w$ is evidently undesirable. Can some modification of the method be applicable to the DE

    $$w'(z) = [1 + w^2(z)]^{1/2}[1 + k^2 w^2(z)]^{1/2}, \qquad 0 < k < 1,$$

    with $w(0) = 0$? This is an elliptic DE satisfied by $-i \operatorname{sn}(iz; k)$.
4.  Consider the equation $w' = 1 + w^2$, $w(0) = 0$. What information is obtainable from Theorems 2.7.1 and 2.7.2 in this case? Does (2.7.27) give reasonable results? Note that $R = \frac{1}{2}\pi$.
5.  Apply the methods of this section to the system

    $$w' = 2z^3 + w^3, \qquad w(0) = 0.$$

    In particular, use $y'(x) < 2R^3 + y^3$ for $0 < x < R$. The gamma function satisfies $\Gamma(p)\Gamma(1 - p) = \pi \operatorname{cosec} p\pi$, which may prove useful.

## 2.8. VARIATION OF PARAMETERS

As usual we are concerned with an equation

$$w' = F(z, w), \qquad (2.8.1)$$

where $(z, w) \mapsto F(z, w)$ is holomorphic in some dicylinder $D$ of $\mathbb{C}^2$. We plan to examine the dependence of the solution on *parameters*.

Parameters are of two kinds: *internal* and *external*. A parameter is internal if its value characterizes a particular solution or possibly a class of solutions but does not occur explicitly in the equation.

As an orientation consider the trivial Riccati equation

$$w' = a + bw^2. \qquad (2.8.2)$$

Here $a$ and $b$ are external parameters. Suppose that the solution

$w(z; z_0, w_0)$ is under consideration. Here $D = \mathbf{C}^2$. Now

$$w(z; 0, 0) = \left(\frac{a}{b}\right)^{1/2} \tan[(ab)^{1/2}z]. \tag{2.8.3}$$

For $a = 0$ the solution is identically zero. For $b = 0$ it reduces to $az$. In both cases (2.8.3) reduces to the indicated solution if we pass to the limit with the parameter in question, keeping $z$ and the other parameter fixed. The formula shows that $w(z; 0, 0)$ is an analytic function of each of the parameters once the determination of the square root has been fixed so that $w(z; 0, 0)$ is single valued. Suppose that both square roots are real positive when $a$ and $b$ are real positive. Now keep $z$ and $b$ fixed, neither being zero, and consider the solution as a function of $a$. It is clearly a meromorphic function of $a$ in the finite complex $a$-plane. Similar results hold if we fix $a$ and $z$ and vary $b$.

On the other hand,

$$w(z; z_0, w_0) = \left(\frac{a}{b}\right)^{1/2} \tan\left\{(ab)^{1/2}(z - z_0) + \arctan\left[\left(\frac{b}{a}\right)^{1/2} w_0\right]\right\}. \tag{2.8.4}$$

This is clearly an analytic function of $z_0$ as well as of $w_0$ when the other varieties are kept fixed.

There are many other internal parameters. Constants of integration play a remarkable role in the advanced theory of nonlinear DE's, as will be seen later. In the present case the general solution of (2.8.2) may be written as

$$\frac{p(\sin pz - Cp \cos pz)}{b(\cos pz + C \sin pz)}, \qquad p = (ab)^{1/2}, \tag{2.8.5}$$

where $C$ is an arbitrary constant of integration. The solution is evidently a fractional linear ($=$ Möbius) transform of $C$, and this property is characteristic of Riccati's equation (see Section 4.1). Again we see that the solution is an analytic function of the internal parameter $C$ for fixed $a$, $b$, and $z$. Many other types of internal parameters could be shown and will be presented in later contexts.

Generally speaking, if the case (2.6.2) is at all typical we would expect the solution of (2.8.1) to be an analytic function of whatever parameters, external or internal, may enter into the problem of determining a solution. This is indeed the case. We obtain numerous results along such lines from Gronwall's lemma; see Theorems 1.6.5 and 1.6.6.

We start by getting information about the dependence of $w(z; z_0, w_0)$ on the initial value $w_0$. Here the basic fact is that, if $F(z, w)$ is holomorphic in $D$, it has a partial derivative $F_w(z, w)$ with respect to $w$ which is also holomorphic in $D$. Shrinking $D$ if necessary, we may assume that the partial derivative is bounded:

$$|F_w(z, w)| \le B \qquad \text{for } (z, w) \in D. \qquad (2.8.6)$$

This implies a Lipschitz condition,

$$|F(z, w_1) - F(z, w_2)| \le B|w_1 - w_2|. \qquad (2.8.7)$$

Furthermore, we have

$$F(z, w_1) - F(z, w_2) = F_w(z, w_1)(w_1 - w_2) + o(|w_1 - w_2|). \qquad (2.8.8)$$

Let us now consider the so-called *variational system* for the solution $w(z; z_0, w_0)$:

$$W'(z) = F_w[z, w(z; z_0, w_0)]W(z), \qquad W(z_0) = 1, \qquad (2.8.9)$$

with the unique solution

$$W(z; z_0, 1) = \exp\left\{\int_{z_0}^{z} F_w[s, w(s; z_0, w_0)] \, ds\right\}. \qquad (2.8.10)$$

Since $D$ is convex, we may take the integrals here and in the following along a straight line.

Consider two solutions, $w(z; z_0, w_0)$ and $w(z; z_0, w_1)$, of (2.8.1). Here of course $(z_0, w_0)$ and $(z_0, w_1)$ are in $D$. Let $D_0$ be a disk in the $z$-plane with center at $z = z_0$ in which the two solutions are holomorphic and bounded. We note first that the two solutions are proximate if $|w_1 - w_2|$ is small. More precisely, we have

**THEOREM 2.8.1**

*There exists a disk $D^*$ in the w-plane such that, if $w_0$ and $w_1$ are both in $D^*$ and $z$ lies in $D_0$, then*

$$|w(z; z_0, w_1) - w(z; z_0, w_0)| \le |w_1 - w_0| \, exp \, [B|z - z_0|]. \qquad (2.8.11)$$

*Proof.* To simplify the formulas we write $w_1(z)$ and $w_0(z)$ for the two solutions. Then

$$w_1(z) - w_0(z) = w_1 - w_0 + \int_{z_0}^{z} \{F[s, w_1(s)] - F[s, w_0(s)]\} \, ds.$$

By the Lipschitz condition (2.8.7) this gives

$$|w_1(z) - w_0(z)| \le |w_1 - w_0| + B\left|\int_{z_0}^{z} |w_1(s) - w_0(s)| \, dr\right|,$$

where $r$ is the arc length on the rectilinear path of integration. We can easily reduce this inequality to the form (1.6.15) by setting

$$f(t) = |w_1(z_0 + t\,e^{i\theta}) - w_0(z_0 + t\,e^{i\theta})|, \qquad \theta = \arg(z - z_0),$$
$$g(t) \equiv |w_1 - w_0|, \qquad \text{and} \qquad K(t) \equiv B.$$

We then get (2.8.11) by substituting in (1.6.18). ■

Much more may be said, however.

**THEOREM 2.8.2**

*The solution* $w(z; z_0, w_0)$ *is a differentiable function of the complex variable* $w_0$ *and hence locally holomorphic. We have*

$$w(z; z_0, w_1) - w(z; z_0, w_0) = W(z; z_0, 1)(w_1 - w_0) + o(|w_1 - w_0|) \qquad (2.8.12)$$

*and*

$$\frac{\partial}{\partial w_0} w(z; z_0, w_0) = exp\left\{\int_{z_0}^{z} F_w[s, w(s; z_0, w_0)]\,ds\right\}. \qquad (2.8.13)$$

*Proof.* We revert to the abridged notation used above. The problem is to show that

$$\Delta(z) \equiv w_1(z) - w_0(z) - W(z)(w_1 - w_0) = o(|w_1 - w_0|). \qquad (2.8.14)$$

Now the second member of this equation reduces to

$$\int_{z_0}^{z} \{F[s, w_1(s)] - F[s, w_0(s)] - (w_1 - w_0)F_w[s, w_0(s)]W(s)\}\,ds.$$

By (2.8.8) this becomes

$$\int_{z_0}^{z} F_w[s, w_0(s)][w_1(s) - w_0(s) - (w_1 - w_0)W(s)]\,ds + o(|w_1 - w_0|).$$

Thus we have

$$|\Delta(z)| \leq o(|w_1 - w_0|) + B\left|\int_{z_0}^{z} |\Delta(s)|\,ds\right|. \qquad (2.8.15)$$

This can also be reduced to the standard form (1.6.15) in an obvious manner and yields the inequality

$$|\Delta(z)| \leq o(|w_1 - w_0|) \exp[B|z - z_0|], \qquad (2.8.16)$$

which implies (2.8.12) and (2.8.13). Note that the difference quotient tends to $W(z; z_0, 1)$, no matter how $w_1$ tends to $w_0$. The analyticity is implied by this fact. ■

This theorem was discovered in 1896 by Ivar Bendixson, using a function-theoretical argument. The use of Gronwall's inequality is of later date. Theorem 2.8.2 is in a certain sense a forerunner of the inequality. The theorem extends to systems of $n$ first order DE's in $n$ unknowns. Here Jacobians come into play and the formulas become fairly complicated. See Halanay (1966); the first extension was given by G. Peano in 1897.

We turn now to the dependence of $w(z; z_0, w_0)$ on the internal parameter $z_0$. We have analogues of Theorems 2.8.1 and 2.8.2 for this case. Since the proofs follow the same lines, the argument will merely be sketched. We assume that

$$|F(z, w)| \leq M, \qquad (z, w) \in D. \tag{2.8.17}$$

Consider two solutions of (2.8.1):

$$w(z; z_1, w_0) \equiv w_1(z), \qquad w(z; z_2, w_0) \equiv w_2(z), \tag{2.8.18}$$

where $z$, $z_1$, and $z_2$ are restricted to a disk $D_0$ with center at $z = z_0$, in which $w_1(z)$ and $w_2(z)$ are holomorphic and bounded.

### THEOREM 2.8.3

*With the stated assumptions*

$$|w(z; z_1, w_0) - w(z; z_2, w_0)| \leq M|z_1 - z_2| \, exp \, [B \, \delta(z)], \tag{2.8.19}$$

*where* $\delta(z) = min \, [|z - z_1|, |z - z_2|]$.

*Proof (sketch).* Suppose that $z$ is nearer to $z_1$ than to $z_2$. Then

$$w_1(z) = w_0 + \int_{z_1}^{z} F[s, w_1(s)] \, ds,$$

$$w_2(z) = w_0 + \int_{z_1}^{z} F[s, w_2(s)] \, ds - \int_{z_1}^{z_2} F[s, w_2(s)] \, ds,$$

where we have used the fact that the integral of $F[s, w_2(s)]$ along the perimeter of a triangle with vertices at $z$, $z_1$, and $z_2$ is zero. Now the last integral in the displayed formulas is less than $M|z_1 - z_2|$ in absolute value. Using the Lipschitz condition (2.8.7), we get

$$|w_2(z) - w_1(z)| \leq M|z_2 - z_1| + B \left| \int_{z_1}^{z} |w_2(s) - w_1(s)| |ds| \right|.$$

By (1.6.18) this gives

$$|w_2(z) - w_1(z)| \leq |z_2 - z_1| \exp \, [B|z - z_1|].$$

Since $\delta(z) = |z - z_1|$ in this case (2.8.19) is verified. ∎

### THEOREM 2.8.4

*With the stated assumptions, let* $W(z) = W(z; z_0, 1)$ *be the solution of* (2.8.9). *Then* $w(z; z_0, w_0)$ *is a differentiable function of the complex parameter* $z_0$ *and*

$$\frac{\partial}{\partial z_0} w(z; z_0, w_0) = - F(z_0, w_0) W(z). \tag{2.8.20}$$

*Proof* (*sketch*). Consider two solutions of (2.8.1), namely, $w_0(z) = w(z; z_0, w_0)$ and $w_1(z) = w(z; z_1, w_0)$, where $z, z_0, z_1$ are in the disk $D_0$. Set

$$D(z) \equiv w_1(z) - w_0(z) + F(z_0, w_0) W(z)(z_1 - z_0). \tag{2.8.21}$$

It is required to prove that $D(z) = o(|z_1 - z_0|)$ for any $z_1$ in $D_0$. We have

$$w_1(z) - w_0(z) = \int_{z_1}^{z} F[s, w_1(s)] \, ds - \int_{z_0}^{z} F[s, w_0(s)] \, ds$$

$$= \int_{z_0}^{z} \{F[s, w_1(s)] - F[s, w_0(s)]\} \, ds - \int_{z_0}^{z_1} F[s, w_1(s)] \, ds.$$

Here the first integral is by (2.8.8)

$$\int_{z_0}^{z} F_w[s, w_0(s)][w_1(s) - w_0(s)] \, ds + E_1.$$

The error term $E_1$ is small by (2.8.8) and (2.8.19), for $E_1$ goes to zero with $|z_1 - z_2|$ faster than

$$\left| \int_{z_0}^{z} |w_1(s) - w_0(s)| \|ds\| \right| \le M|z_1 - z_0| \left| \int_{z_0}^{z} \exp[\delta(s)] \, |ds| \right|,$$

so that $E_1 = o(|z_1 - z_0|)$.

Next, note that

$$W(z) = 1 + \int_{z_0}^{z} F_w[s, w_0(s)] W(s) \, ds.$$

Hence

$$-\int_{z_0}^{z_1} F[s, w_0(s)] \, ds + F(z_0, w_0)(z_1 - z_0)$$

$$= -\int_{z_0}^{z_1} \{F[s, w_0(s)] - F(z_0, w_0)\} \, ds,$$

the absolute value of which does not exceed

$$B \left| \int_{z_0}^{z_1} |w_0(s) - w_0| \|ds| \right| \le B \left| \int_{z_0}^{z_1} |ds| \left| \int_{z_0}^{s} |F[t, w_0(t)]| \|dt| \right| \right| \le \tfrac{1}{2} B^2 |z_1 - z_0|^2.$$

Combining the various loose ends, we get

$$D(z) = \int_{z_0}^{z} F_w[s, w_0(s)] D(s) \, ds + E_2 \qquad \text{with } E_2 = o(|z_1 - z_0|), \tag{2.8.22}$$

and this implies the differentiability of $w(z; z_0, w_0)$ with respect to $z_0$ and gives the value of the derivative as in (2.8.20). ∎

The remaining remarks in this section are devoted to external parameters. We are now concerned with an equation

$$w' = F(z, w, \lambda), \qquad w(z_0, \lambda_0) = w_0, \tag{2.8.23}$$

where $(z, w, \lambda) \mapsto F(z, w, \lambda)$ is holomorphic in some tricylinder $D$ in $\mathbf{C}^3$. We may assume that $D$ is centered at the origin and that

$$F(z, w, \lambda) = \sum_0^\infty \sum_0^\infty \sum_0^\infty c_{ijk} z^i w^j \lambda^k \tag{2.8.24}$$

convergent in $D$. It is then natural to take $z_0 = w_0 = \lambda_0 = 0$, to assume a power series solution

$$w(z, \lambda) = \sum_0^\infty \sum_0^\infty c_{mn} z^m \lambda^n, \tag{2.8.25}$$

and to prove convergence using Cauchy's majorant method. We can find positive numbers $a$, $b$, $c$, $M$ such that

$$|c_{ijk}| a^i b^j c^k \leqslant M, \qquad \forall i, j, k.$$

Hence

$$G(z, w, \lambda) \equiv M \left(1 - \frac{z}{a}\right)^{-1} \left(1 - \frac{w}{b}\right)^{-1} \left(1 - \frac{\lambda}{c}\right)^{-1} \tag{2.8.26}$$

is a majorant of $F(z, w, \lambda)$, and the solution of

$$W' = G(z, W, \lambda), \qquad W(0, 0) = 0 \tag{2.8.27}$$

will be a majorant of (2.8.25). This majorant is easily found to be

$$W(z, \lambda) = b - \left[ b^2 + \frac{2abcM}{c - \lambda} \log\left(1 - \frac{z}{a}\right) \right]^{1/2}. \tag{2.8.28}$$

This is a holomorphic function of $(z, \lambda)$ in a dicylinder in $\mathbf{C}^2$, and, in particular, for a fixed small value of $|z|$ it is a holomorphic function of $\lambda$ in a disk. Thus we have proved ∎

**THEOREM 2.8.5**

*The system* (2.8.25) *with* $z_0 = w_0 = \lambda_0 = 0$ *has a unique solution* (2.8.25) *which is a holomorphic function of* $(z, \lambda)$ *in any dicylinder in* $\mathbf{C}^2$ *where* $W(z, \lambda)$ *is holomorphic.*

This is a powerful theorem and may, in fact, be considered too powerful for applications where the external parameters are normally linear. In such a case the solution would usually be an entire function of $\lambda$

of low order or else a meromorphic function, also of low order. In conclusion let us remark that expansion in powers of a parameter is commonly used, for instance, in celestial mechanics, where Poincaré put this useful device on a rigorous basis in his lectures on this subject in the 1890's.

## EXERCISE 2.8

1. Complete the proofs of Theorems 2.8.3 and 2.8.4.
2. In (2.8.28) for $W(z, \lambda)$ we have $|z| < a$ and $|\lambda| < c$. If $\lambda$ is fixed and $|\lambda| < c$, find the singularities of $W(z, \lambda)$ as a function of $z$. Which singularity is nearest to the origin?
3. The equation $(1 - z)w' = \lambda w$, $w(0) = 1$, $\forall \lambda$, has a solution holomorphic in the unit disk of the $z$-plane for all $\lambda$. For a fixed $z$, $|z| < 1$, verify that the solution is an entire function of $\lambda$. Estimate the maximum modulus, and compare $\log M[r; w(\lambda)]$ with powers of $r = |\lambda|$.
4. Apply Poincaré's method to the equation
$$w'' - (z + \lambda)w = 0, \qquad w(0) = 1, \qquad w'(0) = 0, \quad \forall \lambda.$$
This involves setting $w(z) = w_0(z) + \Sigma_1^\infty w_n(z)\lambda^n$ with $w_0(0) = 1$, $w_0'(0) = 0$, $w_n(0) = w_n'(0) = 0$, $n > 0$. Here $w_0''(z) = zw_0(z)$ and may be assumed to be known. Furthermore, $w_n''(z) = w_{n-1}(z)$, $n > 0$. For a fixed $z = x > 0$, try to estimate $\log M[r; w(\lambda)]$. It should grow as $r^{1/2}$.

## LITERATURE

The presentation is based largely on the author's LODE. Chapter 2 of LODE has a bearing on Sections 2.1–2.6; Chapter 3 goes with Section 2.8. In this treatise essentially only the analytic case is considered. Section 2.7 contains some new material, though the basic idea is inherent in earlier work by the author and others. Further references are:

Banach, S. Sur les opérations dans les ensembles abstraits et leur application aux équations intégrales. *Fundamenta Math.*, **3** (1922), 133–181.

Bendixson, I. Démonstration de l'existence de l'intégral d'une équation aux dérivées partielles linéaires. *Bull. Soc. Math. France*, **24** (1896) 220–225.

Birkhoff, G. D. and O. D. Kellogg. Invariant points in function space. *Trans. Amer. Math. Soc.*, **23** (1922), 96–115.

Brouwer, L. E. J. On continuous vector distributions on surfaces. *Verh. Konikl. Akad. Wet. Amsterdam*, **11** (1909); **12** (1910).

Cacciopoli, R. Un teorema generale sull'essistenza di elementi uniti in una trasformazione funzionale. *Rend. Accad. Lincei Roma*, (6), **11** (1930), 98–115.

Cauchy, A. Mémoire sur l'emploi du nouveau calcul, appelé calcul des limites, dans l'intégration d'un système d'équations différentielles. *Compt. Rend. Acad. Sci. Paris*, **15** (1842); *Oeuvres* (1), **7** (1898), 5–17.

Halanay, A. *Differential Equations: Stability, Oscillations, Time Lags.* Academic Press, New York, 1966. (Re Theorem 2.8.2.)

Lindelöf, E. Démonstration élémentaire de l'existence des intégrales d'un système d'équations différentielles ordinaires. *Acta Soc. Sci. Fenn.*, **21**, No. 7 (1897), 13 pp.

────. Démonstration élémentaire de l'existence des fonctions implicites. *Bull. Sci. Math.*, (2), **23** (1899), 68–75.

Picard, É. Mémoire sur la théorie des équations aux dérivées partielles et la méthode des approximations successives. *J. math. pures appl.*, (4), **6** (1890), 145–210.

Poincaré, H. *Les méthodes nouvelles de la mécanique céleste.* 3 vols. Gauthier-Villars, Paris, 1892, 1893, 1898.

# 3
# SINGULARITIES

In Chapter 2 our study of the equation

$$w'(z) = F[z, w(z)], \qquad w(z_0) = w_0$$

led to the conclusion that there is a unique holomorphic solution whenever the mapping $(z, w) \to F(z, w)$ is holomorphic at $(z_0, w_0)$. In this chapter we shall study what happens at a point where this assumption does not hold. Such a point is a singularity at least for the solution which approaches $w_0$ when $z \to z_0$. But there are other possibilities. Solutions may become infinite as $z \to z_0$ or may not tend to any limit, finite or infinite. Usually the point at infinity is a fixed singularity. We have to examine these various possibilities and try to bring some order into the confusion. Pioneering work in this field was done by Lazarus Fuchs (1833–1902), whose name will be encountered in various later chapters. Here we shall be concerned mostly with the work of Paul Painlevé (1863–1933).

## 3.1. FIXED AND MOVABLE SINGULARITIES

Singularities of DE's are of two types: *fixed* and *movable*. Fixed singularities are external in the sense that they are given position and nature by the DE, and at least the position should be obtainable in a finite number of steps without knowledge of the solutions. Movable singularities, on the other hand, depend on internal parameters, and normally such a point can be put anywhere in the complex plane by manipulating an internal parameter. Shift the parameter and the singularity moves. Thus the equation

$$w' = w^2 - \pi^2 \cot^2 \pi z \qquad (3.1.1)$$

has fixed singularities at the integers and at infinity. The finite singularities are simple poles for some solutions, but the general solution has *transcendental branch points* where infinitely many branches are permuted. The movable singularities are simple poles and may be placed anywhere in the complex plane.

The nature of a movable singularity will depend on the nature of the equation. Thus the equations

$$w' + e^w = 0 \qquad \text{and} \qquad v' + v \log^2 v = 0 \qquad (3.1.2)$$

are satisfied by

$$w = \log \left( \frac{1}{z - c} \right) \qquad \text{and} \qquad v = \exp \left( \frac{1}{z - c} \right), \qquad (3.1.3)$$

respectively. In the first case there is a movable logarithmic branch point; in the second, a movable essential singularity.

In the following discussion such cases will be excluded. In the first order case (but not for systems) this can be achieved by assuming that $F(z, w)$ is a *rational function of $w$ with coefficients which are algebraic functions of $z$*. This restriction makes the number of fixed singularities finite and permits only algebraic movable singularities. During most of our work we shall make the further restriction that the coefficients are single valued; this implies that they may be assumed to be polynomials. We consider equations of the form

$$w' = \frac{P(z, w)}{Q(z, w)} \qquad (3.1.4)$$

with

$$P(z, w) = \sum_{j=0}^{p} A_j(z) w^j, \qquad Q(z, w) = \sum_{k=0}^{q} B_k(z) w^k, \qquad (3.1.5)$$

where the $A$'s and $B$'s are algebraic functions of $z$.

It is appropriate to provide the reader with a brief abstract of the theory of algebraic functions. We shall need some of the properties of such functions for the discussion of the fixed singularities of (3.1.4). For each algebraic function $z \mapsto A(z)$ there is a unique irreducible polynomial in two variables:

$$V(z, u) = p_0(z) u^n + p_1(z) u^{n-1} + \cdots + p_{n-1}(z) u + p_n(z), \qquad (3.1.6)$$

such that $V[z, A(z)] \equiv 0$. Here the $p$'s are polynomials in $z$ without a common factor. Conversely, every such polynomial $V(z, u)$ defines an algebraic function. The zeros of $A(z)$ are the zeros of $p_n(z)$; the infinitudes, the zeros of $p_0(z)$. The point at infinity may be a singular point of $A(z)$. Additional singularities may be furnished by the zeros of the *discriminant equation*

$$D_0(z) = 0. \qquad (3.1.7)$$

If $z = z_0$ is a root of this equation, then the equation

$$V(z_0, u) = 0 \tag{3.1.8}$$

has multiple roots and $z = z_0$ is an algebraic branch point of $A(z)$. Here $D_0(z)$ is the *resultant* of eliminating $u$ between the two equations $V(z, u) = 0$ and $V_u(z, u) = 0$. Since the number of multiple roots is at most equal to the total number of roots, which is $n$, we see that the number of singular points of $A(z)$ is finite and is at most $n + 1 + \deg p_0$.

It should be noted that the algebraic functions of one complex variable form a *field* over C: if $A_1$ and $A_2$ are algebraic functions, then so are their sum, product, and quotients.

We can now start an enumeration of the possible fixed singular points of the DE (3.1.4). The first subset of such points is formed by the singularities of the $A$'s and the $B$'s, say,

$$S_1: s_1, s_2, \ldots, s_\mu, \infty. \tag{3.1.9}$$

This is a finite set. The second subset $S_2$ is the set of points $t_j$, if any, such that $Q(t_j, w)$ is identically zero, say,

$$S_2: t_1, t_2, \ldots, t_\nu. \tag{3.1.10}$$

At $z = t_j$ all the algebraic functions $B_0, B_1, \ldots, B_q$ are zero, and this means that the corresponding polynomials $V_j(z, 0)$ are zero, where $V_j(z, u)$ is the defining polynomial for $B_j(z)$. Now the common roots of a number of polynomials can be found, for instance by the *method of Bezout*. We can affirm, however, that the number of common roots cannot exceed $\min_j (m_j)$, where $m_j$ is the degree of $V_j(z, 0)$.

The third subset,

$$S_3: e_1, e_2, \ldots, e_\lambda, \tag{3.1.11}$$

is formed by the roots of another discriminant $D_3(z)$, the vanishing of which expresses the existence of numbers $w_0$ such that

$$P(e_j, w_0) = 0, \qquad Q(e_j, w_0) = 0, \tag{3.1.12}$$

so that $P/Q$ takes on the indeterminate form $0/0$ at $(z_0, w_0)$. Now the condition for $P(z, w) = 0, Q(z, w) = 0$ to have a common root is the vanishing of a determinant with $p + q$ rows and columns, the entries being the $p + q + 2$ coefficients $A_j$ and $B_k$ with the rows filled out with zeros. Now $D_3(z)$ is an algebraic function of $z$ obtained by expanding the determinant, and as an algebraic function it satisfies a polynomial.

equation $V(z, u) = 0$. Thus the $e$'s are the roots of the polynomial $V(z, 0)$. Hence the set $S_3$ is also finite.

In studying the infinitudes of the solutions of (3.1.4), we use the transformation $w = 1/v$, which transforms (3.1.4) into

$$v' = -\frac{v^2 P(z, 1/v)}{Q(z, 1/v)} = \frac{P_1(z, v)}{Q_1(z, v)}, \tag{3.1.13}$$

where $P_1$ and $Q_1$ are polynomials in $v$ with the $A_j$'s and $B_k$'s as coefficients. The fixed singularities of this equation are also fixed singularities of (3.1.4). The sets $S_1$ and $S_2$ are unchanged, but the result of eliminating $v$ between $P_1 = 0$ and $Q_1 = 0$ is not necessarily the same as that of eliminating $w$ between $P = 0$ and $Q = 0$, since now the common roots of $P_1(z, 0) = 0$ and $Q_1(z, 0) = 0$ are relevant. Thus we have possibly an additional set of fixed singularities,

$$S_4: i_1, i_2, \ldots, i_\kappa. \tag{3.1.14}$$

The set (3.1.14) is a finite one, and the union of the four sets $S_1$–$S_4$ is a finite set $S$. This is the set of potential fixed singularities of (3.1.4).

A point of the set $S$ need not be a singularity of all or any of the solutions, as is illustrated by the following DE's:

$$u' = 1 + z^{3/2}u - z^{1/2}u^2, \qquad w' = \mu w/z. \tag{3.1.15}$$

In both cases $z = 0$ as a singularity of the coefficients is a potential singularity. The first equation has the particular solution $u = z$, which is holomorphic at $z = 0$. The general solution has a branch point of *order 1* there (order 1 means *two* branches). The second equation has the general solution $Cz^\mu$, which is holomorphic at $z = 0$ iff $\mu$ is a positive integer. There is a pole if $\mu$ is a negative integer, an algebraic branch point if $\mu$ is a rational number, and a transcendental critical point in all other cases. Here the nature of the singularity depends essentially on the algebraic nature of the external parameter $\mu$.

This enumeration of potential fixed singularities is due to Paul Painlevé in the 1890's, and we shall see in the next section how important it is for the problem of analytic continuation of the solutions of the equation. Before considering this question, let us mention that Painlevé could look back on a long line of French mathematicians who had contributed to one phase or another of the theory of nonlinear DE's. Besides E. Picard and H. Poincaré special mention should be made of Charles Briot (1817–1882) and Jean Claude Bouquet (1819–1885). In the 1850's they were writing a treatise on *elliptic functions* (the first of its kind). An elliptic function is a doubly periodic meromorphic function. They showed that, if $f$ and $g$ are both elliptic and have the same periods, there exists a polynomial

$P(Z_1, Z_2)$ such that

$$P[f(z), g(z)] \equiv 0.$$

In particular, if $f(z)$ is elliptic so is $f'(z)$, and the two have the same periods so that there exists a polynomial DE,

$$P(w, w') \equiv \sum c_{jk}(w)^j (w')^k = 0, \qquad (3.1.16)$$

which is satisfied by $f(z)$ and, more generally, by $f(z + a)$, where $a$ is an arbitrary constant since (3.1.16) is *autonomous*, i.e., does not contain $z$ explicitly. Similar equations of the second or higher orders also hold for elliptic functions.

Not all equations of this type are satisfied by elliptic functions, and Briot and Bouquet set themselves the task of finding all equations of type (3.1.16) which have elliptic solutions. We shall take up this problem in Chapter 11 for first order equations and in Chapter 12 for second order equations. Briot and Bouquet also studied first order DE's with a fixed singular point at the origin; see Section 11.1.

At this juncture we merely add that the search for elliptic equations is a special case of finding equations of type (3.1.16) whose movable singularities are poles. To this problem we can give at least a partial solution, which permits eliminating a number of equations from the list of suspects; see Section 3.3.

## EXERCISE 3.1

1.  Suppose that $z \mapsto f(z)$ is a solution of (3.1.1) with a simple pole at $z = z_0$. Show that the equation also has a solution with a simple pole at $z = z_0 + n$. Here $z_0$ is an arbitrary complex number, and $n$ is an arbitrary integer. [If $z_0$ is an integer, there exist two distinct solutions which have a simple pole at $z = z_0$. The general solution has a transcendental critical point at each integer involving the powers $(z - z_0)^r$, where $r$ is a root of the equation $r^2 - r = 1$.]

2.  With the aid of the theory of symmetric functions, prove that the sum of two algebraic functions is algebraic.

3.  Solve Problem 2 for the product.

4.  Find the discriminant $D_0(z)$ of (3.1.7), given that

$$V(z, u) = zu^3 + 2z^2u^2 + z + 2.$$

5.  Determine the fixed singularities of the DE

$$(z + w)w' - z + w = 0.$$

Show that only the point at infinity qualifies and that it is a pole of order 2 for all solutions. Show that there are mobile branch points of order 1. (The equation has elementary solutions.)

6. In the equation

$$w' = P_0(z) + P_1(z)w + \cdots + P_n(z)w^n$$

the $P$'s are polynomials in $z$ and $n > 1$. Show that the fixed singular points are the zeros of $P_n(z)$ and the point at infinity.

7. In Problem 6 suppose that the $P$'s are constants. Try to show that, if the equation is satisfied by an entire function $w(z)$, then $w(z)$ must be a constant. Show that there are always singular constant solutions, and describe them. What happens for $n = 1$?

## 3.2. ANALYTIC CONTINUATION; MOVABLE SINGULARITIES

We consider again the system

$$w' = F(z, w) = \frac{P(z, w)}{Q(z, w)}, \qquad w(z_0) = w_0, \qquad (3.2.1)$$

with the unique solution $w(z; z_0, w_0)$. Here $P$ and $Q$ are polynomials in $w$ of degree $p$ and $q$, respectively, and the coefficients are algebraic functions of $z$. Furthermore, $(z, w) \mapsto F(z, w)$ is holomorphic at $(z_0, w_0)$. In particular, $z_0$ does not belong to the set $S$ of fixed singularities.

Starting at $z = z_0$, we continue $w(z; z_0, w_0)$ analytically along a path $C$. We continue $F[z, w(z; z_0, w_0)]$ along the same path. This means that $z$ has to avoid the singularities of the coefficients $A_j$ and $B_k$, as well as points where $Q(z, w)$ could vanish. If $w(z; z_0, w_0)$ admits of analytic continuation, so do its powers of finite order; i.e., both $P[z, w(z; z_0, w_0)]$ and $Q[z, w(z; z_0, w_0)]$ can be continued along any path along which $w[z, z_0, w_0)$ can be continued, but their quotient may possibly impose some conditions on the path $C$. At any rate, under the stated assumptions the law of permanence of functional equations (Theorem 1.11.2) permits us to conclude that the analytic continuation of the solution $w(z; z_0, w_0)$ satisfies the DE all along $C$. Moreover, if $z = z^*$ is a point on $C$ and if $w(z^*; z_0, w_0) = w^*$, there exists a disk, its center at $z^*$, in which $w(z; z_0, w_0)$ coincides with the local solution $w(z; z^*, w^*)$.

We now suppose that $w(z; z_0, w_0)$ has been continued analytically along $C$ from $z = z_0$ to $z = z_1$, and ask what will happen to the solution as $z \to z_1$. There are various possibilities.

1. $w(z; z_0, w_0) \to w_1$ and $F(z, w)$ is holomorphic at $(z_1, w_1)$.
2. $F(z, w)$ is not holomorphic at $(z_1, w_1)$, but $[F(z, w)]^{-1}$ is holomorphic there and $Q(z_1, w) \not\equiv 0$.
3. As in possibility 2, but $Q(z_1, w) \equiv 0$.
4. $P(z, w)$ and $Q(z, w)$ are holomorphic at $(z_1, w_1)$, but both are zero there.

5.  At least one of the functions $P$ and $Q$ is not holomorphic at $(z_1, w_1)$.
6.  As $z \to z_1$, $w \to \infty$.
7.  $w(z; z_0, w_0)$ does not tend to any limit, finite or infinite, as $z \to z_1$.

We list here the results to be proved. In case:

1.  The point $z = z_1$ is not a singularity.
2.  The point is a movable algebraic branch point.
3–5 and 7.  The point $z = z_1$ belongs to the set $S$ of fixed singular points.
6.  The point is a singularity which may be fixed or movable.

Let us start with case 1.

## THEOREM 3.2.1

*If analytic continuation of a given solution of (3.2.1) leads to a point $z_1$ and a function value $w = w_1$ and if $F(z, w)$ is holomorphic at $(z_1, w_1)$, then the local solution $w(zz_1, w_1)$ gives the analytic continuation of the given solution in the disk where $w(z; z_1, w_1)$ is holomorphic.*

*Proof.* In this case a local solution $w(z; z^*, w^*)$ exists for all $z^*$ on $C$, including the point $z = z_1$, where we can take $w^* = w_1$. Now $w(z; z_1, w_1)$ is holomorphic in some disk $|z - z_1| < r$, and this disk contains the end of $C$. If $z^* \in C$ and $|z_1 - z^*| < r$, then at $z = z^*$ we have two solutions, $w(z; z^*, w^*)$ and $w(z; z_1, w_1)$. They are identical provided that they take on the same value at $z = z^*$. In this case $w(z; z_1, w_1)$ is the analytic continuation of $w(z; z^*, w^*)$ and vice versa. Now along $C$ the radius of convergence of the power series representing $w(z; z^*, w^*)$ is a continuous, positive function of $z^*$ and is bounded away from zero. This follows from Cauchy's formula (2.5.12) if we observe that $F(z, w)$ is holomorphic at each point $(z^*, w^*)$ for $z^*$ on $C$ and $w^* = w(z^*; z_0, w_0)$, except possibly at the end point, where $w^* = w_1$. We may assume that $C$ is rectifiable and that

$$C: z = z^*(t), \qquad 0 \le t \le L, \tag{3.2.2}$$

where $L$ is the length of $C$. This is a compact set in the plane. At each point $z^*(t)$ on $C$ there are corresponding quantities $a(t)$, $b(t)$, and $M(t)$. These quantities refer to the local expansion of $F(z, w)$ in powers of $z - z^*(t)$ and $w - w^*(t)$ and may be assumed to vary continuously with $t$. Formula (2.5.12) then asserts that the local solution $w[z; z^*(t), w^*(t)]$ has an expansion which converges at least for

$$|z - z^*(t)| < r(t), \qquad 0 \le t \le L, \tag{3.2.3}$$

where

$$r(t) = a(t)\left\{1 - \exp\left[-\frac{b(t)}{2a(t)M(t)}\right]\right\}. \tag{3.2.4}$$

Now we can cover $C$ by a set of disks defined by (3.2.3), and since $C$ is compact the Heine-Borel theorem ensures the existence of a finite subcovering corresponding to

$$t_0 = 0 < t_1 < t_2 < \cdots < t_n < L.$$

We may assume that the subcovering is so dense that the center of the $(j + 1)$th disk lies in the $j$th one. This implies that the $n$th disk contains $z_1$, so that $w[z; z^*(t_n), w^*(t_n)]$ is holomorphic at $z = z_1$. Since this solution is the result of $n$ successive rearrangements [see Section 1.11, formula (1.11.3)] of the solution $w(z; z_0, w_0)$ at the points $z^*(t_j), j = 1, 2, \ldots, n$, its value at $z = z_1$ is $w_1$, the postulated limit of the solution for analytic continuation along $C$. Thus we have two solutions of the DE at $z = z_1$ with the same value $w_1$ there. Hence they are identical, and the theorem is proved. ∎

This theorem is due to Painlevé and is often called by his name. We shall encounter deeper theorems, however, which also bear his name. We now consider case 2.

**THEOREM 3.2.2**

*If $w(z; z_0, w_0) \to w_1$ as $z \to z_1$, if P and Q are holomorphic at $(z_1, w_1)$, if $P(z_1, w_1) \neq 0$, and if $Q(z_1, w)$ has a zero of exact multiplicity k at $w = w_1$, then $z = z_1$ is an algebraic branch point of order k for $w(z; z_0, w_0)$.*

*Proof.* The assumptions evidently imply that $F(z, w)$ is not holomorphic at $(z_1, w_1)$, while $[F(z, w)]^{-1} \equiv G(z, w)$ is holomorphic and not zero at all points $(z_1, w)$. We now consider the solution of the DE,

$$\frac{dz}{dw} = G(z, w), \qquad z(w_1) = z_1. \tag{3.2.5}$$

This equation has a unique solution by the existence and uniqueness theorems, say,

$$z(w; w_1, z_1) = z_1 + \sum_{n=1}^{\infty} c_n (w - w_1)^n. \tag{3.2.6}$$

Since $G(z_1, w)$ has a multiple zero at $w = w_1$, a certain number of the coefficients $c_n$ are zero. To make this precise, suppose that $G(z, w)$ is presented to us in the form

$$G(z, w) = \sum_{n=0}^{\infty} g_n(w)(z - z_1)^n, \tag{3.2.7}$$

where the functions $w \mapsto g_n(w)$ are holomorphic in one and the same disk, $|w - w_1| < r$. For $z = z_1$ this reduces to $g_0(w)$, so that

$$G(z_1, w) = g_0(w) = \sum g_{m0}(w - w_1)^m. \qquad (3.2.8)$$

Now by assumption $Q(z_1, w)$ has a zero of order $k$ at $w = w_1$, and, since $P(z_1, w_1) \neq 0$, this implies the same property for $G(z_1, w)$. It follows that

$$g_{00} = g_{10} = \cdots = g_{k-1,0} = 0, \qquad g_{k0} \neq 0.$$

We have now from (3.2.5)

$$\sum_{m=1}^{\infty} mc_m(w - w_1)^{m-1} = \sum_{n=0}^{\infty} g_n(w)\left[\sum_{m=0}^{\infty} c_m(w - w_1)^m\right]^n. \qquad (3.2.9)$$

Suppose that in the expansion (3.2.6) the series starts with the power $(w - w_1)^p$. This means that in (3.2.9) we have on the left $(w - w)^{p-1}$ as the lowest power, while on the right $(w - w_1)^k$ furnishes the lowest power. Hence

$$p = k + 1, \qquad (k + 1)c_{k+1} = g_{k0} \neq 0.$$

Thus we have

$$z - z_1 = (w - w_1)^{k+1} \sum_{m=0}^{\infty} c_{k+1+m}(w - w_1)^m. \qquad (3.2.10)$$

Since the power series does not vanish for $w = w_1$, we can extract the $(k + 1)$th root on both sides and obtain

$$(z - z_1)^{1/(k+1)} = (w - w_1) \sum_{m=0}^{\infty} d_{k+1+m}(w - w_1)^m. \qquad (3.2.11)$$

Here $d_{k+1} = (c_{k+1})^{1/(k+1)} \neq 0$, and the power series has a positive radius of convergence by the double series theorem of Weierstrass. We set

$$(z - z_1)^{1/(k+1)} = s, \qquad w - w_1 = t,$$

and want to solve the equation

$$s = t \sum_{m=0}^{\infty} d_{k+1+m} t^m \qquad (3.2.12)$$

for $t$. This we can do with the aid of the implicit function theorem (Theorem 2.6.1), the assumptions of which are satisfied. Hence

$$t = \sum_{j=1}^{\infty} b_j s^j$$

with a positive radius of convergence. Reverting to the old variables, we get

$$w - w_1 = \sum_{j=1}^{\infty} b_j(z - z_1)^{j/(k+1)}. \qquad (3.2.13)$$

This means that the analytic continuation of $w(z; z_0, w_0)$ along $C$ has led to a branch point of order $k$ of the solution where $k + 1$ branches are permuted cyclically. ∎

Here a question arises: In what sense is this a movable singularity, or would it have to rate as a fixed one? To elucidate this point let us examine the conditions for this type of singularity to occur. Give $z$ a value $z^0$, not belonging to the set $S$ of fixed singularities, and examine the equation

$$Q(z^0, w) = \sum_{k=0}^{q} B_k(z^0)w^k = 0. \tag{3.2.14}$$

This is an algebraic equation for $w$ of degree $q$ unless $B_q(z^0) = 0$. Normally there are $q$ roots $w_1^0, w_2^0, \ldots, w_q^0$. If $w^0$ is one of these roots, the event $w(z) \to w^0$ as $z \to z^0$ gives rise to a branch point for $w(z)$ at $z = z^0$, and normally the branch point is of order 1 with two permuting branches. Now the roots of (3.2.14) are continuous (even analytic) functions of $z^0$, so a small change in $z^0$ imposes a small change in $w^0$. This means that there is a solution $w^*(z)$ close to $w(z)$ which has a branch point $z^{0*}$ near to $z^0$, and normally this branch point is also of order 1. In this sense we can speak of movable branch points.

We now consider cases 3–5. The point $z_1$ is a fixed singularity in each of the three cases, and almost anything and everything can happen at such a point. Some examples were given in the preceding section, (3.1.15) and Exercise 3.1. More examples are to be found in Exercise 3.2.

As for case 6, here we set $w = 1/v$. The transformed equation is

$$v' = -v^2 \frac{P(z, 1/v)}{Q(z, 1/v)}, \qquad v(z) \to 0 \quad \text{as } z \to z_1. \tag{3.2.15}$$

Again we assume that $z_1 \not\in S$. There are two distinct subcases according to whether $q + 2 - p$ is $> 0$ or $\leq 0$.

Case 6:1. If $q + 2 > p$, (3.2.15) becomes

$$v' = -v^{q+2-p} \frac{\sum_0^p A_j(z)v^{p-j}}{\sum_0^q B_k(z)v^{q-k}} \equiv \frac{P_1(z, v)}{Q_1(z, v)}. \tag{3.2.16}$$

Here $P_1(z_1, 0) = 0$ and $Q_1(z_1, 0) = B_q(z_1)$, which is $\neq 0$ since $z_1$ is not a fixed singularity. For the same reason all the coefficients $A_j$ and $B_k$ are holomorphic at $z = z_1$. Since it follows that the last member of (3.2.16) is holomorphic at $(z_1, 0)$, (3.2.16) has a unique solution which takes the value 0 at $z = z_1$. By inspection one sees that $v(z) \equiv 0$ is that solution, and this means that case 6:1 cannot occur.

Case 6:2. Suppose first that $q + 2 = p$. Then (3.2.16) takes the form

$$v' = G(z, v), \qquad \text{where } G(z_1, 0) = -\frac{A_p(z_1)}{B_q(z_1)}. \tag{3.2.17}$$

Now $G(z, w)$ is holomorphic at $(z_1, 0)$, so there is a unique solution which vanishes at $z = z_1$. If $A_p(z_1) \neq 0$, this is a simple zero; but if $A_p(z)$ has a zero of multiplicity $m$ at $z = z_1$, then $v(z; z_1, 0)$ has a zero of multiplicity $m + 1$ at $z = z_1$. This means a pole of multiplicity $m + 1$ for $w(z; z_0, w_0)$.

Next, suppose that $q + 2 < p$. Then we replace (3.2.16) by

$$\frac{dz}{dv} = v^{p-q-2}H(v, z), \qquad z(0) = z_1, \qquad (3.2.18)$$

with obvious notation. Here we may assume $H(v, z)$ to be holomorphic at $(0, z_1)$ and different from zero. In this case $A_p(z_1) \neq 0$ since $z_1$ does not belong to $S$, and this guarantees the holomorphism of $H$. Now $H(0, z_1) = 0$ iff $B_q(z_1) = 0$, so our assumption implies that $B_q(z_1) \neq 0$. If this assumption is not satisfied, the following argument has to be modified along the lines used in the proof of Theorem 3.2.2. Under the stated assumption, (3.2.18) has a unique solution of the form

$$z - z_1 = v^{p-q-1}(c_0 + c_1 v + \cdots), \qquad c_0 \neq 0. \qquad (3.2.19)$$

Here we extract the $(p - q - 1)$th root and invert the result to obtain

$$v(z) = \sum_{j=1}^{\infty} d_j (z - z_1)^{j/(p-q-1)}, \qquad d_1 \neq 0. \qquad (3.2.20)$$

Now $w(z)$ is the reciprocal of this, so that

$$w(z) = (z - z_1)^{-1/(p-q-1)} \sum_{i=0}^{\infty} e_i (z - z_1)^{j/(p-q-1)}, \qquad (3.2.21)$$

where $e_0 \neq 0$ and the series has a positive radius of convergence. If $B_q(z_1)$ should be zero and, more generally, $H(v, z_1)$ has a zero of order $k$, then in (3.2.21) $p - q - 1$ should be replaced by $p - q - 1 + k$, as is seen by imitating the argument used in the proof of Theorem 3.2.2. Thus we have proved: ∎

**THEOREM 3.2.3**

*Equation* (3.2.1) *has no movable infinitudes if* $p < q + 2$. *There are movable poles if* $p = q + 2$. *The poles are simple unless accidentally* $A_p(z)$ *vanishes at the point in question. If* $p > q + 2$, *there are movable branch points where the solution becomes infinite. Normally* $p - q - 1$ *branches are permuted at such a point.*

The behavior of the solutions of (3.2.1) as $z \to \infty$ was studied by Pierre Boutroux (1880–1922) in a monograph published in 1908. (French mathematicians are often related: Boutroux was a nephew of Poincaré.) If $p = q + 1$, the integrals are of *exponential growth* (at least if the coefficients are polynomials, the only case considered by Boutroux). If $p \neq q + 1$, the solutions grow as *powers* of $z$, omitting neighborhoods of the infinitudes when $p - q > 1$. We shall take up such questions in Sections 4.6 and 11.2.

## EXERCISE 3.2

1. Discuss the fixed singularities of $w' = z^{1/2} + z^{3/2}w - w^2$.
2. The solution of $w' = z^3 + w^3$, $w(0) = w_0$, has movable branch points. Find the order. If $w_0 > 0$, the branch point nearest to the origin lies on the positive real axis. How does it move if $w_0$ increases?
3. The equation $w' = z^2 + w^2$, $w(0) = 0$, has movable poles (of what order?). If $w_0 > 0$, there are positive poles (why?). How does the pole nearest to the origin move when $w_0$ increases?
4. What fixed singularities, if any, do the equations in Problems 2 and 3 have?
5. How should $w_0$ be chosen in order for the equation in Problem 3 to possess solutions with poles on the imaginary axis?
6. Fill in the missing details in the proof of Theorem 3.2.2.
7. Solve the same Problem for Theorem 3.2.3. In particular, carry out a proof for the case when $B_q(z_1) = 0$ and $H(v, z_1)$ has a zero of order $k$ at $v = 0$.

Show that the following equations have fixed singularities, where for suitable approach the solutions tend to no limit.

8. $w' = z^{-2}w$.
9. $w' = i(1 - z)^{-1}w$.
10. $w' = w$.

## 3.3. PAINLEVÉ'S DETERMINATENESS THEOREM; SINGULARITIES

We have to show that case 7 can occur only at a fixed singular point of the DE. Note that our conclusions have always been based on a pair of values $(z_1, w_1)$, where $w_1 = \lim w(z)$ as $z \to z_1$ and where the limit is finite or infinite. If there is no limit, the point of departure for an attack on the problem is lost. Painlevé was the first to call attention to this difficulty in 1888 and to show that it cannot arise for a DE of type (3.1.4) and (3.1.5) as long as $z_1$ does not belong to the singular set $S$. Here the restrictive assumptions on $P$ and $Q$ are essential, and the result does not hold for systems of first order equations or equations of second or higher order.

## THEOREM 3.3.1

*If the solution $z \mapsto w(z; z_0, w_0)$ of (3.1.4) is continued analytically along a rectifiable arc $C$ from $z = z_0$ to $z = z_1$, avoiding the set $S$ of fixed* *singularities, and if* $z_1 \notin S$, *then the solution tends to a definite limit, finite or infinite, as* $z \to z_1$.

An equivalent and shorter formulation is

**THEOREM 3.3.2**

*If* $F(z, w)$ *is a rational function of* $w$ *with coefficients which are algebraic functions of* $z$, *then the movable singularities of the solutions are poles and/or algebraic branch points.*

*Proof.*   Suppose that the analytic continuation has been achieved along $C$ and that as $z$ approaches $z_1$, not in $S$, the solution tends to no limit. It will be shown that this assumption leads to a contradiction. Mark in the $w$-plane the $q$ roots, $w_1, w_2, \ldots, w_q$, of the equation

$$Q(z_1, w) = 0. \tag{3.3.1}$$

For each $k$ let $\gamma_k$ be a small circle with center at $w_k$. Then there exists a $\rho > 0$ such that for $|z - z_1| < \rho$ the equation

$$Q(z, w) = 0$$

has one and only one root inside each circle $\gamma_k$ and none outside the circular disks. This implies the existence of a constant $M$ such that on the rim of each circle $\gamma_k$ and for $|z - z_1| < \rho$ we have

$$\left| \frac{P(z, w)}{Q(z, w)} \right| < M. \tag{3.3.2}$$

Furthermore, we take a circle $\Gamma$ with center at $z = 0$ and a large radius. We can take $M$ so large that (3.3.2) holds also on $\Gamma$ and $|z - z_1| < \rho$. Let $T$ be the domain in the $w$-plane inside $\Gamma$ and outside all circles $\gamma_k$. Then (3.3.2) will also hold for $w$ in $T$ and $|z - z_1| < \rho$ by the principle of the maximum.

As $z$ describes $C$, we have several possibilities for the behavior of $w(z; z_0, w_0)$. If ultimately $w$ gets outside $\Gamma$ and stays outside, then $w(z; z_0, w_0)$ tends to infinity. If the path traced by the solution ultimately gets inside one of the circles $\gamma_k$ and stays inside, the solution tends to the limit $w_k$. By assumption none of these events takes place. This means that on the arc $C$ there is an infinite sequence of points $\{z_j\}$ such that $w(z_j; z_0, w_0) = W_j$ lies in $T$ and $z_j \to z_1$. If $|z_1 - z_j| < \rho$, then at the point $(z_j, W_j)$ the function $F(z, w)$ is holomorphic and less than $M$ in absolute value. The local solution $w(z; z_j, W_j)$, which coincides with the analytic continuation of $w(z; z_0, w_0)$, is holomorphic in a disk of positive radius $r_j$. The bounded infinite sequence $\{W_j\}$ has at least one limit point, $W_0$ say. Now at $(z_1, W_0)$ there is a local solution $w(z; z_1, W_0)$, holomorphic in a disk $|z - z_1| < r_0$. This disk contains infinitely many points $z_j$ for which $W_j \to W_0$. It follows that, if the analytic continuation of $w(z; z_0, w_0)$ along $C$ is via the function elements $w(z; z_j, W_j)$, using only those values of $j$ for which $W_j \to W_0$, then $\lim w(z; z_0, w_0) = W_0$. Then Theorem 3.2.1 shows that the continuation of our solution is holomorphic at $z = z_1$ and has the definite limit $W_0$. This is a finite quantity, and $F(z, w)$ is holomorphic at $(z_1, W_0)$. This shows that the hypothesis that the analytic continuation tends to no limit as $z \to z_1$ is untenable, and the theorems are proved. ∎

Consider, in particular, the equation

$$w'(z) = P_0(z) + P_1(z)w(z) + \cdots + P_m(z)[w(z)]^m, \qquad (3.3.3)$$

where the $P$'s are polynomials in $z$ and $m > 1$. The coefficients admit $z = \infty$ as the only singularity. Here $Q(z, w) \equiv 1$, so cases 2–5 cannot occur. Case 6 will occur and give movable infinitudes. If $P_m(z_1) \neq 0$, the corresponding expansions take the form

$$w(z) = [-A(z - z_1)]^{-1/(m-1)}\left[1 + \sum_{1}^{\infty} c_j(z - z_1)^{j/(m-1)}\right], \qquad (3.3.4)$$

where $A = (m - 1)P_m(z_1)$. Thus, if $m = 2$, the singularity is a movable pole; if $m > 2$, a movable branch point where the solution becomes infinite and $m - 1$ branches are permuted when $z$ makes a circuit about the point. The fixed singularities of this equation are the point at infinity and the zeros of $P_m(z)$.

The point at infinity is usually a singularity of a higher order: a limit point of poles if $m = 2$, of infinitary branch points if $m > 2$. A $k$-fold zero of $P_m(z)$, say at $z = a$, permits the existence of solutions holomorphic at $z = a$, and also solutions with a branch point where the solution becomes infinite and the order of infinitude is at most $(k + 1)/(m - 1)$. The case $m = 2$ is known as the *Riccati equation*.

We may conclude from the preceding discussion that the equations whose movable singularities are poles are Riccati equations. In view of the importance of this fact we shall give a formal proof in the setting of (3.1.4) and (3.1.5).

**THEOREM 3.3.3**

*Consider the equation*

$$w' = F(z, w) = \frac{P(z, w)}{Q(z, w)}, \qquad (3.3.5)$$

*where P and Q are polynomials in w of degree p and q, respectively, with coefficients which are algebraic functions of z. If no solution of the equation can have a branch point at a nonsingular point of the equation,* then $F(z, w)$ *is a quadratic polynomial in* w, *that is,* (3.3.5) *is a Riccati equation,*

$$w'(z) = A_0(z) + A_1(z)w(z) + A_2(z)[w(z)]^2. \qquad (3.3.6)$$

*Proof.* We start by noting that $Q(z, w)$ must be of degree 0 in $w$. Otherwise, setting $z = z_0$ with $z_0$ not in $S$, we find that the equation $Q(z_0, w) = 0$ has $q$ roots. If $w_0$ is one of the roots, then by Theorem 3.2.2 there is a solution taking on the value $w_0$ at $z = z_0$ and this solution has a branch point there. Hence we must have $q = 0$, and we may take $Q(z, w) \equiv 1$. Thus the equation is of the form (3.3.3), where the coefficients are algebraic functions rather than polynomials. Hence we are dealing with

$$w' = P(z, w) = A_0(z) + A_1(z)w + \cdots + A_p(z)w^p. \qquad (3.3.7)$$

If now $p > 2$, we set $w = 1/v$ and obtain

$$v'(z) = -[A_0(z)v^p + A_1(z)v^{p-1} + \cdots + A_p(z)]v^{2-p}.$$

Here $Q_1(z, v) = v^{p-2}$, which vanishes for $v = 0$ regardless of the value of $z$. Taking a nonsingular value $z_0$ for $z$, we have $A_p(z_0) \neq 0$ and the transformed equation has a solution which vanishes at $z = z_0$ and has a branch point there. Hence the original DE has a solution $w(z) = [v(z)]^{-1}$ with a branch point at a nonsingular point $z_0$ contrary to the assumption. Thus we must have $p \leq 2$. ∎

Riccati's equation has a number of interesting properties, which will be discussed at some length in Chapter 4.

The rest of this section will be devoted to further discussion of singularities, fixed or movable, of (3.1.16), (3.3.3), (3.3.7), and related types of equations. The basic tool is the *test-power test*. This is most effective in the case of movable singularities but can be used also with fixed ones, at least for purposes of orientation. The simple idea is that, if a first order DE has a solution which at some point $z_1$ becomes infinite as

$$a(z - z_1)^{-\alpha} \qquad \text{while } w'(z) \sim -\alpha a(z - z_1)^{-\alpha-1}, \qquad (3.3.8)$$

in both cases up to terms of lower order, then $a$ and $\alpha$ must be so chosen that terms of highest order cancel when (3.3.8) is substituted in the equation. Let us try this out on some simple examples where the facts are known.

For the tangent equation

$$w' = 1 + w^2 \qquad (3.3.9)$$

we find $\alpha = 1$, $a = -1$, and in this case $z = z_1$ is a movable simple pole with residue $-1$. This is indeed the situation for $w(z) = \tan(z - z_1 - \frac{1}{2}\pi)$.

For the Jacobi equation

$$w' = [(1 - w^2)(1 - k^2 w^2)]^{1/2} \tag{3.3.10}$$

we get $\alpha = 1$, $a = \pm 1/k$, values which agree with known facts.

Since these results are rather encouraging, we shall try to apply the method to equations where we do not know the true situation in advance. The Briot-Bouquet equation (3.1.16) is a good testing ground [actually (3.3.10) is a Briot-Bouquet equation]. If we are looking for solutions with prescribed types of singularities and if the test-power test shows that a given equation cannot have such singularities, the test is conclusive. On the other hand, if the test shows that the equation could possibly admit solutions with such and such singularities, an examination of the appropriate series expansion becomes necessary before we can accept the evidence. Thus in the case of the very simple second order nonlinear equation

$$w'' = zw^2 \tag{3.3.11}$$

the test for movable singularities gives $\alpha = 2$, $a = 6$, but the point $z_1$ is not a double pole. See Section 12.4.

We return to (3.1.16). It contains aggregates of the form

$$c_{jk} w^j (w')^k.$$

With each such product we associate a straight line

$$y = (j + k)x + k. \tag{3.3.12}$$

We are interested in the points of intersection of such lines (see Figure 3.1). Their coordinates are rational numbers and candidates for the value of $\alpha$. There are essentially two different cases.

I.   All points of intersection are located in the closed left half-plane.
II.  Some intersections fall in the open right half plane.


**THEOREM 3.3.4**

*In case I the equation can have no movable infinitudes of type* (3.3.8). *In case II, if* $(x_0, y_0)$ *is a point of intersection of lines* (3.3.12) *such that* (i) $x_0 > 0$, $y_0 > 0$, (ii) *the vertical line* $x = x_0$ *has no intersection with the system* (3.3.12) *above the point* $(x_0, y_0)$, *and* (iii) $(x_0, y_0)$ *is the only point of intersection with these properties, then* $\alpha = x_0$ *gives the only admissible movable infinitude of type* (3.3.8). *In case I an intersection in the left half-plane will give movable branch points where the solution is zero, provided that the transformation* $w = 1/v$ *leads to an equation in which the condition under II holds.*

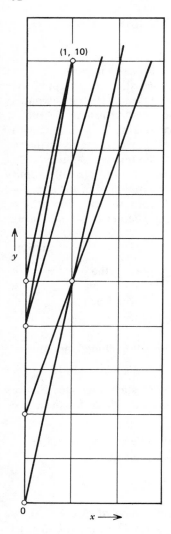

(1, 10)

$y$

0

$x \longrightarrow$

**Figure 3.1**

*Proof.* Suppose that case I is present. We substitute the test powers
(3.3.8) in the DE

$$\sum c_{jk} w^j (w')^k = 0. \tag{3.3.13}$$

If there exists a solution with an infinitude of this type, it must be possible
to determine $\alpha$ (and $a$) so that the terms of highest order are of equal
weight and can be made to cancel. This requires, in particular, that there

be two points, $(j, k)$ and $(m, n)$, such that the equation

$$(j + k)\alpha + k = (m + n)\alpha + n$$

is satisfied by a positive number $\alpha$. This is clearly impossible if all lines (3.3.12) intersect in the left half-plane.

   Suppose now that case II is present. Then there are two lines in the set (3.3.12) which intersect at a point $(x_0, y_0)$ in the open right half-plane. Suppose that these are the lines

$$y = (j + k)x + k \qquad \text{and} \qquad y = (m + n)x + n.$$

If they are to intersect, one line must have the larger $y$-intercept, the other the greater slope. Suppose that $k < n$, $j > m$. Then

$$x_0 = \frac{n - k}{(j - m) - (n - k)}. \tag{3.3.14}$$

This is a positive rational number. Hence for a suitable choice of $a$ the terms

$$(-x_0)^k a^{j+k} c_{jk} (z - z_1)^{-(j+k)x_0 - k} \qquad \text{and} \qquad (-x_0)^n a^{m+n} c_{mn} (z - z_1)^{-(m+n)x_0 - n}$$

are the largest and will cancel. Condition (ii) of Theorem 3.3.4 guarantees that in the interval $(0, x_0)$ all the other lines of the system lie below the line $y = (j + k)x + k$ and that possible intersections in this area need not be considered. Intersections $(x_1, y_1)$ with $x_0 < x_1$ are not excluded, and this is why condition (iii) is needed. Thus we see that (3.3.13) has movable infinitudes.

   For such an infinitude to be a pole, say of order $N$, we must have $(n - k)(N + 1) = (j - m)N$. This implies the existence of an integer $M$ such that

$$j - m = M(N + 1), \qquad n - k = MN. \tag{3.3.15}$$

If no such integers $M$ and $N$ can be found and $x_0 > 0$, the singularity is not polaroid but algebroid. In the case under consideration logarithmic perturbations cannot occur, so a polaroid singularity is a pole and an algebroid singularity is algebraic.

   Let us now return to case I for a moment. If the transformation $w = 1/v$ is applied to (3.3.13), we obtain after multiplication by $v^p$, where $p = \max (j + 2k)$,

$$\sum (-1)^k c_{jk} v^{p-j-2k} (v')^k = 0, \tag{3.3.16}$$

which is of type (3.3.13). Any intersection of lines of system (3.3.12), say $(x_1, y_1)$ with $x_1 < 0$, will correspond to intersections of lines of the system

$$y = (p - j - k)x + k. \tag{3.3.17}$$

In particular, $(x_1, y_1)$ will go into a point $(-x_1, y_2)$ in the first quadrant. If this point satisfies the conditions under case II, then (3.3.16) has a mobile infinitude at $z = z_1$ and (3.3.13) has a solution with a movable zero at $z_1$. If $-x_1$ is a positive integer, this is a regular zero where the solution is holomorphic, otherwise an algebraic branch point. This completes the proof. ■

We have also proved ■

**THEOREM 3.3.5**

*A necessary condition that* (3.3.13) *be satisfied by an elliptic function is that among the exponents one can find two pairs* (j, k) *and* (m, n) *such that the* $x_0$ *defined by* (3.3.14) *is a positive integer which satisfies conditions* (ii) *and* (iii) *of Theorem* 3.3.4.

The various cases encountered above are typified by the following equations:

$$\left.\begin{array}{r} (1 + w^2)w' = 1, \text{ no finite infinitudes, movable zeros;} \\ (w')^2 = 1 + w^3, \text{ movable double poles;} \\ w' = 1 + w^3, \text{ movable branch points, order 1.} \end{array}\right\} \quad (3.3.18)$$

Our theorems leave open the question of what happens if the conditions under case II hold neither for (3.3.13) nor for (3.3.16). Here no singularity of type (3.3.8) seems to be possible. A case in point is the equation

$$(1 + w)(w')^2 - w^2w' + w^2 = 0. \tag{3.3.19}$$

Here we can solve the equation for $w'$, obtaining two roots. One leads to a solution asymptotic to $e^z$, the other to $z$. Here, however, $z$ is an elementary function of $w$, having a logarithmic singularity.

The test-power test can be applied to first order equations with variable coefficients and to equations of higher order. Here it becomes imperative to test the result by other means. Equation (3.3.11) may serve as a warning in this connection.

In a discussion of (3.3.13) it is useful to observe that $w'$ is an algebraic function of $w$, the singularities of which can be read off from the discussion at the beginning of Section 3.1. The question of infinitudes of solutions of (3.3.13) is now reduced to the question of the behavior of $w'$ for large values of $w$. This is a classical problem in the theory of algebraic functions. The method is variously known as *Newton's diagram* or *polygon*, as the *algebraic* or the *analytical triangle*, or as the *method of Puisseux* (after V. A. Puisseux, 1820–1883). It is closely related to the method used above, which led to the line (3.3.12). Here instead we plot the points (j, k) in the (x, y)-plane and draw the least convex polygon which

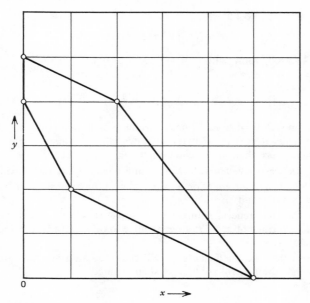

**Figure 3.2**

contains these points. See Figure 3.2, which illustrates the case

$$(w')^5 + w^2(w')^4 + 2(w')^4 - w(w')^2 - w^5 = 0. \tag{3.3.20}$$

The sides of the polygon which face infinity give asymptotic relations of the form

$$w' \sim aw^\beta, \tag{3.3.21}$$

where $-1/\beta$ is the slope of one of the boundary lines. In the figure $\beta = 2$ or $\frac{3}{4}$. The first value corresponds to a solution with simple poles, and the second to a solution $\sim az^4$ for large values of $z$. If so desired, we can elaborate (3.2.21) into a convergent series

$$w' = \sum a_j w^{\beta_j},$$

but (3.3.21) is sufficient for our present purpose. According to the value of $\beta$, we have various possibilities:

1. $\beta = 1 + \delta,\ \delta > 0$, gives

$$w \sim [\delta(z - z_1)]^{-1/\delta}. \tag{3.3.22}$$

This is a mobile infinitude, a pole of order $N$ if $\delta = 1/N$.

2. $\beta = 1,\ w \sim Ce^{az}$.

3.  $0 < \beta < 1$. We have

$$w \sim [(1 - \beta)a(z - z_1)]^{1/(1-\beta)}, \tag{3.3.23}$$

a mobile branch point where $w = 0$.

4.  $\beta = 0$, $w \sim a(z - z_1)$, a mobile zero.

## EXERCISE 3.3

1.  Equation (3.3.3) with $m > 2$ has movable branch points so that the general solution has at least $m - 1$ determinations. Show by an example that it need not have more than $m - 1$ determinations.

2.  Apply the test-power test to (2.7.3), and compare the result with (2.7.22).

3.  Apply the test to the equation $w' = 1 + z^2 w^3$ at the fixed singularity $z = 0$. Try to verify the result.

4.  Show that the general solution of $w' = 1 + z^{1/2} w^2$ is single valued on the Riemann surface of $z^{1/2}$. Try to generalize to other cases involving algebraic coefficients.

5.  Suppose that the coefficients of a Riccati equation are polynomials in $z$. Show that the general solution is single valued.

6.  Verify (3.3.2).

Assuming the validity of the test-power test, apply it to the following DE's:

7.  $(w')^2 = 4(w - e_1)(w - e_2)(w - e_3)$, $e_1 + e_2 + e_3 = 0$.

8.  $w'' = 6w^2 - \frac{1}{2}g_2$.

9.  $w'' = z^{-1}w^2$.

    The equations in Problems 7 and 8 are satisfied by the $\wp$-function of Weierstrass, and the movable singularities are double poles. The equation in Problem 9 is a special Thomas-Fermi equation, and the movable singularities look like double poles but are not. The discrepancy shows up in the sixth term of the proposed Laurent series.

10. In the proof of Theorem 3.3.1 does the assumption that the coefficients $A_j$ and $B_k$ are algebraic functions play an essential role in the argument? Could it be replaced by a less restrictive assumption without jeopardizing the conclusion?

11. Consider the equation

$$w'(z) = \frac{az + bw(z)}{cz + dw(z)},$$

where $a$, $b$, $c$, $d$ are arbitrary fixed constants with $ad - bc \neq 0$. What is the nature of the fixed singularities? Of the movable singularities?

## 3.4.  INDETERMINATE FORMS

Except for a problem in Exercise 3.3, there has been no mention of case 4. This type of fixed singularity has been the object of much work for close to a century and has had a great direct and indirect influence on the development of mathematics. What started with Poincaré in 1878 as a scrutiny of the properties of functions defined by an ordinary DE developed in his fertile mind into a theory of the curves defined by a DE (1879, 1881, 1885, 1886), which laid the foundation of the *geometric theory of differential equations*. Moreover, it led him to the creation of what he called *analysis situs*, now known as *topology*.

A fixed singularity $z = z_0$ exhibits case 4 if there is a point $w = w_0$ such that in the quotient

$$\frac{P(z, w)}{Q(z, w)} \tag{3.4.1}$$

both numerator and denominator are holomorphic at $(z_0, w_0)$, but $P(z_0, w_0) = Q(z_0, w_0) = 0$ so that $P/Q$ becomes an indeterminate form $0/0$ at $(z_0, w_0)$. We place the singularity at $(0, 0)$, and in this chapter we shall consider only the case in which both $P$ and $Q$ are linear homogeneous functions of $z$ and $w$. More general considerations are postponed until Chapter 12. Thus

$$w' = \frac{az + bw}{cz + dw} \quad \text{with } ad - bc \neq 0. \tag{3.4.2}$$

Together with this equation, it is occasionally convenient to consider the system of *autonomous equations*

$$\dot{z}(t) = cz(t) + dw(t), \qquad \dot{w}(t) = az(t) + bw(t), \tag{3.4.3}$$

where $t$ is an arbitrary parameter and the dot indicates differentiation with respect to $t$ (Newton's notation). The word "autonomous" refers to the fact that $t$ does not occur explicitly in the equation. If $[z(t), w(t)]$ is a solution of (3.4.3) and $p$ is an arbitrary number, $[z(t + p), w(t + p)]$ is also a solution. Furthermore,

$$\frac{\dot{w}(t)}{\dot{z}(t)} = \frac{az(t) + bw(t)}{cz(t) + dw(t)}, \tag{3.4.4}$$

so $[z(t), w(t)]$ is a parametric representation of the solution of (3.4.2).

In Chapter 12 we shall be concerned, at least briefly, with the case in which

$$P(z, w) = \sum a_{jk} z^j w^k, \qquad Q(z, w) = \sum b_{jk} z^j w^k. \tag{3.4.5}$$

There is an old tradition in applied as well as pure mathematics of linearizing a difficult problem in the hope that the linear case will give a reasonable approximation of the nonlinear problem. The results so obtained are often misleading, and if the equations involve no linear terms at all the method fails.

After these words of warning let us take a look at (3.4.2), where we want a solution $w(z)$ that tends to zero with $z$. The equations are solvable by elementary means, but it is usually difficult to obtain the desired information from the elementary solution. There are a couple of cases, however, in which explicit solutions of $w$ in terms of $z$ are available and are of interest to our problem. Take $c = 1$, $d = 0$ and the equation

$$w' = a + b\frac{w}{z}. \tag{3.4.6}$$

If $b \neq 1$, the general solution is

$$w = \frac{a}{b-1}z + Cz^b, \tag{3.4.7}$$

while, if $b = 1$, the solution becomes

$$w = z(a \log z - C), \tag{3.4.8}$$

where $C$ is an arbitrary constant. In the second case all solutions go to zero with $z$. In the first case this will be true iff $\mathrm{Re}\,(b) > 0$. If $\mathrm{Re}\,(b) \leq 0$, there is one and only one solution that goes to zero with $z$. These special cases are of interest both for the nature of the solution and for the conditions that have to be imposed on the constants $a$ to $d$.

We now return to the general case in the form of the system (3.4.3). An equivalent form is the matrix-vector DE

$$\dot{\mathbf{v}}(t) = \mathcal{M}\mathbf{v}(t), \tag{3.4.9}$$

with

$$\mathbf{v}(t) = \begin{bmatrix} z(t) \\ w(t) \end{bmatrix}, \qquad \mathcal{M} = \begin{bmatrix} a & b \\ c & d \end{bmatrix}, \tag{3.4.10}$$

and the derivative of a vector is defined as the vector of the derivatives of the components. Here the characteristic values and the characteristic vectors of $\mathcal{M}$ become important (see the end of Section 1.4), and the characteristic equation is

$$\begin{vmatrix} a - \lambda & b \\ c & d - \lambda \end{vmatrix} = 0 \qquad \text{or} \qquad \lambda^2 - (a+d)\lambda + ad - bc = 0. \tag{3.4.11}$$

Since $ad - bc \neq 0$, no root can be zero, i.e., $\mathcal{M}$ is a regular (= nonsingular) matrix and has a unique inverse $\mathcal{M}^{-1}$. The roots of (3.4.11) may be distinct or equal, and the solutions of (3.4.9) are essentially different in the two cases. If the roots are distinct, they are numbered $\lambda_1$, $\lambda_2$ with the convention that $\mathrm{Re}\,(\lambda_1) \geq \mathrm{Re}\,(\lambda_2)$. If the sign of equality should hold here, it is required that $\mathrm{Im}\,(\lambda_1) > \mathrm{Im}\,(\lambda_2)$. If the roots are equal, the common value is $\frac{1}{2}(a + d)$.

Let $\mathfrak{M}_2$ denote the space of all 2-by-2 matrices. We make it into a metric space by defining a norm for $\mathcal{A} = (a_{jk})$:

$$\|\mathcal{A}\| = \max_j (|a_{j1}| + |a_{j2}|). \tag{3.4.12}$$

In this metric

$$\|\mathcal{M}^n\| \leq \|\mathcal{M}\|^n. \tag{3.4.13}$$

We define the exponential function by the power series (with $\mathcal{E}$ the unit matrix)

$$\exp(\mathcal{M}t) = \mathcal{E} + \frac{\mathcal{M}}{1!} t + \frac{\mathcal{M}^2}{2!} t^2 + \cdots + \frac{\mathcal{M}^n}{n!} t^n + \cdots. \tag{3.4.14}$$

Here the sum of the norms is at most $\exp(\|\mathcal{M}\| |t|)$, so the series is absolutely convergent in norm. It is a differentiable function of $t$ with derivative $\mathcal{M} \exp(t\mathcal{M})$, as is seen by (permissible!) term-by-term differentiation of the series (3.4.14).

It follows that the solution of (3.4.9) with the initial value $\mathbf{v}_0$ is

$$\mathbf{v}(t) = \exp(\mathcal{M}t)\mathbf{v}_0. \tag{3.4.15}$$

This is not quite the solution that we want. Indeed, we want a solution which goes to the zero vector when $t$ approaches some limit. What can be achieved in this direction will depend on the characteristic values of $\mathcal{M}$.

If $\lambda_1 \neq \lambda_2$, the normal form of the matrix $\mathcal{M}$ is

$$\mathcal{D} = \begin{bmatrix} \lambda_1 & 0 \\ 0 & \lambda_2 \end{bmatrix}, \tag{3.4.16}$$

and there exists a nonsingular matrix $\mathcal{A}$ such that

$$\mathcal{M} = \mathcal{A}^{-1}\mathcal{D}\mathcal{A} \quad \text{or} \quad \mathcal{D} = \mathcal{A}\mathcal{M}\mathcal{A}^{-1}. \tag{3.4.17}$$

We can transform (3.4.9) into an equivalent equation by multiplying both sides on the left by $\mathcal{A}$ to obtain

$$\mathcal{A}\dot{\mathbf{v}}(t) = \mathcal{A}\mathcal{M}\mathcal{A}^{-1}\mathcal{A}\mathbf{v}(t). \tag{3.4.18}$$

Hence, if we set $\mathcal{A}\mathbf{v}(t) = \mathbf{u}(t)$, then

$$\dot{\mathbf{u}}(t) = \mathcal{D}\mathbf{u}(t). \tag{3.4.19}$$

This has now the solution

$$\mathbf{u}(t) = \exp(\mathcal{D}t)\mathbf{u}_0 \quad \text{with } \mathbf{u}_0 = \mathcal{A}\mathbf{v}_0. \tag{3.4.20}$$

Now the powers of $\mathcal{D}$ are found to be

$$\mathcal{D}^n = \begin{bmatrix} \lambda_1^n & 0 \\ 0 & \lambda_2^n \end{bmatrix},$$

and this shows that the solution may be written as

$$\mathbf{u}(t) = \begin{bmatrix} e^{\lambda_1 t} & 0 \\ 0 & e^{\lambda_2 t} \end{bmatrix} \mathbf{u}_0. \tag{3.4.21}$$

If now the components of $\mathbf{u}(t)$ and $\mathbf{u}_0$ are, respectively, $u_1(t)$, $u_2(t)$ and $u_{01}$, $u_{02}$, we have

$$u_1(t) = u_{01} e^{\lambda_1 t}, \qquad u_2(t) = u_{02} e^{\lambda_2 t}. \tag{3.4.22}$$

Now, if Re $(\lambda_1)$ and Re $(\lambda_2)$ are both positive, $u_1(t)$ and $u_2(t)$ go to zero as $t \to -\infty$. If both signs are negative, $u_1(t)$ and $u_2(t)$ go to zero as $t \to +\infty$. Since in either case

$$\lim \mathbf{u}(t) = \mathbf{0},$$

it follows that

$$\lim \mathbf{v}(t) = \lim \mathcal{A}^{-1} \mathbf{u}(t) = \mathcal{A}^{-1} \lim \mathbf{u}(t) = \mathbf{0} \tag{3.4.23}$$

when $t$ tends to plus or minus infinity, as the case may be. This means that the system (3.4.3) has a solution that goes to zero and hence also that there is a solution of (3.4.2) which exhibits case 4 at $(0, 0)$. For $0 > $ Re $(\lambda_1) > $ Re $(\lambda_2)$ we have

$$w(z) = Cz[1 + o(1)] \tag{3.4.24}$$

as $z \to 0$. Here $C$ is an arbitrary constant since $\mathbf{v}_0$ is an arbitrary vector.

On the other hand, if Re $(\lambda_1) > 0 > $ Re $(\lambda_2)$, we get a different picture. As above, we obtain (3.4.22) for $u_1(t)$ and $u_2(t)$; but now, letting $t \to +\infty$, we get the limit zero for one coordinate and plus or minus infinity for the other, and the same is true for $t \to -\infty$. We notice, however, that

$$\dot{u}_1(t) = \lambda_1 u_1(t), \qquad \dot{u}_2(t) = \lambda_2 u_2(t),$$

and this gives

$$\lambda_2 u_2(t) \dot{u}_1(t) - \lambda_1 u_1(t) \dot{u}_2(t) = 0,$$

whence

$$[u_1(t)]^{-\lambda_2} [u_2(t)]^{\lambda_1} = C. \tag{3.4.25}$$

If the $\lambda$'s are real, $\lambda_1 > 0 > \lambda_2$, this is a system of hyperbola-like curves in the $(u_1, u_2)$-plane, and the coordinate axes are members of the family, the only members that pass through the origin. We can make an orthogonal transformation of the coordinates, carrying $(u_1, u_2)$ into $(x_1, x_2)$ to get

$$[x_1(t) + x_2(t)]^{-\lambda_2} [x_1(t) - x_2(t)]^{\lambda_1} = C_1, \tag{3.4.26}$$

and then the solutions that pass through the origin become

$$x_2 = x_1 \qquad \text{and} \qquad x_2 = -x_1. \tag{3.4.27}$$

We have next the case of a double root, $\lambda_0 = \frac{1}{2}(a + d)$ of (3.4.11). Here we can take as the normal form

$$\mathcal{J} = \begin{bmatrix} \lambda_0 & 1 \\ 0 & \lambda_0 \end{bmatrix}. \qquad (3.4.28)$$

Again there is a nonsingular matrix $\mathcal{A}$ such that

$$\mathcal{M} = \mathcal{A}^{-1}\mathcal{J}\mathcal{A} \qquad \text{or} \qquad \mathcal{J} = \mathcal{A}\mathcal{M}\mathcal{A}^{-1}. \qquad (3.4.29)$$

Here

$$\mathcal{J}^n = \begin{bmatrix} \lambda_0^n & \lambda_0^{n-1} \\ 0 & \lambda_0^n \end{bmatrix},$$

and this gives

$$\exp(\mathcal{J}t) = \begin{bmatrix} \exp(\lambda_0 t) & t\exp(\lambda_0 t) \\ 0 & \exp(\lambda_0 t) \end{bmatrix}, \qquad (3.4.30)$$

whence

$$u_1(t) = (u_{01} + u_{02}t)e^{\lambda_0 t}, \qquad u_2(t) = u_{02}e^{\lambda_0 t}. \qquad (3.4.31)$$

If $\text{Re}(\lambda_0) \neq 0$, all solutions go to $(0, 0)$ as $t \to +\infty$ or $-\infty$. If $\text{Re}(\lambda_0) = 0$, the solutions will still go to zero if $t \to \infty$ along a suitably chosen ray in the complex $t$-plane.

We say that two vector-matrix equations are *equivalent*:

$$\dot{\mathbf{y}}(t) = \mathcal{M}_1\mathbf{y}(t) \qquad \text{equivalent to} \qquad \dot{\mathbf{v}}(t) = \mathcal{M}_2\mathbf{v}(t) \qquad (3.4.32)$$

if the matrices $\mathcal{M}_1$ and $\mathcal{M}_2$ are *equivalent* in the sense of matrix algebra, i.e., there is a regular matrix $\mathcal{B}$ such that

$$\mathcal{B}\mathcal{M}_1 = \mathcal{M}_2\mathcal{B}. \qquad (3.4.33)$$

This means that $\mathcal{M}_1$ and $\mathcal{M}_2$ have the same characteristic values, and all equations in the equivalence class have a singularity of type 4 at $(0, 0)$.

## EXERCISE 3.4

1. Prove that equivalent matrices have the same spectra.
2. If
$$\mathcal{M} = \begin{bmatrix} 1 & -1 \\ 2 & 3 \end{bmatrix}$$
   find characteristic values and vectors.
3. Prove (3.4.13).
4. Verify the statements concerning the convergence properties of (3.4.14) and the differentiability of the series.
5. Verify (3.4.21) and (3.4.22).
6. Verify (3.4.24).

7. If $\mathscr{B}$ is an orthogonal matrix (i.e., its row vectors are orthogonal as well as the column vectors) and $\mathscr{B}$ was used in passing from (3.4.25) to (3.4.26), show that $\mathscr{B}\mathscr{A}v(t)$ is a solution of a vector-matrix DE equivalent to (3.4.9).

8. Verify (3.4.30) and (3.4.31).

9. Show that (3.4.7) goes into (3.4.8) when $b \mapsto 1$ and $C$ is chosen as an appropriate function of $h = b - 1$.

10. If $t$ is eliminated between the two equations in (3.4.32) and $w = v_1(t)$, $z = v_2(t)$, show that a result is obtained similar to (3.4.8).

11. Solve $w' = (z/w)^2$, which exhibits case 4 at $(0, 0)$.

## LITERATURE

As a general reference for this chapter see the author's LODE, Section 12.1.

For the theory of algebraic functions see the author's AFT, Vol. II, Chapter 12. For details consult:

Bliss, G. A. *Algebraic Functions.* Colloquium Publications, American Mathematical Society, New York, 1933.

Painlevé's Stockholm Lectures are not easily available but were published as:

Painlevé, P. *Leçons sur la théorie analytique des équations différentielles, professées à Stockholm, 1895.* Hermann, Paris, 1897.

See also:

Boutroux, P. *Leçons sur les fonctions définies par les équations différentielles du premier ordre.* Gauthier-Villars, Paris, 1908.

For the theory of elliptic functions, see AFT, Vol. II, Chapter 13. The Briot-Bouquet publications referred to are:

Briot, Ch. and J. Cl. Bouquet. *Théorie des fonctions elliptiques.* 2nd ed. Paris, 1875.

———. Intégrations des équations différentielles au moyen des fonctions elliptiques. *J. École Polytech.*, **21**: 36 (1856).

For the geometric theory of DE's see:

Lefschetz, S. *Differential Equations: Geometric Theory.* Interscience, New York, 1957; 2nd ed., 1963 and the literature cited there.

Further references are:

Fuchs, L. Über Differentialgleichungen deren Integrale feste Verzweigungspunkte besitzen. *Sitzungsber. Akad., Berlin* (1884), 699–720.

Poincaré, H. Sur un théorème de M. Fuchs. *Acta Math.*, **7** (1885), 1–32.

# 4
# RICCATI'S EQUATION

In the beginning there was a Riccati—to wit, Conto Jacopo Riccati (1676–1754). In the *Acta Eruditorum* (1723, pp. 502–574, and Supplementum VIII, 1724, pp. 68–74) he treated a DE which Daniel Bernoulli (*ibid.*, 1725, pp. 473–475) reduced to the form

$$y'(x) + ay^2(x) + bx^m = 0.$$

It is closely related to Bessel's equation (Section 6.3).

This is the origin of what later became known as Riccati's equation. Its unique character in the theory of first order nonlinear DE's and its importance for first and second order linear DE's became clear much later. In this chapter we shall discuss various aspects of the theory at some length. We shall treat the elementary theory, the nature of the singularities, the dependence of the solutions on internal parameters, and some geometric consequences of the situation. The equation has some extremal properties in the class of all nonlinear first order DE's expressed by the Malmquist-Wittich-Yosida theorem. This calls for a sketch of the Nevanlinna value distribution theory for meromorphic functions. This theory plays an important role in the present chapter and later ones.

## 4.1. CLASSICAL THEORY

We take the Riccati equation in the form

$$y'(z) = f_0(z) + f_1(z)y(z) + f_2(z)[y(z)]^2, \tag{4.1.1}$$

where the coefficients are holomorphic functions of $z$ in the domain under consideration. The equation is closely related to the linear second order DE

$$w''(z) + P(z)w'(z) + Q(z)w(z) = 0, \tag{4.1.2}$$

for if we set

$$y(z) = -[f_2(z)]^{-1}w'(z)[w(z)]^{-1}, \tag{4.1.3}$$

then $w(z)$ will be a solution of

$$w''(z) - \{f_2'(z)[f_2(z)]^{-1} + f_1(z)\}w'(z) + f_1(z)f_2(z)w(z) = 0. \quad (4.1.4)$$

This is of the form (4.1.2) with

$$P(z) = -f_2'(z)[f_2(z)]^{-1} - f_1(z), \qquad Q(z) = f_0(z)f_2(z). \quad (4.1.5)$$

This is the passage from the Riccati equation to a second order linear DE. But we can of course reverse the process and get a Riccati equation corresponding to a given second order linear equation. Suppose that the given equation is (4.1.2), where $P$ and $Q$ are holomorphic in some domain. Here we set

$$y(z) = -\frac{w'(z)}{w(z)} \quad (4.1.6)$$

and obtain the Riccati equation

$$y'(z) = -Q(z) + P(z)y(z) + [y(z)]^2. \quad (4.1.7)$$

It is clear that the correspondence between the two classes of equations (4.1.1) and (4.1.2) is not one to one. To a given Riccati equation corresponds infinitely many linear second order equations, and vice versa.

The case where $f_0(z) \equiv 0$ is a special Bernoulli equation and hence reducible to a first order linear DE, which is solvable by quadratures. The equation

$$y'(z) = f(z)y(z) + g(z)[y(z)]^n, \quad (4.1.8)$$

where $n \neq 0$ and 1 but is not necessarily an integer, is reduced to

$$v'(z) = (1-n)[f(z)v(z) + g(z)] \quad (4.1.9)$$

via the substitution

$$v(z) = [y(z)]^{1-n}. \quad (4.1.10)$$

Equation (4.1.8) was proposed by Jacob Bernoulli (uncle of the Daniel Bernoulli mentioned above) in 1695 and was solved by his brother Johann and by C. W. Leibniz in 1697, using different methods.

Since the solution $v(z; z_0, v_0)$ of the linear equation

$$v'(z) + P_1(z)v(z) = R(z) \quad (4.1.11)$$

is

$$v(z) = v_0 \exp\left[-\int_{z_0}^z P_1(s)\,ds\right] + \int_{z_0}^z R(s)\exp\left[\int_s^z P_1(t)\,dt\right]ds, \quad (4.1.12)$$

it follows that the solution of

$$y'(z) = f_1(z)y(z) + f_2(z)[y(z)]^2 \qquad (4.1.13)$$

is

$$y(z) = \left\{ v_0 \exp\left[ -\int_{z_0}^z f_1(s)\, ds \right] \right.$$
$$\left. -\int_{z_0}^z f_2(s) \exp\left[ \int_s^z f_1(t)\, dt \right] ds \right\}^{-1}. \qquad (4.1.14)$$

This is of the form

$$\frac{1}{v_0 C(z) + D(z)}, \qquad (4.1.15)$$

i.e., a fractional linear transformation operating on the internal parameter $v_0$. As will be shown in the next section, the general solution of Riccati's equation can always be written as a fractional linear (= Möbius) transformation of an internal parameter with coefficients which are functions of $z$ and are expressible in terms of particular solutions of the equation.

A Riccati equation is formally invariant under a Möbius transformation operating on the dependent variable. Take an arbitrary set of four constants, $a$, $b$, $c$, $d$, with $ad - bc \neq 0$ and substitute

$$y = \frac{a + bv}{c + dv} \qquad (4.1.16)$$

in (4.1.1). Then $v$ satisfies the Riccati equation

$$(bc - ad)v' = (c^2 f_0 + acf_1 + a^2 f_2)$$
$$+ [2cdf_0 + (ad + bc)f_1 + 2abf_2]v$$
$$+ (d^2 f_0 + bdf_1 + b^2 f_2)v^2. \qquad (4.1.17)$$

In particular, $1/y$ satisfies

$$-v' = f_2(z) + f_1(z)v + f_0(z)v^2. \qquad (4.1.18)$$

We recall from Chapter 3 that the movable singularities of solutions of a Riccati equation are simple poles. Since the zeros of $y$ are poles of $v$, the solution of (4.1.18), it is seen that the zeros of $y(z)$, not located at a fixed singularity, are also simple.

A Riccati equation may have a rational function as a solution. Thus

$$y = \frac{2z}{z^2 - 1} \qquad \text{satisfies} \qquad -y' = \frac{y}{z(z^2 - 1)} + \tfrac{1}{2}y^2. \qquad (4.1.19)$$

**EXERCISE 4.1**

1. Verify (4.1.12).
2. Interpret $D(z)$ in (4.1.15).

3. Show that the general solution of (4.1.19) may be written as

$$y = 2z(z^2 - 1)^{-1/2}[(z^2 - 1)^{1/2} - C]^{-1}.$$

Each particular solution with $C \neq 0$ has two simple poles on the Riemann surface of $(z^2 - 1)^{1/2}$, and in this case $z = \pm 1$ are simple branch points of the solution. The special case $C = 0$ is the one figuring in (4.1.19).

4. The equation $y' = 1 - y^2$ has a solution which becomes infinite as $z = x$ decreases to zero. Show that $xy(x) \to 1$. Show that this solution is positive and decreases to $+1$ as $x \to +\infty$. Does the equation have any constant solutions?

5. [H. Wittich] Show that the function $z \mapsto z^{-1/2} \tan z^{1/2}$ satisfies the equation

$$y' = \frac{1}{2z} - \frac{1}{2z} y + \tfrac{1}{2}y^2.$$

Find the general solution. Discuss the distribution of the poles.

6. [H. Wittich] Show that the equation

$$y' = (z^3 - 1)^2 z^{-4} + (z^3 + 2)(z^4 - z)^{-1}y + y^2$$

has a general solution of the form

$$y(z) = (z - z^{-2}) \tan (\tfrac{1}{2}z^2 + z^{-1} + C).$$

7. The function $z \mapsto \tan z$ is a solution of $y' = 1 + y^2$. Apply the transformation (4.1.10) to this equation, and choose the constants $a, b, c, d$ in such a manner that the resulting equation becomes $v' = 1 + v^2$. Use this device to obtain the addition theorem for the tangent function.

8. The equation in Problem 7 has two constant solutions, $y = +i$ and $y = -i$. Use the existence and uniqueness theorems to show that $\tan z$ assumes the values $+i$ and $-i$ nowhere in the finite plane.

9. Let $y_1, y_2, y_3, y_4$ be four distinct solutions of

$$y' - f_0 - f_1 y - f_2 y^2 = 0.$$

Substitution of these particular values of $y$ leads to a system of four linear homogeneous equations which are known to have the nontrivial solution 1, $-f_0, -f_1, -f_2$. Discuss the resulting determinant condition. Compare Problem 4.2:1.

## 4.2. DEPENDENCE ON INTERNAL PARAMETERS; CROSS RATIOS

We shall show that the solution of

$$y'(z) = f_0(z) + f_1(z)y(z) + f_2(z)[y(z)]^2, \qquad y(z_0) = y_0, \qquad (4.2.1)$$

is a fractional linear function of $y_0$ with coefficients which are analytic functions of $z$. Compare formula (4.1.15) for the special case when $f_0(z) \equiv 0$.

In the complex plane we mark the fixed singularities of the equation, say $\zeta_1, \zeta_2, \ldots, \zeta_m$. These points are joined to the point at infinity by cuts which have no finite points in common. Let $T$ stand for the plane less the cuts, so that $T$ is a domain. Let $z_0 \in T$, and choose $y_0$ arbitrarily. This gives a unique solution $y(z; z_0, y_0)$ of (4.2.1). If $y_0$ is finite, this is a holomorphic function of $z$ in some disk $|z - z_0| < r$. It can be continued analytically along any rectifiable arc $C$ in $T$. Since there are no movable branch points, the continuation is analytic save for poles.

Let $z_1 \in C$, and consider $y(z_1; z_0, y_0) \equiv y_1$, where $y_1$ is either a definite finite number or infinity. Thus $y(z_1; z_0, y_0)$ is analytic save for poles in the finite $y_0$-plane, so that $y_1$ is a meromorphic function of $y_0$. But this holds also in the extended plane, for $y(z; z_0, \infty) = [v(z; z_0, 0)]^{-1}$, where $v$ is the solution of (4.1.18). Now a function which is meromorphic in the extended plane is a rational function, so that

$$y_1 = y(z_1; z_0, y_0) = R_1(y_0). \tag{4.2.2}$$

But here we can interchange the roles of $z_0$ and $z_1$. We consider

$$y(z; z_1, y_1), \tag{4.2.3}$$

i.e., the solution which takes the value $y_1$ at $z = z_1$. This solution can be continued analytically along $C$ from $z = z_1$ to $z = z_0$, and since every solution is single valued in $T$ we have

$$y(z_0; z_1, y_1) = y_0. \tag{4.2.4}$$

Here $y(z_0; z_1, y_1)$ is a meromorphic function of $y_1$ in the extended $y_1$-plane, i.e., a rational function of $y_1$, so that

$$y_0 = R_2(y_1). \tag{4.2.5}$$

Now, if a mapping $u \to v$ of the extended $u$-plane onto the extended $v$-plane is rational and has a rational inverse, it is one to one and is given by a Möbius transformation

$$v = \frac{au + b}{cu + d}, \qquad ad - bc \neq 0. \tag{4.2.6}$$

Applying this to the case on hand, we get

**THEOREM 4.2.1**

*The solution* $y(z; z_0, y_0)$ *is the result of applying a Möbius transformation to* $y_0$:

$$y(z; z_0, y_0) = \frac{y_0 A(z) + B(z)}{y_0 C(z) + D(z)}. \tag{4.2.7}$$

Now a fractional linear transformation is determined by three of its transforms, uniquely up to a common factor. Let the transformation be given by (4.2.6). Give $u$ four complex values, $u_1$, $u_2$, $u_3$, $u_4$, and let the corresponding values of $v$ be $v_1$, $v_2$, $v_3$, $v_4$. We introduce the *cross ratio*

$$\mathbf{R}(z_1, z_2, z_3, z_4) = \frac{z_1 - z_3}{z_1 - z_4} : \frac{z_2 - z_3}{z_2 - z_4}, \tag{4.2.8}$$

and find after a lengthy but elementary computation that

$$\mathbf{R}(v_1, v_2, v_3, v_4) = \mathbf{R}(u_1, u_2, u_3, u_4). \tag{4.2.9}$$

This implies that the cross ratio of four solutions of (4.2.1) is a constant. To see this we give $y_0$ four values, $y_{01}$, $y_{02}$, $y_{03}$, $y_{04}$, in (4.2.7) and denote the corresponding solutions by $y_1$, $y_2$, $y_3$, $y_4$, so that

$$\mathbf{R}(y_1, y_2, y_3, y_4) = \mathbf{R}(y_{01}, y_{02}, y_{03}, y_{04}). \tag{4.2.10}$$

This means that, knowing three solutions of a Riccati equation, $y_1$, $y_2$, $y_3$, we know all solutions. The general solution may be written as

$$y(z) = \frac{[y_1(z) - y_3(z)]y_2(z) - C[y_2(z) - y_3(z)]y_1(z)}{[y_1(z) - y_3(z)] - C[y_2(z) - y_3(z)]}, \tag{4.2.11}$$

where $C$ is an arbitrary constant (theorem due to Eduard Weyr, 1875). We note that the solution is a fractional linear function of the internal parameter $C$.

There is another way of tackling this problem which is also quite instructive. It is based on the representation of the solution of a Riccati equation in terms of solutions of an associated linear second order DE; see (4.1.1)–(4.1.5). The discussion will involve several concepts which belong to the next chapter but undoubtedly are familiar to the reader from the elementary theory of DE's. Let $w_1(z)$ and $w_2(z)$ be two linearly independent solutions of (4.1.4). Then the general solution of this equation is

$$w(z) = C_1 w_1(z) + C_2 w_2(z), \tag{4.2.12}$$

where $C_1$ and $C_2$ are arbitrary constants. This means that the general solution of the Riccati equation is given by

$$-\frac{1}{f_2(z)} \cdot \frac{C_1 w_1'(z) + C_2 w_2'(z)}{C_1 w_1(z) + C_2 w_2(z)}.$$

Actually the solution does not depend on two arbitrary constants; only their ratio, $C_1/C_2 = C$, is significant. Hence

$$y(z) = -\frac{1}{f_2(z)} \frac{C w_1'(z) + w_2'(z)}{C w_1(z) + w_2(z)}. \tag{4.2.13}$$

In this setting the general solution is a linear fractional transform of $C$. We see that the zeros of $y(z)$ are the zeros of

$$w_2'(z) + Cw_1'(z), \tag{4.2.14}$$

while the poles are the zeros of

$$w_2(z) + Cw_1(z). \tag{4.2.15}$$

Since both zeros and poles of $y(z)$ are simple, we conclude that the functions (4.2.14) and (4.2.15) have simple zeros.

In (4.2.13) it is permissible to let $C$ become infinite. The limit exists and is a solution of the Riccati equation.

We have seen that the general solution of the Riccati equation is a fractional linear transform of an internal parameter with coefficients which are functions of $z$. Conversely, let $A(z)$, $B(z)$, $C(z)$, $D(z)$ be four functions holomorphic in a domain of the $z$-plane. Let $K$ be an arbitrary parameter, and form the family of functions

$$z \mapsto y_K(z) = \frac{A(z) + KB(z)}{C(z) + KD(z)}. \tag{4.2.16}$$

These functions satisfy a Riccati equation, for we have

$$y' = \{(A'C - AC') + [(B'C + A'D) - (AD' + BC')]K \\ + (B'D - BD')K^2\}(C + DK)^{-2}.$$

We solve (4.2.16) for $K$ to obtain

$$K = \frac{Cy - A}{B - Dy}.$$

Furthermore,

$$(C + DK)^{-1} = \frac{B - Dy}{BC - AD}.$$

Substitution now gives

$$(AD - BC)^2 y' = (A'C - AC')(B - Dy)^2 \\ + [BC' - B'C + AD' - A'D](A - Cy)(B - Dy) \\ + (B' - BD')(A - Cy)^2, \tag{4.2.17}$$

which clearly is a Riccati equation.

## EXERCISE 4.2

1. In Problem 4.1:8 it was found that a certain determinant had to be zero. Show that this determinant is the numerator in the derivative of the cross ratio $R_x(y_1, y_2, y_3, y_4)$ and hence that the cross ratio is a constant.

2.  The symmetric group on four letters has 24 elements. If the group operates on the cross ratio of these four, how many different cross ratios are obtained?
3.  For what values of $C$ in (4.2.11) are the initial solutions $y_1$, $y_2$, $y_3$ obtained? Verify the formula.
4.  Verify that the only rational functions with a rational inverse are the fractional linear functions.
5.  The Schwarzian derivative

$$\{w, z\} = \frac{w'''}{w'} - \frac{3}{2}\left(\frac{w''}{w'}\right)^2$$

is a differential invariant of the group of fractional linear transformations acting on $w$, i.e.,

$$\{W, z\} = \{w, z\} \qquad \text{if } W = \frac{aw + b}{cw + d}, \quad ad - bc \neq 0,$$

where $a$, $b$, $c$, $d$ are constants. Verify.

6.  Show that the third order nonlinear DE

$$\{w, z\} = P(z)$$

is a Riccati equation for the logarithmic derivative of $w'(z)$.

## 4.3.  SOME GEOMETRIC APPLICATIONS

Cross ratios are a fundamental concept of projective geometry. Let $l_1$ and $l_2$ be two lines in $\mathbf{R}^3$, the points on $l_1$ being given by a parameter $s$, and those on $l_2$ by a parameter $t$. Let $A$, $B$, $C$, $D$ be any four constants with $AD - BC \neq 0$. The mapping

$$s \to t = \frac{As + B}{Cs + D} \tag{4.3.1}$$

defines a one-to-one correspondence between the points of the two lines, and this correspondence is the most general *projective relation*. Such a relation is characterized by the fact that the cross ratio of four points on $l_1$ is the same as the cross ratio of the four corresponding points on $l_2$. Thus, if the points on $l_1$ are given by the values $s_1$, $s_2$, $s_3$, $s_4$ of the parameter $s$ and if the corresponding points on $l_2$ are given by $t = t_1, t_2, t_3, t_4$, then

$$\mathbf{R}(s_1, s_2, s_3, s_4) = \mathbf{R}(t_1, t_2, t_3, t_4). \tag{4.3.2}$$

For the Riccati equation (4.2.1) we have seen that the cross ratio of four solutions, $y$, $y_1$, $y_2$, $y_3$, is a constant:

$$\mathbf{R}(y, y_1, y_2, y_3) = \mathbf{R}(y_0, y_{01}, y_{02}, y_{03}) \equiv K, \tag{4.3.3}$$

where $y_0$, $y_{01}$, $y_{02}$, $y_{03}$ are the initial values of the solutions at some point $z_0$, not a fixed singularity. By (4.2.11) this gives

$$y = \frac{(y_1 - y_3)y_2 - K(y_2 - y_3)y_1}{(y_1 - y_3) - K(y_2 - y_3)}, \tag{4.3.4}$$

and $K$ is an arbitrary constant. We note that the general solution $y(x)$ is a rational function—more precisely, a fractional linear function of each of three arbitrarily chosen particular solutions.

Thus we see that Riccati's equation leads to *projectivities*. Conversely, various problems in differential geometry lead to Riccati equations. We shall discuss two such problems.

### The Formulas of Frenet

Let $\Gamma$ be a rectifiable curve in 3-space, $s$ its *arc length*, $R = R(s)$ the *radius of curvature*, $T = T(s)$ the *radius of torsion*. Let $X$, $Y$, $Z$ be running coordinates. Let the orientation of the axes be such that an ordinary right-hand screw placed along the $Z$-axis will advance in the positive $Z$-direction when rotated from the positive $X$-direction to the positive $Y$-direction. Let $x_1(s)$, $x_2(s)$, $x_3(s)$ be the *moving trihedral* of unit vectors taken along the *tangent*, the *principal normal*, and the *binormal* of $\Gamma$ at $s$. The trihedral is supposed to be oriented in the same manner as the coordinate axes. Let $\Gamma$ be given by the vector equation

$$\mathbf{x} = \mathbf{g}(s). \tag{4.3.5}$$

Then the *tangent vector* is $\mathbf{x}_1(s) = \mathbf{g}'(s)$, which is of length 1. Furthermore, $\mathbf{g}''(s)$ is the *curvature vector*. It is not normalized (i.e., of length 1), but $\mathbf{x}_2(s) = \mathbf{g}''(s)\|\mathbf{g}''(s)\|^{-1}$ is the principal normal. The binormal is a unit vector orthogonal to the tangent and the principal normal and properly oriented.

The formulas of Frenet we can interpret as simultaneous first order vector DE's satisfied by the moving trihedral. They are named after Jean Frédéric Frenet (1816–1900), who discovered them (in component form) in 1847. They were rediscovered by Alfred Serret (1819–1885) in 1850. The vector equations are as follows:

$$\mathbf{x}_1'(s) = \frac{\mathbf{x}_2(s)}{R(s)}, \qquad \mathbf{x}_2'(s) = -\frac{\mathbf{x}_1(s)}{R(s)} + \frac{\mathbf{x}_3(s)}{T(s)}, \qquad \mathbf{x}_3'(s) = -\frac{\mathbf{x}_1(s)}{T(s)}. \tag{4.3.6}$$

Let the components of the vector $\mathbf{x}_j(s)$ be $x_{jk}(s)$, and introduce the unitary matrix

$$\mathcal{U}(s) \equiv \begin{bmatrix} x_{11}(s) & x_{12}(s) & x_{13}(s) \\ x_{21}(s) & x_{22}(s) & x_{23}(s) \\ x_{31}(s) & x_{32}(s) & x_{33}(s) \end{bmatrix} \tag{4.3.7}$$

We can multiply (4.3.6) on the left or on the right by an arbitrary matrix, and the resulting vector equations are valid as long as the original equations are. In particular, we can get the system of three scalar DE's for the first column of (4.3.7):

$$x'_{11}(s) = \frac{x_{21}(s)}{R(s)}, \qquad x'_{21}(s) = -\frac{x_{11}(s)}{R(s)} + \frac{x_{21}(s)}{T(s)}, \qquad x'_{31}(s) = -\frac{x_{21}(s)}{T(s)}. \qquad (4.3.8)$$

Since the sum of the squares of the elements in a column of $\mathcal{U}(s)$ equals one, we can set

$$\begin{aligned} \frac{x_{11} + ix_{21}}{1 - x_{31}} &= \frac{1 + x_{31}}{x_{11} - ix_{21}} = u, \\[2mm] \frac{x_{11} - ix_{21}}{1 - x_{31}} &= \frac{1 + x_{31}}{x_{11} + ix_{21}} = -\frac{1}{v}. \end{aligned} \qquad (4.3.9)$$

From these relations we get

$$x_{11} = \frac{1 - uv}{u - v}, \qquad x_{21} = i\frac{1 + uv}{u - v}, \qquad x_{31} = \frac{u + v}{u - v}. \qquad (4.3.10)$$

If these expressions are substituted in (4.3.8), we obtain the DE

$$\frac{du}{ds} = -\frac{1}{2}\left[\frac{1}{R(s)} - \frac{i}{T(s)}\right] - \frac{1}{2}\left[\frac{1}{R(s)} + \frac{i}{T(s)}\right]u^2, \qquad (4.3.11)$$

and the same Riccati equation is satisfied by $v$. Hence $u$ and $v$ are distinct solutions of the same Riccati equation.

Now if a solution $u(s)$ of (4.3.11) has been found, then

$$v(s) = -\overline{[u(s)]}^{-1} \qquad (4.3.12)$$

is found to satisfy (4.3.11). Here the bar indicates, as usual, the complex conjugate. Knowing $u(s)$ and $v(s)$, we can compute the first column of $\mathcal{U}(s)$. The same DE is found if we apply this procedure to the second or the third column.

This discussion implies that a rectifiable space curve is completely determined by initial conditions: (i) for $s = s_0$ the point $(x_0, y_0, z_0)$ is on the curve, and at this point the moving trihedral is given; (ii) the radii of curvature and of torsion are given functions of the arc length $s$; (3) $R(s)$ and $T(s)$ are differentiable functions of $s$ with bounded derivatives. Then the existence theorems of Chapter 2 applied to (4.3.11) show that the matrix $\mathcal{U}(s)$ is uniquely determined and hence also the curve itself. This observation and this method date back to the Norwegian mathematician Marius Sophus Lie (1842–1899) in 1882.

Relations of the form

$$R = F_1(s), \qquad T = F_2(s) \tag{4.3.13}$$

are known as the *natural equations* of a curve.

The problem of determining a surface in $R^3$ from a knowledge of its fundamental forms,

$$E \, (du)^2 + 2F \, du \, dv + G \, (dv)^2,$$
$$L \, (du)^2 + 2m \, du \, dv + N \, (dv)^2, \tag{4.3.14}$$

also leads ultimately to the integration of an ordinary Riccati equation. The formulas are too complicated to be given here.

## Asymptotic Lines of a Ruled Surface

As a second differential geometric application of Riccati's equation we take the problem of finding the *asymptotic lines on a ruled surface S* in $R^3$. The surface is given by a vector equation

$$\mathbf{x} = \mathbf{G}(u, v) \tag{4.3.15}$$

in terms of two parameters, $u$ and $v$. In the case of a ruled surface this takes the form

$$\mathbf{x} = \mathbf{G}(u) + v\mathbf{H}(u). \tag{4.3.16}$$

Here the system $u = C$ is the family of the *rectilinear generators* (the "rulers"). The *asymptotic lines are the curves which annihilate the second fundamental form*. In the present case $N$ is identically zero, $M$ is a function of $u$ alone, and $L$ is quadratic in $v$. The form reduces to

$$du \, (L \, du + 2M \, dv) = 0.$$

Setting $du = 0$, we get the generators $u = C$ as one set of asymptotic lines, while the second set satisfies the Riccati equation

$$\frac{dv}{du} = A + Bv + Cv^2, \tag{4.3.17}$$

where the coefficients are functions of $u$ alone. We may assume the coefficients to be continuous. In this case the right-hand side satisfies a Lipschitz condition with respect to $v$, and through each point $(u_0, v_0)$ passes a unique solution of (4.3.17). If $v_1$, $v_2$, $v_3$, $v_4$ are four solutions of (4.3.17), their cross ratio is a constant. This says that, if the four solutions are real curves, they cut each rectilinear generator in four points having a constant cross ratio. In turn, this says that the rectilinear generators are cut by the second family of asymptotic lines in projective point sets.

In the case of a hyperboloid of one sheet the second set of asymptotic lines is also a family of straight lines, and it is a known fact that the two sets of rectilinear generators cut each other in projective point sets.

We refrain from further geometric applications of Riccati's equation.

### EXERCISE 4.3

1. Two vectors in $R^3$ are orthogonal if their dot product $x \cdot y = \Sigma_1^3 x_i y_i = 0$. Find the components of the binormal vector, knowing that it is orthogonal to the tangent and principal normal vectors.

### 4.4. ABSTRACT OF THE NEVANLINNA THEORY, I

In the 1920's the brothers Frithiof and Rolf Nevanlinna, pupils of Ernst Lindelöf, developed an extensive theory of the *value distribution* of entire and meromorphic functions, aimed at sharpening the theorem of Picard. In 1933 the Japanese mathematician Kôsaku Yosida gave an elegant application of this theory to nonlinear DE's by presenting a new proof and extensions of a striking theorem by Johannes Malmquist (1882–1952) dating from 1913. A systematic study of the implications of the Nevanlinna theory for DE's was undertaken by Hans Wittich, starting in 1950. We shall give some of the results of Wittich and Yosida in Section 4.6 and in later chapters. It is clear that such a program requires at least a brief review of the salient points in the value distribution theory. The basic concepts and formulas are given in the present section; the proximity function of the logarithmic derivative, the second fundamental theorem, and the defect relations, in the next section.

It is customary to start with the theorem of Jensen [J. L. V. W. Jensen, (1859–1925), Danish mathematician and telephone engineer]. We shall however, go further back and begin with a modified form of Cauchy's integral which was a useful tool in the Hadamard theory of entire functions of finite order.

Let $z \mapsto G(z)$ be a function holomorphic in the disk $|z| \leq R < \infty$, and set $G(z) = U(z) + i V(z)$, where $U$ and $V$ are real potential functions. We have then

$$G(z) = -\overline{G(0)} + \frac{1}{\pi i} \oint \frac{U(t)}{t - z} \, dt, \tag{4.4.1}$$

$$G(z) = \overline{G(0)} + \frac{1}{\pi} \oint \frac{V(t)}{t - z} \, dt. \tag{4.4.2}$$

Here the path of integration is the circle $|t| = R$ described in the positive sense, and the bar indicates the complex conjugate. The proof is left to the reader.

Now let $a$ be a complex number, $0 < |a| < R$, and form the *Möbius function*:

$$Q(z; a, R) = \frac{R(z - a)}{R^2 - \bar{a}z}. \tag{4.4.3}$$

It is holomorphic in the extended plane except for a simple pole at $z = R^2 \bar{a}^{-1}$; it maps the whole plane into itself in a one-to-one conformal manner. Here $z = a$ goes into $w = 0$, $z = R^2 \bar{a}^{-1}$ into $w = \infty$. The circle $|z| = R$ goes into the unit circle $|w| = 1$; the interior (exterior) of the first circle goes into the interior (exterior) of the second circle.

Suppose now that $z \mapsto f(z)$ is meromorphic in $|z| \leq R$ and has neither zeros nor poles at $z = 0$ or on the circle $|z| = R$. Let the zeros be

$$a_1, a_2, \ldots, a_j \qquad \text{with } |a_\alpha| = r_\alpha, \tag{4.4.4}$$

and the poles

$$b_1, b_2, \ldots, b_k \qquad \text{with } |b_\beta| = p_\beta. \tag{4.4.5}$$

Here multiple zeros or poles are counted as often as indicated by their multiplicity. We now construct a function $z \mapsto H(z)$ which has neither zeros nor poles in $|z| \leq R$ and for which $|H(z)| = |f(z)|$ on $|z| = R$. This we can do by multiplying $f(z)$ by a product of $Q$-functions and their reciprocals so as to eliminate the zeros and poles of $f$ without affecting the absolute value on $|z| = R$. We take

$$H(z) = f(z) \prod_{\alpha=1}^{j} [Q(z; a_\alpha, R)]^{-1} \prod_{\beta=1}^{k} Q(z; b_\beta, R), \tag{4.4.6}$$

which clearly has all the desired properties.

Next let

$$L(z) = \log H(z). \tag{4.4.7}$$

This function is holomorphic in $|z| \leq R$ once we have chosen the value of the logarithm at some particular point. Since $H(z) \neq 0, \infty$ in the disk, its logarithm is single valued and hence holomorphic. We now use (4.4.1) with $G(z) = L(z)$. Here $U(z) = \log |H(z)|$, $V(z) = \arg H(z)$, and $|H(z)| = |f(z)|$ on $|z| = R$. Hence

$$\log H(z) = -\overline{\log H(0)} + \frac{1}{\pi i} \oint \frac{\log |f(t)|}{t - z} \, dt, \tag{4.4.8}$$

and by (4.4.6)

$$\log f(z) = -\overline{\log H(0)} + \frac{1}{\pi i} \oint \frac{\log |f(t)|}{t - z} \, dt$$
$$+ \sum_{\alpha=1}^{j} \log Q(z; a_\alpha, R) - \sum_{\beta=1}^{k} \log Q(z; b_\beta, R). \tag{4.4.9}$$

The logarithms in the left member and in the sums have not been defined. For the time being, this information is immaterial and any ambiguity will disappear when derivatives are taken with respect to $z$. At this juncture we set $z = 0$ and obtain

$$\log f(0) = -\overline{\log H(0)} + \frac{1}{\pi} \int_0^{2\pi} \log |f(R e^{i\theta})| \, d\theta$$
$$+ \sum_{\alpha=1}^{j} \log \left(-\frac{a_\alpha}{R}\right) - \sum_{\beta=1}^{k} \log \left(-\frac{b_\beta}{R}\right).$$

Here we equate the real parts and get Jensen's formula,

$$\frac{1}{2\pi} \int_0^{2\pi} \log |f(R e^{i\theta})| \, d\theta = \log |f(0)| + \sum_{\alpha=1}^{j} \log \frac{R}{r_\alpha} - \sum_{\beta=1}^{k} \log \frac{R}{p_\beta}. \qquad (4.4.10)$$

Now we are ready for Rolf Nevanlinna. He got his first fundamental theorem by the following argument. For $t$ real, set

$$t^+ = \max (t, 0). \qquad (4.4.11)$$

Then

$$\log |f| = \log^+ |f| - \log^+ \frac{1}{|f|}$$

and

$$\frac{1}{2\pi} \int_0^{2\pi} \log |f(r e^{i\theta})| \, d\theta = m (r, \infty; f) - m (r, 0; f). \qquad (4.4.12)$$

The expressions in the right member are the *proximity functions* of $f$ with respect to the values $\infty$ and $0$. If $a \neq \infty$, we set

$$m (r, a; f) = \frac{1}{2\pi} \int_0^{2\pi} \log^+ \frac{1}{|f(r e^{i\theta}) - a|} \, d\theta \qquad (4.4.13)$$

while

$$m (r, \infty; f) = \frac{1}{2\pi} \int_0^{2\pi} \log^+ |f(r e^{i\theta})| \, d\theta. \qquad (4.4.14)$$

To cope with the sums in the right member of (4.4.10) we introduce *enumerative functions*. Let $n(s, a; f)$ be the number of times that $f(z) = a$ for $|z| \leq s$, multiple roots counted with the proper multiplicity. If $f(0) \neq a$, we set

$$N(r, a; f) = \int_0^r n(s, a; f) \frac{ds}{s}. \qquad (4.4.15)$$

This can also be written as a Riemann-Stieltjes integral,

$$\int_0^r \log \frac{r}{s} \, d_s n (s, a; f), \qquad (4.4.16)$$

and this observation shows that the difference of the two sums in (4.4.10) equals

$$N(R, \infty; f) - N(R, 0; f).$$

This gives

$$[m(R, 0; f) + N(R, 0; f)] - [m(R, \infty; f) + N(R, \infty; f)] = \log |f(0)|. \qquad (4.4.17)$$

If $f(z)$ is meromorphic in the finite plane, $R$ may be taken arbitrarily large. The expression

$$T(r; f) = m(r, \infty; f) + N(r, \infty; f) \qquad (4.4.18)$$

is then defined for all $r > 0$. It is the *characteristic function* of $f$ in the sense of Nevanlinna. It measures the *affinity* of $f$ for the value $\infty$, for $N(r, \infty; f)$ measures the frequency of the poles and $m(r, \infty; f)$ gets its contribution from the arcs of $|z| = r$, where $|f(z)|$ is large. Now (4.4.17) states that

$$T\left(r; \frac{1}{f}\right) - T(r; f) = \log |f(0)|. \qquad (4.4.19)$$

In other words the affinity of $f$ for the value 0 differs from its affinity for the value $\infty$ by a constant. Replacing $f$ by $f - a$, $a$ finite, we see that the affinity for all values is essentially the same:

$$T(r; f - a) = T(r; f) + O(\log r). \qquad (4.4.20)$$

For fixed $a$, the deviation is normally $O(1)$, but the larger $O(\log r)$ is required if $f(0) = a$.

Similar formulas hold if $f(z)$ is single valued and meromorphic for $0 < r_0 \leqslant |z| < \infty$ and is different from zero and infinity on $|z| = r_0$. The proximity functions are unchanged but with $r_0 < r$. In the enumerative functions the lower limit of integration is $r_0$ instead of zero. $T(r; f)$ is still defined by (4.4.18), and (4.4.20) holds.

These various relations constitute Nevanlinna's *first fundamental theorem*.

Among the various inequalities satisfied by $T(r; f)$ we list the following:

$$T(r; f_1 f_2 \ldots f_m) \leqslant \sum_{j=1}^{m} T(r; f_j), \qquad (4.4.21)$$

$$T\left(r; \sum_{j=1}^{m} f_j\right) \leqslant \sum_{j=1}^{m} T(r; f_j) + \log m. \qquad (4.4.22)$$

A Möbius transformation applied to a meromorphic function $f$ leads to another meromorphic function $F$, and the difference of $T(r; f)$ and $T(r; F)$ is at most $O(\log r)$.

A meromorphic function $f$ such that $T(r;f) = O(\log r)$ forces $f$ to be rational. To see this, set

$$\liminf \frac{T(r;f)}{\log r} = c. \qquad (4.4.23)$$

Here $c > 0$ unless $f$ is a constant. It is desired to prove that $c$ is finite and equal to a positive integer iff $f$ is a rational function, in which case "lim inf" becomes "lim."

To perceive this, suppose that $f$ has a total number of $m$ poles in the finite plane and that $f(z) \sim Az^n$ for large values of $z$. Then $n(r, \infty; f) = m$ for $r > p_m$, so that $N(r, \infty; f)$ lies between $m \log r$ and $m \log r + C$ and hence $\lim N(r, \infty; f)/(\log r) = m$. If $n > 0$, then $m(r, \infty; f) = n \log r + O(1)$, so that

$$\lim \frac{T(r;f)}{\log r} = m + n, \qquad (4.4.24)$$

where the term $n$ is omitted if $n < 1$.

Conversely, suppose that $0 < c < \infty$, and it is to be proved that $f$ is a rational function. Let $a$ be an arbitrary complex number or infinity. Suppose that $f(0) \neq a$, and consider

$$N(r, a; f) = \int_0^r n(s, a; f) \frac{ds}{s} > \int_{r^{1/k}}^r n(s, a; f) \frac{ds}{s},$$

where $k$ is any real number $\geq 2$. The last member is at least

$$\frac{k-1}{k} n(r^{1/k}, a; f) \log r.$$

Here $(k-1)/k \geq \frac{1}{2}$, and $r^{1/k}$ sweeps the interval $(1, r^{1/2})$ as $k$ goes from two to infinity. By assumption there exists a sequence $\{r_n\}$, $r_n \uparrow \infty$, such that $T(r_n; f) < 2c \log r_n$. Since every $r$ is ultimately included in some interval $(1, r_n^{1/2})$, it follows that $n(r, a; f) < 4c$ for all $a$ and all large $r$. The only meromorphic function that could and does possess such a property is a rational function.

In addition to the Nevanlinna characteristic function we have to consider the *spherical characteristic* $T^0(r; f)$, introduced in 1929 by Tatsujiro Shimizu in Japan and Lars Ahlfors in Finland independently of each other. Consider the mapping defined by $z \mapsto f(z)$ of the disk $|z| \leq s$ on the $w$-plane; it is the germ of a Riemann surface $R_s$ which covers parts of the $w$-plane several times; in fact, $n(s, a; f)$ measures how many times the particular value $w = a$ is covered. We make a stereographic projection of $R_s$ on a sphere of radius $\frac{1}{2}$, tangent to the $w$-plane at the origin. The area of the projection is

$$\pi A(s;f) = \int_{|z| \le s} \frac{|f'(z)|^2 \, d\omega}{[1 + |f(z)|^2]^2}, \qquad (4.4.25)$$

where $d\omega$ is the surface element in the $z$-plane. We now define

$$T^0(r;f) = \int_0^r A(s;f) \frac{ds}{s}. \qquad (4.4.26)$$

This function is strictly increasing and its second derivative with respect to $\log r$ is positive, so that $T^0(r;f)$ is *logarithmically convex*. Here this property is evident; the Nevanlinna characteristic has the same property, but in that case the proof is fairly intricate. The logarithmic convexity is equivalent to the inequality

$$T^0(r^c s^{1-c}; f) \le cT(r;f) + (1-c)T(s;f). \qquad (4.4.27)$$

It turns out that *the difference between the two characteristics is a bounded function of* r, *so they are interchangeable.*

*Order* and *type* of meromorphic functions are defined by means of the characteristic. Thus the order of $f$ is

$$\rho(f) = \limsup \frac{\log T^0(r;f)}{\log r}, \qquad (4.4.28)$$

and if $0 < \rho < \infty$ the type is

$$\tau(f) = \limsup r^{-\rho} T^0(r;f). \qquad (4.4.29)$$

The type is *normal* if $0 < \tau(f) < \infty$, *minimal* if $\tau(f) = 0$, *maximal* if $\tau(f) = +\infty$. In the case of entire functions this notion of order agrees with the classical definition in terms of the maximum modulus,

$$\rho(f) = \limsup \frac{\log \log M(r;f)}{\log r}. \qquad (4.4.30)$$

The notion of type is also the same (i.e., normal, minimal, maximal), but in the case of normal type the numerical value will depend on which definition is used.

For entire or meromorphic functions of finite order $\rho$ the three integrals

$$\int_1^\infty s^{-\alpha} d_s n(s, a; f), \qquad \int_1^\infty s^{-\alpha-1} n(s; a; f) \, ds,$$
$$\int_1^\infty s^{-\alpha-1} N(s, a; f) \, ds \qquad (4.4.31)$$

are *equiconvergent* and do converge for any $\alpha > \rho$. If the roots of the equation $f(z) = a$ form the set $\{z_n(a)\}$ and $|z_n(a)| = r_n(a)$, the convergence of the first integral implies convergence of the series

$$\sum_{n=1}^{\infty} [r_n(a)]^{-\alpha}. \tag{4.4.32}$$

The *exponent of convergence* of the series is that number $\sigma$ such that the series converges for all $\alpha > \sigma$ and diverges for all $\alpha < \sigma$, while $\alpha = \sigma$ may yield either convergence or divergence. In the present case $\sigma \leq \rho$, and it may be shown that $\sigma < \rho$ can hold for at most two values of $a$. These convergence properties clearly imply that the series

$$\sum_{n=1}^{\infty} T^0[r_n(a); f]^{-\alpha} \tag{4.4.33}$$

has a convergence exponent not exceeding one and this holds also if $f$ is of infinite order. If in the integrals (4.4.31) the factor $s^{-\alpha}$ is replaced by $[T^0(s; f)]^{-\alpha}$, integrals are obtained for which similar equiconvergence theorems hold.

## EXERCISE 4.4

1. It is desired to prove (4.4.1) and (4.4.2). To this end note that on the circle $|z| = R$ the function $z \mapsto G(z)$ is a Fourier power series. Prove that the Fourier coefficients are

$$\frac{1}{2\pi} \int_0^{2\pi} G(R e^{i\varphi}) e^{-im\varphi} \, d\varphi = \begin{cases} 0, & m < 0, \\ c_m R^m, & m \geq 0, \end{cases} \quad \text{if } G(z) = \sum_{m=0}^{\infty} c_m z^m.$$

   Furthermore, $c_0 + \bar{c}_0 = 2U(0)$, $c_0 - \bar{c}_0 = 2iV(0)$. Complete the proof.

2. The function $z \mapsto \cos z$ is entire. Determine $n(r, 0; f)$, $m(r, 0; f)$, $m(r, \infty; f)$, and $T(r; f)$. Find order and type.

3. Solve problem 2 for the logarithmic derivative $-\tan z$.

4. Verify the stated properties of $Q(z, a, R)$.

5. Complete the proof of (4.4.10).

6. Prove the properties of $\log^+$ which justify (4.4.21) and (4.4.22).

7. If $u \mapsto F(u)$ is a fractional linear transformation on $u$ with constant coefficients, show that $T[r; F(f)] = T(r; f) + O(\log r)$.

8. Verify the statements made about (4.4.31) and (4.4.32). Use the function $e^z$ to show that the exponent of convergence in (4.4.32) can be less than one for two values of $a$.

9. Why does the condition $n(r, a; f) < 4c$ for all $a$ and all large $r$ imply that a meromorphic function $f$ is rational?

10. The generalized equiconvergence theorem mentioned at the end of the section involves the following items:

$$\int_1^{\infty} [T^0(s; f)]^{-\alpha} d_s n(s, a; f), \qquad \int_1^{\infty} [T^0(s; f)]^{-\alpha} n(s, a; f) \frac{ds}{s},$$

$$\int_1^\infty [T^0(s;f)]^{-\alpha} N(s,a;f) \frac{ds}{s}, \qquad \sum_{n=1}^\infty \{T[r_n(a);f]\}^{-\alpha}.$$

To prove equiconvergence, start with the second integral.

11.  Show that the Laplacian of $\log[1+|f(z)|^2]$ equals $4|f'(z)|^2[1+|f(z)|^2]^{-2}$.

12.  Temporarily, let us say that a value $w = a$ is *exceptional* for a meromorphic function $f$ if

$$\limsup N(r, a; f)[T^0(r;f)]^{-1} = \eta(a) < 1.$$

It is a Picard value if $\eta(a) = 0$. Show that for $z \mapsto \int_0^z \exp(-s^2)\,ds$ there are three exceptional values, $\pm\frac{1}{2}\sqrt{\pi}, \infty$. Find the values of $\eta(a)$.

13.  An exceptional value may be *asymptotic* in the sense that $f(z) \to a$ as $z \to \infty$ in some sector of the $z$-plane. Show that this is the case for the function in Problem 12, and find the sectors. (To be an exceptional value, however, it is neither necessary nor sufficient to be asymptotic.)

## 4.5.  ABSTRACT OF THE NEVANLINNA THEORY, II

In this section we continue the study of the Nevanlinna theory. The main object will be the proximity functions of the logarithmic derivative of a meromorphic function, but the second fundamental theorem and the defect relations will also be presented.

The starting point is (4.4.9):

$$\log f(z) = -\overline{\log H(0)} + \frac{1}{\pi i} \oint \frac{\log|f(t)|}{t-z}\,dt$$
$$+ \sum_{\alpha=1}^j \log Q(z, a_\alpha, R) - \sum_{\beta=1}^k \log Q(z, b_\beta, R). \qquad (4.5.1)$$

The logarithms in the left member and in the sums may be specified by analytic continuation, but actually the chosen determinations are immaterial for our purposes since we are going to take derivatives, thus removing the ambiguities. We may clearly differentiate the integral with respect to $z$ under the integral sign and obtain

$$\frac{f'(z)}{f(z)} = \frac{R}{\pi} \int_0^{2\pi} \frac{\log|f(Re^{i\varphi})|}{(t-z)^2} e^{i\varphi}\,d\varphi$$
$$+ \sum_1^j \left(\frac{1}{z-a_\alpha} + \frac{\bar{a}_\alpha}{R^2 - \bar{a}_\alpha z}\right) - \sum_1^k \left(\frac{1}{z-b_\beta} + \frac{\bar{b}_\beta}{R^2 - \bar{b}_\beta z}\right). \qquad (4.5.2)$$

Here $z = re^{i\theta}$ is different from all zeros and poles of $z \mapsto f(z)$ in the disk, and $r < R$. We aim to estimate this expression with the ultimate view of obtaining an upper bound for the proximity function $m(r, \infty; f'/f)$. Let $d$ be the distance of $z$ from the nearest zero or pole, so that $|z - a_\alpha| \geq d$,

$|z - b_\beta| \geq d$. Then

$$\frac{\bar{a}}{R^2 - \bar{a}z}$$

has its greatest value when $\bar{a}z$ is real positive, so that

$$\left|\frac{\bar{a}}{R^2 - \bar{a}z}\right| \leq \frac{R}{R^2 - rR} = \frac{1}{R - r}.$$

The total number of terms in the sums is $n(R, 0; f) + n(R, \infty; f)$, so that the sums contribute at most

$$\left(\frac{1}{d} + \frac{1}{R - r}\right)[n(R, 0; f) + n(R, \infty; f)]$$

to the estimate. The integral gives at most

$$\frac{2R}{(R - r)^2}[m(R, 0; f) + m(R, \infty; f)].$$

All told, we get

$$\left|\frac{f'(z)}{f(z)}\right| \leq \frac{2R}{(R - r)^2}[m(R, 0; f) + m(R, \infty; f)]$$

$$+ \left(\frac{1}{d} + \frac{1}{R - r}\right)[n(R, 0; f) + n(R, \infty; f)]. \qquad (4.5.3)$$

Let $\delta$ be any small positive number, $d \sim [T^0(r; f)]^{-1-\delta}$, and $R = r + d$. We have then

$$\left|\frac{f'(z)}{f(z)}\right| \leq 2(r + d)d^{-2}[m(r + d, 0; f) + m(r + d, \infty; f)]$$

$$+ 2d^{-1}[n(r + d, 0; f) + n(r + d, \infty; f)]. \qquad (4.5.4)$$

Here $z = re^{i\theta}$, and values of $r$ must be excluded around each $p_n$ and $r_n$. Since $d = d(r)$ may be chosen as a decreasing function of $r$, the excluded values may be adjusted so that the total length of the excluded intervals is dominated by the sum of the convergent series,

$$2 \sum_{n=1}^{\infty} \{[T^0(p_n; f)]^{-1-\delta} + [T^0(r_n; f)]^{-1-\delta}\}. \qquad (4.5.5)$$

So far, we have been able to keep the discussion perfectly general, but now we shall assume that $z \mapsto f(z)$ is a meromorphic function of *finite* order $\rho$. Let $\alpha$ be any number $> \rho$. Then

$$m(r + d, 0; f) + m(r + d, \infty; f) \leq 2T^0(r + d; f) + O(\log r)$$

$$\leq 2(r + d)^\alpha + O(\log r) < 3r^\alpha \qquad (4.5.6)$$

for sufficiently large values of $r$. This crude estimate is also valid for $[n(r + d, 0; f) + n(r + d, \infty; f)]$, as may be concluded from the convergence of the first integral under (4.4.31). Assuming $r + d < 2r$ and combining, we get an upper bound

$$12r^{1+(3+2\delta)\alpha} + 6r^{(2+\delta)\alpha} < 18r^{1+(3+2\delta)\alpha}.$$

This gives

**THEOREM 4.5.1**

*If* $z \mapsto f(z)$ *is a meromorphic function of finite order* $\rho$ *and if* $\gamma$ *is any number* $\gamma > 1 + 3\rho$, *then*

$$\left|\frac{f'(z)}{f(z)}\right| \leq 18|z|^{\gamma} \tag{4.5.7}$$

*for* $|z|$ *outside of intervals of finite total length.*

This estimate is probably very poor, but it is good enough for our main purpose, which is stated by

**THEOREM 4.5.2**

*Under the assumptions of Theorem* 4.5.1

$$m\left(r, \infty; \frac{f'}{f}\right) = O(log\ r) \tag{4.5.8}$$

*for* r *outside the exceptional intervals which are of finite total length.*

As a matter of fact, all values are admissible when $\rho(f)$ is finite. For a function of infinite order we have instead

$$m\left(r, \infty; \frac{f'}{f}\right) = O(\log r) + O[\log T^0(r; f)], \tag{4.5.9}$$

but now for $r$ outside a set of finite logarithmic measure [i.e., a set $E$ such that $\int_E (dr/r) < \infty$]. We shall not need these refinements, but we do need

**THEOREM 4.5.3**

*If* f *is of finite order and if* $m(r, \infty; f) = O(log\ r)$, *then* $m(r, \infty; f') = O(log\ r)$.

This is so because

$$m(r, \infty; f') = m\left(r, \infty; \frac{f'}{f}f\right) \leq m\left(r, \infty; \frac{f'}{f}\right) + m(r, \infty; f) + \log 2 = O(\log r).$$

We have now all the material that is needed for Wittich's proof of the

Malmquist-Yosida theorem. But we have also what is required for a proof of the second fundamental theorem and the defect relations for the case in which $\rho(f) < \infty$. Here we can get along without putting restrictions on $m(r, \infty; f)$. The following argument is based on Wittich (1955, pp. 17–18).

It is a question of estimating $T^0(r; f')$ from above and from below when $f$ is a meromorphic function of finite order. We introduce a new set of enumerative functions,

$$N_1(r, a; f) = \int_0^r n_1(s, a; f) \frac{ds}{s}, \qquad f(0) \neq a, \qquad (4.5.10)$$

where the enumeration extends over the multiple roots of the equation $f(z) = a$, but a $k$-fold root is counted only $k - 1$ times. Hence

$$N(r, \infty; f') = 2N(r, \infty; f) - N_1(r, \infty; f) \leqslant 2N(r, \infty; f). \qquad (4.5.11)$$

Thus, omitting reference to infinity when no ambiguity is likely, we have

$$\begin{aligned}
T^0(r; f') &= N(r; f') + m(r; f') = 2N(r; f) - N_1(r; f) + m(r; f') \\
&= 2N(r; f) - N_1(r; f) + m\left(r; \frac{f'}{f} f\right) \\
&\leqslant 2N(r; f) - N_1(r; f) + m\left(r; \frac{f'}{f}\right) + m(r; f) + \log 2 \\
&\leqslant T^0(r; f) + N(r; f) - N_1(r; f) + O(\log r)
\end{aligned}$$

and finally

$$T^0(r; f') \leqslant 2T^0(r; f). \qquad (4.5.12)$$

The estimate from below is harder to get but yields a number of valuable results. Let $a_1, a_2, \ldots, a_q$ be $q$ $(> 1)$ complex numbers and form the polynomial

$$P(w) = \prod_1^q (w - a_j).$$

Here we replace $w$ by $f(z)$ and get a composite meromorphic function $H(z) = P[f(z)]$. We have

$$\frac{1}{P(w)} = \sum_{j=1}^q \frac{A_j}{w - a_j}. \qquad (4.5.13)$$

Since $m(r; 1/H) \leqslant m(r; 1/f') + m(r; f'/H)$, it is seen that $T^0(r; f') \leqslant N(r; 1/f') + m(r; 1/H) - m(r; f'/H) + O(1)$. By (4.5.13)

$$m\left(r; \frac{f'}{H}\right) = m\left(r; \sum_{j=1}^q \frac{A_j f'}{f - a_j}\right) = \sum_{j=1}^q m\left(r; \frac{f'}{f - a_j}\right) + O(1) = O(\log r)$$

by Theorem 4.5.2. The first fundamental theorem gives

$$m\left(r;\frac{1}{H}\right) = N(r;H) - N\left(r;\frac{1}{H}\right) + m(r;H) + O(1).$$

Furthermore, we have

$$m(r;H) = qm(r;f) + O(1),$$

so that

$$m\left(r;\frac{1}{H}\right) = qN(r;f) - \sum_{j=1}^{q} N(r,a_j;f) + qm(r;f) + O(1).$$

It is now seen that

$$T^0(r;f') \leqslant N\left(r;\frac{1}{f'}\right) + qN(r;f) - \sum_{j=1}^{q} N(r,a_j;f) + qm(r;f) - O(\log r)$$

$$= N\left(r;\frac{1}{f}\right) + \sum_{j=1}^{q} m(r,a_j;f) + O(\log r),$$

or

$$N\left(r;\frac{1}{f'}\right) + \sum_{j=1}^{q} m(r,a_j;f) - O(\log r) \leqslant T^0(r;f')$$

$$\leqslant N(r;f') + m(r;f) + O(\log r). \qquad (4.5.14)$$

We can now dispense with the services of $T^0(r;f')$ and are then left with the inequality

$$\sum_{j=1}^{q} m(r,a_j;f) \leqslant m(r,\infty;f) + N(r;f') - N\left(r;\frac{1}{f'}\right) + O(\log r). \qquad (4.5.15)$$

Here we add $\Sigma\, N(r,a_j;f)$ on both sides to obtain

$$qT^0(r;f) \leqslant \sum_{j=1}^{q} N(r,a_j;f) + m(r,\infty;f) + 2N(r,\infty;f)$$

$$- N_1(r,\infty;f) - N\left(r,\infty;\frac{1}{f}\right)$$

or

$$(q-2)T^0(r;f) \leqslant \sum_{1}^{q} N(r,a_j;f) - N_1(r,\infty;f) - m(r,\infty;f)$$

$$- N\left(r,\infty;\frac{1}{f'}\right) + O(\log r)$$

and finally

$$(q-2)T^0(r;f) \leqslant \sum_{1}^{q} N(r,a_j;f) - N_1(r,\infty;f) + O(\log r). \qquad (4.5.16)$$

This inequality is the *second fundamental theorem* of R. Nevanlinna.

Here $q$ is any integer $> 1$. The case $q = 2$ is obviously trivial. The $a_j$'s are any distinct complex numbers, infinity being permitted. An alternative form of the inequality is

$$\sum_1^q m(r, a_j; f) < 2T^0(r; f) - N_1(r, \infty; f) + O(\log r). \qquad (4.5.17)$$

From (4.5.17) we get

$$\sum_{j=1}^q \liminf \frac{m(r, a_j; f)}{T^0(r; f)} + \liminf \frac{N_1(r, \infty; f)}{T^0(r; f)} \leq 2. \qquad (4.5.18)$$

Set

$$\liminf \frac{m(r, a_j; f)}{T^0(r; f)} = 1 - \limsup \frac{N(r, a_j; f)}{T^0(r; f)} = \delta(a_j), \qquad (4.5.19)$$

$$\liminf \frac{N_1(r, \infty; f)}{T^0(r; f)} = \Phi. \qquad (4.5.20)$$

These quantities are called the *defect* of $a_j$ and the *total algebraic ramification* of $f$, respectively. "Ramification" refers to the Riemann surface defined by the mapping $f(z) = w$. Formula (4.5.20) then gives the *defect relation*,

$$\sum_1^q \delta(a_j) + \Phi \leq 2, \qquad (4.5.21)$$

for any finite value of $q$. This implies that $\delta(a) \neq 0$ for at most a countable set of values of $a$. Although meromorphic functions with any prescribed finite number of defective values are known, no example with an infinite number is known as yet.

It is clear from (4.5.19) that the *maximal defect that any value can have is one*, and (4.5.21) shows that *there can be at most two values of maximal defect one*. This is Picard's theorem. The function $z \mapsto \tan z$ shows that this number can be reached.

One can also introduce a local *ramification index*, $\vartheta(a)$:

$$\vartheta(a) = \liminf \frac{N_1(r, a; f)}{T^0(r; f)}. \qquad (4.5.22)$$

This index measures the frequency of algebraic branch points located over $w = a$ on the Riemann surface. The value $a$ is *completely ramified* if the equation $f(z) = a$ has no simple roots. This means that the roots are at least double roots and the corresponding ramification index $\geq \frac{1}{2}$. Since the *sum of the ramification indices is at most two*, it follows that a *meromorphic function can have at most four completely ramified values*. There are functions having four completely ramified values: the ℘-function of Weierstrass is perhaps the simplest and the best known (see Problem

3.3:7). This is the solution of the DE

$$[w']^2 = 4(w - e_1)(w - e_2)(w - e_3), \tag{4.5.23}$$

where $e_1 + e_2 + e_3 = 0$ and the solution which has a pole at the origin is chosen. Here the four values $e_1$, $e_2$, $e_3$, and $\infty$ are completely ramified.

The theorem on complete ramification has an important bearing on algebraic geometry and uniformization. Here a classical theorem due to Picard (1887) asserts that *an algebraic plane curve does not admit of a parametric representation in terms of meromorphic functions if its genus* (= *deficiency*) *is* $p > 1$. Equivalently, an algebraic relation $F(z, w) = 0$ cannot be uniformized globally by functions meromorphic in the finite plane if the genus exceeds one. The genus of an algebraic curve is a positive integer or zero invariant under birational transformations of the coordinates. An algebraic curve of order $N$ can have at most $\frac{1}{2}(N - 1) \times (N - 2)$ singularities (nodes and cusps). If there are actually $n$ nodes and $c$ ordinary cusps, the deficiency is $p = \frac{1}{2}(N - 1)(N - 2) - n - c$, and this is the genus.

For the curves

$$y^2 = a \prod_{j=1}^{3} (x - a_j) \qquad \text{and} \qquad y^2 = b \prod_{k=1}^{4} (x - b_k), \tag{4.5.24}$$

where the $a_j$'s and $b_k$'s are distinct, we have $\frac{1}{2}(N - 1)(N - 2)$ equal to one and three, respectively, and $n + c$ is zero in the first case and two in the second, so that $p = 1$ in both cases. If two of the numbers $a_j$ or two of the $b_k$ should coincide, then $p = 0$ and parametric representations in terms of rational functions of $t$ are available. If the $a_j$'s are distinct, a preliminary affine transformation of the coordinates will reduce the equation to the form

$$v^2 = 4(u - e_1)(u - e_2)(u - e_3) \tag{4.5.25}$$

with $e_1 + e_2 + e_3 = 0$, and we can set $u = \wp(t)$, $v = \wp'(t)$.

For the second equation in (4.5.24) the Jacobi normal form,

$$v^2 = (1 - u^2)(1 - k^2 u^2), \tag{4.5.26}$$

may be more appropriate. Here a fractional linear transformation can be used to reduce the equation in question to the form of (4.5.26), where $k$ is an essential parameter depending on the original configuration. After the reduction a parametric representation $u = \operatorname{sn}(t)$, $v = \operatorname{cn}(t) \operatorname{dn}(t)$ in terms of the Jacobi functions is available. Thus both equations under (4.5.24) admit of parametric representations by functions meromorphic in the finite plane.

So much for the *elliptic* case $p = 1$. For the *hyperelliptic* case we have two alternative normal forms,

$$y^2 = a \prod_{j=1}^{2p+1} (x - a_j) \quad \text{and} \quad y^2 = b \prod_{k=1}^{2p+2} (x - b_k). \quad (4.5.27)$$

These curves are of genus $p$, and any algebraic curve of genus $p$ can be transformed into one or the other of the normal forms by a *birational transformation*, i.e.,

$$x = R_1(u, v), \quad y = R_2(u, v), \quad u = R_3(x, y), \quad v = R_4(x, y), \quad (4.5.28)$$

where the $R$'s are rational functions of the indicated variables. Now, if the coordinates $x$, $y$ of the first curve under (4.5.27) could be represented globally by, let us say, $x = f_1(t)$, $y = f_2(t)$, where $f_1$ and $f_2$ are meromorphic functions of $t$ in the finite plane, then

$$[f_2(t)]^2 = \prod_{j=1}^{2p+1} [f_1(t) - a_j], \quad (4.5.29)$$

and the values $a_1, a_2, \ldots, a_{2p+1}, \infty$ must be completely ramified for $f_1(t)$ in order to make $f_2(t)$ single valued. But if $p > 1$, we cannot have $2p + 2$ completely ramified values. This proves Picard's theorem. The curves in question may be uniformized; however, $x = A_1(t)$, $y = A_2(t)$, where $A_1$ and $A_2$ are *automorphic* functions which are meromorphic but have only a restricted domain of existence with a natural boundary.

We return briefly to the main theme of this section. We plan to apply the formulas to functions which are single valued and meromorphic in a neighborhood of infinity, say, for $|z| \geq R > 0$. The main difference between this case and that of $R = 0$ resides in the definition of the enumerative and proximity functions, the area function $A(s; f)$ and the characteristics where now the lower limit of integration is $R$ instead of zero. We can justify this extension by noting that $f(z)$ may be split into the sum of two functions $f(z) = f_1(z) + f_2(z)$, where $f_1$ is holomorphic for $|z| \geq R$ and zero at infinity, while $f_2$ is meromorphic in the finite plane. Our previous considerations apply unchanged to $f_2$, and in forming the relevant integrals for $f$ with $|z| > R$ the contributions coming from $f_1$ are at most $O(\log r)$.

## EXERCISE 4.5

1. Justify (4.5.5).
2. Justify Theorem 4.4.1.
3. How should (4.5.10) be modified if the equation $f(z) = a$ has a multiple root at $z = 0$?

4.  If $f = \tan z$, find the limit of the quotient of $N(r, \infty; f')$ and $N(r, \infty; f)$ as $r \to \infty$. Compare (4.5.11).

5.  Does $z \mapsto \cos z$ have any defective values? Are there any completely ramified values? Compute $\Phi$ of (4.5.20).

6.  Consider $z \mapsto \tan z$ for defective values. Are there any ramified values?

7.  Estimate $m(r, a; f)$ for $f = \tan z$. The values $+i$ and $-i$ require special treatment.

8.  For an entire function, $T(r; f) < \log M(r; f) + O(1)$. Why? What happens to this inequality if $f$ is meromorphic and becomes infinite as $r \uparrow R$ and $|z| = R$ contains one or more poles?

9.  Show that the mean square modulus

$$M_2(r; f) = \left\{ \frac{1}{2\pi} \int_0^{2\pi} |f(re^{i\theta})|^2 \, d\theta \right\}^{1/2}$$

is piecewise logarithmically convex and has at most one minimum for $p_n < r < p_{n+1}$. [*Hint:* Use the Laurent series for $f(z)$ in the annulus under consideration.]

10. Could the DE $(w')^2 = a\prod_{j=1}^{2p+1}(w - a_j)$ for $p > 1$ be satisfied by a function meromorphic in some domain?

11. Given the curve $y^2 = a(x - a_1)(x - a_2)(x - a_3)$, how should the constants $c$, $d$, $e$ be chosen so that the transformation $x = cs + d$, $y = et$ take the curve into the normal form (4.5.25)?

12. Find a change of variables which takes the second equation in (4.5.24) into the Jacobi normal form (4.5.26).

13. Determine the genus of the ellipse, and find a parametric representation by (i) rational functions, and (ii) trigonometric functions.

14. Solve Problem 13 for the hyperbola.

15. The cubic $y^2 = a(x - b)^2(x - c)$, $c \neq b$, has a double point at $x = b$. Find a rational representation by laying a pencil of straight lines $y = t(x - b)$ through the double point.

## 4.6. THE MALMQUIST THEOREM AND SOME GENERALIZATIONS

The Malmquist theorem of 1913 may be formulated as

**THEOREM 4.6.1**

*Let* $R(z, w)$ *be a rational function of* $z$ *and* $w$. *Suppose that the DE*

$$w' = R(z, w) \tag{4.6.1}$$

*has a solution* $z \mapsto w(z)$ *which is a transcendental meromorphic function with infinitely many poles. Then* $R(z, w)$ *is a polynomial in* $w$ *of degree* 2, *i.e., the DE is a Riccati equation*

$$w'(z) = R_0(z) + R_1(z)w(z) + R_2(z)[w(z)]^2, \tag{4.6.2}$$

*where the R's are rational functions of* z.

*Proof.* In the main we shall follow Wittich's proof (1954) except for the trivial but useful Lemma 4.6.1, which may possibly be new in this connection and appears to be essential for the generalizations.

As in Chapter 3 let

$$R(z, w) = \frac{P(z, w)}{Q(z, w)}, \tag{4.6.3}$$

where

$$P(z, w) = \sum_{j=0}^{p} p_j(z)w^j, \tag{4.6.4}$$

$$Q(z, w) = \sum_{k=0}^{q} Q_k(z)w^k, \tag{4.6.5}$$

where the $P_j$'s and the $Q_k$'s are polynomials in z.

We have assumed that the postulated solution has infinitely many poles. The case in which there is only a finite number would lead to the conclusion that (4.6.2) is actually linear. This was shown by C. C. Yang (1971) using Yosida's results.

Now the existence of a single distant pole, distant in the sense that $P_p(z)Q_q(z) \neq 0$, implies that

$$p = q + 2 \tag{4.6.6}$$

and that the pole is simple. For suppose that $z = z_0$ is a distant pole of $w(z)$ of multiplicity $\alpha$ and leading coefficient $c_0$. Then

$$R(z, w) = \frac{P_p(z_0)}{Q_q(z_0)} c_0^{p-q}(z - z_0)^{(q-p)\alpha}[1 + o(1)], \tag{4.6.7}$$

while

$$w'(z) = -\alpha c_0(z - z_0)^{-\alpha-1}[1 + o(1)]. \tag{4.6.8}$$

The DE requires that

$$(p - q - 1)\alpha = 1. \tag{4.6.9}$$

Since $\alpha$ is a positive integer, we must have $\alpha = 1$ and (4.6.6) holds.

We can then reduce the DE to the form

$$w' = a_2(z)w^2 + a_1(z)w + a_0(z) + \frac{P_1(z, w)}{Q(z, w)} \tag{4.6.10}$$

by dividing $P(z, w)$ by $Q(z, w)$. The $a$'s are rational functions of z, and the degree of $P_1(z, w)$ with respect to $w$ is at most $q - 1$. Here a further simplification can be made. We may assume that $a_2(z) \equiv 1$, for if this is not the case at the outset, the substitution $v(z) = a_2(z)w(z)$ leads to a DE for $v$ of the same type as (4.6.10), where now the coefficient of $v^2$ is identically one. If this simplification has already been made, we are dealing with the DE

$$w' = w^2 + a_1(z)w + a_0(z) + \frac{P_1(z, w)}{Q(z, w)}. \tag{4.6.11}$$

We shall now study the solution $w(z)$ in the neighborhood of a pole. Write

$$\frac{w'}{w^2} = 1 + \frac{a_1(z)}{w} + \frac{a_0(z)}{w^2} + \frac{P_1}{w^2 Q}. \tag{4.6.12}$$

Our first objective is to find conditions on $z$ and $w(z)$ so that the right member of this equation is of the form $1 + h(z)$, where $|h(z)| < \frac{1}{2}$. Now the coefficients are rational functions which behave as integral powers of $z$ for large values of $|z|$. Let $R$ be a large positive number, and consider the set

$$S = [z; |z| > R, |w(z)| > |z|^g], \tag{4.6.13}$$

where $g$ will be determined in a moment. Now for $z \in S$ we can find numbers $a, b, c$ such that

$$\left| \frac{a_1(z)}{w(z)} \right| < |z|^{a-g}, \qquad \left| \frac{a_2(z)}{[w(z)]^2} \right| < |z|^{b-2g},$$

$$\frac{|P_1(z, w)|}{|[w(z)]^2 Q(z, w)|} < |z|^{c-3g}. \tag{4.6.14}$$

Now choose $g$ subject to the two conditions

$$g > \max [a, \tfrac{1}{2}b, \tfrac{1}{3}c], \tag{4.6.15}$$

$$\max [R^{a-g}, R^{b-2g}, R^{c-3g}] < \tfrac{1}{6}. \tag{4.6.16}$$

The principle of the maximum shows that, with this choice of $g$ and for $z \in S$, each of the three quotients in (4.6.14) is less than $\frac{1}{6}$. Thus for $z \in S$

$$\frac{w'(z)}{[w(z)]^2} = 1 + h(z) \qquad \text{with } |h(z)| < \tfrac{1}{2}. \tag{4.6.17}$$

If now $z_1$ and $z_2$ are in $S$ and can be joined by a line segment also in $S$, then

$$\frac{1}{w(z_1)} - \frac{1}{w(z_2)} = \int_{z_1}^{z_2} [1 + h(s)] \, ds, \tag{4.6.18}$$

where the absolute value of the integral lies between $\frac{1}{2}$ and $\frac{3}{2}$ of $|z_1 - z_2|$. In particular, if $z = z_0$ is a pole of $w(z)$ and if we take $z_1 = z_0$ and let $z$ be any point in $S$ "visible" from $z_0$, then

$$\tfrac{2}{3}|z - z_0|^{-1} < |w(z)| < 2|z - z_0|^{-1}. \qquad (4.6.19)$$

It is clear that every pole of $w(z)$ belongs to $S$ and also some neighborhoods of the poles. It is necessary to subject $S$ to a closer scrutiny. It is an open, unbounded set. It contains a countable number of maximal components, $S_j$ say, i.e., open connected subsets not contained in any larger open connected subset. Each $S_j$ contains at least one pole, and if $S_j$ is convex, so that every point of $S_j$ is visible from any other point in $S_j$, then $S_j$ contains one and only one pole. The same conclusion holds if any two points of $S_j$ may be joined by a path in $S_j$ of length less than twice the distance between the points, for then the right-hand side of (4.6.18) cannot be zero.

We now define a polar neighborhood for the point $z_n$ as the set

$$U(z_n) = \{z ; |z - z_n| < \tfrac{1}{2}|z_n|^{-g}\}. \qquad (4.6.20)$$

The problem is to show that $U(z_n) \subset S$. This requires some delicate considerations. If all the points of the circle $|z - z_n| = |z_n|^{-g}$ satisfy $|w(z)| > |z|^g$, then by the theorem of the maximum the inequality also holds inside the circle and the whole disk is in $S$ and so does $U(z_n)$. On the other hand, if the inequality does not hold everywhere on the circle, we can find a concentric circle of a smaller radius passing through a point $z = t$ such that $|w(z)| > |z|^g$ in the smaller disk and $|w(t)| = |t|^g$. Thus the smaller disk is a polar neighborhood of the pole $z_n$ and lies in $S$. It is required to show that it contains $U(z_n)$. Here (4.6.18) may be used, where we take $z_1 = z_n$ and $z_2 = t$. Thus

$$|t - z_n| > \tfrac{2}{3}|w(t)|^{-1} = \tfrac{2}{3}|t|^{-g},$$

and we have to show that $|t - z_n| > \tfrac{1}{2}|z_n|^{-g}$. Now $|t| \le |z_n| + |t - z_n| < |z_n| + |z_n|^{-g}$ so that

$$|t - z_n| > \tfrac{2}{3}\{|z_n| + |z_n|^{-g}\}^{-g} = \tfrac{2}{3}|z_n|^{-g}\{1 + |z_n|^{-g-1}\}^{-g}.$$

We shall show that the last factor exceeds $\tfrac{3}{4}$ for all $g > 0$ and all $|z_n| > e$. Now

$$(1 + x^{-g-1})^{-g}$$

is an increasing function of $x$ for positive $g$ and $x$, so it takes its minimum for $e \le x \le \infty$ at $x = e$. Thus we have to show that

$$(1 + e^{-g-1})^{-g} > \tfrac{3}{4} \quad \text{or} \quad 1 + e^{-g-1} < (\tfrac{4}{3})^{1/g} = (1 + \tfrac{1}{3})^{1/g}.$$

Now for $1 \le g$ the last member exceeds $1 + 3^{-1/g}$, and we have

$$e^{-g-1} < 3^{-1/g} \quad \text{for } 1 \le g.$$

On the other hand, if $0 < g < 1$ we have

$$\tfrac{1}{g} \log \tfrac{4}{3} > e^{-g-1} > \log (1 + e^{-g-1})$$

so the required inequality is true for all $g > 0$ and $|z_n| \geqslant e$. It follows that all polar neighborhoods with $|z_n| \geqslant e$ are subsets of $S$.

Polar neighborhoods can have no points in common, for if they did then one pole would be "visible" from the other in $S$, and this contradicts (4.6.18).

Suppose that at a certain point $z = z_0$ we have $|w(z_0)| \geqslant |z_0|^{g+1}$. This implies that $z_0$ belongs to a polar neighborhood. By the argument given above we may show that the disk

$$D = \{z ; |z - z_0| < \tfrac{1}{2}|z_0|^{-g}\} \qquad (4.6.21)$$

belongs to $S$. Then for any $z$ in $D$

$$\frac{1}{w(z)} = - \int_{z_0}^{z} [1 + h(s)] \, ds + \frac{1}{w(z_0)}.$$

Here the two expressions involving $z$ are holomorphic in $D$, and the integral has a simple zero at $z = z_0$. On the rim of $D$

$$\left| \int_{z_0}^{z} [1 + h(s)] \, ds \right| > \tfrac{1}{2}|z - z_0| = \tfrac{1}{4}|z_0|^{-g},$$

while $|w(z_0)|^{-1} \leqslant |z_0|^{-g-1} < \tfrac{1}{4}|z_0|^{-g}$ if $|z_0| > 4$. We may now invoke the theorem of Rouché, which says that $|w(z)|^{-1}$ has the same number of zeros in $D$ as the integral has, i.e., just one simple zero. This in turn implies that $w(z)$ has a simple pole in $D$, proving the assertion. See Lemma 5.6.1.

Outside $S$ we have

$$|w(z)| < |z|^{g}, \qquad (4.6.22)$$

i.e., outside the polar neighborhoods the solution grows at most as a power of $z$. This is in agreement with a general theorem of Pierre Boutroux (1908) for equations of type (4.6.1) with $p > q + 1$.

Our next task is a question of polar statistics. We have to estimate the characteristic function $T(r; w)$, which requires estimating $n(r, \infty; w)$ and $m(r, \infty; w)$. The first of these functions gives the number of poles of the solution $z \mapsto w(z)$ in the disk $|z| < r$, while the second receives its main contributions from those arcs of the circle $|z| = r$ which belong to polar neighborhoods of $w(z)$. The whole problem reduces to a geometric problem of counting polar neighborhoods. Since they do not intersect, the question becomes one of closest packing: *How many small disks of diameter* $r^{-g}$ *can be placed with their centers on a circle of radius* $r$ *when no overlapping is permitted?* The answer is given by the trivial but useful

**LEMMA 4.6.1**

*On the circle* $|z| = r > R$ *there are at most*

$$2\pi r^{g+1}[1 + o(1)] \qquad (6.4.23)$$

*poles, while in the disk* $|z| \leqslant r$ *the number of poles is at most*

$$4r^{2g+2}[1 + o(1)]. \qquad (4.6.24)$$

*Proof.* The first estimate is obvious, since this is the solution of the closest packing problem. The disk $|z| \leqslant r$ presents a slightly different problem. The diameter of a polar neighborhood with center at $z = z_0$ with $|z_0| = s$ is $s^{-g}$, a decreasing function of $s$. This means that the smallest of the small disks are those centered on the rim of the big disk. Thus we get an upper bound for the packing of the polar neighborhoods by finding how many disks with diameter $r^{-g}$ can be packed into a disk of diameter $2r$. Taking the quotient of the areas, we get (4.6.24). ∎ This gives

**LEMMA 4.6.2**

$$N(r, \infty; w) < \frac{2}{g+1} r^{2g+2}[1 + o(1)],$$

*for this is the integral of* $n(s, \infty; w)/s$.

**LEMMA 4.6.3**

$$m(r, \infty; w) = O(\log r) \qquad and \qquad T(r; w) < r^{2g+2}$$

*so that* $w(z)$ *is of finite order not exceeding* $2g + 2$.

*Proof.* The crux of the matter is to estimate the proximity function,

$$m(r, \infty; w) = \frac{1}{2\pi} \int_0^{2\pi} \log^+ |w(re^{i\theta})| \, d\theta. \qquad (4.6.25)$$

Now, for $z$ not belonging to a polar neighborhood, we have $|w(z)| < |z|^g$ so the contribution to (4.6.25) from such arcs is $O(\log r)$. It is thus obvious that major contributions to (4.6.25) can come only from the polar neighborhoods which happen to have points in common with the circle $|z| = r$, and the biggest contributions should come from polar neighborhoods centered on the circle. Suppose that $N(z_0)$ is such a neighborhood. There is a pole at $z = z_0$ with residue $-1$. Then $N(z_0)$ subtends an angle $\omega$ of opening $r^{-g-1}[1 + o(1)]$ at the origin. If $\theta$ is the angle which the bisector of $\omega$ makes with the radius vector, then on the intersection $\gamma$ of $N(z_0)$

with $|z| = r$ we have

$$w(z) < 2[r|\theta|]^{-1}, \tag{4.6.26}$$

so the arc $\gamma$ makes a contribution less than

$$\frac{1}{\pi}(g+2)r^{-g-1}\log r.$$

Since no more than $2\pi r^{g+1}[1 + o(1)]$ polar neighborhoods can be centered on $|z| = r$, we get

$$m(r, \infty; w) < 2(g+2)\log r \tag{4.6.27}$$

and the lemma is proved, since $T(r; w) = N(r, \infty; w) + m(r, \infty; w)$. ∎

We recall that for a meromorphic function of finite order, (4.6.27) implies

$$m(r, \infty; w') = O(\log r). \tag{4.6.28}$$

Incidentally our estimate for the order, $2g + 2$, is about twice the correct order, as we shall see. At this stage all we need is that the order is finite.

We are now ready for the last step in the proof of Theorem 4.6.1. Set

$$\frac{P_1(z, w)}{Q(z, w)} = F[z, w(z)] = F(z), \tag{4.6.29}$$

so that

$$F(z) = w'(z) - [w(z)]^2 - a_1(z)w(z) - a_0(z). \tag{4.6.30}$$

Every pole of $w(z)$ is a zero of $F(z)$, since $P_1$ is of lower degree in $w$ than $Q$. It is clear that $F(z)$ is holomorphic for all values of $z$ with $|z| > R$. Now for $|z| > R$

$$m(r, \infty; F) \leqslant m(r, \infty; w') + 3m(r, \infty; w) + O(\log r) = O(\log r).$$

The first fundamental theorem gives

$$N\left(r, \infty; \frac{1}{F}\right) = T(r; F) + O(\log r) = m(r, \infty; F) + O(\log r)$$

since $N(r, \infty; F) = 0$. This implies

$$N\left(r, \infty; \frac{1}{F}\right) = O(\log r). \tag{4.6.31}$$

From this we conclude that $F$ cannot actually depend on $w(z)$, for if it did then the infinitely many poles of $w$ are also poles of $1/F$ and $N(r, \infty; 1/F)$ would be of the same order of magnitude as $N(r, \infty; w)$. If $F$ is independent of $w$ and is a rational function of $z$ alone, it can be incorporated with $a_0(z)$. It follows that the equation is actually a Riccati DE, and the Malmquist-Wittich-Yosida theorem is proved. ∎

## COROLLARY

*If* (4.6.1) *is not a Riccati equation but has a meromorphic solution, then the solution is a rational function of* $z$.

The following result due to Wittich is of interest in this connection.

## THEOREM 4.6.2

*A meromorphic solution of* (4.6.1) *which is of order* $<\frac{1}{2}$ *is a rational function.*

*Proof.* We may assume that the equation is Riccati, say,

$$w' = R_0(z) + R_1(z)w + R_2(z)w^2,$$

since otherwise the assertion follows from the corollary of Theorem 4.6.1. We may assume that $R_2(z) \not\equiv 0$, for otherwise $w(z)$ must be rational. We may also assume that $w(z)$ has infinitely many poles, since otherwise again $R_2(z) \equiv 0$ and $w(z)$ is rational. With at most a finite number of exceptions the poles are simple. Next we remove the first order term and make the coefficient of the quadratic term equal to one. All this can be done by setting

$$w = \frac{u}{R_2} - \frac{1}{2}\frac{R_1}{R_2} - \frac{1}{2}\frac{R_2'}{R_2^2}, \tag{4.6.32}$$

which gives a result of the form

$$u' = A(z) + u^2, \tag{4.6.33}$$

where $z \mapsto A(z)$ is a rational function of $z$. We can then find a rational function $R(z)$ such that $v(z) = u(z) - R(z)$ is meromorphic in the finite plane with infinitely many poles, all simple and of residue $-1$. Such a function must be the logarithmic derivative of an entire function. We define

$$v(z) = -\frac{E'(z)}{E(z)}$$

and find that $z \mapsto E(z)$ satisfies

$$E'' - 2RE' + (A + R^2 - R')E = 0. \tag{4.6.34}$$

Now $v(z)$ is of finite order (say by Lemma 4.6.3); hence $E$ must also be of finite order so that

$$E(z) = e^{h(z)}P(z).$$

By a theorem due to Jacques Hadamard $h(z)$ is a polynomial in $z$ and $P(z)$ is a canonical product for the zeros of $E(z)$, i.e., the poles of $v(z)$;

$P(z)$ also satisfies a second order linear DE,

$$P'' + 2(h' - R)P' + (A + R^2 - R' - 2Rh' + (h')^2 + h'')P = 0. \qquad (4.6.35)$$

Now an entire function which satisfies a linear second order DE cannot have an order $< \frac{1}{2}$, and the function $\cos(z^{1/2})$, which is of order $\frac{1}{2}$ and satisfies $zw'' + \frac{1}{2}w' + \frac{1}{4}w = 0$, shows that the lower limit for the order can be reached. This is under the assumption that the coefficients are rational functions of $z$. For such questions consult Section 5.4. Now $N(r, \infty; w) = N(r, \infty; 1/P) + O(\log r)$, so that $w(z)$ must be of the same order as $P(z)$, i.e., $\geq \frac{1}{2}$. This proves the theorem. ∎

For a Riccati equation with polynomial coefficients

$$w' = P_0(z) + P_1(z)w + P_2(z)w^2, \qquad (4.6.36)$$

H. Wittich has given a direct, elegant proof that the meromorphic solutions are of finite order. This follows from a use of the spherical characteristic (4.4.26) with

$$\pi A(r; w) = \iint_{|t| \leq r} \frac{|w'(t)|^2}{[1 + |w(t)|^2]^2} s \, ds \, d\theta, \qquad t = s e^{i\theta}, \qquad (4.6.37)$$

and by (4.6.36)

$$|w'|^2 \leq |P_0|^2 + \cdots + |P_2|^2 |w|^4;$$

and for $j = 0, 1, \ldots, 4$,

$$\frac{|w|^j}{(1 + |w|^2)^2} < 1$$

so that $A(r; w) < Kr^k$ and hence also $T^0(r; w) < (K/k)r^k$. Thus the order of the solution is at most $k$, and the assertion is proved. It is worth while observing that $k = 2 \max [\deg |P_j|, j = 0, 1, 2] + 2$. This estimate is independent of that of Lemma 4.6.2. Wittich has also attacked the order problem from a different angle. The following theorem will be proved in Section 5.4.

### THEOREM 4.6.3

*If the Riccati equation*

$$w' = R_0(z) + R_1(z)w + w^2 \qquad (4.6.38)$$

*has rational coefficients, if the limits*

$$\lim_{z \to \infty} R_0(z)|z|^{-\alpha} \qquad and \qquad \lim_{z \to \infty} R_1(z)|z|^{-\beta} \qquad (4.6.39)$$

*exist and are neither zero nor infinity, if the equation has a solution* w(z) *which is single valued and meromorphic outside a circle* $|z| = R$, *and if the solution has infinitely many poles, then* $w(z)$ *is of finite order* $\rho[w]$, *which is an integral multiple of* $\frac{1}{2}$, *and*

$$\rho[w] = 1 + max\ [\beta, \tfrac{1}{2}\alpha]. \tag{4.6.40}$$

Note that $\alpha$ and $\beta$ are integers, and at least one of the inequalities $\beta > -1, \alpha > -2$ must hold. The reason for this restriction is that a transformation of type (4.1.6), which leads from the Riccati equation to a second order linear DE, must lead to an equation with an irregular-singular point at infinity.

Theorem 4.6.2 is of course a corollary of this result.

In his 1933 paper K. Yosida proved several generalizations of Malmquist's theorem. We shall state some of these.

**THEOREM 4.6.4**

*If* $R(z, w)$ *is a rational function of* $z$ *and* $w$, *and if the DE*

$$(w')^n = R(z, w) \tag{4.6.41}$$

*has a transcendental meromorphic solution, then* $R(z, w)$ *must be a polynomial in* $w$ *of degree not exceeding* $2n$. *More precisely,* $p = n[1 + (1/\alpha)]$.

Here the test-power test shows that

$$\alpha(p - q - n) = n, \tag{4.6.42}$$

where, as usual, $p$ is the degree in $w$ of the numerator of $R(z, w)$ while $q$ is the degree of the denominator. It is seen that $\alpha$, the order of the poles, is a divisor of $n$. Since $\alpha$ is at least one and at most $n$, one gets

$$n + 1 \leqslant p - q \leqslant 2n. \tag{4.6.43}$$

On the other hand, Yosida derives an inequality from the Nevanlinna theory, which in the present case takes the form $2n \geqslant p$. He also finds $2n \geqslant 2n + q$, so that $q = 0$.

In principle we can use the argument of Wittich to prove this result. We shall sketch the case $n = 2$. Before doing so, let us remark that there is a great difference between the cases $n = 1$ and $n > 1$. A Riccati equation with polynomial coefficients will always have transcendental meromorphic solutions with poles clustering at infinity. This need not be true, however, for equations

$$(w')^2 = \sum_{j=0}^{4} P_j(z)w^j. \tag{4.6.44}$$

Here the equations

$$(w')^2 = z^k(4w^3 - 1) \qquad (4.6.45)$$

are instructive. The solution is

$$\wp\left(\frac{2z^{k/2+1}}{k+2} + C; 0, 1\right), \quad k \neq -2; \qquad \wp(\log z + C; 0, 1), \quad k = -2. \qquad (4.6.46)$$

If the solution is to be single valued, $k$ must be a nonnegative even integer. If $k < -2$, the poles do not cluster at infinity even if $k$ is an even integer.

We now return to the DE (4.6.41) with $n = 2$. Here we have

$$(p - q - 2)\alpha = 2 \qquad (4.6.47)$$

with two solutions in terms of integers: (i) $\alpha = 1$, $p = q + 4$, and (ii) $\alpha = 2$, $p = q + 3$.

We consider only the first case, in which all distant poles are simple. Proceeding as above, we get

$$(w')^2 = \sum_{j=0}^{4} a_j(z)w^j + \frac{P_1(z, w)}{Q(z, w)}, \qquad (4.6.48)$$

where the $a_j(z)$ are rational functions of $z$, and the degree of $P_1$ with respect to $w$ is at least one unit less than that of $Q$. Here we assume that $a_4(z) \equiv 1$. If this were not the case at the outset, we could make a change of the dependent variable by setting

$$v(z) = [a_4(z)]^{1/2} w(z). \qquad (4.6.49)$$

It should be noted, however, that to preserve the rational character of the coefficients in the transformed equation, $[a_4(z)]^{1/2}$ must be rational and the result for (4.6.45) suggests that $a_4(z)$ must be the square of a polynomial.

Assuming $a_4(z) \equiv 1$, we can define a number $g$ and corresponding polar neighborhoods as in the case $n = 1$. Suppose that $|z| > R$, and set

$$S = [z; |w(z)| > |z|^g]. \qquad (4.6.50)$$

For $|z| > R$ and $z \in S$ we have

$$\left|\frac{a_j(z)}{[w(z)]^{4-j}}\right| \leq |z|^{a_j-(4-j)g}, \qquad \frac{|P_1[z, w(z)]|}{|Q[z, w(z)][w(z)]^4|} \leq |z|^{\beta-5g}. \qquad (4.6.51)$$

Choose $g$ so that

$$\max_j [\alpha_j - (4 - j)g] < 0, \qquad j = 0, 1, 2, 3 \quad \text{and} \quad \beta - 5g < 0,$$
$$\max [R^{\alpha_j-(4-j)g}, R^{\beta-5g}] < 0.1. \qquad (4.6.52)$$

With this choice of $g$ and for $z \in S$ it is seen that

$$\left(\frac{w'}{w}\right)^2 = 1 + h_0(z) \qquad \text{with } |h_0(z)| < \tfrac{1}{2}. \qquad (4.6.53)$$

Here we extract the square root, taking $1^{1/2} = +1$, to obtain

$$\frac{w'}{w^2} = 1 + h(z) \qquad \text{with } |h(z)| < \tfrac{1}{4}, \qquad (4.6.54)$$

whence

$$\tfrac{4}{3}|z - z_0|^{-1} < |w(z)| < \tfrac{4}{3}|z - z_0|^{-1} \qquad (4.6.55)$$

if $z_0$ is a pole of $w(z)$. The positive square root corresponds to a pole with residue $-1$; the negative, to a pole with residue $+1$. Both are possible, and the estimate is the same in the two cases.

We can define polar neighborhoods by (4.6.20), and they will be subsets of $S$. Actually we could replace the factor $\tfrac{1}{2}$ by a somewhat larger number, but that fact is immaterial. The polar statistics carry over unchanged, and we find that $w(z)$ is of finite order, $< 2g + 2$. The proof is then completed as in the case $n = 1$. This proves the case $n = 2$, $\alpha = 1$. The procedure for the case $\alpha = 2$ is the same. In principle the method works for any value of $n$ and any admissible $\alpha$, so that $p = n[1 + (1/\alpha)]$. Furthermore, it is assumed that the coefficient of the highest power of $w$ is the $n$th power of a polynomial.

Yosida (1933a) also considered more complicated nonlinear DE's of first and higher orders. We refer to his paper for further results. See also a paper by A. A. Gol'dberg (1956). Yosida's results have been extended by Ilpo Laine (1971–1974) and Chung-Chun Yang (1972). They allow the coefficients to be transcendental entire or meromorphic functions of lower order than that of the solution under consideration. (The author is indebted to Professors H. Hochstadt, I.-C. Hsu, and H. Wittich and to Dr. Yang for numerous bibliographical references on these questions.) The methods here used would not seem capable of coping with such a situation, at least not in its full generality. Meromorphic solutions would normally be of infinite order, but this difficulty can be surmounted in some cases at least. In particular, the entire function case is amenable. If the coefficients are actually meromorphic with infinitely many poles, there will be infinitely many fixed singularities and the existence of meromorphic solutions is doubtful. See Problems 16–22 below.

## EXERCISE 4.6

1. Find the value of $c_0$ in (4.6.7) and (4.6.8).
2. The discussion in the text naturally applies to a Riccati equation. Consider

$w' = 1 + z^2 w + zw^2$. Remove the coefficient $z$ of $w^2$, and find the resulting coefficients $a_0$ and $a_1$. Find safe bounds for $R$ and $g$.

3.  If $h > 0$, is $(1 + h)^{1/2} < 1 + \frac{1}{2}h$? More generally, if $g$ is a real number, $g > 1$, is $(1 + h)^{1/g} < 1 + h/g$?

4.  Verify the calculation that justifies defining a polar neighborhood by (4.6.20).

5.  If $w' = P_0 + P_1 w + P_2 w^2 + P_3 w^3$, the $P$'s being polynomials, how should the dependent variable be transformed so that (i) the quadratic term disappears from the equation, and (ii), in addition, the coefficient of the cubic term becomes identically one?

6.  The equation $(w')^2 = 4w^3$ has rational solutions. Find them, and find a change of variables $z = cs$, $w = ay$ which leaves the equation invariant.

7.  Solve Problem 6 for $w'' = 4w^3$.

8.  Suppose that $P_0$, $P_1$, $P_2$ are polynomials in $z$. Show that the equation

$$w' = P_0(z) + P_1(z)w + P_2(z)w^2 - z^2 w^3$$

cannot be satisfied by a rational function unless it be a polynomial.

9.  Show that the DE

$$w' = 4 + 3z^2 + z^4 - zw^3$$

has a particular rational solution. Show that the general solution has movable algebraic singularities, and determine their nature.

10.  Verify the calculations that led to (4.6.27).

11.  Determine $A(z)$ in (4.6.33).

12.  Why must $v(z)$ be the logarithmic derivative of an entire function?

13.  How is the estimate $k \leqslant 2 \max (\deg P_j) + 2$ obtained in the discussion of (4.6.37)?

14.  Verify (4.6.46).

15.  If the change of variable (4.6.49) is applied to (4.6.48), find the transformed equation.

The equation

$$w' = f(z)(1 + w^2)$$

is integrable by quadratures. Discuss the nature of the singularities if $f(z)$ equals

16.  $e^z$.

17.  $z^k$, $k \neq -1$.

18.  $z^{-1}$.

19.  $2(1 + z^2)^{-1}$.

20.  $\sec^2 z$.

21.  Which of the preceding equations possess solutions that are single valued in a neighborhood of infinity and have the point at infinity as the only cluster points of its poles?

22.  If $f(z)$ is an entire function, show that

$$T(r; w) \leqslant \exp{[rT^2(r; f)]},$$

using (4.6.37), for instance.

## LITERATURE

Sections C.1 and 12.1 of the author's LODE have a bearing on this chapter.

Frenet's formulas and the moving trihedral are discussed in Sections 15.4 and 15.5 of another book of the author's:

Hille, E. *Analysis*, Vol. II, Ginn-Blaisell, Waltham, Mass., 1966.

For connections between the Riccati equation and differential geometry see p. 186 (Vol. 1) and pp. 54 and 1196 (Vol. 3) of:

Strubecker, Karl. *Differentialgeometrie*. 3 vols. Walther de Gruyter, Berlin, 1964.

The Nevanlinna theory is discussed at some length in Chapter 14 of Vol. II of the author's AFT. See also:

Hayman, W. *Meromorphic Functions*. Oxford University Press, 1964.

Nevanlinna, Rolf. *Le théorème de Picard-Borel et la théorie des fonctions méromorphes*. Gauthier-Villars, Paris, 1929; Chelsea, New York, 1973.

Sario, Leo and Kiyoshi Noshiro, *Value Distribution Theory*. Van Nostrand, Princeton, N.J., 1966.

The Malmquist theorem is discussed in:

Bieberbach, Ludwig. *Theorie der gewöhnlichen Differentialgleichungen*. Grundlehren, No. 86. Springer-Verlag, Berlin, 1953, pp. 86–101.

Wittich, Hans. *Neuere Untersuchungen über eindeutige analytische Funktionen*. Ergebnisse of Math., N.S., No. 8. Springer-Verlag, Berlin, 1955, pp. 73–80; 2nd ed., 1968.

Further references for Section 4.6 are:

Boutroux, Pierre. *Leçons sur les fonctions défines par les équations différentielles du premier ordre*. Gauthier-Villars, Paris, 1908, pp. 47–54.

Gol'dberg, A. A. On single-valued solutions of first-order differential equations. *Ukrain. Mat. Žurnal*, 8:3 (1956), 254–261. (Russian; NRL Translation 1224, 1970.)

Hille, Einar. Finiteness of the order of meromorphic solutions of some non-linear ordinary differential equations. *Proc. Roy. Soc. Edinburgh*, (A) 72:29 (1973/74), 331–336.

Laine, Ilpo. On the behaviour of the solutions of some first order differential equations. *Ann. Acad. Sci. Fenn.*, Ser. A, I, **497** (1971), 26 pp.

———. Admissible solutions of Riccati differential equations. *Publ. Univ. Joensuu.*, Ser. B, 1 (1972).

———. Admissible solutions of some generalized algebraic differential equations. *Publ. Univ. Joensuu.*, Ser. B, **10** (1974), 6 pp.

Malmquist, Johannes. Sur les fonctions à un nombre fini des branches définies par les équations différentielles du premier ordre. *Acta Math.*, **36** (1913), 297–343.

——. Sur les fonctions à un nombre fini de branches satisfaisantes à une équation différentielle du premier ordre. *Acta Math.* **42** (1920), 433–450.

Wittich, Hans. Ganze transzendente Lösunger algebraischer Differentialgleichungen. *Math. Ann.*, **122** (1950), 221–234.

——. Zur Theorie der Riccatischen Differentialgleichung. *Math. Ann.*, **127** (1954), 433–450.

——. Einige Eigenschaften der Lösungen von $w' = a(z) + b(z)w + c(z)w^2$. *Arch. Math.*, **5** (1954), 226–232.

——. Zur Theorie linearer Differentialgleichungen im Komplexen. *Ann. Acad. Sci. Fenn.*, Ser. A, I, Math., **379** (1966), 19 pp.

Yang, Chung-Chun. On deficiencies of differential polynomials. I, II. *Math. Zeit.*, **116** (1970), 197–204; **125** (1972), 107–112.

——. A note on Malmquist's theorem on first-order differential equations. *NRL-MRC Conference on Ordinary Differential Equations*, edited by L. Weiss, Academic Press, New York, 1972, pp. 597–607. Also *Yokohama Math. J.*, **20** (1972), 115–125.

——. On meromorphic solutions of generalized algebraic differential equations. *Annal. Mat. pura appl.*, (IV) **91**, (1972), 41–52.

Yosida, Kôsaku. A generalization of a Malmquist's theorem. *Japan. J. Math.*, **9** (1933), 253–256.

——. A note on Riccati's equation. *Proc. Phys.-Math. Soc. Japan*, (3) **15**, (1933), 227–237.

——. On the characteristic function of a transcendental meromorphic solution of an algebraic differential equation of the first order and the first degree. *Proc. Phys.-Math. Soc. Japan*, (3) **15**, (1933), 337–338.

——. On algebroid solutions of ordinary differential equations. *Japan. J. Math.*, **10** (1934), 119–208.

## ADDENDUM

The modified Weierstrass DE

$$(w')^2 = [f'(z)]^2(w - e_1)(w - e_2)(w - e_3)$$

is satisfied by $w(z) = \wp[f(z); 2\omega_1, 2\omega_3]$. The reader is urged to discuss the nature of the singularities of $w(z)$ and the existence of meromorphic solutions if $f(z)$, e.g., is a power of $z$. The case $f'(z) = (z^2 - 1)^{-1/2}$, $\omega_1 = \frac{1}{2}$, $\omega_3 = \pi i$, is treated by Stevens B. Bank and Robert P. Kaufman in a forthcoming paper. Here the finite fixed singularities are only apparent, $w(z)$ is single valued and meromorphic in the finite plane, of order zero, and $T(r; w) = 0[\log^2 r]$. Verify! S. Bank has many important papers on nonlinear DE's but seems to pass by the Malmquist theorem cycle.

# 5

# LINEAR DIFFERENTIAL EQUATIONS: FIRST AND SECOND ORDER

In this chapter and the next four we shall be concerned with linear DE's. It is appropriate to start with the first and second order cases which have been considered here and there in preceding chapters. The second order case is the most important for the applications and exhibits most of the phenomena that we aim to study in the $n$th order case. It also presents fewer algebraic difficulties.

Our previous existence theorems apply at nonsingular points. All singularities are fixed; they are the singularities of the coefficients and the zeros of the coefficient of the highest derivative plus the point at infinity. The singularities are of two types: *regular* and *irregular*. In the case of a regular singular point, expansions valid in a full neighborhood of the point will be given. They involve ordinary power series with multipliers which are fractional powers and/or logarithms. Similar expansions involving Laurent series could be given at the irregular singularities, if any, but they are normally of little use in studying the analytical properties of the solution. They will not be discussed here. Instead we devote considerable space to questions of growth and asymptotic integration. There is also a brief mention of the group of the equation.

## 5.1. GENERAL THEORY: FIRST ORDER CASE

An $n$th order DE is *linear and homogeneous* if it is the sum of terms of the form $P_k(z)w^{(n-k)}(z)$ equated to zero; thus

$$P_0(z)w^{(n)}(z) + P_1(z)w^{(n-1)}(z) + \cdots + P_n(z)w(z) = 0. \qquad (5.1.1)$$

It is often convenient to write $L(z, w)$ or simply $L(w)$ for the left member. The linearity implies that *the sum of two solutions is a solution, and a constant multiple of a solution is also a solution*, for

$$L(C_1w_1 + C_2w_2) = C_1L(w_1) + C_2L(w_2) = 0.$$

The general solution of (5.1.1) involves $n$ arbitrary constants and is a linear homogeneous function of the constants. The initial-value problem

$$w(z_0) = w_0, \qquad w'(z_0) = w_1, \qquad \ldots, \qquad w^{(n-1)}(z_0) = w_{n-1} \quad (5.1.2)$$

has a unique solution when $z = z_0$ does not belong to the fixed singularities of the equation.

The equation

$$L(w) = Q(z) \qquad (5.1.3)$$

is said to be *linear and nonhomogeneous*. Its solution is of the form $W(z) + V(z)$, where $W(z)$ is the general solution of (5.1.1) and $V(z)$ is a particular solution of (5.1.3). Any particular solution will do unless some selective condition has to be satisfied. Thus in the case of

$$w' + w = z \qquad (5.1.4)$$

there is one and only one solution which is a polynomial, namely, $z - 1$, and there is one and only one solution which is zero for $z = 0$, namely, $e^{-z} + z - 1$.

Let us consider the first order case which figured in Section 4.1. The equation

$$w'(z) = F_0(z) + F_1(z)w(z) \qquad (5.1.5)$$

can be solved by "quadratures," i.e., by explicit integration. To achieve this, set $w = uv$. Then

$$(uv)' = uv' + vu' = F_0 + F_1uv.$$

A condition may be imposed on one of the factors. Take

$$v' = F_1v \qquad \text{so that} \qquad v(z) = \exp\left[\int F_1(s)\,ds\right] \qquad (5.1.6)$$

and $vu' = F_0$. Hence

$$u(z) = C + \int F_0(t)\exp\left[-\int F_1(s)\,ds\right]dt.$$

Suppose that it is desired to find $w(z; z_0, w_0)$, the solution that takes on the value $w_0$ when $z = z_0$. This gives

$$w(z; z_0, w_0) = \exp\left[\int_{z_0}^{z} F_1(s)\,ds\right]$$

$$\times \left\{ w_0 + \int_{z_0}^{z} F_0(t)\exp\left[-\int_{z_0}^{t} F_1(s)\,ds\right]dt \right\}. \qquad (5.1.7)$$

This expresses the fact that $w(z; z_0, w_0)$ is the sum of the solution of the homogeneous equation which is $w_0$ at $z = z_0$ plus the particular solution of the nonhomogeneous equation which is zero for $z = z_0$.

This discussion has been of a somewhat formal nature, but if $z \mapsto F_0(z)$ and $z \mapsto F_1(z)$ are holomorphic with a common domain of holomorphy $D$, to which $z_0$ belongs, every step makes sense and is justified. The integrations may be performed along any path $C$ in $D$, and the resulting solution is locally holomorphic. It is holomorphic in the large if $D$ is simply connected. The solution cannot have any movable singularities, as is obvious from its form. The fixed singularities are those of $F_0$ and $F_1$ plus the point at infinity.

We shall scrutinize in more detail the function $v(z)$ defined by (5.1.6). Suppose that $F_1(z)$ is single valued and holomorphic except for isolated singularities, one of which is located at $z = a$. Here we have a Laurent series

$$F_1(z) = \sum_{-\infty}^{\infty} F_{1n}(z - a)^n. \tag{5.1.8}$$

The question is now, What happens to $w(z; z_0, w_0)$ in a neighborhood of $z = a$. Here the first question is: What happens when $z$ describes a closed path $\Gamma$ in the positive sense around $z = a$, leaving all other singularities, if any, on the outside? Suppose that we start at a point $z = z_1$ on $\Gamma$ with a function element $v_0(z)$, and we return to $z_1$ with a function element $v_1(z)$, possibly distinct from $v_0(z)$. But both elements are solutions of the equation $v' = F_1v$, all the solutions of which are constant multiples of an arbitrarily chosen solution, not identically zero. It follows that there is a number $\omega$ such that $v_1(z) = \omega v_0(z)$ and $\omega \neq 0$. Set

$$\rho = \frac{\log \omega}{2\pi i}, \tag{5.1.9}$$

where, to start with, the determination of the logarithm is left open. This means that $\omega$ is determined only modulo 1. The power $(z - a)^\rho$ is multiplied by

$$e^{2\pi i\rho} = \omega$$

as $z$ describes $\Gamma$, and

$$V(z) = (z - a)^{-\rho}v(z)$$

is single valued in the neighborhood of $z = a$ and as such may be represented by a Laurent series convergent in a disk $0 < |z - a| < R$, so that

$$v(z) = (z - a)^\rho \sum_{-\infty}^{\infty} b_n(z - a)^n. \tag{5.1.10}$$

Here there are essentially two distinct possibilities. If $z = a$ is actually a singular point of $F_1(z)$, some powers with negative exponents occur in (5.1.8) and there are two possibilities: (i) a finite number, and (ii) infinitely many, of such powers. In the first case $F_1(z)$ has a pole; in the second, an essential singular point. We have a similar situation for $V(z)$. We now want to know what condition $F_1(z)$ must satisfy in order that $V(z)$ tends to a finite definite limit as $z \to a$. If this is the case, then

$$v(z) = (z - a)^\rho [b_0 + b_1(z - a) + \cdots],    \tag{5.1.11}$$

where now $\rho$ is a definite real or complex number. To find what condition this imposes on $F_1$, we note that $v(z)$ is the logarithmic derivative of $F_1(z)$, so that

$$F_1(z) = [\log v(z)]' = \frac{\rho}{z - a} + \frac{b_1 + 2b_2(z - a) + \cdots}{b_0 + b_1(z - a) + b_2(z - a)^2 + \cdots}$$
$$= (z - a)^{-1}[c_{-1} + c_0(z - a) + c_1(z - a)^2 + \cdots].    \tag{5.1.12}$$

Thus $F_1$ must have a simple pole at $z = a$ with residue $c_{-1} = \rho$.

If this condition is satisfied, we say that $z = a$ is a *regular-singular point* of (5.1.6). The first study of singularities of linear DE's was made by Lazarus Fuchs in 1866 and was followed by the work of G. Frobenius (1849–1917) in 1873. Fuchs referred to such a singularity as a *Bestimmtheitsstelle*; later German terminology is *schwach singuläre Stelle*, and the term *stark singuläre Stelle* is used for an *irregular-singular point*. For (5.1.6) such a singularity would correspond to a pole of higher order than the first of $F_1(z)$ or an essential singular point. A point of the second kind is $z = 0$ for the equation

$$w' = z^{-2}w    \qquad \text{with } w(z) = C \exp\left(-\frac{1}{z}\right).    \tag{5.1.13}$$

The origin is evidently an essential singular point of the solution which goes to zero as $z \to 0$ in the sector $|\arg z| < \frac{1}{2}\pi$, and to infinity in $|\arg(-z)| < \frac{1}{2}\pi$. Furthermore, $w(z)$ takes on every value $\neq 0$ and $\infty$ in $\epsilon$-sectors centered on the imaginary axis. This is typical for irregular singularities of finite rank or grade, terms which will be defined in Section 5.4. The DE

$$w' = e^z w    \qquad \text{with } w(z) = C \exp(e^z)    \tag{5.1.14}$$

has $z = \infty$ as an irregular-singular point of infinite rank and exhibits a correspondingly more complicated behavior at infinity. Instead of sectors we have now to consider half-strips.

The equation (5.1.6) belongs to the *Fuchsian class* if there is only a finite number of singular points, each of which is a regular-singular point including the point at infinity. This forces $F_1(z)$ to be a rational function with simple poles, say,

$$F_1(z) = \frac{P(z)}{\prod\limits_1^n (z - a_j)}. \tag{5.1.15}$$

In order that the point at infinity also be regular-singular, the degree of $P(z)$ cannot exceed $n - 1$. Let the partial fraction expansion of $F_1(z)$ be

$$F_1(z) = \sum_{j=1}^n \frac{r_j}{z - a_j}$$

so that

$$v(z) = \prod_{j=1}^n (z - a_j)^{r_j}. \tag{5.1.16}$$

### EXERCISE 5.1

1. When is $v(z)$ as defined by (5.1.16) single valued?
2. What does it look like at $z = \infty$? What conditions should the exponents satisfy in order that $v(z)$ be single valued in a neighborhood of $z = \infty$?
3. Verify the statements concerning the behavior of the solutions of (5.1.13). Show also that in the disks $|z - i2^{-n}| < 2^{-n-2}$, $n = 0, 1, 2, \ldots$, the solution takes on every value except zero and infinity infinitely often [Gaston Julia, 1924].
4. Discuss limits and value distributions for the solutions of (5.1.14). There are two Picard values.
5. Suppose that $F_1(z)$ has two simple poles only, one at $z = a$, the other at $z = b$. Suppose that $F_0(z)$ is a polynomial in $z$ of degree not exceeding two. Find the nature of $w(z; z_0, w_0)$ at $z = a$.
6. Discuss the equation $w' = i(z - 1)^{-1}w$. Here $z = 1$ is a regular-singular point.
7. Explain the conditions for belonging to the Fuchsian class.

### 5.2. GENERAL THEORY: SECOND ORDER CASE

It was observed in Section 4.1 that there is a close connection between Riccati's equation,

$$y'(z) = F_0(z) + F_1(z)y(z) + F_2(z)[y(z)]^2, \tag{5.2.1}$$

and the linear second order DE,

$$w''(z) + P(z)w'(z) + Q(z)w(z) = 0. \tag{5.2.2}$$

Setting

$$y(z) = -\frac{1}{F_2(z)} \frac{w'(z)}{w(z)}, \tag{5.2.3}$$

we proceed from (5.2.1) to a DE of type (5.2.2), while setting

$$-\frac{w'(z)}{w(z)} = y(z) \qquad (5.2.4)$$

takes (5.2.2) into a Riccati equation. As a matter of fact the theories of these two types of equations are essentially equivalent.

Now Riccati's equation has no movable branch points, but it does have movable poles. The first property clearly carries over to the second order linear case, but what happens to the movable poles in the transit? Suppose that $y(z)$ has a pole at $z = a$, not one of the fixed singularities, i.e., the zeros of $F_2$, the singularities of $F_0$, $F_1$, $F_2$, and the point at infinity. Thus $F_2(a) \neq 0$, so that

$$y(z) = -\frac{1}{F_2(a)} \frac{1}{(z-a)} [1 + a_1(z-a) + a_2(z-a)^2 + \cdots]. \qquad (5.2.5)$$

This follows from setting $y = 1/v$ and finding the solution of the corresponding Riccati equation (4.1.16), which has a zero at $z = a$. Now $F_2(z)$ is holomorphic at $z = a$ and has a convergent power series expansion there. It follows that

$$-\int F_2(t) y(t)\, dt = \int [(t-a)^{-1} + b_0 + b_1(t-a) + \cdots]\, dt$$

$$= \log (z-a) + P_1(z-a),$$

where $P_1(u)$ is a power series in $u$. Since $w(z)$ is the exponential function of this expression, it is of the form

$$w(z) = c(z-a)P_2(z-a),$$

where $P_2(u)$ is a power series in $u$ and $P_2(0) = 1$. Convergence follows from the double series theorem of Weierstrass. Thus a pole of $y(z)$ becomes a zero of $w(z)$, and this function is holomorphic at $z = a$.

With a slight change of notation, suppose that $w_1(z)$ and $w_2(z)$ are solutions of the DE

$$P_0(z)w''(z) + P_1(z)w'(z) + P_2(z)w(z) = 0, \qquad (5.2.6)$$

and that $w_2(z)$ is not a constant multiple of $w_1(z)$. It is assumed that $z$ is restricted to a domain $D$, where the coefficients $P_j$ are holomorphic and $P_0(z) \neq 0$. We have then two more equations:

$$P_0(z)w_1''(z) + P_1(z)w_1'(z) + P_2(z)w_1(z) = 0,$$
$$P_0(z)w_2''(z) + P_1(z)w_2'(z) + P_2(z)w_2(z) = 0.$$

These equations, together with (5.2.6), form a system of three linear homogeneous equations which are satisfied by $P_0(z)$, $P_1(z)$, $P_2(z)$ not all identically zero. It follows that the determinant of the system

$$\begin{vmatrix} w'' & w' & w \\ w_1'' & w_1' & w_1 \\ w_2'' & w_2' & w_2 \end{vmatrix} \equiv 0 \tag{5.2.7}$$

must be identically zero. Next, the coefficients $P_j$ are proportional to the minors of the first row, i.e.,

$$P_0 : P_1 : P_2 = \begin{vmatrix} w_1' & w_1 \\ w_2' & w_2 \end{vmatrix} : -\begin{vmatrix} w_1'' & w_1 \\ w_2'' & w_2 \end{vmatrix} : \begin{vmatrix} w_1'' & w_1' \\ w_2'' & w_2' \end{vmatrix}. \tag{5.2.8}$$

This shows that it must be possible to find a system $w_1$, $w_2$ such that the first minor is not zero and is a holomorphic function of $z$ in $D$. Such systems were introduced by Fuchs in 1868 under the name *fundamental systems*.

For the purpose of orientation, suppose that the first minor is identically zero so that

$$w_1 w_2' - w_2 w_1' \equiv 0.$$

We may assume that $w_1 \neq 0$; then division by $w_1^2$ gives

$$\frac{d}{dz}\left(\frac{w_2}{w_1}\right) = 0$$

or $w_2/w_1 = c$, a constant. In this case $w_2(z)$ is a constant multiple of $w_1(z)$, so that $w_1$ and $w_2$ are linearly dependent [see (1.1.1) and below]. Thus for a fundamental system we must choose linearly independent solutions of (5.2.6). Now, if $z_0$ is not a singular point, the initial conditions:

$$\begin{aligned} w_1(z_0) &= 0, & w_2(z_0) &= 1, \\ w_1'(z_0) &= 1, & w_2'(z_0) &= 0, \end{aligned} \tag{5.2.9}$$

determine uniquely a pair of solutions which cannot be constant multiples of each other. We shall see in a moment that the first minor is nowhere zero in $D$ if it is not zero at $z = z_0$.

If $w_1$, $w_2$ form a fundamental system, the DE can be recovered from (5.2.8) by expanding the determinant (5.2.7) according to the elements of the first row. This gives

$$-\frac{P_1}{P_0} = \frac{w_1 w_2'' - w_2 w_1''}{w_1 w_2' - w_2 w_1'}, \qquad \frac{P_2}{P_0} = \frac{w_1' w_2'' - w_2' w_1''}{w_1 w_2' - w_2 w_1'}. \tag{5.2.10}$$

Here the numerator of the first quotient is the derivative of the denominator. Let $W(z; w_1, w_2)$ denote the denominator. It is called the *Wronskian* after the Polish mathematician known as Wronski, whose real name appears to have been Josef Maria Hoene (1778–1853). From

$-P_1/P_0 = W'/W$ it follows that

$$W(z) = W(z_0) \exp\left[ - \int_{z_0}^{z} \frac{P_1(t)}{P_0(t)}\, dt \right], \qquad (5.2.11)$$

a formula due to Niels Henrik Abel (1802–1829). It occurs in one of his five memoirs in Vol. 2 (1827) of Crelle's *Journal* (= *Journal der reine und angewandte Mathematik*, dubbed *J. d. unangewandte Math.* by Crelle's hecklers). In (5.2.11) we may integrate along any rectifiable path in $D$. If $W(z_0) \neq 0$, the same is true of $W(z)$ everywhere in $D$. On the other hand, if $W(z_0) = 0$, then $W(z)$ is identically zero in $D$. Thus we see that (5.2.9) determines a fundamental system.

Going back to (5.2.7), we assume that $w_1$, $w_2$ form a fundamental system and that $w$ is any solution of (5.2.6). Then the fact that the determinant is identically zero implies that the first row is linearly dependent on the second and third rows. Hence there exist two constants, $C_1$ and $C_2$, such that

$$w(z) = C_1 w_1(z) + C_2 w_2(z). \qquad (5.2.12)$$

We say that $n$ functions, $F_1$, $F_2$, ..., $F_n$, are *linearly dependent* if $n$ constants, $C_1$, $C_2$, ..., $C_n$, exist, not all zero, such that

$$C_1 F_1(z) + C_2 F_2(z) + \cdots + C_n F_n(z) \equiv 0 \qquad (5.2.13)$$

in the common domain of definition of the functions. They are *linearly independent* if such a relation can hold iff all the $C$'s are zero. In this terminology any three solutions of a linear second order DE are linearly dependent, while two solutions which form a fundamental system are linearly independent.

The solutions are locally holomorphic in any domain $D$ in which the quotients $P_1/P_0$ and $P_2/P_0$ are holomorphic. We come now to the question of what happens in a neighborhood of a singular point, and start with the problem of analytic continuation along a closed path $\Gamma$ surrounding an isolated singularity $z = a$. We start at $z = z_1$ with a fundamental system $w_1$, $w_2$. Analytic continuation along $\Gamma$ back to $z_1$ will lead to a system $w_1^*$, $w_2^*$ which is still fundamental. Since any three solutions are linearly dependent, there are constants $C_{jk}$ such that

$$\begin{aligned} w_1^* &= C_{11} w_1 + C_{12} w_2, \\ w_2^* &= C_{21} w_1 + C_{22} w_2. \end{aligned} \qquad (5.2.14)$$

These equations must be solvable for $w_1$, $w_2$ so that

$$C_{11} C_{22} - C_{12} C_{21} \neq 0.$$

This means that the matrix

$$\mathcal{M} = \begin{bmatrix} C_{11} & C_{12} \\ C_{21} & C_{22} \end{bmatrix} \tag{5.2.15}$$

is nonsingular. This matrix $\mathcal{M}$ governs the analytic continuation of the fundamental system $(w_1, w_2)$ along the path $\Gamma$. Using the notation of the matrix calculus, we may write

$$\mathbf{v}^* = \mathcal{M}\mathbf{v}, \tag{5.2.16}$$

where $\mathbf{v}$ and $\mathbf{v}^*$ are the column vectors

$$\mathbf{v} = \begin{bmatrix} w_1 \\ w_2 \end{bmatrix}, \qquad \mathbf{v}^* = \begin{bmatrix} w_1^* \\ w_2^* \end{bmatrix}. \tag{5.2.17}$$

Here the *transit matrix* $\mathcal{M}$ obviously depends on the choice of the original fundamental system. However, changing to another fundamental system $(w_0, w_{00})$ merely replaces $\mathcal{M}$ by a *similar matrix*, i.e., a matrix $\mathcal{M}^*$ such that there exists a nonsingular matrix $\mathcal{B}$ with $\mathcal{M}^* = \mathcal{B}\mathcal{M}\mathcal{B}^{-1}$, for if $(w_0, w_{00})$ is a fundamental system there must exist a nonsingular matrix $\mathcal{A}$ such that

$$\begin{bmatrix} w_0 \\ w_{00} \end{bmatrix} = \mathcal{A} \begin{bmatrix} w_1 \\ w_2 \end{bmatrix} \quad \text{or} \quad \begin{bmatrix} w_1 \\ w_2 \end{bmatrix} = \mathcal{A}^{-1} \begin{bmatrix} w_0 \\ w_{00} \end{bmatrix}.$$

In the second equality we operate on the left by $\mathcal{M}$ to obtain

$$\mathcal{M} \begin{bmatrix} w_1 \\ w_2 \end{bmatrix} = \begin{bmatrix} w_1^* \\ w_2^* \end{bmatrix} = \mathcal{M}\mathcal{A}^{-1} \begin{bmatrix} w_0 \\ w_{00} \end{bmatrix}.$$

Here we operate with $\mathcal{A}$ on the left in the last two members to obtain

$$\mathcal{A} \begin{bmatrix} w_1^* \\ w_2^* \end{bmatrix} = \mathcal{A}\mathcal{M}\mathcal{A}^{-1} \begin{bmatrix} w_0 \\ w_{00} \end{bmatrix} = \begin{bmatrix} w_0^* \\ w_{00}^* \end{bmatrix}$$

as the result of the circuit applied to $(w_0, w_{00})$. Here the matrices $\mathcal{M}$ and $\mathcal{A}\mathcal{M}\mathcal{A}^{-1}$ are clearly similar in the sense of matrix theory.

By (1.4.14) the characteristic equation of $\mathcal{M}$ is

$$\det [\mathcal{M} - \omega\mathcal{E}] = 0, \tag{5.2.18}$$

where $\mathcal{E}$ is the 2-by-2 unit matrix. This gives

$$\omega^2 - (C_{11} + C_{22})\omega + C_{11}C_{22} - C_{12}C_{21} = 0. \tag{5.2.19}$$

Since

$$\mathcal{A}\mathcal{M}\mathcal{A}^{-1} - \omega\mathcal{E} = \mathcal{A}\mathcal{M}\mathcal{A}^{-1} - \mathcal{A}\omega\mathcal{E}\mathcal{A}^{-1} = \mathcal{A}(\mathcal{M} - \omega\mathcal{E})\mathcal{A}^{-1}$$

and the determinant of a product of matrices equals the product of the determinants of the factors, and since $\det (\mathcal{A}^{-1}) = [\det (\mathcal{A})]^{-1}$, the characteristic equations of similar matrices are identical. This shows also that the characteristic roots of (5.2.18) are numerical invariants associated with the path $\Gamma$. Since $\det(\mathcal{M}) \neq 0$, the roots are different from zero.

There are two cases: (i) the roots $\omega_1$, $\omega_2$ are distinct, and (ii) the equation has a double root. In case (i) we can find two linearly independent solutions, $w_0$ and $w_{00}$, which are multiplied by a constant when $z$ describes $\Gamma$, for among the matrices similar to $\mathcal{M}$ there is also the diagonal matrix

$$\mathcal{D} = \begin{bmatrix} \omega_1 & 0 \\ 0 & \omega_2 \end{bmatrix}. \tag{5.2.20}$$

$\mathcal{D}$ is clearly similar to $\mathcal{M}$, and we can find a regular matrix $\mathcal{A}$ such that

$$\mathcal{D} = \mathcal{A}^{-1}\mathcal{M}\mathcal{A}. \tag{5.2.21}$$

We start at $z = z_1$ with the fundamental system $w_1$, $w_2$, and apply successively the matrices $\mathcal{A}$, $\mathcal{M}$, and $\mathcal{A}^{-1}$. The result is a fundamental system $w_0$, $w_{00}$, and a circuit along $\Gamma$ takes it into

$$\mathcal{D} \begin{bmatrix} w_0 \\ w_{00} \end{bmatrix} = \begin{bmatrix} \omega_1 w_0 \\ \omega_2 w_{00} \end{bmatrix}. \tag{5.2.22}$$

Although the fundamental system which transforms in this manner is not uniquely determined, the multipliers are unique and are the two roots of (5.2.19).

We can now discuss the analytical nature of the multiplicative solutions along the lines pursued in Section 5.1 in the discussion of $v(z)$. We set

$$\rho_1 = \frac{\log \omega_1}{2\pi i}, \qquad \rho_2 = \frac{\log \omega_2}{2\pi i}, \tag{5.2.23}$$

again leaving the logarithms arbitrary up to multiples of $2\pi i$. It is found that

$$w_0(z) = (z - a)^{\rho_1} V_1(z), \qquad w_{00}(z) = (z - a)^{\rho_2} V_2(z). \tag{5.2.24}$$

Here $V_1$ and $V_2$ are single valued in the neighborhood of $z = a$ and may be represented by convergent Laurent series in $z - a$. It will be determined later under what circumstances the Laurent series may be replaced by ordinary power series.

Let us first settle the case in which the characteristic equation has a double root $\omega_1$. A solution $w_0(z)$ can obviously be found, which is multiplied by $\omega_1$ when $z$ describes $\Gamma$. Take any solution $w_{00}$ which is linearly independent of $w_0$, and suppose that describing $\Gamma$ takes $w_{00}$ into

$$w_{00}^* = \beta w_0 + \gamma w_{00}.$$

Thus the fundamental system $w_0$, $w_{00}$ has the transit matrix

$$\mathcal{M}_0 = \begin{bmatrix} \omega_1 & 0 \\ \beta & \gamma \end{bmatrix}$$

with respect to the path $\Gamma$. Its characteristic equation is

$$\begin{vmatrix} \omega_1 - \omega & 0 \\ \beta & \gamma - \omega \end{vmatrix} = (\omega_1 - \omega)(\gamma - \omega) = 0.$$

Since $\mathcal{M}_0$ and $\mathcal{M}$ are similar matrices, we must have $\gamma = \omega_1$ and

$$w_{00}^* = \beta w_0 + \omega_1 w_{00}.$$

It follows that the quotient $w_{00}/w_0$ is transformed into

$$\frac{\beta}{\omega_1} + \frac{w_{00}}{w_0}$$

as $z$ describes $\Gamma$. This means that the quotient has an *additive period* $\beta/\omega_1$. Now the function

$$z \mapsto \frac{\beta}{\omega_1} \frac{1}{2\pi i} \log(z - a)$$

has the same property, since the logarithm increases by $2\pi i$ along the circuit. Hence

$$\frac{w_{00}}{w_0} - \frac{\beta}{\omega_1} \frac{1}{2\pi i} \log(z - a) = V_2(z),$$

where $V_2(z)$ is single valued in a neighborhood of $z = a$ and may be expanded in a convergent Laurent series in terms of $z - a$. Thus in the case of a double root of (5.2.19) a fundamental system can be found of the form

$$w_0(z) = (z - a)^{\rho_1} V_1(z),$$

$$w_{00}(z) = (z - a)^{\rho_1}\left[ V_3(z) + \frac{\beta}{\omega_1} \frac{1}{2\pi i} V_1(z) \log(z - a) \right], \qquad (5.2.25)$$

where

$$V_3(z) = V_1(z) V_2(z)$$

and the $V$'s are convergent Laurent series in $z - a$.


## EXERCISE 5.2

1. If the Wronskian of two locally holomorphic functions $F_1$ and $F_2$ vanishes in a set $S$ with a limit point in the common domain of holomorphism of $F_1$ and $F_2$, show that the Wronskian vanishes identically in $D$ and $F_2 = CF_1$, $C$ a constant.

2. A 2-by-2 matrix $z \mapsto \mathcal{M}(z)$ is holomorphic in a domain $D$ iff the four entries $m_{jk}(z)$ are holomorphic there. Suppose that $\det [\mathcal{M}(z)]$ vanishes in a set $S$ with the properties as in Problem 1. Show that $\mathcal{M}(z)$ is algebraically singular in $D$, i.e., has no inverse anywhere in $D$.

3. If $\cos z^{1/2}$ and $\sin z^{1/2}$ form a fundamental system for a linear second order DE, find the equation.

4. Solve Problem 3 for the system $z^{1/2}$, $z^{1/2} \log z$.

5. The DE's in Problems 3 and 4 have $z = 0$ and $z = \infty$ as only singularities. Find the transit matrix for a circuit about the origin in the positive sense. Find the characteristic equations and its roots. The singularities are obviously regular singular. How would you choose the $\rho$'s?

## 5.3. REGULAR-SINGULAR POINTS

In Section 5.2 it was shown that at an isolated singularity $z = a$ of $P_1/P_0$ and/or $P_2/P_0$ it is possible to find a fundamental system of the form (5.2.24) if the roots of the characteristic equation are distinct, and of the form (5.2.25) if they are equal.

The next problem is to determine if the singularity is *regular singular* or *irregular singular*. In the first case the functions $z \mapsto V_j(z)$ are holomorphic at $z = a$ for a suitable choice of the $\rho_j$'s. This problem can be solved with the aid of formulas (5.2.10), which express the coefficients of the DE in terms of a fundamental system. The question is now to find necessary and sufficient conditions at $z = a$ for the presence of a regular-singular point at $z = a$.

**THEOREM 5.3.1**

*A point* $z = a$ *is a regular-singular point of the DE*

$$w''(z) + P(z)w'(z) + Q(z)w(z) = 0 \tag{5.3.1}$$

*iff* (i) *at least one of the coefficients* P *and* Q *has an isolated singularity at* $z = a$, (ii) P(z) *either is regular or has a simple pole at* $z = a$, *and* (iii) Q(z) *either is regular or has a pole of order* $\leq 2$.

*Proof.* One can proceed directly from (5.2.10), but a slight modification saves labor. We write the fundamental system in the form

$$\begin{aligned}
w_1(z) &= (z - a)^{\rho_1} V_1(z), \\
w_2(z) &= (z - a)^{\rho_2} V_2(z) + A(z - a)^{\rho_1} V_1(z) \log (z - a)
\end{aligned} \tag{5.3.2}$$

with the following conventions: (i) $A = 0$ if $\rho_1 \not\equiv \rho_2 \pmod 1$, while (ii) if $\rho_1 \equiv \rho_2$, then $A = \delta = (\beta/\omega_1)/(2\pi i)$ and $V_2 = V_3$ in our previous notation.

Now

$$P(z) = -\frac{W'(z; w_1, w_2)}{W(z; w_1, w_2)} = -\frac{d}{dz} \log\left[ w_1^2 \frac{d}{dz}\left(\frac{w_2}{w_1}\right)\right], \qquad (5.3.3)$$

$$Q(z) = -\frac{w_1''}{w_1} - P(z)\frac{w_1'}{w_1}, \qquad (5.3.4)$$

where the second relation follows from the fact that $w_1$ is a solution of (5.3.1). In (5.3.3) and (5.3.4) the $\rho$'s are now definite complex numbers and the $V$'s are ordinary power series with $V_1(a) = V_2(a) = 1$. Now

$$\frac{w_2}{w_1} = A \log(z - a) + (z - a)^{\rho_2 - \rho_1} V_4(z),$$

where $V_4 = V_2/V_1$ and $V_4(a) = 1$ if $A = 0$. In any case the derivative equals

$$\frac{A}{z - a} + (\rho_2 - \rho_1)(z - a)^{\rho_2 - \rho_1 - 1}[1 + O(|z - a|)].$$

Next

$$w_1^2 \frac{d}{dz}\left(\frac{w_2}{w_1}\right) = A(z - a)^{2\rho_1 - 1}[1 + O(|z - a|)]$$
$$+ (\rho_2 - \rho_1)(z - a)^{\rho_1 + \rho_2 - 1}[1 + O(|z - a|)].$$

Now, if $A = 0$, $\rho_1 \neq \rho_2$, the logarithm of the last member is

$$(\rho_1 + \rho_2 - 1)\log(z - a) + O(|z - a|),$$

and its derivative equals

$$\frac{\rho_1 + \rho_2 - 1}{z - a} + O(1) = -P(z). \qquad (5.3.5)$$

This expression is also valid if $A \neq 0$ and $\rho_1 \equiv \rho_2$.

Next we have

$$Q(z) = -\frac{w_1''}{w_1} - P(z)\frac{w_1'}{w_1} = -\frac{w_1''}{w_1'}\frac{w_1'}{w_1} - P(z)\frac{w_1'}{w_1},$$

where $P(z)$ is given by (5.3.5) and

$$\frac{w_1'}{w_1} = \frac{\rho_1}{z - a} + O(1).$$

We take $\rho_1 \neq 0$ for the time being. Then

$$w_1'(z) = \rho_1(z - a)^{\rho_1 - 1}[1 + O(|z - a|)],$$
$$\log w_1'(z) = (\rho_1 - 1)\log(z - a) + O(1),$$

so that

$$\frac{w_1''(z)}{w_1'(z)} = \frac{\rho_1 - 1}{z - a} + O(1).$$

Combining and substituting in (5.3.4), we get

$$Q(z) = \frac{c}{(z-a)^2} + O(|z-a|^{-1}), \qquad c = \rho_1\rho_2, \qquad (5.3.6)$$

which is seen to be true also for $\rho_1 = 0$. This proves that the conditions of the theorem are necessary and shows also that

and
$$\lim (z-a) P(z) = -(\rho_1 + \rho_2 - 1)$$
$$\lim (z-a)^2 Q(z) = \rho_1\rho_2 \qquad (5.3.7)$$

are necessary conditions.

To prove sufficiency we write the DE in the form

$$(z-a)^2 w''(z) + (z-a)p(z)w'(z) + q(z)w(z) = 0, \qquad (5.3.8)$$

following the precept of G. Frobenius. Here $p(z)$ and $q(z)$ are holomorphic in some neighborhood of $z = a$. The quadratic equation

$$f_0(s) = s(s-1) + p(a)s + q(a) = 0 \qquad (5.3.9)$$

is known as the *indicial equation* at $z = a$, and by (5.3.7) its roots are $\rho_1$ and $\rho_2$. It is required to show that (5.3.8) has at least one solution of the form $(z-a)^\rho V(z-a)$ and two such solutions if $\rho_1$ and $\rho_2$ are incongruent (modulo 1). To simplify the notation we take $a = 0$. We also assume that the singularities of $p(z)$ and $q(z)$ lie outside the unit circle. If this is not the case at the outset, an affine transformation $z \mapsto az + b$ will lead to the desired situation.

We have thus the equation

$$z^2 w''(z) + z p(z)w'(z) + q(z)w(z) = 0 \qquad (5.3.10)$$

with

$$p(z) = \sum_{n=0}^{\infty} p_n z^n, \qquad q(z) = \sum_{n=0}^{\infty} q_n z^n. \qquad (5.3.11)$$

Here we substitute

$$w(z) = \sum_{n=0}^{\infty} c_n z^{n+\rho}, \qquad (5.3.12)$$

where $\rho$ and the coefficients $c_n$ are to be determined. Here various formal processes are involved, as we know from the discussion in Section 2.4. The series (5.3.12) is differentiated twice termwise, the formal first derivative is multiplied by the series for $p(z)$, using Cauchy's product theorem, and similarly the series for $q(z)$ and for $w(z)$ are multiplied; the three series are then combined, terms with equal powers of $z$ are collected, and their coefficients are equated to zero. All this is justified if we are actually dealing with absolutely convergent series. We have to prove the convergence a posteriori; nothing in our previous work tells us that the series are convergent, although this is a very plausible guess. First we have to find the series, however. The indicated procedure leads to an infinite set of conditional equations:

$$c_0 f_0(\rho) = 0,$$
$$c_1 f_0(\rho + 1) + c_0 f_1(\rho) = 0,$$
$$c_2 f_0(\rho + 2) + c_1 f_1(\rho + 1) + c_0 f_0(\rho) = 0,$$ (5.3.13)
$$\cdots$$
$$c_k f_0(\rho + k) + c_{k-1} f_1(\rho + k - 1) + \cdots + c_0 f_k(\rho) = 0,$$
$$\cdots.$$

Here $f_0(\rho)$ is given by (5.3.9) and

$$f_j(\rho) = \rho p_j + q_j, \qquad 0 < j. \qquad (5.3.14)$$

We may assume that $c_0 = 1$. Then the first condition under (5.3.13) requires that $f_0(\rho) = 0$, so that $\rho$ has to be one of the roots of the indicial equation (5.3.11). The latter has two roots, $\rho_1$ and $\rho_2$, where the numbering is such that Re $(\rho_1) \leqslant$ Re $(\rho_2)$. After $\rho$ has been chosen, (5.3.13) determine successively $c_1, c_2, \ldots, c_k, \ldots$, provided that $f_0(\rho + k) \neq 0$ for all $k > 0$. This will certainly be the case for $\rho = \rho_2$, the root with the greater real part. For $\rho = \rho_1$ there are four possibilities: (i) if $\rho_2 - \rho_1$ is not zero or a positive integer, the procedure works also for $\rho = \rho_1$; (ii) if $\rho_2 - \rho_1 = k > 0$, we come to an impasse unless accidentally

$$c_{k-1} f_1(\rho_1 + k - 1) + \cdots + c_0 f_k(\rho_1) = 0, \qquad (5.3.15)$$

in which case $c_k$ may be chosen arbitrarily and all coefficients $c_n$ with $n > k$ are uniquely determined in terms of the preceding coefficients: in this case we obtain a one-parameter family of solutions corresponding to $\rho = \rho_1$, the parameter being $c_k$; (iii) $\rho_2 - \rho_1 = k > 0$, and (5.3.15) does not hold; in this case and in (iv) $\rho_1 = \rho_2$ only one solution is obtainable in this manner and some other device is needed to produce a second linearly independent solution. Let us first settle the convergence question.

Suppose that in case (i) or (ii) we have computed a formal solution. It is desired to prove that it is convergent in some punctured neighborhood of the origin. Here the assumed properties of $p(z)$ and $q(z)$ come into play. Since these functions are holomorphic in a disk $|z| < r$, where $r > 1$, we can find constants $K > 0, R > 1$ such that for all $n$

$$|p_n| < K(n+1)^{-1} R^{-n}, \qquad |q_n| < K(n+1)^{-1} R^{-n}. \qquad (5.3.16)$$

Furthermore, it is possible to find a $d > 0$ such that

$$|f_0(\rho + n)| > d(n+1)^2. \qquad (5.3.17)$$

This is to hold for $\rho = \rho_1$ and $\rho_2$ for all $n > 0$ in case (i) and for $n > k$ in case (ii). The assumption (5.3.16) implies that

$$|f_j(\rho + k)| < \frac{\sigma + k + 1}{j + 1} R^{-j}, \qquad j > 0, \, k \geq 0, \tag{5.3.18}$$

where $\sigma = \max(|\rho_1|, |\rho_2|)$. Now

$$|c_n| \, |f_0(\rho + n)| \leq |c_{n-1}| \, |f_1(\rho + n - 1)| + \cdots + |c_0| \, |f_n(\rho)|.$$

Suppose that an integer $m$ and a constant $M$ have been found such that

$$|c_j| \leq \frac{M}{j + 1} R^{-j}, \quad j = 0, 1, \ldots, m - 1. \tag{5.3.19}$$

Such a choice can always be made, but here we want to show that if $m$ and $M$ satisfy certain conditions, (5.3.19) holds also for $j = m$ and hence, by induction, for all $j$.

In fact, from (5.3.17)–(5.3.19) it follows that

$$|c_m| < \frac{KM}{d(m + 1)^2} R^{-m} \left\{ \frac{1}{2} \frac{\sigma + m}{m} + \frac{1}{3} \frac{\sigma + m - 1}{m - 1} + \cdots + \frac{1}{m + 1} \frac{\sigma + 1}{1} \right\}.$$

Here the fraction $(\sigma + k)/k$ decreases as $k$ increases, so for $k$ between 1 and $m$ it has its largest value $(\sigma + 1)$ for $k = 1$. Hence

$$|c_m| < \frac{KM(\sigma + 1)}{d(m + 1)^2} R^{-m} \left\{ \frac{1}{2} + \frac{1}{3} + \cdots + \frac{1}{m + 1} \right\}.$$

Here the quantity between the braces is $< \log(m + 1)$. Hence

$$|c_m| < \frac{K(\sigma + 1)}{d(m + 1)} \frac{\log(m + 1)}{m + 1} M R^{-m}, \tag{5.3.20}$$

and (5.3.19) will hold also for $j = m$ provided $m$ is chosen so that

$$\frac{K(\sigma + 1)}{d} \frac{\log(m + 1)}{m + 1} < 1. \tag{5.3.21}$$

Now $d$, $K$, $\sigma$ are fixed numbers, so that this condition can always be satisfied. Once $m$ has been chosen, we can find an $M$ such that (5.3.19) is satisfied for $j < m$. A finite number of conditions are involved so that $M$ can be found. Once this has been accomplished, complete induction shows that (5.3.19) holds for all $j$. Thus the formal solution is absolutely convergent for $0 < z < R$, and the formal solution is then the actual solution.

In case (i) we have then two convergent series solutions, and they are linearly independent for

$$\frac{w_2(z)}{w_1(z)} = z^{\rho_2-\rho_1} \frac{V_2(z)}{V_1(z)} = z^{\rho_2-\rho_1} V_3(z), \tag{5.3.22}$$

where the $V_j$'s are absolutely convergent power series and $V_j(0) = 1$. If $\mathrm{Re}\,(\rho_1) < \mathrm{Re}\,(\rho_2)$ the quotient tends to zero with $z$, while if $\mathrm{Re}\,(\rho_1) = \mathrm{Re}\,(\rho_2)$ there is no limit. In neither case can the quotient be a constant, so the two solutions are linearly independent.

The same argument applies in case (ii), the only difference being that the integer $m$ of (5.3.19) must also satisfy $m > k$.

In cases (iii) and (iv), $\rho_2 - \rho_1 = k$, a nonnegative integer. Here we use a method for reducing the order of a linear DE by one unit. In our case, $n = 2$, this means that the reduced equation is of order 1 and the discussion in Section 5.1 applies. We know one solution, $w_2(z)$, corresponding to $\rho = \rho_2$. For the moment let $u(z)$ denote the known solution, and set

$$w = u \int v \, dt, \tag{5.3.23}$$

where $v$ is to be found. Then

$$w' = u' \int v \, dt + uv, \qquad w'' = u'' \int v \, dt + 2u'v + uv'.$$

Thus $v$ satisfies the equation

$$\frac{v'(z)}{v(z)} = -2 \frac{u'(z)}{u(z)} - \frac{p(z)}{z}, \tag{5.3.24}$$

so that

$$\log v = -2 \log u - \int \frac{p(t)}{t} \, dt$$

or

$$v(z) = C[u(z)]^{-2} \exp\left[-\int^z \frac{p(t)}{t} \, dt\right]. \tag{5.3.25}$$

Here we take $u(z) = w_2(z) = z^{\rho_2} V_2(z)$, and note that the second factor in (5.3.25) equals

$$z^{-p_0} V(z) \qquad \text{with} - p_0 = \rho_1 + \rho_2 - 1. \tag{5.3.26}$$

Here $V$ and $V_2$ are convergent power series, and $V(0) = 1$. In the case of $V(z)$ we use the fact that $\exp[P(z)]$ is a convergent power series if $P(z)$ has this property, this by virtue of the Weierstrass double series theorem. Hence

$$v(z) = z^{-2\rho_2+\rho_1+\rho_2-1} V_0(z) = z^{-k-1} V_0(z),$$

where $V_0(z)$ is also a convergent power series and $V_0(0) = 1$. Suppose that

$$v(z) = z^{-k-1} \sum_{n=0}^{\infty} b_n z^n. \qquad (5.3.27)$$

Then

$$\int v(t)\, dt = {\sum}' \frac{b_n}{n-k} z^{n-k} + b_k \log z,$$

where the prime indicates that $n \neq k$. Hence

$$w_1(z) = w_2(z) \sum_{n=0}^{\infty}{}' \frac{b_n}{n-k} z^{n-k} + b_k w_2(z) \log z$$

or

$$w_1(z) = z^{\rho_1} V_1(z) + b_k w_2(z) \log z. \qquad (5.3.28)$$

This qualifies as the second solution in the fundamental system where $w_2(z)$ is the first component. It should be noted that normally $b_k \neq 0$. If $b_k = 0$, case (ii) is present instead of case (iii). Furthermore, if $k = 0$, so that the indicial equation has a double root, $b_0 \neq 0$. Thus logarithms always occur in case (iv). In addition, it should be noted that $w_2(z)$ is a multiplier of $\log z$. This completes the proof of Theorem 5.3.1. ∎

## EXERCISE 5.3

1.  The equation $z^2 w'' - 3zw' + 4w = 0$ has $z^2$ as a solution. Find a second linearly independent solution, using (5.3.23).

2.  If $a$ and $b$ are constants, what is the indicial equation of $z^2 w'' + azw' + bw = 0$? If the roots $\rho_1$ and $\rho_2$ are distinct, then

    $$\frac{z^{\rho_1} - z^{\rho_2}}{\rho_1 - \rho_2}$$

    is a solution. What is the limit of this solution as $\rho_2 \to \rho_1$? Is the limit a solution of the limiting equation, and if so why?

3.  The equation $z^2 w'' + (z^2 - k^2 + \tfrac{1}{4})w = 0$ is closely related to Bessel's equation. Find the indicial equation at $z = 0$ and its roots. They differ by an integer if $2k$ is zero or an integer. If $k = 0$ or a positive integer, a logarithm is present in the general solution, but not if $k = n + \tfrac{1}{2}$. The latter statement is trivial if $n = 0$, $k = \tfrac{1}{2}$. Verify it for $n = 1$, $k = \tfrac{3}{2}$, given that $(z \cos z - \sin z)/z$ is a solution.

4.  Prove that when the indicial equation has a double root the second solution always involves a logarithm. Suppose that $z$ describes a closed circuit in the positive sense about the origin (an isolated regular-singular point). Find the transit matrix.

5.  Verify (5.3.16).

6.  Justify (5.3.17).

7.  Apply the transformation (5.3.23) to a third order linear DE.

## 5.4. ESTIMATES OF GROWTH

Our study of irregular-singular points will proceed along three different lines which complement each other. The singularity is placed at infinity. In this section we try to estimate the rate of growth of a solution when rates of growth for the coefficients are known, the equation being

$$w'' + P(z)w' + Q(z)w = 0. \tag{5.4.1}$$

In Section 5.5 we shall consider asymptotics on the real line for real coefficients, and in Section 5.6 we shall use a method of asymptotic integration in sectors of the complex plane.

The coefficients are supposed to be single valued and holomorphic for $0 \leqslant R < |z| < \infty$, and at least one of the functions

$$z P(z) \quad \text{and} \quad z^2 Q(z) \tag{5.4.2}$$

has a singularity at infinity. It will be assumed that the solution $w(z)$ under consideration is single valued for $|z| > R$ and that it behaves as an entire function of $z$ for $|z| > R$. With the notation borrowed from the Nevanlinna theory, we define the *order* of $w(z)$ to be

$$\rho(w) = \limsup_{r \to \infty} \frac{\log T(r; w)}{\log r}, \tag{5.4.3}$$

where $R < r$ and

$$T(r; w) = m(r, \infty; r) = \frac{1}{2\pi} \int_0^{2\pi} \log^+ |w(r e^{i\theta})| \, d\theta. \tag{5.4.4}$$

Furthermore, when $w(z)$ ranges over the solutions of (5.4.1), we define the *grade* of the singularity of the equation as

$$g = \sup \rho(w). \tag{5.4.5}$$

Note that the grade is not the same as what is usually referred to as the *rank* of an irregular singular point: the rank is an integer (or infinity), whereas the grade may very well be a fraction. In fact, for an $n$th order linear DE the grade must be an integral multiple of $1/n$, as first observed by the Swedish mathematician Anders Wiman (1865–1959) in 1916.

Although the conditions on $P(z)$ and $Q(z)$ stated above guarantee that $z = \infty$ is an irregular-singular point, they do not exclude, for instance, a polynomial solution. Thus the equation

$$w'' - z Q(z)w' + Q(z)w = 0 \tag{5.4.6}$$

is satisfied by $w(z) = z$ for any choice of $Q(z)$.

An early attempt to estimate the grade of the irregular singularity for the case of an $n$th order linear DE with polynomial coefficients is due to

Helge von Koch (1870–1924) in 1918, followed a year later by Oscar Perron (1880–), who removed an $\epsilon$ from the estimate. If a personal note may be permitted, the author heard von Koch present his solution in a course on infinite determinants at the University of Stockholm in 1917–1918. It was infinitely complicated. We shall sketch a proof here; the general theorem for $n$th order equations will be given in Section 9.4.

**THEOREM 5.4.1**

*Suppose that*

$$P(z) = z^p \, V_1(z), \qquad Q(z) = z^q \, V_2(z) \tag{5.4.7}$$

*where we set* $p = -\infty$ *if* $P(z) \equiv 0$. *The functions* $V_1$ *and* $V_2$ *are assumed to be holomorphic for* $|z| > R$ *and to tend to finite limits* $\neq 0$ *as* $z \to \infty$. *Set*

$$g_0 = 1 + max \ (p, \tfrac{1}{2}q). \tag{5.4.8}$$

*Then the grade of* $w(z)$ *is at most* $g_0$, *and there are bounds* $B = B(w)$, *and* $K$ *independent of* $w$, *such that*

$$|w(r \, e^{i\theta})| \leqslant B(w) \ exp \ (Kr^{g_0}). \tag{5.4.9}$$

*Proof.* The idea of the proof is to reduce the second order linear DE to a linear first order vector-matrix equation, to take norms, and to cast the result in a form to which Gronwall's lemma applies Theorem 1.6.5. To carry out this program we set

$$w_1 = z^\alpha w, \qquad w_2 = z^\beta w', \tag{5.4.10}$$

where $\alpha$ and $\beta$ are real numbers to be chosen. The result of the substitution is

$$w_1' = \frac{\alpha}{z} w_1 + z^{\alpha-\beta} w_2,$$

$$w_2' = -z^{\beta-\alpha} Q(z) w_1 + \left[\frac{\beta}{z} - P(z)\right] w_2.$$

This may be written as a vector-matrix equation,

$$v'(z) = \mathcal{A}(z) \, v(z), \tag{5.4.11}$$

with obvious notation. The norm of $\mathcal{A}(z)$ is taken to be

$$\|\mathcal{A}(z)\| = \max \ [|a_{11}(z)| + |a_{12}(z)|, |a_{21}(z)| + |a_{22}(z)|]$$

and

$$\|v(z)\| = \max \ [|w_1(z)|, |w_2(z)|].$$

The problem is now to choose $\alpha$ and $\beta$ in such a manner that $\|\mathcal{A}\|$ is as small as possible for large values of $|z|$, taking into account the assumption (5.4.7) on the behavior of $P$ and $Q$ for large $|z|$. It is clear that the behavior of $a_{12}(z) = z^{\alpha-\beta}$ and $a_{21}(z) = -z^{\beta-\alpha}Q(z)$ is decisive. They become approximately equal if $\alpha - \beta = \frac{1}{2}q \geqslant -\frac{1}{2}$. Some simplification is gained by choosing $\alpha = \frac{1}{2}q$, $\beta = 0$. The final choice leads to the matrix

$$\mathcal{A}(z) = \begin{bmatrix} \frac{1}{2}q/z & z^{q/2} \\ -z^{-q/2}Q(z) & -P(z) \end{bmatrix}, \tag{5.4.12}$$

the norm of which is $O(r^\gamma)$, $\gamma = \max(p, \frac{1}{2}q)$ and $r = |z|$.

From the resulting equation we then get

$$\mathbf{v}(z) = \mathbf{v}(z_0) + \int_{z_0}^{z} \mathcal{A}(s)\,\mathbf{v}(s)\,ds. \tag{5.4.13}$$

Here $R < |z_0| < |z|$, $\arg z_0 = \arg z$, $0 \leqslant \arg z_0 < 2\pi$, and $|z_0| = r_0$. After integrating radially from $z_0$ to $z$, we can find constants $C$ and $M$ such that

$$\|\mathbf{v}(z_0)\| < C, \qquad \|\mathcal{A}(z)\| \leqslant Mr^\gamma, \qquad z = re^{i\theta}. \tag{5.4.14}$$

Then if $\|\mathbf{v}(z)\| = f(r)$ we have the inequality

$$f(r) < C + M \int_{r_0}^{r} t^\gamma f(t)\,dt, \tag{5.4.15}$$

which is of the form (1.6.13). This implies that

$$f(r) < C + C \int_{r_0}^{r} Mt^\gamma \exp\left(\int_{t}^{r} Mu^\gamma\,du\right) dt,$$

which simplifies to

$$f(r) < C \exp\left[\frac{M}{\gamma+1}(r^{\gamma+1} - r_0^{\gamma+1})\right]. \tag{5.4.16}$$

Since $\gamma + 1 = g_0$ and $\max[|w_1(z)|, |w_2(z)|] < f(r)$, it follows that

$$|w(re^{i\theta})| < B(w)r^{-q/2} \exp\left(\frac{M}{g_0}r^{g_0}\right), \tag{5.4.17}$$

where

$$B(w) = C \exp\left(-\frac{M}{g_0}r_0^{g_0}\right).$$

This is (5.4.9) in a sharper form. If the factor $r^{-q/2}$ is omitted, we obtain an estimate for $|w'(z)|$. This proves Theorem 5.4.1. ∎

This is a best possible result in various ways. Thus, if the coefficients $P$ and $Q$ are polynomials, we have always $g = g_0$ and $g_0$ is the normal order of a transcendental entire solution. Note that there may be a polynomial solution in addition to the transcendental solution. Thus consider (5.4.6), where $Q(z)$ is a polynomial of exact degree $q$. Here $g_0 = q + 2$, and using the order reducing transformation (5.3.23) with $u = z$ it is seen that the order of the transcendental solution is indeed $q + 2$.

The following observation illustrates one sense in which the result is optimal:

## THEOREM 5.4.2

*If* $g = \frac{1}{2}k$, k *a positive integer, then there exist a second order linear DE for which* $z = \infty$ *is an irregular-singular point of grade* g *and a solution* w(z) *such that*

$$\lim |z|^{-g} \log |w(z)| > 0 \qquad (5.4.18)$$

*for a radial approach to infinity, a finite number of directions excepted.*

*Proof.* It suffices to produce the example

$$w(z) = \frac{1}{2}[\exp (z^g) + \exp (-z^g)], \qquad (5.4.19)$$

which is clearly of order g and satisfies the DE

$$w'' - \frac{g-1}{z} w' - g^2 z^{2g-2} w = 0. \qquad (5.4.20)$$

Here $2g - 2 = k - 2$ is an integer and $p = -1, \frac{1}{2}q = g - 1$, so that $g = 1 + \max (p, \frac{1}{2}q)$. The exceptional directions in (5.4.18) are $\arg z = (m/k)\pi, m = 1, 2, \ldots, k$. These rays divide the plane into $k$ sectors in the interior of which either Re $(z^g)$ or Re $(-z^g)$ goes to $+\infty$. ∎

The orders of the transcendental entire functions which satisfy second order linear DE's with polynomial coefficients have been investigated by K. Pöschl (1953) and H. Wittich (1952). With our previous notation, let $p \geq q + 1$. Then all transcendental solutions are of the order $1 + p = g_0$. If $p \leq \frac{1}{2}q$, all transcendental solutions are of the order $1 + \frac{1}{2}q = g_0$. Deviation from this pattern can occur only if $\frac{1}{2}q < p \leq q$. Here $g_0 = 1 + p$, and there are always solutions of this order; under certain circumstances, however, a lower order $q - p + 1$ may also be present.

In the case last mentioned the solution of lower order admits $w = 0$ as a Picard value. If this is understood in the strict sense to mean that $w_1(z) \neq 0$ for all $z$, then $w_1(z) = \exp [B(z)]$, where $B(z)$ is a polynomial, say of degree $d$. Taking this observation as our point of departure, we can construct infinitely many linear second order DE's which have $\exp [B(z)]$ as particular solutions, while the order of the general solution is $> d$. Let $C(z)$ be any polynomial of degree $m$. Then $w_1(z)$ satisfies the equation

$$w'' - [B'(z) + C(z)]w' - [B''(z) - B'(z)C(z)]w = 0. \qquad (5.4.21)$$

A linearly independent solution is given by

$$w_2(z) = \exp [B(z)] \int_0^z \exp \left[ \int_0^t C(s) \, ds - B(t) \right] dt. \qquad (5.4.22)$$

If now $m + 1 > d$, the order of $w_2$ is $m + 1$. Here $p = m$, $q = m + d - 1 > m$, so that

$$\tfrac{1}{2}q = \tfrac{1}{2}(m + d - 1) < m = p < m + d - 1 = q.$$

Thus the stated conditions are satisfied. Furthermore, $g_0 = m + 1$ and $\rho(w) = g_0$ except for constant multiples of $w_1(z)$.

Using Theorem 5.4.1, we can now give the proof of Theorem 4.6.3, omitted in the preceding chapter.

*Proof of Theorem* 4.6.3.   Given the Riccati equation

$$w' = R_0(z) + R_1(z)w + w^2, \tag{5.4.23}$$

where $R_0$ and $R_1$ are rational functions of $z$ such that

$$R_0(z) = O(|z|^\alpha), \qquad R_1(z) = O(|z|^\beta) \quad \text{as } z \to \infty, \tag{5.4.24}$$

it is supposed that (5.4.23) has a solution $w(z)$, single valued and meromorphic for $|z| > R \geq 0$ and with infinitely many poles clustering at infinity. It is to be shown that $w(z)$ is of finite order, $\rho(w)$, which is a positive multiple of $\tfrac{1}{2}$—more precisely,

$$\rho(w) = 1 + \max(\beta, \tfrac{1}{2}\alpha). \tag{5.4.25}$$

We restrict ourselves to the case in which $R_0$ is a polynomial of degree $\alpha$ and $R_1$ a polynomial of degree $\beta$.

Set

$$w = -\frac{W'}{W},$$

so that

$$W'' + R_1(z)W' + R_0(z)W = 0 \tag{5.4.26}$$

and $-w$ is the logarithmic derivative of $W$. Now $W$ is an entire function of $z$ of finite order, the order being $g_0$, the right member of (5.4.25). We know that there is only one possibility for a lower order: if $\tfrac{1}{2}\alpha < \beta \leq \alpha$, there is a solution of (5.4.26) of order $\alpha - \beta + 1 < g_0$. We may disregard such a solution, however, since it has zero as a Picard value and its logarithmic derivative can have only a finite number of poles. Thus $\rho[W] = g_0$. We have now

$$m(r, \infty; w) = m\left(r, \infty; \frac{W'}{W}\right) = O(\log r)$$

and

$$N(r, \infty; w) = N(r, 0; W) = T(r; W) + O(\log r) = O(r^{g_0}),$$

so that

$$T(r; w) = O(r^{g_0}),$$

as asserted. ∎

We return now to Theorem 5.4.1. If $P$ or $Q$ is an entire transcendental function instead of a polynomial, we may expect the transcendental solutions of (5.4.1) to be of infinite order. This is indeed the case. Equation (5.4.6) shows that such a situation does not preclude the presence of a polynomial solution.

### THEOREM 5.4.3

*If in* (5.4.1) *either* P *or* Q *is an entire transcendental function while the other is a polynomial, then every transcendental solution of* (5.4.1) *is an entire function of infinite order. This is not necessarily true, however, if both* P *and* Q *are entire.*

*Proof.* Suppose that $Q$ is transcendental while $P$ is a polynomial. The proof is an application of the Nevanlinna theory. If the assertion were false, there is a solution $w(z)$ which is an entire function of finite order so that $T(r; w) = O(r^\rho)$ and $T(r; w) = m(r, \infty; w)$. Then

$$Q(z) = -\frac{w''}{w} - P(z)\frac{w'}{w} = -\frac{w''}{w'}\frac{w'}{w} - P(z)\frac{w'}{w}, \qquad (5.4.27)$$

and a consideration of proximity functions gives

$$m(r, \infty; Q) \leq m\left(r, \infty; \frac{w''}{w'}\right) + 2m\left(r, \infty; \frac{w'}{w}\right) + m(r, \infty; P) \quad (5.4.28)$$

with an error of at most $O(\log r)$. Here each of the terms on the right is $O(\log r)$, and this leads to a contradiction since $T(r; Q) = m(r, \infty; Q)$ goes to infinity faster than any constant multiple of $\log r$.

To get a counter example for the case in which both $P$ and $Q$ are entire transcendental functions, we use (5.4.21). If $B(z)$ is a polynomial in $z$ of degree $d$ and if $C(z)$ is entire transcendental, both $B'(z) + C(z)$ and $B''(z) - B'(z)C(z)$ are entire and $w_1(z)$ is of finite order while $w_2(z)$ is of infinite order. ∎

The following theorem due to H. Wittich (1948) may be regarded as a converse of Theorem 5.4.3.

### THEOREM 5.4.4

*In* (5.4.1) *suppose that* P *and* Q *are entire functions, and suppose that the equation has a fundamental system* $w_1(z)$, $w_2(z)$, *where* $w_1$ *and* $w_2$ *are entire functions of order* $\rho_1$ *and* $\rho_2$, *respectively. Then* P *and* Q *are polynomials.*

*Proof.* Let $W(z) = W(z; w_1, w_2)$ be the Wronskian of the system. This is an entire function of order $\rho = \max(\rho_1, \rho_2)$. But

$$P(z) = -\frac{d}{dz}[\log W(z)].$$

It follows that

$$m(r, \infty; P) = m\left(r, \infty; \frac{W'}{W}\right) = O(\log r). \tag{5.4.29}$$

Now $P$ is an entire function, so this gives $T(r; P) = O(\log r)$ and $P$ must be a polynomial. From (5.4.27) we then obtain the fact that $Q$ is also a polynomial. ∎

**THEOREM 5.4.5**

*For the DE*

$$w'' = Q(z)w, \tag{5.4.30}$$

*where $Q(z)$ is an entire function, if it is known that the equation has a solution $w(z)$ which is an entire function of finite order, then $Q(z)$ is a polynomial.*

*Proof.* We have

$$m(r, \infty; Q) = m\left(r, \infty; \frac{w''}{w}\right) = m\left(r, \infty; \frac{w''}{w'}\frac{w'}{w}\right) = O(\log r),$$

which again implies that $Q$ is a polynomial. ∎

In his investigations Wittich has made extensive use of the so-called *central index* and the corresponding *maximal term* in the Maclaurin series of an entire function,

$$E(z) = \sum_{n=0}^{\infty} a_n z^n. \tag{5.4.31}$$

Let $r = |z| > 0$ be fixed. Since the terms $|a_j|r^j$ tend to zero as $j \to \infty$, there is a largest member of the sequence $\{|a_j|r^j\}$. If there are several contestants for this distinction, the one with the highest index $j$ is designated as the maximal term, $m(r) = m(r; E)$, and its index $j$ is the *central index* $\nu(r) = \nu(r; E)$. These notions were introduced by A. Wiman (1914–1916) and amplified by Georges Valiron in 1923. Wiman showed that there is a close relation between the maximal term and the maximal modulus of an entire function. Let $M(r; E) = \max |E(r e^{i\varphi})|$; then

$$m(r; E) < M(r; E) < m(r; E)[\log m(r; E)]^{1/2+\delta}, \qquad 0 < \delta, \tag{5.4.32}$$

and

$$\nu(r; E) < [\log m(r; E)]^{1/2+\delta}, \tag{5.4.33}$$

where the first part of (5.4.32) holds for all $r$, while the second half together with (5.4.33) holds for $r$ outside an exceptional set of finite logarithmic measure, i.e., a set where the total variation of $\log r$ is finite. These relations extend also to the derivatives of $E(r)$. Thus, if $\nu_j(r; E)$ is the central index of $E^{(j)}(z)$ and $M_j(r) = M(r; E^{(j)})$, then, for $r$ outside sets of finite logarithmic measure,

$$\nu_j(r) = \nu(r)[1 + \epsilon_j(r)], \qquad 0 < \epsilon_j(r) = O[\nu(r)^{1/2+\delta}], \qquad (5.4.34)$$

$$M_j(r) = \left[\frac{\nu(r)}{r}\right]^j M(r)[1 + o(1)]. \qquad (5.4.35)$$

We shall not prove any of these statements; the interested reader is referred to Wittich (1955, pp. 9–10) and the literature therein cited.

At this juncture we call attention to another of Wiman's observations. In the case of the DE (5.4.30), where $Q(z)$ is a polynomial of degree $d$, it is found that

$$\log M(r; w) \sim \nu(r; w) \sim \int_0^r [Q(t)]^{1/2} dt \qquad (5.4.36)$$

up to terms of lower order. Suppose now that we are given an $n$th order linear DE

$$w^{(n)} + P_1(z)w^{(n-1)} + \cdots + P_n(z)w = 0, \qquad (5.4.37)$$

where the coefficients are polynomials in $z$ (or, more generally, functions holomorphic for $0 < R < |z| < \infty$ with poles at infinity). Consider the algebraic equation

$$W^n + P_1(z)W^{n-1} + \cdots + P_n(z) = 0. \qquad (5.4.38)$$

If the $P$'s are polynomials, this equation defines an algebraic function normally with $n$ determinations

$$W_1(z), W_2(z), \ldots, W_n(z) \qquad (5.4.39)$$

at infinity. According to Wiman, the corresponding functions

$$\exp\left[\int^z W_1(s)\, ds\right], \ldots, \exp\left[\int^z W_n(s)\, ds\right] \qquad (5.4.40)$$

are approximate solutions of (5.4.37). Details were worked out in an Uppsala dissertation (1924) by Mogens Matell.

## EXERCISE 5.4

1.  Verify the statements concerning (5.4.21). In particular, prove that the order of $w_2(z)$ is infinite when $C(z)$ is an entire transcendental function, assuming that the coefficients of the Maclaurin series are nonnegative. Prove that in

this case

$$w_2(r) > [C(r)]^{-1} \left\{ \exp \left[ \int_0^r C(t)\, dt \right] - 1 \right\}.$$

2. Consider the equation $w'' = z^2 w$ with $w(0) = 1$, $w'(0) = 0$. Find its Maclaurin series. Compute $\nu(r; w)$ from the series, and show that $\log M(r; w) = \frac{1}{2} r^2 + O(\log r)$.

3. Let $Q_1(t)$ and $Q_2(t)$ be positive and continuous for $0 \le t < \infty$. Consider the three DE's

$$u''(t) - Q_1(t)u(t) = 0, \qquad v''(t) - Q_2(t)v(t) = 0,$$
$$w''(t) - [Q_1(t) + Q_2(t)]w(t) = 0,$$

with $u(0) = v(0) = w(0) = 1$, $u'(0) = a$, $v'(0) = b$, $w'(0) = c$, where $a$, $b$, $c$ are positive and $c < a + b$. Show that

$$u(t)v(t) > w(t), \qquad t > 0.$$

4. Determine other solutions of (5.4.6) besides $w = z$.

5. In Theorem 5.4.4 what bounds can be obtained for the degrees of the polynomials $P$ and $Q$?

6. In (5.4.21) take $B(z) = \frac{1}{2} z^2$, $C(z) = z^3$ and determine the functions $W_1(z)$ and $W_2(z)$ of (5.4.38). Compare $\exp [\int^z W_1(t)\, dt]$ and $\exp [\int^z W_2(t)\, dt]$ with the known solutions of the DE.

7. For the equation $zw'' + \frac{1}{2} w' + \frac{1}{4} w = 0$ we have $R = 0$, $p = q = -1$, and $g_0 = \frac{1}{2}$. Show that $\cos z^{1/2}$ is a particular solution which is of order $\frac{1}{2}$. Find the general solution.

   In the following six problems the function $t \mapsto G(t)$ is positive and continuous on $[0, \infty)$. It is desired to study the solutions $w(t)$ of the DE $w''(t) = G(t)w(t)$, where $w(t)$ is real and not identically zero. A useful tool is the identity

$$[w(t)w'(t)]_a^b = \int_a^b [w'(t)]^2\, dt + \int_a^b G(t)[w(t)]^2\, dt.$$

8. Verify the identity.

9. Show that $w(t)w'(t)$ can be zero at most once in $[0, \infty)$.

10. From the point $(0, 1)$ draw the graphs of the solutions corresponding to $w(0) = 1$, $w'(0) = a$, $-\infty < a < \infty$. Show that no two graphs can have a point in common except for the starting point. Show that any solution with $a \ge 0$ becomes infinite with $t$.

11. Suppose that $a < 0$. Show that the solutions separate into three mutually exclusive classes: (i) solutions with a positive minimum, which go to infinity with $t$; (ii) a solution which goes to a finite limit $L$, $0 \le L < 1$; (iii) solutions which have a positive zero and go to $-\infty$ when $t \to \infty$. There is only one solution in class (ii), and its graph separates those of class (i) from those of class (iii).

12. Show that, if there is a solution such that $\lim w(t) = 1$ and $\lim tw'(t) = 0$, then $tG(t) \in L(0, \infty)$.

13. Show the existence of a solution such that $w'(t)$ and $[G(t)]^{1/2}w(t)$ are both $L_2(0, \infty)$.

### 5.5. ASYMPTOTICS ON THE REAL LINE

Before starting to study this section, the reader should familiarize himself with Problems 5.4:8–13. In this section we plan an inquiry into the behavior for large positive values of $t$ of the real solutions of the DE

$$w''(t) = G(t)w(t), \qquad (5.5.1)$$

where $G(t)$ is positive and continuous in $[0, \infty)$. The results of Exercise 5.4 indicate that the general solution becomes infinite with $t$ but that an exceptional solution tends to a finite limit as $t \to +\infty$. In this setting the exceptional solution is known as a *subdominant*, while the general solution, which involves two arbitrary constants, is termed the *dominant*.

Normally a subdominant tends to zero as $t \to +\infty$. The case in which $tG(t) \in L(0, \infty)$ is exceptional. Here the dominant satisfies for some $a$ the conditions

$$\lim_{t \to \infty} \frac{w_1(t)}{t} = a \neq 0, \qquad \lim_{t \to \infty} w_1'(t) = a, \qquad (5.5.2)$$

while, if suitably normalized, the subdominant satisfies

$$\lim_{t \to \infty} w_0(t) = 1, \qquad \lim_{t \to \infty} t w_0'(t) = 0. \qquad (5.5.3)$$

Conversely, the existence of such solutions implies that $tG(t) \in L(0, \infty)$. Actually this is true under the weaker condition that $G(t)$ ultimately keeps a constant sign.

We now assume that $tG(t) \notin L(0, \infty)$. Define $w_1(t)$ by $w_1(0) = 1$ and $w_1'(0) = 0$. Then $w_1$ is increasing, and (5.5.2) is replaced by

$$\lim \frac{w_1(t)}{t} = +\infty, \qquad \lim w_1'(t) = +\infty. \qquad (5.5.4)$$

We now define a subdominant solution by

$$w_0(t) = w_1(t) \int_t^\infty [w_1(s)]^{-2} \, ds. \qquad (5.5.5)$$

the integral clearly exists, since $w_1(t)$ ultimately exceeds $at$ for any positive $a$. Furthermore, $w_0(t)$ is decreasing since

$$w_0'(t) = w_1'(t) \int_t^\infty [w_1(s)]^{-2} \, ds - [w_1(t)]^{-1}$$

$$< \int_t^\infty w_1'(s)[w_1(s)]^{-2} \, ds - [w_1(t)]^{-1} = 0. \qquad (5.5.6)$$

This implies that $w_0(t)$ decreases to a nonnegative limit as $t \to +\infty$. The limit cannot be positive, for then $tG(t)$ would be $L(0, \infty)$, contrary to assumption.

The subdominant solution also has the property that $w_0'(t)$ and $[G(t)]^{1/2}w(t)$ both are $L_2(0, \infty)$. It is clear that no dominant solution can satisfy this condition. On the other hand, the identity

$$w_0(t)w_0'(t) - w_0(a)w_0'(a) = \int_a^t [w_0'(s)]^2 \, ds + \int_a^t G(s)[w_0(s)]^2 \, ds \quad (5.5.7)$$

shows that the left member tends to the limit $- w_0(a)w_0'(a)$ as $t \to +\infty$. It follows that each of the integrals must then tend to a finite limit $< - w_0(a)w_0'(a)$ as $t \to \infty$, and this proves the stated integrability properties.

We now set

$$M(t, w) = \int_0^t [w'(s)]^2 \, ds, \qquad N(t, w) = \int_0^t G(s)[w(s)]^2 \, ds, \quad (5.5.8)$$

$$L(t, w) = M(t, w) + N(t, w). \quad (5.5.9)$$

As a curiosity we note that

$$M(t, w) \equiv N(t, w) \qquad \text{iff } w(t) = w(0) \exp \left\{ \int_0^t [G(s)]^{1/2} \, ds \right\}. \quad (5.5.10)$$

For a solution of (5.5.1) this can hold iff $G(t)$ is a constant. For a general continuous and positive $G(t)$ it can hold only asymptotically. Actually, for dominant solutions and under mild restrictions on $G$ it holds in the weak sense

$$\lim_{t \to \infty} \frac{M(t, w)}{N(t, w)} = 1. \quad (5.5.11)$$

Let $G(t)$, in addition to being positive and continuous, also be strictly increasing. The classification of the solutions given in Problem 5.4:11 implies the existence of (i) solutions with a zero, (ii) solutions with an extremum (maximum or minimum), (iii) dominant solutions with $w(t)w'(t) \neq 0$ for $0 \leq t$, (iv) subdominant solutions.

We now transfer the problem to a study of the Riccati equation

$$y'(t) = G(t) - [y(t)]^2 \qquad \text{with } y(t) = \frac{w'(t)}{w(t)}. \quad (5.5.12)$$

The integral curves of this equation also exhibit tetrachotomy. If $w(t)$ belongs to class (i) and has a zero at $t = a$, then the $y$-curve of (i) has a vertical asymptote for $t = a$ and $y(t) < 0$ to the left of the asymptote and $y(t) > 0$ to the right of the asymptote. If $w(t)$ belongs to class (ii) and $w'(t_0) = 0$, then $y(t) < 0$ for $t < t_0$, positive for $t_0 < t$. These $y$-curves form class (ii). If $w(t)$ belongs to class (iii), the $y$-curves of (iii) have positive ordinates for all $t \geq 0$. Finally, there is one and only one $y$-curve of class (iv):

$$C_0: y = y_0(t), \tag{5.5.13}$$

which corresponds to the subdominant solution $w_0(t)$ of (5.5.5). This curve lies in the fourth quadrant and separates the curves of class (i) from those of class (ii). All $y$-curves below $C_0$ have vertical asymptotes; they escape from the lower half-plane by going to $-\infty$ as $t \uparrow a$ and reappear at the upper end of the asymptote in the upper half-plane, where they spend the rest of their existence. See Figure 5.1.

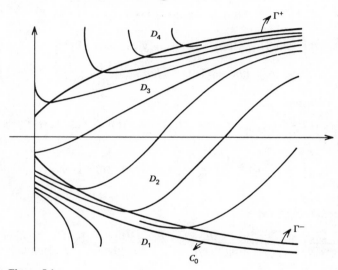

**Figure 5.1**

We shall examine the $y$-curves for large values of $t$, aiming to show that $y_0(t)$ is approximately $-[G(t)]^{1/2}$ for large values of $t$, while for all other $y$-curves $y$ is approximately $+[G(t)]^{1/2}$. In a $(t, y)$-plane we mark the curves

$$\Gamma^+: y = [G(t)]^{1/2}, \qquad \Gamma^-: y = -[G(t)]^{1/2}. \tag{5.5.14}$$

These curves, together with the positive $t$-axis, divide the plane into four regions, $D_1$–$D_4$. Figure 5.1 gives a schematic view of the regions and some of the integral curves. The figure brings out the fact, already clear from the preceding discussion, that all curves, with the sole exception of $C_0$, ultimately enter $D_3$, where $y'(t) > 0$. The curves must stay there, for $\Gamma^+$ is an *entrance boundary* where the slope is zero. Since $G(t)$ is strictly monotone, an integral curve cannot cross from $D_3$ into $D_4$. This means that ultimately

$$y(t) < [G(t)]^{1/2}. \tag{5.5.15}$$

We shall estimate the difference between the two sides

$$D(t) = [G(t)]^{1/2} - y(t). \qquad (5.5.16)$$

**THEOREM 5.5.1**

*If* G(t) *is continuous, positive, and strictly increasing, and if* y(t) *is a solution of* (5.5.12), *increasing and positive for* $0 \le c < t$, *then*

$$\int_c^t D(s)\, ds < \tfrac{1}{4} log\ G(t) - \tfrac{1}{2} log\ y(c), \qquad (5.5.17)$$

$$8 \int_c^t D(s)[log\ G(s)]^{-2}\, ds < [log\ y(c)]^{-1} \qquad if\ y(c) > 1, \qquad (5.5.18)$$

$$\lim_{t \to \infty} sup\ D(t) \le \tfrac{1}{2} \lim_{t \to \infty} sup\ \frac{G'(t)}{G(t)}. \qquad (5.5.19)$$

*Proof.* We have

$$y'(t) = G(t) - [y(t)]^2 = \{[G(t)]^{1/2} + y(t)\}\, D(t) > 2y(t)D(t)$$

so that

$$\int_c^t D(s)\, ds < \frac{1}{2} \int_c^t \frac{y'(s)}{y(s)}\, ds = \tfrac{1}{2} \log y(t) - \tfrac{1}{2} \log y(c)$$

$$< \tfrac{1}{4} \log G(t) - \tfrac{1}{2} \log y(c),$$

as asserted. Next, if $y(c) > 1$,

$$\frac{1}{\log y(c)} = \int_c^\infty \frac{y'(t)\, dt}{y(t)[\log y(t)]^2} > 2 \int_c^\infty \frac{D(t)\, dt}{[\log y(t)]^2} > 8 \int_c^\infty \frac{D(t)\, dt}{[\log G(t)]^2},$$

and this is (5.5.18).

For the proof of (5.5.19) we add the assumption that $G(t)$ has a continuous derivative everywhere; the existence of a derivative almost everywhere follows from the assumed strict increase. Under this assumption $D(t)$ is differentiable and satisfies a Riccati equation

$$D'(t) + 2[G(t)]^{1/2}D(t) - [D(t)]^2 = \tfrac{1}{2}G'(t)[G(t)]^{-1/2}. \qquad (5.5.20)$$

Now $0 < D(t) < [G(t)]^{1/2}$, since $y(t)$ satisfies this inequality. It follows that

$$[G(t)]^{1/2}D(t) - [D(t)]^2 > 0,$$

so that

$$D'(t) + [G(t)]^{1/2}D(t) < \tfrac{1}{2}G'(t)[G(t)]^{1/2}.$$

The left member of this inequality becomes an exact derivative upon multiplication by

$$E(t) = \exp \left\{ \int_c^t [G(s)]^{1/2}\, ds \right\}.$$

Integration gives

$$D(t)E(t) - D(c) < \frac{1}{2} \int_c^t \frac{G'(s)}{G(s)} \, dE(s)$$

$$\leq \frac{1}{2} \sup_{c \leq s \leq t} \frac{G'(s)}{G(s)} [E(t) - 1], \qquad (5.5.21)$$

whence

$$D(t) < D(c)[E(t)]^{-1} + \frac{1}{2} \sup_{c \leq s \leq t} \frac{G'(s)}{G(s)} \{1 - [E(t)]^{-1}\}.$$

If here $G'(s)/G(s)$ is unbounded, its supremum is $+\infty$ and (5.5.19) is trivially true. On the other hand, if the fraction is bounded, we note that $c$ can be taken as arbitrarily large, $E(c) = 1$, and $E(t) \uparrow \infty$ with $t$. Now the supremum is arbitrarily close to the lim sup, and again (5.4.19) follows and Theorem 5.5.1 is proved. ■

Various comments and corollaries are in order.

### COROLLARY 1

*If* G *is increasing and has a continuous derivative, if*

$$\lim_{t \to \infty} G'(t)[G(t)]^{-3/2} = 0, \qquad (5.5.22)$$

*and if* w(t) *is a dominant solution of* (5.5.1), *then*

$$\lim_{t \to \infty} [G(t)]^{-1/2} \frac{w'(t)}{w(t)} = 1. \qquad (5.5.23)$$

*Proof.* If $y(t)$ is the corresponding solution of (5.5.12), the problem is to prove that

$$y(t)[G(t)]^{-1/2} = 1 - D(t)[G(t)]^{-1/2}$$

tends to one as $t$ becomes infinite, and this is equivalent to showing that $D(t)[G(t)]^{-1/2}$ tends to 0. Now from (5.5.21) one obtains that

$$D(t)[G(t)]^{-1/2}E(t) < D(c)[G(t)]^{-1/2} + \frac{1}{2}[G(t)]^{-1/2} \int_c^t \frac{G'(s)}{G(s)} \, dE(s)$$

$$< D(c)[G(t)]^{-1/2} + \frac{1}{2} \int_c^t \frac{G'(s)}{[G(s)]^{3/2}} \, dE(s), \qquad (5.5.24)$$

since $G(s)$ is strictly increasing. Then (5.5.22) implies that the integral is $o[E(t)]$ and thus proves (5.5.23). ■

### COROLLARY 2

*Under the assumptions of Corollary* 1 (5.5.11) *holds.*

*Proof.* Apply the rule of L'Hospital to the quotient $M/N$. The quotient of the derivatives is

$$\frac{[w'(t)]^2}{G(t)[w(t)]^2},$$

and this tends to one by (5.5.23).  ∎

### COROLLARY 3

*Under the original assumptions on* G(t) *and for all dominant solutions with* $w(c) > 0$, $w'(c) > 0$

$$w(t) > [w(c) \, w'(c)]^{1/2}[G(t)]^{-1/4} \exp \int_c^t [G(s)]^{1/2} \, ds. \qquad (5.5.25)$$

*Proof.*  This is a consequence of (5.5.17), for we have

$$\frac{w'(t)}{w(t)} = [G(t)]^{1/2} - D(t),$$

whence

$$\log \frac{w(t)}{w(c)} = \int_c^t [G(s)]^{1/2} \, ds - \int_c^t D(s) \, ds$$

$$> \int_c^t [G(s)]^{1/2} \, ds - \tfrac{1}{4} \log G(t) + \tfrac{1}{2} \log y(c),$$

and this implies (5.5.25).  ∎

An asymptotic relation of the form

$$w(t) = [1 + o(1)][G(t)]^{-1/4} \exp \left\{ \int_0^t [G(s)]^{1/2} \, ds \right\} \qquad (5.5.26)$$

holds if, for instance, $G$ has continuous first and second order derivatives and

$$[G(t)]^{-1/4} \frac{d^2}{dt^2} \{[G(t)]^{-1/4}\} \in L(0, \infty). \qquad (5.5.27)$$

This follows from an application of the Liouville transformation. See the next section.

The subdominant is harder to pin down, but we have the following results.

### THEOREM 5.5.2

*If* G(t) *is continuous, positive, and strictly increasing, then the subdominant solution of* (5.5.12) *has these properties*:

1.  $y_0(t) < -[G(t)]^{1/2}$ *for all* t.
2.  $1 + [G(t)]^{1/2}[y_0(t)]^{-1} \in L(0, \infty)$.

3. $\lim_{t\to\infty} \sup y_0(t)[G(t)]^{-1/2} = -1$.

4. *The measure of the set where* $y_0(t)[G(t)]^{-1/2} < -1 - \epsilon$ *is finite for every* $\epsilon > 0$.

*Proof.* Assertion 1 is that the graph $C_0$ of $y = y_0(t)$ lies in the region $D_1$. It is certainly located in the lower half-plane for

$$y_0(t) = \frac{w_0'(t)}{w_0(t)},$$

where $w_0(t)$ is defined by (5.5.5) and has a negative derivative. In $D_2$ the integral curves have positive slopes and are increasing so that such a curve ultimately crosses the positive $t$-axis and gets into the upper half-plane. All this is out of the question for $C_0$. Hence $y = y_0(t)$ stays in $D_1$, and point 1 is verified.

Set $y_0(t) = -z_0(t)$, and note that

$$z_0'(t) = [z_0(t)]^2 - G(t),$$

whence

$$\int_0^t \frac{z_0'(s)}{[z_0(s)]^2} \, ds = \int_0^t \left\{ 1 - \frac{G(s)}{[z_0(s)]^2} \right\} ds.$$

Both integrands are positive, and as $t$ tends to infinity, the left member tends to the finite limit $[z_0(0)]^{-1}$. Thus the right member must also tend to a finite limit, as asserted in point 2. Now a bounded continuous positive function in $L(0, \infty)$ can exceed a given $\epsilon > 0$ only on a set of finite measure. The factor $1 + [G(t)]^{1/2}|y_0(t)|^{-1}$ lies between 1 and 2, so it must be the other factor, $1 - [G(t)]^{1/2}|y_0(t)|^{-1}$, that is small on the average. This is expressed by points 3 and 4. ∎

**COROLLARY**

*We have*

$$\lim_{t\to\infty} \sup \frac{w_0'(t)}{[G(t)]^{1/2} w_0(t)} = -1, \tag{5.5.28}$$

*and the set of values* t *for which the quotient is less than* $-1 - \epsilon$ *is of finite measure.*

**EXERCISE 5.5**

1. Suppose that $w_0(t) > 0$ is a subdominant solution of (5.5.1), and form

$$w(t) = w_0(t) \int_0^t [w_0(s)]^{-2} \, ds.$$

Show that $w'(t) > 0$ and that $w(t)$ is a dominant solution.

2.  If $t \mapsto f(t)$ is positive and continuous together with its first and second order derivatives, find the equation of type (5.5.1) which is satisfied by $\exp[f(t)]$.

3.  Suppose that $G$ and $H$ are continuous increasing functions and $0 < G(t) < H(t)$. Consider the systems

    $$w'' = G(t)w, \qquad W'' = H(t)W, \quad \text{with } 0 \leqslant w(0) = W(0),\ 0 \leqslant w'(0) < W'(0).$$

    Show that $w(t) < W(t)$ for $0 < t$.

4.  The function $t \mapsto [G(t)]^{-1/4} \exp\{\int_0^t [G(s)]^{1/2}\, ds\}$ satisfies the DE $w'' = Q(t)w$. Find $Q(t)$ if $G(t)$ is positive and continuous, together with its first and second order derivatives.

5.  If $5[G'(t)]^2 > 4G(t)G''(t)$, show that (5.5.1) has a dominant solution satisfying

    $$w(t) < [G(t)]^{-1/4} \exp\left\{\int_0^t [G(s)]^{1/2}\, ds\right\}.$$

    [*Hint:* Use Problems 3 and 4. For the opposite inequality see (5.5.25).]

6.  If $Q$ is defined as in Problem 4 and if $G(t) = t^k$, $0 \leqslant k$ an integer, find $Q(t) - G(t)$. Show that the condition of Problem 5 is satisfied, and give the corresponding estimates for dominant solutions.

7.  Let $t \mapsto F(t)$ be positive and continuous. The equation $w''(t) + F(t)w(t) = 0$ is said to be *oscillatory* in $(a, b)$ if every real solution has at least one zero in $(a, b)$. If $c > 0$, is every solution of $w'' + ct^{-2}w = 0$ oscillatory in $(0, \infty)$?

8.  If $F(t) > M > 0$ for all $t > 0$, prove that any interval of length $\pi M^{-1/2}$ contains at least one zero of any given real solution.

9.  Why is it necessary to specify "real" in these cases?

10.  Given the equation $y' = G(t) - y^3$, $0 \leqslant t$, let $\Gamma$ be the curve $\Gamma: y = [G(t)]^{1/3}$. Here $G$ is supposed to be positive, continuous, strictly increasing, and unbounded. Discuss the integral curves in a way similar to that used for (5.5.12).

11.  If a solution exists for $t > a > 0$ but becomes infinite as $t$ decreases to $a$, show that $y(t)(t - a)^{1/2}$ tends to a limit, and find this limit.

12.  Show that each solution is ultimately less than $[G(t)]^{1/3}$. Set $y(t) = [G(t)]^{1/3} - D(t)$, and show that $D(t) \in L(a, \infty)$ for large $a$.

13.  If $\lim_{t \to \infty} G'(t)[G(t)]^{-5/3} = 0$, show that

    $$\lim_{t \to \infty} y(t)[G(t)]^{-1/3} = 1.$$

## 5.6.  ASYMPTOTICS IN THE PLANE

We now return to the case of analytic coefficients and will be concerned with a linear second order DE with the point at infinity as an irregular-singular point normally of finite grade. The method to be used is a reduction of the equation to a quasi-Bessel form with the aid of the *transformation of Liouville* (Joseph Liouville, 1809–1882), dating from 1837. This has a strong smoothing effect and leads to a transformed equation which is a perturbed sine equation. It applies to formally *self-adjoint* equations:

$$[K(z)w'(z)]' + G(z)w(z) = 0. \tag{5.6.1}$$

We can use it for the equation

$$y'' + P_1(z)y' + Q_1(z)y = 0, \tag{5.6.2}$$

where the coefficients are polynomials in $z$ or, more generally, functions holomorphic for $|z| > R$ except for poles at infinity. Condition (5.4.2) is supposed to be satisfied; it is the necessary condition on the coefficients in order for $z = \infty$ to be irregular singular. It is convenient, however, to reduce the equation to a form where the first derivative is missing. This is done by setting

$$y(z) = w(z) \exp\left[-\frac{1}{2}\int^z P_1(s)\,ds\right], \tag{5.6.3}$$

which leads to

$$w''(z) + Q(z)w(z) = 0, \tag{5.6.4}$$

where

$$Q(z) = Q_1(z) - \tfrac{1}{4}[P_1(z)]^2 - \tfrac{1}{2}P_1'(z). \tag{5.6.5}$$

Note that, if $P_1$ and $Q_1$ are polynomials, so is $Q$. Moreover, the grade of the singularity of (5.6.4) is the same as that of (5.6.2), namely,

$$g = 1 + \max(p_1, \tfrac{1}{2}q_1) \tag{5.6.6}$$

with obvious notation. Note that the multiplier of $w(z)$ in (5.6.3) is of the order $p_1 + 1 \leq g$.

New variables are introduced in (5.6.4) by setting

$$Z = \int^z [Q(s)]^{1/2}\,ds, \qquad W = [Q(z)]^{1/4}w. \tag{5.6.7}$$

The result is of the form

$$W'' + [1 - F(Z)]W = 0, \tag{5.6.8}$$

where differentiation is with respect to $Z$. Various expressions are available for $F(Z)$. Setting

$$[G(z)]^{1/4} = H(Z), \tag{5.6.9}$$

we get

$$F(Z) = \frac{H''(Z)}{H(Z)}. \tag{5.6.10}$$

In terms of the original variable $z$ we have

$$F(Z) = \frac{1}{4} \frac{Q''(z)}{[Q(z)]^2} - \frac{5}{16} \frac{[Q'(z)]^2}{[Q(z)]^3}, \qquad (5.6.11)$$

where now differentiation is with respect to $z$. An alternative expression is

$$F(Z) = -[Q(z)]^{-1/4} \frac{d^2}{dz^2} \{[Q(z)]^{-1/4}\}. \qquad (5.6.12)$$

Compare (5.5.27).

As an illustration consider the case in which $Q(z)$ is a polynomial of degree $m$, say,

$$Q(z) = a_0 z^m + \cdots. \qquad (5.6.13)$$

Then for $|z| > R$

$$Z(z) = (a_0)^{1/2} \frac{2 z^{(m/2)+1}}{m+2} [1 + o(1)]. \qquad (5.6.14)$$

Here $R$ is the absolute value of the zero of $Q(z)$ furthest away from the origin. The remainder in (5.6.14) is an ordinary power series in $1/z$ without a constant term if $m$ is odd, while if $m$ is even, $m = 2k$, the $(k+1)$th term can be a constant times $z^{-k-1} \log z$, the other terms being negative powers of $z$. For simplicity it will be assumed here and in the following discussion that no logarithms occur. Then $Z(z)$ will have a pole at $z = \infty$ of order $k+1$ if $m = 2k$, and a branch point if $m$ is odd. Furthermore,

$$F(Z) = -\frac{m(m+4)}{16a_0} z^{-m-2} [1 + O(z^{-1})]. \qquad (5.6.15)$$

Now, if there is no logarithmic term in (5.6.14), we can invert the series and expand $z$ in fractional powers of $Z$. Set

$$u = [\tfrac{1}{2}(m+2) a_0^{-1/2} Z]^{2/(m+2)}. \qquad (5.6.16)$$

Then

$$z = u[1 + O(u^{-1})], \qquad (5.6.17)$$

and, finally,

$$F(Z) = -\frac{m(m+4)}{4(m+2)^2} Z^{-2} [1 + O(Z^{-2/(m+2)})]. \qquad (5.6.18)$$

The important thing here is that $Z^2 F(Z)$ is bounded at infinity and that $Z^{-2}$ is integrable out to infinity. This motivates a study of the solutions of the DE (5.6.8) under suitable assumptions on $F(Z)$, which are patterned on the particular case just considered. Although not the most general that could be handled, they are sufficient for the cases of interest here.

Hypothesis F.  It is assumed that $z \mapsto F(z)$ is an analytic function such that:

1.  $F(z)$ is holomorphic in a domain $D$ extending to infinity and located on the Riemann surface of $\log z$. $D$ contains a sector $S$: $-2\pi < \arg z < 2\pi$, $R < |z|$. For $\theta$ such that $-2\pi \leq \theta \leq 2\pi$ we denote by $D_\theta$ that part of $S$ where, if $z \in S$, the ray $z + r e^{i\theta}$ is in $S$.

2.  $\lim_{r \to \infty} r F(r e^{i\theta}) = 0$.

3.  For each $z$ in $D_\theta$ the integral

$$\int_0^\infty |F(z + r e^{i\theta})| \, dr \qquad (5.6.19)$$

exists, and the set of all such integrals is bounded for all admissible values of $z$ and $\theta$, the least upper bound being $M$.

This hypothesis holds if, e.g., $F(z)$ is a rational function of $z$ having its poles inside the circle $|z| = R$ and having a zero at infinity of at least the second order. It is also satisfied by the function $Z \mapsto F(Z)$ obtained in (5.6.18) from (5.6.13) by the Liouville transformation.

The investigation proceeds by a number of steps.

**THEOREM 5.6.1**

*Let* w(z) *be a solution not identically zero of the DE*

$$w''(z) + [1 - F(z)]w(z) = 0, \qquad (5.6.20)$$

*where F satisfies Hypothesis F and* w(z) *is defined in* $D_0$. *Then there exists a solution* $w_0(z)$ *of the sine equation*

$$w''(z) + w(z) = 0 \qquad (5.6.21)$$

*such that for all* z = x + iy *in* $D_0$

$$|w(x + iy) - w_0(x + iy)| \leq M(y)\left\{ exp\left[ \int_x^\infty |F(s + iy)| \, ds \right] - 1 \right\}, \qquad (5.6.22)$$

*where* $M(y) = max_s |w_0(s + iy)|$.

*Proof.* Choose a point $x_0 > R$ on the real axis in $D_0$. Let $z$ be any point in $D_0$. There is then a $\theta$, $-\frac{1}{2}\pi < \theta < \frac{1}{2}\pi$, such that $x_0$ and $z$ both belong to $D_\theta$. We can then join $x_0$ and $z$ by a path $\Gamma$ composed of, at most, two straight line segments, one parallel to the real axis, the other parallel to the ray $\arg z = \theta$. See Figure 5.2. No matter how $\theta$ is chosen subject to the stated conditions, we have

$$\int_\Gamma |F(s)| \, |ds| \leq 2M,$$

and by condition F(3) this holds uniformly for $z$ in $D_0$.

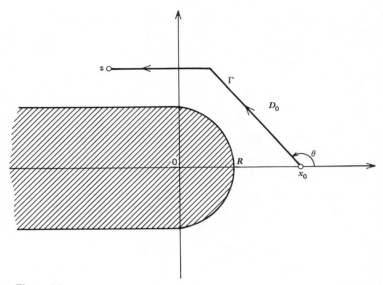

**Figure 5.2**

Next, there is a unique solution $w_1(z)$ of (5.6.21) such that

$$w_1(x_0) = w(x_0), \qquad w_1'(x_0) = w'(x_0). \tag{5.6.23}$$

We have then, if $\Gamma$ is oriented from $x_0$ to $z$,

$$w(z) = w_1(z) + \int_\Gamma \sin(z - t)F(t)w(t)\,dt. \tag{5.6.24}$$

This is verified by twofold differentiation with respect to $z$ combined with the initial conditions (5.4.23). Let $D_0(c)$ be that part of $D_0$ whose distance from the real axis is at most $c > 0$. Suppose now that $|w_1(z)| \leq K$, $|\sin(t - z)| \leq K$ for $t$ and $z$ in $D_0(c)$. Then

$$|w(z)| \leq |w_1(z)| + \int_\Gamma |\sin(t - z)|\,|F(t)|\,|w(t)|\,|dt|$$

or

$$|w(z)| \leq K + K \int_\Gamma |F(t)|\,|w(t)|\,|dt|.$$

If now $t$ is expressed in terms of arc length on $\Gamma$, it is seen that the inequality is one to which Gronwall's lemma applies (Theorem 1.6.5), so that

$$|w(z)| \leq K \exp \left[ K \int_\Gamma |F(t)| \, |dt| \right] \leq K \exp (2KM). \qquad (5.6.25)$$

Thus $|w(z)|$ is uniformly bounded in $D_0(c)$.

Let us now form

$$w(z) - \int_z^\infty \sin (t - z) F(t) w(t) \, dt \equiv w_0(z). \qquad (5.6.26)$$

Here the integral is taken along the line $t = z + r$, $0 \leq r < \infty$. The integral exists by F(3) plus (5.6.25), which is valid as long as $z$ is confined to $D_0(c)$. It follows that $w_0(z)$ is defined and holomorphic in $D_0$ and bounded in any $D_0(c)$. Twofold differentiation shows that $w_0(z)$ is a particular solution of (5.5.21). Thus

$$w(z) = w_0(z) + \int_z^\infty \sin (t - z) F(t) w(t) \, dt. \qquad (5.6.27)$$

This is a singular Volterra integral equation which has a unique solution, namely, the solution of (5.6.20), from which we started. Here we have

$$|w(x + iy) - w_0(x + iy)| \leq \int_x^\infty |F(s + iy)| \, |w(s + iy)| \, dy,$$

and this implies (5.6.22) by Gronwall's lemma.

The Volterra equation may be solved by the method of successive approximations, and the estimate

$$|w_n(z) - w_{n-1}(z)| \leq M(y) \frac{1}{n!} \left[ \int_z^\infty |F(t)| \, dt \right]^n \qquad (5.6.28)$$

shows that the approximations converge rapidly to $w(z)$. ∎

In this manner we get a correspondence between the solutions of (5.6.20), on the one hand, and those of (5.6.21), on the other. It is a one-to-one correspondence which is type preserving. Now (5.6.21) has solutions of three different types:

1. Solutions which go exponentially to zero in the upper half-plane, type $e^{iz}$.
2. Solutions which go exponentially to zero in the lower half-plane, type $e^{-iz}$.
3. Oscillatory solutions, type $\sin (z - z_0)$.

Once a particular $w_0(z)$ has been chosen, it is to be expected that the corresponding solution $w(z)$ will inherit the asymptotic properties of $w_0(z)$, and this is indeed the case.

**THEOREM 5.6.2**

*Equation (5.6.20) has a unique solution* $E^+(z)$ *asymptotic to* $e^{iz}$ *in the sector* $-\pi < \arg z < 2\pi$ *in the sense that*

$$|E^+(z)e^{-iz} - 1| \le exp\left[\int_z^\infty |F(t)|\,|dt|\right] - 1,\qquad (5.6.29)$$

*where the path of integration is* $\arg(t-z)$, *equal to*

$$0 \text{ for } z \in D_0,\qquad \tfrac{1}{2}\pi \text{ for } z \in D_{\pi/2},\qquad \text{and}\qquad \pi \text{ for } z \in D_\pi.$$

We recall that, e.g., $D_{\pi/2}$ is that part of $S$ for which $z \in S$ implies that the points $z + ir \in S$ for any positive $r$.

*Proof.*    Set

$$E^+(z)e^{-iz} \equiv U(z).\qquad (5.5.30)$$

Here $U(z)$ satisfies the integral equation

$$U(z) = 1 + \frac{1}{2i}\int_z^\infty [e^{2i(t-z)} - 1]F(t)U(t)\,dt,\qquad (5.6.31)$$

where the path of integration is $\arg(t - z) = 0$. Since the kernel satisfies

$$\left|\frac{1}{2i}[e^{2i(t-z)} - 1]\right| \le 1,$$

Gronwall's lemma applies and gives

$$|U(z) - 1| \le exp\left[\int_0^\infty |F(z + r)|\,dr\right] - 1\qquad (5.6.32)$$

for all $z$ in $D_0$. This is the first inequality listed above. It shows that $U(z)$ is bounded in $D_0$ since the integral cannot exceed $M$, by F(3).

For the next step it is necessary to swing the path of integration through an angle of 90°. This is permitted for $z \in D_0 \cap D_{\pi/2}$. The integral taken along a quarter-circle with center at $z$ and radius $\rho$ plus two radii, one parallel to the real and the other to the imaginary axis, is obviously zero. The integrand is bounded uniformly with respect to $\rho$ and is $o(\rho^{-1})$ as $\rho \to \infty$, by F(2). It follows that the integral along the curvilinear part tends to zero as $\rho \to \infty$ and

$$\int_z^{z+\infty} = \int_z^{z+i\infty}.$$

The resulting integral equation is valid for all $z$ in $D_{\pi/2}$. Since now the kernel is $< \frac{1}{2}$ in absolute value along the new path of integration, we have the preliminary inequality

$$|U(z) - 1| \leqslant \frac{1}{2} \int_z^{z+i\infty} |F(t)| \, |U(t)| \, |dt|,$$

and by Gronwall's lemma

$$|U(z) - 1| \leqslant \exp\left[\frac{1}{2} \int_z^{z+i\infty} |F(t)| \, |dt|\right] - 1. \tag{5.6.33}$$

This is actually a sharper inequality than what is stated in the theorem for this case.

Again we can turn the path of integration through an angle of 90°, provided that $z \in D_{\pi/2} \cap D_\pi$. The discussion follows the same lines as above, and for any $z \in D_\pi$ we have a new integral equation,

$$U(z) = 1 + \frac{1}{2i} \int_z^{z-\infty} [e^{2i(t-z)} - 1] F(t) U(t) \, dt, \tag{5.6.34}$$

valid for all $z$ in $D_\pi$, and the resulting estimate

$$|U(z) - 1| \leqslant \exp\left[\int_z^{z-\infty} |F(t)| \, |dt|\right] - 1. \tag{5.6.35}$$

This is the last assertion.

No further turns of the path of integration are permitted, since the kernel of (5.6.31) is not bounded in the lower half-plane. This completes the proof. ■

The theorem asserts that

$$|U(z) - 1| \leqslant e^M - 1 \tag{5.6.36}$$

for $z$ in $D_0 \cup D_{\pi/2} \cup D_\pi$, and $M$ is the bound postulated by F(3). This estimate is rather poor if $M$ is large, but it may be sharpened by trimming the domain along the edges. This can obviously be done in such a manner that $M$ can be replaced by a smaller number. Set

$$I(z, \theta) = \int_0^\infty |F(z + r e^{i\theta})| \, dr, \qquad 0 \leqslant \theta \leqslant \pi.$$

By suitable trimming we can obtain the fact that in the resulting region $D^+$

$$\min I(z, \theta) < \log 2.$$

If this has been done, then for $z \in D^+$

$$|U(z) - 1| < 1 \qquad \text{and} \qquad U(z) \neq 0. \tag{5.6.37}$$

The same method can be used to prove that

$$|E^-(z) \, e^{iz} - 1| \leqslant \exp\left[\int_z^\infty |F(t)| \, |dt|\right] - 1 \tag{5.6.38}$$

in the sector $-2\pi < \arg z < \pi$ for suitable choice of the path of integration. Again, by trimming the sector along the edges, we may obtain a domain $D^-$ in which

$$|E^-(z)\, e^{iz} - 1| < 1 \qquad \text{and} \qquad E^-(z) \neq 0.$$

This result will be referred to as **THEOREM 5.6.3**. The proof is left to the reader.

The case in which $w_0(z)$ is oscillatory requires several steps.

**THEOREM 5.6.4**

*Equation* (5.6.20) *has a solution* $w_1(z)$ *which is asymptotic to* $\sin (z - z_0)$, $z_0 = x_0 + iy_0$, *in* $D_0$ *so that*

$$[\cosh (y - y_0)]^{-1} |w_1(z) - \sin (z - z_0)| \leq \exp \left[ \int_0^\infty |F(z + s)|\, ds \right] - 1. \quad (5.6.39)$$

*There is also a solution* $w_2(z)$, *which is asymptotic to* $\sin (z - z_0)$ *in* $D_\pi$ *in the sense that*

$$[\cosh (y - y_0)]^{-1} |w_2(z) - \sin (z - z_0)| \leq \exp \left[ \int_0^\infty |F(z - s)|\, ds \right] - 1. \quad (5.6.40)$$

Normally $w_1(z) \neq w_2(z)$. Both solutions are oscillatory, $w_1$ in the right half-plane, and $w_2$ in the left. Furthermore, $w_1(z)$ may be continued analytically into $D_\pi$ as well as into $D_{-\pi}$, and the continuations may be expected to be oscillatory in both regions. In particular, $w_1(z)$ is oscillatory in $D_\pi$ if $y_0$ is sufficiently large positive, and in $D_{-\pi}$ if $y_0$ is sufficiently large negative.

*Remark.* Here we say that a solution is *oscillatory* in a region which extends to infinity if it has infinitely many zeros in the region. The next theorem will provide estimates of the number of zeros.

For the proof we need the theorem of Rouché (Eugène Rouché, 1832–1910), which may be stated as follows:

**LEMMA 5.6.1**

*Let the rectifiable simple closed curve* C *bound a simply connected domain* D. *Let* f, g, h *be functions holomorphic in* $D \cup C$ *such that* (i) $h = f + g$, *and* (ii) $|f(z)| > |g(z)|$ *everywhere on* C. *Then* f(z) *and* h(z) *have the same number of zeros in* D.

The proof is left to the reader.

*Proof of Theorem 5.6.4.* Formula (5.6.39) is read off directly from

(5.6.22) since in the present case $M(y) = \cosh(y - y_0)$. The existence and properties of $w_2(z)$ follow from the analogue of Theorem 5.6.1 for the region $D_\pi$. A proof is obtained by replacing $z$ by $-z$, which preserves the form of the equation.

The analytic continuation of $w_1(z)$ from $D_0$ to the adjacent regions $D_\pi$ and $D_{-\pi}$ clearly exists. Let us consider $D_\pi$. By the analogue of Theorem 5.6.1 there exists a solution $w_0(z)$ of (5.6.21) to which the continuation is asymptotic in $D_\pi$. Here $w_0(z)$ may be expected to be oscillatory in $D_\pi$. Now $w_0(z)$ cannot be a constant multiple of $e^{iz}$, for this would make $w_1(z)$ bounded in the upper half-plane, contradicting (5.6.39). We can exclude the possibility that $w_0(z)$ is a constant multiple of $e^{-iz}$ if $w_1(z)$ is oscillatory in the intersection of $D_0$, $D_\pi$, and $D^+$, and this would be the case if $y_0$ is large positive. Similar considerations hold for the continuation of $w_1(z)$ into $D_{-\pi}$ and for the continuations of $w_2(z)$. ∎

We shall now discuss the zeros of

$$w_1(z) = \sin(z - z_0) + R(z) \tag{5.6.41}$$

in the right half-plane, where

$$|R(z)| \leqslant \cosh(y - y_0) \left\{ \exp\left[ \int_0^\infty |F(z + s)|\, ds \right] - 1 \right\}$$

for $z \in D_0$. Let $\delta$ be given, $0 < \delta < \tfrac{1}{2}\pi$, and let $H_\delta$ be a half-plane $\operatorname{Re}(z) > \delta_0$, such that $z \in D_0$ and

$$\exp\left[ \int_0^\infty |F(z + s)|\, ds \right] < 1 + \frac{\sin \delta}{\cosh \delta} \tag{5.6.42}$$

is satisfied. We assume that $z_0$ and $H_\delta$ are such that $z_0 - \delta \in H_\delta$, while $z_0 - \pi + \delta$ does not. We now introduce a set of square boxes $Q_{n,\delta}$, where

$$|x - x_0 - n\pi| < \delta, \qquad |y - y_0| < \delta, \qquad z_0 = x_0 + iy_0, \tag{5.6.43}$$

and $n = 0, 1, 2, \ldots$.

We aim to prove that each box contains one and only one zero of $w_1(z)$ and that there are no other zeros in the half-plane $H_\delta$. On the boundary of $Q_{n,\delta}$ we have $|\sin(z - z_0)| \geqslant \sin \delta$, the lower bound being reached for $z = z_0 + n\pi \pm \delta$. On the other hand, the estimate for $|R(z)|$ on the boundary reaches its maximum at the four vertices of the square, and there $|R(z)| < \sin \delta$. This shows that the conditions of Lemma 5.6.1 are satisfied by

$$f(z) = \sin(z - z_0), \qquad g(z) = R(z)$$

on the perimeter of $Q_{n,\delta}$. Now $\sin(z - z_0)$ has a simple zero in the square, so $w_1(z)$ has one and only one zero in each square.

Next, it is required to show that there are no zeros in $H_\delta$ outside the squares. We have

$$|\sin(z - z_0)|^2 = \sin^2(x - x_0) + \sinh^2(y - y_0).$$

Thus, if $|y - y_0| > \delta$, we get $|\sin(z - z_0)| > \sinh|y - y_0|$, while

$$|R(z)| < \cosh(y - y_0)\frac{\sin \delta}{\cosh \delta} < \cosh(y - y_0)\tanh \delta.$$

Since $\sinh|y - y_0| > \cosh(y - y_0)\tanh \delta$ for $|y - y_0| > \delta$, it is seen that $|\sin(z - z_0)| > |R(z)|$, so that $w_1(z)$ cannot be zero outside the strip $|y - y_0| < \delta$ for $z$ in $H_\delta$. The rectangles between the squares remain for examination. But here we have $|\sin(z - z_0)| > \sin \delta$ and $|R(z)| < \sin \delta$ in any one of the rectangles. Thus we have proved

**THEOREM 5.6.5**

*The solution* $w_1(z)$ *is oscillatory in the half-plane* $H_\delta$ *defined above. It has one and only one zero in each of the squares* $Q_{n,\delta}$, $n = 0, 1, 2, \ldots$, *and no zeros in* $H_\delta$ *outside the squares.*

Let us now return to our starting point, the DE (5.6.4), to which we applied the Liouville transformation and obtained (5.6.8), where $F(Z)$ is given by any one of the three expressions (5.6.10)–(5.6.12). In the case where $Q(z)$ is a polynomial of degree $m$ as in (5.6.13), we have to scrutinize the conformal mapping defined by the Abelian integral

$$Z(z) = \int^z [Q(s)]^{1/2}\, ds = (a_0)^{1/2}\frac{2z^{(m/2)+1}}{m + 2}[1 + o(1)]. \qquad (5.6.44)$$

If $m$ is odd, the remainder is an ordinary power series in $1/z$; if $m$ is even, $m = 2k$, there may be a term of the form $z^{-k-1}\log z$. The function $z \mapsto Z(z)$ maps partial neighborhoods of $z = \infty$ onto partial neighborhoods of $Z = \infty$. To apply the preceding theory we must determine what regions in the $z$-plane correspond to approximate half-planes in the $Z$-plane.

To illustrate let us consider the DE

$$w''(z) + (a^4 - z^4)w(z) = 0, \qquad 0 < a. \qquad (5.6.45)$$

A singular boundary value problem was considered by Edward Charles Titchmarsh (1899–1963) in 1946 for this equation. It involved subdominant solutions, and Titchmarsh conjectured that such a solution had only real (finite in number) and purely imaginary (infinitely many) zeros. This can be proved as follows (Hille, 1966). Here for $|z| > a$ and up to an additive constant,

$$Z(z) = \frac{i}{3} z^3 \left[ 1 + \sum_{n=1}^{\infty} b_n \left( \frac{a}{z} \right)^{4n} \right], \tag{5.6.46}$$

whence

$$z = (-3iZ)^{1/3} \left[ 1 + \sum_{n=1}^{\infty} c_n a^{4n} (-3iZ)^{-4n/3} \right]. \tag{5.6.47}$$

There are six regions $D_1$–$D_6$ in the $z$-plane which correspond to approximate upper or lower half-planes. These are sketched in Figure 5.3, and the

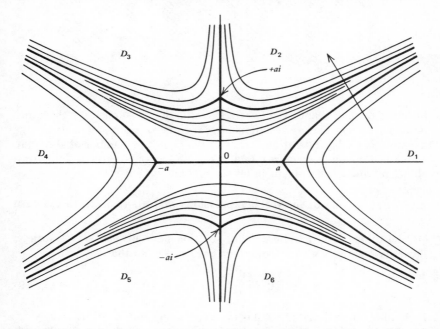

**Figure 5.3.** At the four points $a, -a, ai, -ai$ the curve tangents make angles of 120° with each other. The six curve bundles have asymptotes that make 30°, 90°, 150°, 210°, 270°, and 330° angles with the horizontal.

"half-planes" corresponding to $D_1$ and $D_2$ are indicated in Figure 5.4. These two "half-planes" are connected along the real $Z$-axis from $\frac{1}{4} a^3 B(\frac{3}{2}, \frac{1}{4})$ to $+\infty$, where $B(p, q)$ is the Euler beta function. There is a subdominant solution $E_k(z)$ for each $D_k$, and the solution $E_1(z)$ is

$$E_1(z) = z^{-1} \exp\left(-\tfrac{1}{3} z^3\right)[1 + o(1)] \tag{5.6.48}$$

in $D_1$. By Theorem 5.6.2 this solution has the same asymptotic form in the three domains $D_6$, $D_1$, and $D_2$ and has only a finite number of zeros, if any,

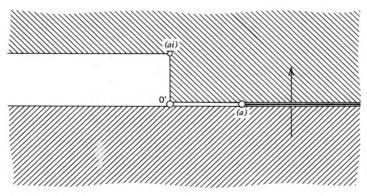

**Figure 5.4**

in any right half-plane, $x > b > 0$. The situation is similar for the other domains, $D_k$. Oscillatory solutions have their zeros normally in six strings asymptotic to the separation curves of the $D_k$'s. The number of zeros (all simple) of absolute value $\leq r$ in a string is $(3\pi)^{-1}r^3[1 + o(1)]$. Thus for a solution which is dominant in all six regions we have

$$N(r, 0; w) = \frac{2}{3\pi} r^3[1 + o(1)]. \tag{5.6.49}$$

On the other hand, a subdominant solution normally lacks two strings, those associated with the boundary lines of $D_k$, so that

$$N(r, 0; E_k) = \frac{4}{9\pi} r^3[1 + o(1)]. \tag{5.6.50}$$

It should be noted that (5.6.45) is invariant under the substitution $z \to -z$. The Titchmarsh functions $T_m(z)$ are the solutions of the singular boundary value problem

$$T_m(x) \in L_2(-\infty, \infty), \qquad m = 1, 2, 3, \ldots \tag{5.6.51}$$

and correspond to a sequence of spectral values $a = a_m$, where $0 < a_m < a_{m+1} \uparrow \infty$. The integrability condition shows that $T_m(z)$ must be subdominant in $D_1$ and in $D_4$. Furthermore, $T_m(z)$ is an even or odd function of $z$ according as $m$ is an even or odd integer. The symmetry forces the two remaining strings to lie on the imaginary axis outside the interval $(-ia_m, +ia_m)$. Problem 5.4:9 shows that $T_m$ can have no zeros on the real axis outside the interval $(-a_m, +a_m)$. That there are no zeros off the axes is indicated in Problem 8.4:20. We have

$$N(r, 0; T_m) = \frac{2}{9\pi} r^3 [1 + o(1)].$$                  (5.6.52)

For $T_m$ the value 0 has the defect $\frac{2}{3}$, while for an ordinary $E_k$ the defect is only $\frac{1}{3}$.

## EXERCISE 5.6

1. Verify (5.6.6).
2. Verify (5.6.7)–(5.6.12).
3. Verify (5.6.13)–(5.6.18).
4. Show that

$$\int_0^\infty |z + r|^{-2} \, dr = \begin{cases} \dfrac{\arg z}{y}, & z = x + iy, \ y \neq 0. \\ \dfrac{1}{x}, & y = 0, \ x > 0. \end{cases}$$

5. Discuss $\int_0^\infty |z + re^{i\theta}|^{-2} \, dr$.
6. Gronwall's lemma was invoked repeatedly in the proofs of Theorems 5.6.1 and 5.6.2. Show that it applies, and verify the inequalities.
7. If $F(z) = z^{-2}$, how is the "trimming" to be done in (5.6.38) for the corresponding solution? Here $U(z)$ is a power of $z$ times a Bessel-Hankel function of order $\frac{1}{2}\sqrt{5}$.
8. Prove the theorem of Rouché. [*Hint*: $h = f(1 + g/f)$. If $z$ describes $C$, where is $1 + g/f$ located?]
9. Verify the various inequalities between $|\sin(z - z_0)|$ and $|R(z)|$ used in the proof of Theorems 5.6.4 and 5.6.5.
10. If $Q(z) = z$, determine the domains $D_k$. Show that the equation is invariant under a positive rotation about the origin through an angle of 120°. What does this imply for the solutions? If $D_1$ is the sector containing the positive real axis, how does a dominant (subdominant) solution behave in $D_1$? Determine $N(r, 0; w)$ and $N(r, 0; E_k)$. What is the defect of the value 0 for $E_k$?
11. Verify the formulas relating to the Titchmarsh equation. How does the beta function enter the problem?
12. What is the asymptotic form of $E_4$ in $D_4$?
13. If $k$ is a positive integer, the DE

$$w''(z) - [k^2 z^{2k-2} + k(k-1)z^{k-2}]w(z) = 0$$

exhibits $2k$ sectors. Show that there is a solution for which $w = 0$ is a Picard value.
14. Discuss $Q(z) = a^m - z^m$. There are $m + 2$ sectors and corresponding subdominants. For $m > 1$ there may be values of $a$ for which a subdominant lacking more than two strings exists. If $m$ is even, there is a Titchmarsh problem.

## 5.7. ANALYTIC CONTINUATION; GROUP OF MONODROMY

Consider the DE

$$w''(z) + P(z)\, w'(z) + Q(z)\, w(z) = 0, \tag{5.7.1}$$

where $P$ and $Q$ are single-valued analytic functions of $z$.

**DEFINITION 5.7.1**

*If* $z = z_0$ *is a point where* P *and* Q *are holomorphic, we define the Mittag-Leffler star* $A(z_0; P, Q)$ *to be the set of all points* $z_1$ *in the finite plane such that* P(z) *and* Q(z) *are holomorphic at all points of the straight line segment joining* $z_0$ *with* $z_1$, *including the end points.*

Thus, if $z_2$ lies on the boundary of $A(z_0; P, Q)$, either $z_2$ is singular for at least one of the functions $P$ and $Q$ or analytic continuation from $z_0$ in the direction of $z_2$ encounters a singularity before reaching $z_2$. As an example take $P(z) = (1 - z)^{-1}$, $Q(z) = (1 + z)^{-2}$, $z_0 = 0$. Here $A(0; P, Q)$ has as its boundary the line segments $(-\infty, -1]$ and $[+1, \infty)$, and only the points $z = -1$ and $z = +1$ are actual singularities.

Suppose now that $w_1(z)$, $w_2(z)$ is a fundamental system of (5.7.1) at $z = z_0$, and form the *Wronskian*

$$W(z; w_1, w_2) = \begin{vmatrix} w_1(z) & w_2(z) \\ w_1'(z) & w_2'(z) \end{vmatrix}, \tag{5.7.2}$$

which is different from zero at $z = z_0$ and in some neighborhood thereof. We introduce also the *Wronskian matrix*

$$\mathcal{W}(z; w_1, w_2) = \begin{bmatrix} w_1(z) & w_2(z) \\ w_1'(z) & w_2'(z) \end{bmatrix}. \tag{5.7.3}$$

This is an element of $\mathfrak{M}_2$, the algebra of 2-by-2 matrices; it is nonsingular at $s = z_0$, and we shall see that it is nonsingular in $A(z_0; P, Q)$.

We shall need some additional notation and terminology. An $n$-by-$n$ matrix $\mathcal{A}(z) = (a_{jk}(z))$ is an element of the matrix algebra $\mathfrak{M}_n$. We often find it convenient to say *the a matrix* $\mathcal{A}$ *has the property* P *if the elements* $a_{jk}$ *have the property.* We shall use this convention if $P$ stands for analyticity, continuity, differentiability, holomorphy, integrability, measurability, or positivity.

We introduce a metric in $\mathfrak{M}_n$ by defining a norm of $\mathcal{A}$ by

$$\|\mathcal{A}\| = \max_j \left[ |a_{j1}| + |a_{j2}| + \cdots + |a_{jn}| \right]. \tag{5.7.4}$$

With this definition the norm of a product does not exceed the product of the norms. Under this norm $\mathfrak{M}_n$ is a complete metric space and is actually a Banach algebra. The algebra has a unit element, the matrix $(\delta_{jk})$ of norm 1.

A simple calculation shows that the Wronskian matrix satisfies

$$\mathcal{W}'(z) = \mathcal{F}(z)\,\mathcal{W}(z), \tag{5.7.5}$$

where $\mathcal{F}$ is the matrix

$$\mathcal{F}(z) = \begin{bmatrix} 0 & 1 \\ -Q(z) & -P(z) \end{bmatrix} \tag{5.7.6}$$

with determinant $Q(z)$. This could be expected from (5.2.11). Now $\mathcal{F}(z)$ is holomorphic in $A(z_0; P, Q)$ by the definition of the star and the definition of holomorphy for a matrix. It is to be expected that the property will carry over to the solution of (5.7.5), and this is indeed the case.

**THEOREM 5.7.1**

*If $\mathcal{W}(z)$ is the solution of (5.7.5) with initial value $\mathcal{W}(z_0)$, a nonsingular matrix, at $z = z_0$, then the unique solution is holomorphic in $A(z_0; P, Q)$.*

*Proof.* This follows from the argument used in Section 2.3, the method of successive approximations, with minor changes. First let $A_0(z_0)$ be a *star domain* with respect to $z = z_0$ contained in $A(z_0; P, Q)$ but bounded and closed. Then there exist constants $\rho$ and $M$ such that

$$\|\mathcal{F}(z)\| \leq M, \qquad |z - z_0| \leq \rho, \qquad z \in A_0.$$

Set $\mathcal{W}(z_0) = \mathcal{W}_0$, and define the successive approximations by

$$\mathcal{W}_n(z) = \mathcal{W}_0 + \int_{z_0}^{z} \mathcal{F}(s)\,\mathcal{W}_{n-1}(s)\,ds. \tag{5.7.7}$$

This makes sense in $A_0$ and shows that the approximations are holomorphic in $A_0$. Moreover, the usual estimates and the classical induction argument show that

$$\|\mathcal{W}_n(z) - \mathcal{W}_{n-1}(z)\| \leq \|\mathcal{W}_0\| \frac{1}{n!} (\rho M)^n,$$

so that the series

$$\mathcal{W}_0 + \sum_{1}^{\infty} [\mathcal{W}_n(z) - \mathcal{W}_{n-1}(z)] \tag{5.7.8}$$

converges uniformly in $A_0$ to an element of $\mathfrak{M}_2$ which is holomorphic in $A_0$. Since clearly

$$\mathcal{W}(z) = \mathcal{W}_0 + \int_{z_0}^{z} \mathcal{F}(s)\,\mathcal{W}(s)\,ds, \tag{5.7.9}$$

it is seen that $\mathcal{W}(z)$ is a solution of (5.7.5) satisfying the initial condition. Uniqueness is proved in the usual manner. The extension to $A(z; P, Q)$ is obvious. ∎

The same method may be used to show that the system

$$\mathcal{V}'(z) = -\mathcal{V}(z)\,\mathcal{F}(z), \qquad \mathcal{V}(z_0) = \mathcal{V}_0 \tag{5.7.10}$$

has a unique solution holomorphic in $\mathbf{A}(z_0;\,P,\,Q)$. We shall use this fact to prove

**THEOREM 5.7.2**

*The solution* $\mathcal{W}(z;\,z_0,\,\mathcal{W}_0)$ *of* (5.7.5) *is a regular* (= *nonsingular*) *element of* $\mathfrak{M}_2$ *iff* $\mathcal{W}_0$ *is regular.*

*Proof.* The following argument is due to W. A. Coppel. Let us first take $\mathcal{W}_0 = \mathcal{E}_2$, the unit element of $\mathfrak{M}_2$. Then

$$\mathcal{W}'(z) = \mathcal{F}(z)\,\mathcal{W}(z), \qquad \mathcal{W}(z_0) = \mathcal{E}_2,$$
$$\mathcal{V}'(z) = -\mathcal{V}(z)\,F(z), \qquad \mathcal{V}(z_0) = \mathcal{E}_2.$$

Here $\mathcal{W}(z)$ and $\mathcal{V}(z)$ are holomorphic in $\mathbf{A}(z_0;\,P,\,Q)$, and

$$\mathcal{V}(z)\,\mathcal{W}'(z) + \mathcal{V}'(z)\,\mathcal{W}(z) \equiv 0. \tag{5.7.11}$$

Now, if a matrix product has a derivative which is identically zero, the product is a constant matrix and

$$\mathcal{V}(z)\,\mathcal{W}(z) \equiv \mathcal{E}_2$$

since this holds for $z = z_0$. Thus $\mathcal{W}(z)$ has $\mathcal{V}(z)$ as left inverse everywhere in $\mathbf{A}(z_0;\,P,\,Q)$. On the other hand, since $\mathcal{W}(z_0) = \mathcal{E}_2$ the matrix $\mathcal{W}(z)$ must have a two-sided inverse in some neighborhood of $z = z_0$, and in this neighborhood

$$[\mathcal{W}(z)]^{-1} = \mathcal{V}(z), \qquad \mathcal{V}(z)\,\mathcal{W}(z) = \mathcal{W}(z)\,\mathcal{V}(z) = \mathcal{E}_2. \tag{5.7.12}$$

Now the last relation is a functional equation satisfied by $\mathcal{V}(z)$ and $\mathcal{W}(z)$, and, by the law of permanency of functional equations, it must hold as long as $\mathcal{W}(z)$ and $\mathcal{V}(z)$ can be continued analytically along the same path—in particular, everywhere in $\mathbf{A}(z_0;\,P,\,Q)$. ∎

If the initial matrix is $\mathcal{W}_0$, a nonsingular matrix, we have

$$\mathcal{W}(z;\,z_0,\,\mathcal{W}_0) = \mathcal{W}(z;\,z_0,\,\mathcal{E}_2)\,\mathcal{W}_0, \tag{5.7.13}$$

for both sides are solutions of (5.7.5) and have the same initial value at $z = z_0$. Here both factors on the right have inverses; hence the left member does also.

**COROLLARY 1**

*A fundamental system defined at* $z = z_0$ *remains a fundamental system everywhere in* $\mathbf{A}(z_0;\,P,\,Q)$.

This of course agrees with our previous remarks concerning the Wronskian (determinant).

## COROLLARY 2

*For any choice of points* $z$ *and* $z_1$ *in* $A(z_0; P, Q)$

$$\mathcal{W}(z; z_0, \mathcal{E}_2) = \mathcal{W}(z; z_1, \mathcal{E}_2) \, \mathcal{W}(z_1; z_0, \mathcal{E}_2). \tag{5.7.14}$$

This follows immediately from (5.7.13).

We can now tackle the problem of analytic continuation of the solutions of the DE (5.7.5). It is clearly sufficient to solve this problem for $\mathcal{W}(z; z_0, \mathcal{E}_2)$. We shall drop the subscript 2 and simply write $\mathcal{E}$ for the unit element. Now let $z_1 \in A(z_0; P, Q)$, $z_1 \neq z_0$, and set $\mathcal{W}(z_1; z_0, \mathcal{E}) = \mathcal{W}_1$. There is then a unique solution $\mathcal{W}(z; z_1, \mathcal{E})$ which is defined in $A(z_1; P, Q)$ and where

$$\mathcal{W}(z; z_0, \mathcal{E}) = \mathcal{W}(z; z_1, \mathcal{E}) \mathcal{W}_1 \tag{5.7.15}$$

for $z \in A(z_0; P, Q) \cap A(z_1; P, Q)$. The right member of this equation gives the analytic continuation of $\mathcal{W}(z; z_0, \mathcal{E})$ in $A(z_1; P, Q)$. Given a finite set of points $\{z_k\}$ with

$$z_k \in A(z_{k-1}; P, Q),$$

we obtain a sequence of analytic continuations of the form

$$\mathcal{W}(z; z_k, \mathcal{E}) \, \mathcal{W}_{k-1} \cdots \mathcal{W}_2 \mathcal{W}_1, \tag{5.7.16}$$

which represent the original solution in $A(z_k; P, Q)$ for $k = 1, 2, \ldots, m$ for analytic continuation along the polygonal line with vertices at $z_0, z_1, \ldots, z_m$.

Suppose that $z_0 \in A(z_m; P, Q)$. Thus, after describing a closed polygon with vertices $z_0, z_1, \ldots, z_m, z_0$, we arrive at $z = z_0$ with the value

$$\mathcal{W}_{m+1} \mathcal{W}_m \cdots \mathcal{W}_2 \mathcal{W}_1,$$

which may very well be distinct from the original value $\mathcal{E}$. Since each $\mathcal{W}_k$ is regular, so is the product. Consider now the set of all closed rectifiable curves $C$ beginning and ending at $z = z_0$ along which $\mathcal{W}(z; z_0, \mathcal{E})$ may be continued analytically. Here we may limit ourselves to polygonal lines, which, however, may be self-intersecting. This is no restriction. On the other hand, we shall assume that $\mathcal{F}(z)$ has only a finite number of singularities, poles or essential singularities. This is a restriction, but since the most interesting case is that in which $\mathcal{F}(z)$ is a rational function, i.e., $P$ and $Q$ are rational scalar functions, the limitation is not serious.

$C$ is to be given an *orientation*. If $C$ is a simple closed polygon, the *Jordan curve theorem* asserts that $C$ separates the plane into an *interior* of $C$ and an *exterior*. These parts have $C$ as a common boundary.

Moreover, any polygonal line which joins a point in the interior with a point in the exterior must intersect $C$ at least once—in any case, an odd number of times. If $z = a$ is a point not on $C$ and if $z$ describes $C$ once, $\arg(z - a)$ has an increment which is a multiple of $2\pi$, say, $2\pi n$. The integer $n$ is known as the *index* or *winding number* of $C$ with respect to $a$. Here $n = 0$ if $a$ is exterior to $C$. If $C$ is simple and if $a$ lies in the interior of $C$, then $n = +1$ or $-1$, depending on how we describe $C$. We say that the orientation of $C$ is positive if $n = +1$; otherwise, negative. To distinguish, we denote the curve with the positive orientation by $C$ and let $-C$ stand for the corresponding curve with the opposite orientation.

If $C$ is self-intersecting, we use a parametric representation $z = z(t)$, $0 < t < 1$, $z(0) = z(1) = z_0$. Here $z(t)$ is continuous and piecewise linear. This induces an orientation of $C$ in which $z(t_1)$ precedes $z(t_2)$ if $t_1 < t_2$. This will be accepted as the positive orientation of $C$ provided that it agrees with the positive orientation of the least simple subpolygon of $C$ that begins and ends at $C$, where the positive orientation is defined by the index. If the two disagree, we change the orientation of $C$.

Now with every curve $C$ there is associated a *transit matrix* $\mathcal{W}(C)$ such that analytic continuation along $C$ carries

$$\mathcal{W}(z; z_0, \mathscr{E}) \qquad \text{into } \mathcal{W}(z; z_0, \mathscr{E})\,\mathcal{W}(C), \qquad (5.7.17)$$

while continuation along $-C$ carries

$$\mathcal{W}(z; z_0, \mathscr{E}) \qquad \text{into } \mathcal{W}(z; z_0, \mathscr{E})[\mathcal{W}(C)]^{-1}. \qquad (5.7.18)$$

The set of all these transit matrices forms a group $\mathfrak{G}$, known as the *group of monodromy* of the equation. It is clear that the unit matrix is an element of $\mathfrak{G}$. It may be shown that, if a path $C$ can be contracted to $z_0$ without passing over a singularity, $\mathcal{W}(C) = \mathscr{E}$. Furthermore, for every matrix $\mathcal{W}(C)$ there is an inverse matrix $[\mathcal{W}(C)]^{-1} = \mathcal{W}(-C)$. If we describe first $C_1$ and then $C_2$, $\mathcal{W}(z; z_0, \mathscr{E})$ will be multiplied on the right by $\mathcal{W}(C_2)\mathcal{W}(C_1)$, where order is essential. The number of elements of $\mathfrak{G}$ is countable (possibly finite) if the number of singularities is finite. With each path, there is associated a continuum of equivalent paths $\Gamma$ for which $\mathcal{W}(C) = \mathcal{W}(\Gamma)$. Two paths are equivalent if $C_1 - C_2$ bounds a region containing no singular points. Thus the paths break up into a countable (possibly finite) number of equivalence classes.

The assumption that the DE has only a finite number of singularities implies that the group is *finitely generated*, i.e., $\mathcal{W}(C)$ can be written as an aggregate of positive and negative powers of the *generators* (at most one generator for each singular point), where the order is essential and powers of the same generator cannot be collected. Finally the group is independent of the choice of $z_0$. *There is only one abstract monodromy group.* The various realizations are *isomorphic.*

## EXERCISE 5.7

1. Determine the star $A(0; P, Q)$ if $P = (1 - z^3)^{-1}$ and $Q(z) = \tanh z$.

2. If norms are defined by (5.7.4), verify that the norm of a product is at most equal to the product of the norms. Is $\|\mathscr{A}^n\|$ necessarily equal to $\|\mathscr{A}\|^n$?

3. Let $\{\mathscr{A}_m\}$ be a sequence of matrices in $\mathfrak{M}_n$. What is the relation between convergence in the sense of the metric and convergence of the entries $a_{m;j,k}$?

4. Consider the matrix DE (5.7.10). What second order linear DE do $v_1(z)$ and $v_2(z)$ satisfy?

5. Show that, if $\mathscr{A}$ is a matrix in $\mathfrak{M}_n$ and if $\|\mathscr{A} - \mathscr{E}\| = k < 1$, then $\mathscr{A}$ has a two-sided inverse given by the convergent matrix series $\mathscr{E} + \sum_1^\infty (\mathscr{E} - \mathscr{A})^n$.

6. Try to justify the appeal to the law of permanency of functional equations made in the proof of Theorem 5.7.2. The law is valid for scalar functions, but here we are concerned with matrices.

7. Consider the equation

$$\mathscr{W}'(z) = \mathscr{A}(1 - z)^{-1}\mathscr{W}(z), \qquad \text{where } \mathscr{A} = \begin{bmatrix} 0 & a \\ a & 0 \end{bmatrix}.$$

A fundamental solution (algebraically regular) can be found as a linear combination of $(z - 1)^a$ and $(z - 1)^{-a}$ with matrix coefficients. Find the latter (not unique).

8. Show that the group of monodromy of the equation in Problem 7 has a single generator. If $a$ is an integer, the solution is single valued and the group reduces to the identity (the unit element). If $a$ is not an integer, the group is *cyclic* (all elements are powers of the generator). The group is finite iff $a$ is rational. Verify.

9. In Problem 7 replace $\mathscr{A}$ by the *nilpotent* matrix

$$\mathscr{B} = \begin{bmatrix} 0 & 1 \\ 0 & 0 \end{bmatrix}.$$

Find a solution matrix and the group. (A matrix is nilpotent if its powers are the zero matrix from a certain point on.)

10. Suppose that (5.7.5) has singularities at $z = 0, 1, \infty$, and nowhere else. Take $z_0 = \frac{1}{2}$, and lay loops around zero and one. Let $\mathscr{W}(C_j) = \mathscr{W}_j, j = 1, 2$. Find a path $C$ such that

$$\mathscr{W}(C) = \mathscr{W}_1 (\mathscr{W}_0)^{-1} (\mathscr{W}_1)^{-1} \mathscr{W}_0.$$

Draw a figure. Such *double loops* play an important role in the integration of linear DE's by definite integrals. See the next chapter.

## LITERATURE

This chapter corresponds to Appendix B of the author's LODE, supplemented with various selections from Chapters 6, 7, and 9. All these have extensive bibliographies to which the reader is referred. Also see, in particular, Chapters 4 and 5 of:

Coddington, E. A. and Norman Levinson. *Theory of Ordinary Differential Equations.* McGraw-Hill, New York, 1955.

Section 5.1 is partly based on pp. 66–90 of:

Schlesinger, Ludwig, *Differentialgleichungen.* 2nd ed. Sammlung Schubert, Vol. 13. Göschen, Leipzig, 1904.

For Sections 5.2 and 5.3, pp. 91–121 of the same treatise served as inspiration, together with pp. 108–138 of:

Bieberbach, Ludwig, *Theorie der gewöhnlichen Differentialgleichungen.* Grundlehren, No. 66. Springer-Verlag, Berlin, 1953.

Some references for Section 5.4 are:

Koch, H. von. Un théorème sur les intégrales irrégulières des équations différentielles linéaires et son application au problème de l'intégration. *Ark Mat., Astr., Fys.*, 13, No. 15 (1918), 18 pp.

Matell, M. *Asymptotische Eigenschaften gewisser linearer Differentialgleichungen.* Uppsala, 1924, 67 pp.

Perron, O. Über einen Satz des Herrn Helge von Koch über die Integrale linearer Differentialgleichungen. *Math. Zeit.*, 3 (1919), 161–174.

Pöschl, Klaus. Zur Frage des Maximalbetrages der Lösungen linearer Differentialgleichungen zweiter Ordnung mit Polynomkoeffizienten. *Math. Ann.*, 125 (1953), 344–349.

Valiron, G. *Lectures on the General Theory of Integral Functions.* Edouard Privat, Toulouse, 1923; Chelsea, New York, 1949.

Wiman, Anders. Über den Zusammenhang zwischen dem Maximalbetrage einer analytischen Funktion und dem grössten Glied der zugehörigen Taylorschen Reihe. *Acta Math.*, 37 (1914), 305–326.

――――. Über den Zusammenhang zwischen dem Maximalbetrage einer analytischen Funktion und dem grössten Betrage bei gegebenem Argument der Funktion. *Acta Math.*, 41 (1916), 1–28.

Wittich, Hans. Ganze transzendente Lösungen algebraischer Differentialgleichungen. *Gött. Nachrichten*, (1952), 277–288.

――――. *Neuere Untersuchungen über eindeutige analytische Funktionen.* Ergebnisse d. Math, N. S., No. 8. Springer-Verlag, Berlin, 1955; 2nd ed., 1968.

――――. Zur Theorie linearer Differentialgleichungen im Komplexen. *Ann. Acad. Sci. Fenn.*, Ser. A, I, Math., 379 (1966).

Section 5.5 is based essentially on LODE, Chapter 9. The results go back to:

Wiman, Anders. Über die reellen Lösungen der linearen Differentialgleichungen zweiter Ordnung. *Ark. Mat., Astr., Fys.*, **12**, No. 14 (1917), 22 pp.

A survey of the field is given by:

Coppel, W. A. Asymptotic solutions of second order linear differential equations. *MRC Technical Summary Report*, No. 555, March 1965.

Section 5.6 is based on Section 7.4 of LODE. It goes back to work done by the author in 1920–1924. The use of Gronwall's lemma is a late interpolation. For the Titchmarsh problem see Chapters VII and VIII of:

Titchmarsh, E. C. *Eigenfunction Expansions Associated with Second Order Differential Equations*. Clarendon Press, Oxford, 1946. (Page 147 for the conjecture.)

Section 5.7 is related to LODE, Sections 6.1 and 6.2.

# 6

# SPECIAL SECOND ORDER LINEAR DIFFERENTIAL EQUATIONS

E. Kamke published in 1944 a magnificent monograph on DE's entitled *Differentialgleichungen, Lösungsmethoden und Lösungen*. Of a total of 666 pages, he devoted over 200 to special second order linear equations and listed some 450 equations. Our projected coverage is much more modest. There are some equations which must be included in any account. Others are mentioned because at some time or other the author had occasion to study them; the reader who is not satisfied with the selection here is referred to Kamke.

We include the hypergeometric equation, the equations of Bessel, Laplace (as a sample of the confluent hypergeometric case), and Legendre, the equations of elliptic and of parabolic cylinders, and some odds and ends.

## 6.1. THE HYPERGEOMETRIC EQUATION

No other DE has such a venerable history as the hypergeometric equation: Euler, Gauss, Kummer, and Riemann all contributed to the theory, and fresh contributions are still being offered. It is an equation that appears in many situations, and the theory has a high degree of formal elegance. It is connected with conformal mapping, difference equations, continued fractions, and automorphic functions, to mention only a few.

Equations of the Fuchsian class with only two singularities lead to trivial Euler equations. Three singularities lead to the hypergeometric case. Here we have a choice between the normal form of Gauss:

$$z(1-z)w'' + [c - (a+b+1)z]w' - abw = 0, \qquad (6.1.1)$$

where the singular points are 0, 1, ∞ and the solutions are expressible in terms of the hypergeometric series,

$$F(a, b, c; z) = 1 + \sum_{n=1}^{\infty} \frac{(a)_n (b)_n \, z^n}{(1)_n (c)_n}, \qquad (6.1.2)$$

where $(p)_n = p \, (p+1) \, (p+2) \cdots (p+n-1)$, and the alternative: the Riemann $P$-function,

$$P \left\{ \begin{matrix} a & b & c \\ \lambda_1 & \lambda_2 & \lambda_3 \ z \\ \mu_1 & \mu_2 & \mu_3 \end{matrix} \right\}, \qquad \sum_{k=1}^{3} (\lambda_k + \mu_k) = 1. \qquad (6.1.3)$$

The $P$-function has three singular points, $a, b, c$; the roots of the indicial equations are $\lambda_1, \mu_1$ at $z = a$, $\lambda_2, \mu_2$ at $z = b$, and $\lambda_3, \mu_3$ at $z = c$, with the consistency relation requiring that the sum of the roots be one. The coefficients of the equation are uniquely determined by these data. This form of the equation dates from 1857. Like most of the things that Riemann touched, this led to a Riemann problem: if the singular points and the monodromy group are given, does there always exist a DE of the Fuchsian class with these singularities and this group? Answers to this question were given by G. D. Birkhoff (1884–1944), David Hilbert (1862–1943), and Josef Plemelj (1873–1967). See Section 9.4.

We return to the normal type of Gauss. Although Gauss published an exhaustive discussion of the hypergeometric series in 1812, his work on the DE itself appeared only after his death.

The DE has three singular points at $z = 0, 1$, and ∞ with the corresponding roots of the indicial equations

$$0, 1-c; 0, c-a-b; a, b. \qquad (6.1.4)$$

Note that logarithmic singularities are to be expected

at $z = 0$     if $c$ is an integer,

at $z = 1$     if $a + b - c$ is an integer, and

at $z = \infty$   if $a - b$ is an integer.

At $z = 0$ and for $c$ not equal to zero or a negative integer,

$$w_{01}(z) = F(a, b, c; z) \qquad (6.1.5)$$

is the holomorphic solution. The series converges for $|z| < 1$, i.e., up to the nearest singularity. We shall see later what happens on the unit circle. The

Mittag-Leffler star $A(0; w_{01})$ of this solution is the finite plane less the line segment $[1, \infty)$. Analytic continuation is possible along any path that does not pass through any of the singular points. A closed path, starting and ending at $z = 0$ and surrounding $z = 1$, would normally lead to a determination of $w_{01}(z)$ which has a singularity at $z = 0$ because the second solution at the origin has a singular point there. To find the second solution, we may use the fact that there is a set of 24 transformations of the variables which carry a hypergeometric equation into such an equation with a different set of parameters. These 24 solutions were found in 1834 by Eduard Kummer (1810–1893). One of these transformations is $w = z^{1-c}u$, which gives a hypergeometric equation for $u$ with parameters

$$a - c + 1, \qquad b - c + 1, \qquad 2 - c. \tag{6.1.6}$$

This gives the second solution at $z = 0$ as

$$w_{02}(z) = z^{1-c} F(a - c + 1, b - c + 1, 2 - c; z). \tag{6.1.7}$$

Here $c$ is not an integer, and the two solutions are obviously linearly independent.

It is of some interest to detour at this point with a view to discussing very briefly the corresponding matrix equation of the Fuchsian class with singularities at $z = 0, 1, \infty$. It is

$$\mathcal{W}'(z) = \left[ \frac{\mathcal{A}}{z} + \frac{\mathcal{B}}{1 - z} \right] \mathcal{W}(z), \tag{6.1.8}$$

where $\mathcal{A}$ and $\mathcal{B}$ are constant matrices. This equation is formally invariant under the transformations of the *anharmonic group* $\mathfrak{A}$ with the six elements

$$s = z, \qquad 1 - z, \qquad \frac{1}{z}, \qquad \frac{1}{1 - z}, \qquad \frac{z}{z - 1}, \qquad \frac{z - 1}{z}. \tag{6.1.9}$$

Each of these transformations leads to an equation of type (6.1.8) though the matrix "parameters" are changed according to fairly simple laws. Actually the group $\mathfrak{A}$ enters also into the transformation of the coefficient "vector" (a 2-vector with matrix entries). The point of departure is a set of six 2-by-2 matrices which form a group under matrix multiplication. They are as follows:

$$\mathcal{E} = \begin{bmatrix} 1 & 0 \\ 0 & 1 \end{bmatrix}, \qquad \mathcal{S} = \begin{bmatrix} 0 & -1 \\ -1 & 0 \end{bmatrix}, \qquad \mathcal{T} = \begin{bmatrix} -1 & 1 \\ 0 & 1 \end{bmatrix},$$

$$\mathcal{S}\mathcal{T} \equiv \begin{bmatrix} 0 & -1 \\ 1 & -1 \end{bmatrix}, \qquad \mathcal{T}\mathcal{S} = \begin{bmatrix} -1 & 1 \\ -1 & 0 \end{bmatrix}, \qquad \mathcal{T}\mathcal{S}\mathcal{T} = \begin{bmatrix} 1 & 0 \\ 1 & -1 \end{bmatrix}. \tag{6.1.10}$$

There are various identities:

$$\mathcal{S}^2 = \mathcal{T}^2 = \mathcal{E}, \qquad (\mathcal{S}\mathcal{T})^3 = (\mathcal{T}\mathcal{S})^3 = \mathcal{E}, \qquad \mathcal{T}\mathcal{S}\mathcal{T} = \mathcal{S}\mathcal{T}\mathcal{S}. \qquad (6.1.11)$$

We shall also need a notion of *conjugacy*. If $\mathcal{U} \in \mathfrak{A}$, set

$$\mathcal{U}^* = \mathcal{T}\mathcal{S}\mathcal{T}\mathcal{U}\mathcal{T}\mathcal{S}\mathcal{T}. \qquad (6.1.12)$$

Then, if $\mathcal{U} \in \mathfrak{A}$ and

$$\mathcal{U} = \begin{bmatrix} a & b \\ c & d \end{bmatrix},$$

we define

$$\mathcal{U}(z) = \frac{az + b}{cz + d}, \qquad \mathcal{U}\begin{bmatrix} \mathcal{A} \\ \mathcal{B} \end{bmatrix} = \begin{bmatrix} a\mathcal{A} + b\mathcal{B} \\ c\mathcal{A} + d\mathcal{B} \end{bmatrix}. \qquad (6.1.13)$$

Here $a$, $b$, $c$, $d$ have values in the set $-1, 0, 1$, while $\mathcal{A}$, $\mathcal{B}$ are the matrix coefficients of (6.1.8). We then have

### THEOREM 6.1.1

*The transformations*

$$s = \mathcal{U}(z), \qquad \begin{bmatrix} \mathcal{A}^* \\ \mathcal{B}^* \end{bmatrix} = \mathcal{U}^*\begin{bmatrix} \mathcal{A} \\ \mathcal{B} \end{bmatrix} \qquad (6.1.14)$$

*carry* (6.1.8) *into*

$$\mathcal{W}'(s) = \left[\frac{\mathcal{A}^*}{s} + \frac{\mathcal{B}^*}{1 - s}\right]\mathcal{W}(s). \qquad (6.1.15)$$

The proof is left to the reader.

Thus if we know one solution of (6.1.8) we know six. Unfortunately the transformation theory of the classical DE is not quite that simple. Of the six anharmonic transformations on the independent variable, only the identity and $s = 1 - z$ lead to hypergeometric equations. Thus the reflection $s = 1 - z$ gives a new equation with parameters

$$a, b, a + b + 1 - c.$$

This gives a fundamental system at $z = 1$:

$$w_{11}(z) = F(a, b, a + b + 1 - c; 1 - z), \qquad (6.1.16)$$

$$w_{12}(z) = (1 - z)^{c-a-b}F(c - a, c - b, c + 1 - a - b; 1 - z). \qquad (6.1.17)$$

For the other solutions it is necessary to proceed as in (6.1.7) by setting

$$w = z^\delta(1 - z)^\epsilon u \qquad (6.1.18)$$

and choosing $\delta$ and $\epsilon$ properly. We list the following cases:

1.  Solutions at infinity: set $s = 1/z$, $\delta = -a$, $\epsilon = 0$. The new parameters

are

$$a, \qquad a - c + 1, \qquad a - b + 1,$$

and the solution

$$w_{\infty 1}(z) = z^{-a}F\left(a, a - c + 1, a - b + 1; \frac{1}{z}\right). \tag{6.1.19}$$

Now (6.1.1) does not change if $a$ and $b$ are interchanged so that a second solution at infinity is given by

$$w_{\infty,2}(z) = z^{-b}F\left(b, b - c + 1, b - a + 1; \frac{1}{z}\right). \tag{6.1.20}$$

One of the solutions breaks down if $a - b$ is an integer. If not, they are obviously linearly independent and form a fundamental system at infinity. These series converge for $|z| > 1$, while the series (6.1.16) and (6.1.17) converge for $|z - 1| < 1$. We can also get solutions converging for $|z - 1| > 1$. Of more interest are solutions converging in a half-plane, either $\text{Re}\,(z) < \frac{1}{2}$ or $\text{Re}\,(z) > \frac{1}{2}$.

2. Set $s = (z - 1)/z$, $\delta = -a$, $\epsilon = 0$. The parameters are $a$, $a - c + 1$, $a + b + 1 - c$, and the solutions are

$$w_{1,3}(z) = z^{-a}F\left(a, a - c + 1, a + b + 1 - c; \frac{z - 1}{z}\right), \tag{6.1.21}$$

$$w_{1,5}(z) = z^{-b}F\left(b, b - c + 1, a + b + 1 - c; \frac{z - 1}{z}\right). \tag{6.1.22}$$

The numbering refers to the fact that both solutions are holomorphic at $z = 1$, where they have the value 1. This means that

$$w_{1,1}(z) = w_{1,3}(z) = w_{1,5}(z), \tag{6.1.23}$$

one of Kummer's 24 relations. The solutions with subscripts 1, 3 and 1, 5 give analytic continuation of $w_{1,1}(z)$ in the half-plane $\text{Re}\,(z) > \frac{1}{2}$. Similar analytic continuations may be obtained for $w_{1,2}(z)$ in the half-plane:

$$w_{1,4}(z) = z^{b-c}(1 - z)^{c-a-b}F\left(c - b, 1 - b, 1 + c - a - b; \frac{z - 1}{z}\right), \tag{6.1.24}$$

$$w_{1,6}(z) = z^{a-c}(1 - z)^{c-a-b}F\left(c - a, 1 - a, 1 + c - a - b; \frac{z - 1}{z}\right), \tag{6.1.25}$$

giving another Kummer relation, $w_{1,2} = w_{1,4} = w_{1,6}$. For $w_{0,1}$ and $w_{0,2}$ we can get representations in the half-plane $\text{Re}\,(z) < \frac{1}{2}$ in a similar manner.

Thus all the solutions of (6.1.1) are expressible in terms of hypergeometric series in one of the six "anharmonic" $z$-variables multiplied by a power of $z$ or/and a power of $1 - z$. The parameters are linear functions of $a, b, c$, and so are the exponents of the powers. It is clear that the solutions are analytic functions of the parameters, as well as of $z$. This implies further that in the transit formulas such as

$$w_{0,1}(z) = A\, w_{1,1}(z) + B\, w_{1,2}(z) \qquad (6.1.26)$$

the coefficients $A, B$ are also analytic functions of $a, b, c$—in fact, meromorphic functions, for we can differentiate (6.1.26) with respect to $z$ and set $z = \frac{1}{2}$ in the result as well as in (6.1.26). We have now two linear equations for $A$ and $B$ with locally holomorphic coefficients. Actually this gives

$$A = \frac{W(w_{01}, w_{12})}{W(w_{11}, w_{12})} \qquad (6.1.27)$$

in terms of the Wronskians of the solutions evaluated at $z = \frac{1}{2}$. Now a Wronskian can become zero iff the two solutions become linearly dependent for some particular choice of the parameters; it becomes infinite (since $z = \frac{1}{2}$ is not a singular point) if one of the solutions becomes identically infinity. This representation shows that $A$ is a meromorphic function of the parameters and tells us something about the zeros and poles of $A$. Thus $A$ has poles at the points $c = 0, -1, -2, \ldots$ because $w_{01}$ has poles at such points. Furthermore, $c - a - b = 0, -1, \ldots$ are poles because $w_{11}$ and $w_{12}$ become linearly dependent. The positive integers congruent to $c - a - b$ drop out of consideration since $w_{11}(\frac{1}{2})$ becomes infinite. Finally, if $c - a$ or $c - b$ is zero or a negative integer, then $w_{01}$ and $w_{12}$ become linearly dependent, so these values should be zeros of $A$. All these observations will be found to be correct once we get the explicit formula for $A$. We note, however, that for $\mathrm{Re}\,(c - a - b) > 0$ we have

$$A = \lim_{z \to 1} F(a, b, c\,;z) \equiv f(a, b, c) \qquad (6.1.28)$$

as follows from (6.1.26), since $w_{11}(z) \to 1$ and $w_{12}(z) \to 0$. This function of three parameters enters into all the basic transit relations; hence the finding of $f(a, b, c)$ becomes paramount.

Let us observe in passing that the hypergeometric series (6.1.2) is absolutely convergent for $|z| < 1$ for any choice of the parameters except $c$ equal to zero or a negative integer. The series is absolutely convergent for $|z| = 1$ provided $\mathrm{Re}\,(c - a - b) > 0$. It converges but not absolutely for $|z| = 1, z \ne 1$, as long as $-1 < \mathrm{Re}\,(c - a - b) \le 0$ and diverges on $|z| = 1$ if $\mathrm{Re}\,(c - a - b) \le -1$.

**LEMMA 6.1.1**

*For Re* $(c - a - b) > 0$ *we have*

$$f(a, b, c) = \frac{\Gamma(c)\Gamma(c - a - b)}{\Gamma(c - a)\Gamma(c - b)}. \qquad (6.1.29)$$

*Proof.* We note that the gamma quotient on the right is the simplest meromorphic function having poles and zeros at the points indicated in the discussion above, but this is no proof. We shall use a device which is worked out in much more detail in Sections 7.3 and 7.4. The gamma quotient suggests bringing in the Euler beta function:

$$\int_0^1 t^{x-1}(1-t)^{y-1}\,dt = \frac{\Gamma(x)\Gamma(y)}{\Gamma(x+y)}, \tag{6.1.30}$$

valid for $\operatorname{Re}(x) > 0$, $\operatorname{Re}(y) > 0$. If now

$$g(u) = \sum_{n=0}^{\infty} a_n u^n, \qquad |u| < 1, \tag{6.1.31}$$

then for $|z| < 1$

$$\begin{aligned}
G(z) &= \int_0^1 t^{x-1}(1-t)^{y-1} g(zt)\,dt \\
&= \sum_{n=0}^{\infty} a_n z^n \int_0^1 t^{x+n-1}(1-t)^{y-1}\,dt \\
&= \sum_{n=0}^{\infty} a_n \frac{\Gamma(x+n)\Gamma(y)}{\Gamma(x+y+n)} z^n
\end{aligned} \tag{6.1.32}$$

by uniform convergence. Here we set

$$x = a, \qquad y = c - a, \qquad g(u) = (1-a)^{-b}.$$

Since

$$(1-u)^{-b} = \sum_{n=0}^{\infty} \frac{(b)_n}{(1)_n} u^n,$$

we get

$$\begin{aligned}
G(z) &= \Gamma(c-a) \sum_{n=0}^{\infty} \frac{(b)_n \Gamma(a+n)}{(1)_n \Gamma(c+n)} z^n \\
&= \frac{\Gamma(c-a)\Gamma(a)}{\Gamma(c)} F(a, b, c; z),
\end{aligned}$$

so that

$$w_{01}(z) = \frac{\Gamma(c)}{\Gamma(c-a)\Gamma(a)} \int_0^1 t^{a-1}(1-t)^{c-a-1}(1-zt)^{-b}\,dt. \tag{6.1.33}$$

Here the convergence of the integral requires that $\operatorname{Re}(a) > 0$, $\operatorname{Re}(c-a) > 0$. If in addition $\operatorname{Re}(c-a-b) > 0$, we can let $z \to 1$ from the left, and the Lebesgue theorem on dominated convergence shows that the resulting integral converges. Thus we get

$$\begin{aligned}
f(a, b, c) &= \frac{\Gamma(c)}{\Gamma(c-a)\Gamma(a)} \int_0^1 t^{a-1}(1-t)^{c-a-b-1}\,dt \\
&= \frac{\Gamma(c)}{\Gamma(c-a)\Gamma(a)} \frac{\Gamma(a)\Gamma(c-a-b)}{\Gamma(c-b)},
\end{aligned}$$

which is the required result. In (6.1.33) and the following formula we may interchange $a$ and $b$.  ∎

We have found the expression for $A$ in (6.1.26). We can now also find $B$. If in this relation $z$ decreases to zero along the real axis, then $w_{0,1} \to 1$, $w_{1,1} \to f(a, b, a + b - c + 1)$ while $w_{1,2} \to f(c - a, \ c - b, \ c - a - b + 1)$, and simplification gives

$$B = \frac{\Gamma(c)\Gamma(a + b - c)}{\Gamma(a)\Gamma(b)}, \tag{6.1.34}$$

again a meromorphic function of $a, b, c$. This is rather important, for $A(a, b, c)$ and $B(a, b, c)$ are defined everywhere in $\mathbf{C}^3$, certain lower dimensional varieties excepted. We can then resort to the law of permanency of functional equations to conclude that the expressions for $A$ and $B$ are independent of the previously assumed restrictions on the parameters.

The same method may be used to find the other transit relations. All that one needs is convergent series expansions valid in a domain with two singular points on the boundary. Here the half-plane solutions are useful. A second method based on integral representations of one type or another will be given in Chapter 7.

**EXERCISE 6.1**

Verify the following relations:

1.  $(d/dz)F(a, b, c; z) = (ab/c)F(a + 1, b + 1, c + 1; z)$.
2.  $(1 - z)^{-a} = F(a, b, b; z)$.
3.  arc sin $z = zF(\tfrac{1}{2}, \tfrac{1}{2}, \tfrac{3}{2}; z^2)$.
4.  arc tan $z = zF(1, \tfrac{1}{2}, \tfrac{3}{2}; -z^2)$.
5.  $\lim\limits_{a \to \infty} F(a, 1, 1; z/a) \equiv e^z$.

6.  The elliptic normal integrals of Jacobi are

$$K = \int_0^1 (1 - s^2)^{-1/2}(1 - k^2 s^2)^{-1/2}\, ds, \qquad iK' = \int_1^{1/k} (1 - s^2)^{-1/2}(1 - k^2 s^2)^{-1/2}\, ds,$$

where $k \neq -1, 0, +1, \infty$. Show that

$$K = \tfrac{1}{2}\pi F(\tfrac{1}{2}, \tfrac{1}{2}, 1; k^2), \qquad K' = \tfrac{1}{2}\pi F(\tfrac{1}{2}, \tfrac{1}{2}, 1; 1 - k^2).$$

7.  Prove the invariance of (6.1.8) under the anharmonic group.
8.  Verify that the matrices (6.1.10) form a group under matrix multiplication, and verify (6.1.9).
9.  Prove Theorem 6.1.1.
10. Determine the set of self-conjugate matrices in the set (6.1.8).
11. The characteristic values of the anharmonic matrices are roots of unity. What roots occur, and how are they reflected in (6.1.9)?

12. Construct a fundamental system with convergent series expansions in $|z - 1| > 1$. Express these solutions in terms of $w_{\infty,1}$ and $w_{\infty,2}$.

13. Verify (6.1.23) and (6.1.27).

14. If $a = \frac{1}{4}$, $b = 1$, $c = \frac{1}{2}$, show that $w_{02}(z) = w_{12}(z) = w_{\infty 1}(z)$ is an algebraic function. Find this function. Are there any other algebraic solutions of the equation?

15. With the parameters as in Problem 14, show that $q(z) = w_{02}(z)/w_{01}(z)$ maps the upper $z$-half-plane on the interior of a curvilinear triangle (possibly degenerate) bounded by straight lines and circular arcs. Describe the triangle, and find the sides and the angles.

16. The equation $w^2 + 2w - z = 0$ has a root that goes to zero with $z$. Express the root as a hypergeometric function.

17. Show also that the equation $w^3 + 3w - z = 0$ is satisfied by $w = \frac{1}{3}zF(\frac{1}{3}, \frac{2}{3}, \frac{3}{2}; -\frac{1}{4}z^2)$, $|z| < 2$.

18. What are the singularities of the function $z \mapsto w(z)$ of the equation in Problem 17, and what is their nature? Use a transit relation to express the solution in a neighborhood of $z = 2i$.

19. Generalize to arbitrary trinomial equations.

20. Show that the hypergeometric equation may be written as

$$z(\vartheta + a)(\vartheta + b)w = \vartheta(\vartheta - 1 + c)w, \qquad \vartheta = z\frac{d}{dz}.$$

## 6.2.  LEGENDRE'S EQUATION

This is the DE

$$(1 - z^2)\, w''(z) - 2z\, w'(z) + a(a + 1)\, w(z) = 0 \tag{6.2.1}$$

or

$$\frac{d}{dz}\left[(1 - z^2)\frac{dw}{dz}\right] + a(a + 1)w = 0. \tag{6.2.2}$$

It is named after the French mathematician Adrien Marie Legendre (1752–1833), who considered the case $a = n$, an integer, in 1785. Here the equation has a polynomial solution, the $n$th *Legendre polynomial*, $P_n(z)$. The substitution $u = \frac{1}{2}(1 - z)$ reduces the equation to hypergeometric form with parameters $a + 1, -a, 1$. The roots of the indicial equations are

$$0, 0; 0, 0: a + 1, -a.$$

Thus at both $z = -1$ and $+1$ logarithms must appear.
The holomorphic solution at $z = +1$ is

$$P_a(z) = F[a + 1, -a, 1; \tfrac{1}{2}(1 - z)]. \tag{6.2.3}$$

If $a$ is an integer, $a = n$, the hypergeometric series breaks off and a polynomial of degree $n$ results. The equation is unchanged if $z$ is replaced by $-z$ and $a + \frac{1}{2}$ by $-(a + \frac{1}{2})$. In particular,

$$P_n(z) = (-1)^n P_n(-z).$$

The second solution at $z = +1$, which involves a logarithm, may be written as

$$P_{a,1}(z) = P_a(z) \log\left[\tfrac{1}{2}(1-z)\right] + \sum_{n=1}^{\infty} \frac{(a+1)_n(-a)_n}{(1)_n(1)_n}$$

$$\times \left[ \sum_{j=0}^{n-1} \left( \frac{1}{a+1+j} + \frac{1}{-a+j} - \frac{2}{j+1} \right) \right] [\tfrac{1}{2}(1-z)]^n, \quad (6.2.4)$$

as the reader may verify.

A rather useful analytic continuation of $w_{01}(z)$, i.e., (6.2.3), is furnished by the representation

$$P_a(z) = [\tfrac{1}{2}(1-z)]^{-1-a} F\left( a+1, -a, 1; \frac{z-1}{z+1} \right) \quad (6.2.5)$$

valid in the right half-plane, the analogue of $w_{1,3}(z)$.

Formula (6.2.4) involves a pair of finite sums,

$$\sum_{j=0}^{n-1} \left( \frac{1}{a+1+j} - \frac{1}{j+1} \right) + \sum_{j=0}^{n-1} \left( \frac{1}{-a+j} - \frac{1}{j+1} \right) \equiv D_n(a). \quad (6.2.6)$$

It is seen that each individual difference is $O(j^{-2})$, so that as $n \to \infty$ each sum converges to a finite limit when $a$ is not an integer. It is not surprising that the limits are expressible in terms of the logarithmic derivative of the gamma function:

$$\Psi(x) = \frac{\Gamma'(x)}{\Gamma(x)} = -C - \frac{1}{x} - \sum_{n=1}^{\infty} \left( \frac{1}{x+n} - \frac{1}{n} \right), \quad (6.2.7)$$

where $C$ is the Euler-Mascheroni constant $0.57721\ldots$. This gives

$$\lim_{n\to\infty} D_n(a) = \Psi(a+1) + \Psi(-a) - 2C \equiv D(a).$$

This leads ultimately to a transit formula

$$P_a(z) = \sin(\pi a)[D(a)P_a(-z) + P_{a,1}(-z)]. \quad (6.2.8)$$

The solutions $P_a(z)$ and $P_{a,1}(z)$ are linearly independent and form a fundamental system. At infinity we have the fundamental system

$$Q_a(z) \quad \text{and} \quad Q_{-1-a}(z), \quad (6.2.9)$$

where

$$Q_a(z) = \frac{\Gamma(\tfrac{1}{2})\Gamma(a+1)}{\Gamma(a+\tfrac{3}{2})} (2z)^{-1-a} F(\tfrac{1}{2}a + \tfrac{1}{2}, \tfrac{1}{2}a + 1, a + \tfrac{3}{2}; z^{-2}). \quad (6.2.10)$$

The transit formula from $P_a$ to the $Q$'s has the elegant form

$$P_a(z) = \tan(\pi a)[Q_a(z) - Q_{-1-a}(z)]. \tag{6.2.11}$$

Here it is assumed that $2a$ is not an integer.

The rest of this section is devoted to the properties of Legendre polynomials. The set $\{P_n(z)\}$ is an *orthogonal system* for the interval $(-1, 1)$, and

$$w_n(t) = (n + \tfrac{1}{2})^{1/2} P_n(t) \tag{6.2.12}$$

is an *orthonormal system*. Moreover, this sytem is *complete* in $L_2(-1, +1)$. This property may be defined in several different but equivalent ways: (i) the only element of $L_2(-1, 1)$ whose Legendre-Fourier coefficients

$$f_n = \int_{-1}^{+1} f(t) P_n(t)\, dt \tag{6.2.13}$$

are zero for all $n$ is the zero element; (ii) the finite linear combinations of the $P_n$'s are dense in $L_2(-1, 1)$; and (iii) the *closure relation*

$$\sum_{n=0}^{\infty} (n + \tfrac{1}{2})|f_n|^2 = \int_{-1}^{1} |f(s)|^2\, ds \tag{6.2.14}$$

holds for all $f$ in $L_2(-1, 1)$.

If $z$ is not real and located in $(-1, 1)$, then $t \mapsto (z - t)^{-1}$ belongs to $L_2(-1, 1)$, and the corresponding Legendre-Fourier coefficients,

$$\frac{1}{2} \int_{-1}^{1} \frac{p_n(t)}{z - t}\, dt = Q_n(z), \tag{6.2.15}$$

are obviously analytic functions of $z$. That the integral equals $Q_n(z)$ was proved by F. Neumann (1848). We have then

$$\frac{1}{z - t} = \sum_{n=0}^{\infty} (2n + 1) P_n(t) Q_n(z), \tag{6.2.16}$$

to start with, in the sense of $L_2$-convergence. But the terms of the series tend to zero so rapidly that there is absolute convergence not merely for $-1 \le t \le +1$ but also in the complex plane for $t$ inside an ellipse with foci at $-1$ and $+1$ and passing through $z$. Moreover, any function holomorphic inside such an ellipse may be represented by a series of Legendre polynomials absolutely convergent inside the ellipse.

**EXERCISE 6.2**

1. Prove that

$$P_n(z) = \frac{2^{-n}}{n!} \frac{d^n}{dz^n} [(z^2 - 1)^n].$$

2. Prove the orthogonality relations

$$\int_{-1}^{1} P_m(t)P_n(t)\, dt = \delta_{mn}(n + \tfrac{1}{2})^{-1}.$$

3. If $y(t)$ is a real-valued solution of (6.2.1), prove that, if

$$F(t) = (1 - t^2)[y'(t)]^2 + a(a + 1)[y(t)]^2, \qquad y(1) = 1,$$

then $-1 < y(t) < +1$ on $(0, 1)$. In particular, this holds for $a = n$, $y(t) = P_n(t)$.

4. Show that $P_n(t)$ has only real simple zeros all in $(-1, 1)$.

5. Show that $P_a(z) > 0$ for $1 < x < \infty$, $a$ real.

6. Let $a = -\tfrac{1}{2} + ki$, and consider $P_{-1/2+ki}(z)$. These are the *conical harmonics* (German: *Kegelfunktionen*) of E. H. Heine (of Heine-Borel theorem fame). Show that these functions are real for $z = t > -1$ and positive on $(-1, 1)$. They arise in potential-theoretical problems involving cones.

7. Let the $z$-plane be cut along the real axis from $-1$ to $+1$, and set $u = z + (z^2 - 1)^{1/2}$, where the square root is real positive when $z$ is real and $> 1$. Show that the curve $|u| = r > 1$ is the ellipse passing through the point $z$ and having its foci at $-1$ and $+1$. Show that

$$\limsup_{n\to\infty} |P_n(z)|^{1/n} = |u|.$$

8. Use these results to discuss the convergence of the series (6.2.16).

9. Let $f(t)$ be holomorphic in an ellipse with foci at $-1, +1$. Verify the stated convergence properties of the Legendre series of $f(t)$. Show that on the ellipse of convergence there is always a singular point.

10. Verify (6.2.4).

11. Verify (6.2.5).

12. Since $\mathrm{Re}\,(c - a - b) = 0$ for Legendre's equation, $P_a(x)$ cannot tend to a finite limit as $x$ decreases to $-1$. Instead it becomes infinite as a multiple of $\log[\tfrac{1}{2}(1 - x)]$. Determine the multiple and derive (6.2.8).

13. Present a similar discussion at $z = \infty$ with a view toward finding (6.2.11).

14. For fixed $z$ the Legendre function $P_a(z)$ satisfies a linear second order *difference equation* with respect to $a$:

$$(a + 1)P_{a+1}(z) - (2a + 1)zP_a(z) + aP_{a-1}(z) = 0.$$

Verify for integral values of $a$, knowing that $P_0(z) \equiv 1$, $P_1(z) = z$.

15. All polynomials $P_{2n+1}(z)$ evidently have $z = 0$ as a root. Show that no number can be a common root of $P_n(z)$ and $P_{n+1}(z)$, and that, if $z = z_0$ is a common root of $P_n(z)$ and $P_{n+2}(z)$, then $z_0 = 0$.

## 6.3.  BESSEL'S EQUATION

Few scientists have been able to contribute to as many fields as Gauss did, but it was as an astronomer that he made his living and the few pupils that he had were trained as astronomers and became good ones. But it was no accident that they also gained fame for mathematical contributions. August Ferdinand Möbius (1790–1868) is a case in point: one volume of his collected works is on astronomy, the remaining three deal with mathematics. Friedrich Wilhelm Bessel (1784–1848) is another. In 1824 he studied the perturbations in the orbits of the planets and in this connection developed the theory of the DE

$$z^2 w'' + z w' + (z^2 - a^2) w = 0, \tag{6.3.1}$$

the parameter $a$ being zero or a positive integer. G. Schlömilch (1857) and R. Lipschitz (1859) proposed to name the equation after Bessel. Special instances of (6.3.1), however, were known long before Bessel. In this connection we can mention the contributions of Jacob Bernoulli (1703), Daniel Bernoulli (1725, 1732), L. Euler (1764), J. L. Lagrange (1769), and J. Fourier (1822). These were investigations into applied mathematics, involving such varied topics as oscillations of heavy chains, vibrations of circular membranes, elliptic motion, and heat conduction in cylinders.

There is no doubt that Bessel functions have enjoyed wide popularity and been of great use. During the 250 years that have elapsed since the start, the literature has grown enormously. G. N. Watson's *Theory of Bessel Functions*, the modern standard treatise on the subject, has 36 pages of bibliography.

Equation (6.3.1) has a regular-singular point at $z = 0$ with roots of the indicial equation equal to $a$ and $-a$. The point at infinity is irregular singular of grade $1 = 1 + \sup(-1, 0)$.

The roots of the indicial equation differ by $2a$, and as long as $2a$ is not an integer, the two functions $J_a(z)$ and $J_{-a}(z)$ form a fundamental system with

$$J_a(z) = \sum_{k=0}^{\infty} \frac{(-1)^k}{\Gamma(k+1)\Gamma(a+k+1)} \left(\frac{z}{2}\right)^{2k+a}. \tag{6.3.2}$$

Here the series converge for $0 < |z| < \infty$.

If $a = n + \frac{1}{2}$, the roots of the indicial equation differ by $2n + 1$ and logarithmic terms could be expected but are actually missing. For $n = 0$ we have

$$J_{1/2}(z) = \left(\frac{2}{\pi z}\right)^{1/2} \sin z, \qquad J_{-1/2}(z) = \left(\frac{2}{\pi z}\right)^{1/2} \cos z \tag{6.3.3}$$

with no logarithms. It may be shown that

$$J_{n+1/2}(z) = A_n(z) \cos z + B_n(z) \sin z, \tag{6.3.4}$$

where $A_n$ and $B_n$ are polynomials in $z^{-1/2}$.

On the other hand, if $a$ is zero or an integer, logarithms do appear, and in addition to the Bessel function $J_a(z)$ of the first kind we have also to consider *Bessel functions of the second kind*. Various definitions are commonly used; they differ only by constant multiples of $J_n(z)$. If $a$ is not an integer, set

$$Y_a(z) = \operatorname{cosec}(\pi a)[J_a(z) \cos(\pi a) - J_{-a}(z)]. \qquad (6.3.5)$$

This is obviously a solution of (6.3.1). We now let $a \to n$ and find that a limit exists, denoted by $Y_n(z)$. It is a Bessel function of the second kind of order $n$. A tedious calculation gives

$$Y_n(z) = -\sum_{k=0}^{n-1} \frac{(n-k-1)!}{k!} (\tfrac{1}{2}z)^{-n+2k}$$
$$+ \sum_{k=0}^{\infty} \frac{(-1)^k}{k!(n+k)!} (\tfrac{1}{2}z)^{n+2k} \Big[ 2\log(\tfrac{1}{2}z) + 2C - \sum_1^{n+k} \frac{1}{m} - \sum_1^k \frac{1}{m} \Big], \qquad (6.3.6)$$

where $C$, as usual, is the Euler constant.

The function $(a, z) \mapsto J_a(z)$ satisfies a number of functional equations with respect to $a$ and/or $z$. Note, in particular, the second order *difference equation*

$$zJ_{a+1}(z) - 2aJ_a(z) + zJ_{a-1}(z) = 0. \qquad (6.3.7)$$

We also note

$$J'_a(z) = \tfrac{1}{2}[J_{a-1}(z) - J_{a+1}(z)]. \qquad (6.3.8)$$

The difference equation shows that the sequence $\{J_n(x)\}$ for $x$ real $>0$ forms a so-called *Sturmian chain*; i.e., if there is an $n$ and an $x_0 > 0$ such that $J_n(x_0) = 0$, then $J_{n-1}(x_0)J_{n+1}(x_0) < 0$. In particular, two consecutive $J_n$'s cannot have a positive root in common.

Equation (6.3.1) is formally real if $z$ is purely imaginary and $a$ is real, for if we set $z = iy$, $0 < y$, the result is

$$y^2 \frac{d^2 w}{dy^2} + y \frac{dw}{dy} - (y^2 + a^2)w = 0. \qquad (6.3.9)$$

For $a$ real, two real solutions of this equation are $I_a(y)$ and $I_{-a}(y)$, where

$$I_a(y) = \sum_{k=0}^{\infty} \frac{(\tfrac{1}{2}y)^{a+2k}}{\Gamma(k+1)\Gamma(a+k+1)}, \qquad (6.3.10)$$

and this differs from $J_a(iy)$ by a constant multiplier. These solutions form a fundamental system for (6.3.9) unless $a$ equals zero or a positive integer, in which case logarithms enter and a second solution may be found along the lines indicated above.

There is a Laurent series expansion

$$\exp\Big[\tfrac{1}{2}z\Big(t - \frac{1}{t}\Big)\Big] = \sum_{-\infty}^{\infty} J_n(z)t^n. \qquad (6.3.11)$$

Here the left member may be considered as a *generating function* of the system $\{J_n(z)\}$. The left member is the product of $\exp(\frac{1}{2}zt)$ and $\exp(-\frac{1}{2}zt^{-1})$. To derive (6.3.11) one expands the factors in exponential series, multiplies termwise, and rearranges according to powers of $t$. The expansion was found by Schlömilch and implies a number of important relations and inequalities.

Take first $z = iy$, $t = -i$, which gives

$$e^y = I_0(y) + 2 \sum_{n=1}^{\infty} I_n(y) \tag{6.3.12}$$

convergent for all $y$. If $y > 0$, all terms are positive and we get

$$I_0(y) < e^y, \qquad I_n(y) < \tfrac{1}{2}e^y. \tag{6.3.13}$$

If $y < 0$, replace $y$ by $|y|$ in the inequalities. Next take $t = e^{i\theta}$ to obtain

$$e^{iz \sin \theta} = \sum_{-\infty}^{\infty} J_n(z) e^{in\theta} \tag{6.3.14}$$

convergent for all $z$ and real $\theta$. It follows that $J_n(z)$ is a Fourier coefficient,

$$J_n(z) = \frac{1}{2\pi} \int_{-\pi}^{+\pi} e^{i(z \sin \theta - n\theta)} \, d\theta. \tag{6.3.15}$$

Here we can take the "formally real part" to obtain

$$J_n(z) = \frac{1}{\pi} \int_0^{\pi} \cos(z \sin \theta - n\theta) \, d\theta, \tag{6.3.16}$$

which happens to be Bessel's definition of $J_n(z)$. Note that the "formally imaginary part" is zero.

Now, if $z = x > 0$, it is seen that

$$|J_n(x)| < 1, \qquad \forall x, \forall n. \tag{6.3.17}$$

Next observe that the function $\theta \mapsto e^{iz \sin \theta} \in L_2(-\pi, \pi)$, so that for $z = x + iy$ one obtains

$$\int_{-\pi}^{\pi} |e^{iz \sin \theta}|^2 \, d\theta = \int_{-\pi}^{\pi} e^{-2y \sin \theta} \, d\theta = 2\pi J_0(2iy) = 2\pi I_0(2y).$$

The closure relation for the orthogonal system $\{e^{ni\theta}\}$ then gives

$$I_0(2y) = \sum_{-\infty}^{\infty} |J_n(x + iy)|^2. \tag{6.3.18}$$

This yields

$$|J_0(x + iy)| \leq e^{|y|}, \qquad |J_n(x + iy)| \leq 2^{-1/2} e^{|y|}, \qquad 0 < n. \tag{6.3.19}$$

Note that (6.3.18) suggests the identity

$$[J_0(z)]^2 + 2 \sum_1^\infty [J_n(z)]^2 \equiv 1 \qquad (6.3.20)$$

for all $z$. By (6.3.18) the relation is true for $z$ real, $y = 0$, and the series (6.3.20) is a uniformly convergent series of holomorphic functions, so the sum is holomorphic. Now, if a holomorphic function is identically one on the real axis, it is identically one in its domain of existence—in this case, the finite plane.

Integral representations of Bessel functions will be considered later. Here we have only to mention *Bessel functions of the third kind*, introduced by Hermann Hankel (1839–1873), a brilliant mathematician and classical scholar, in 1869. Besides the Bessel functions, his name is attached to an integral for the reciprocal of the gamma function and a class of determinants. He codified the field postulates and wrote a history of early mathematics; most of this work was published after his untimely death.

To get the proper background for the Bessel-Hankel functions we have to go back to Section 5.6. The substitution

$$w(z) = z^{-1/2} W(z) \qquad (6.3.21)$$

takes (6.3.1) into

$$W''(z) + \left(1 - \frac{a^2 - \frac{1}{4}}{z^2}\right) W(z) = 0, \qquad (6.3.22)$$

an equation which satisfies Hypothesis F. From the discussion in Section 5.6 we know of the existence of a solution $E_a^+(z)$, which is asymptotic to $e^{iz}$ in the sector $-\pi < \arg z < 2\pi$, and a solution $E_a^-(z)$ asymptotic to $e^{-iz}$ in the sector $-2\pi < \arg z < \pi$. These solutions correspond to the Bessel-Hankel functions $H_a^{(1)}(z)$ and $H_a^{(2)}(z)$; we have

$$H_a^{(1)}(z) = C_1(a)z^{-1/2}E_a^+(z), \qquad H_a^{(2)}(z) = C_2(a)z^{-1/2}E_a^-(z), \qquad (6.3.23)$$

and the constants are chosen so that

$$J_a(z) = \tfrac{1}{2}[H_a^{(1)}(z) + H_a^{(2)}]. \qquad (6.3.24)$$

Thus $J_a(z)$ must be an oscillatory solution of (6.3.1), and it may be shown that

$$J_a(z) = \left(\frac{2}{\pi z}\right)^{1/2} \cos\left[z - \tfrac{1}{2}\pi(a + \tfrac{1}{2})\right] + O(|z|^{-3/2}). \qquad (6.3.25)$$

### EXERCISE 6.3

The first four problems deal with the expansion of the Cauchy kernel $(t - z)^{-1}$ in terms of Bessel functions $J_n(z)$. This is in analogy with (6.2.16) and one of the many expansions of this kind.

1. Prove that

$$e^{zu} = J_0(z) + \sum_{n=1}^{\infty} J_n(z)\{[u + (u^2 + 1)^{1/2}]^n + [u - (u^2 + 1)^{1/2}]^n\},$$

   and show that the series is absolutely convergent for all $z$ and $u$. Prove that the coefficient of $J_n(z)$ is actually a polynomial in $u$ of degree $n$.

2. For Re $(t) > 0$ show that

$$O_n(t) = \frac{1}{2} \int_0^{\infty} e^{-tu}\{[u + (u^2 + 1)^{1/2}]^n + [u + (u^2 + 1)^{1/2}]^n\} \, du$$

   exists as a polynomial of degree $n$ in $1/t$.

3. Show that the series

$$O_0(t)J_0(z) + 2 \sum_{n=1}^{\infty} O_n(t)J_n(z)$$

   converges absolutely for $|z| < |t|$ and uniformly for $|z| \leqslant a < b \leqslant |t|$.

4. Multiply both sides of the identity in Problem 1 by $e^{-tu}$, and integrate with respect to $u$ from zero to infinity. By Problems 2 and 3 the right member equals the series in Problem 3, so that

$$\frac{1}{t - z} = O_0(t)J_0(z) + 2 \sum_{n=1}^{\infty} O_n(t)J_n(z).$$

   Show that the procedure is justified, say for $0 < z < t$, and then extend it to $0 < |z| < |t|$ by analytic continuation.

5. Carl Neumann found in 1867 the expansion

$$f(z) = \sum_{n=0}^{\infty} a_n J_n(z), \qquad a_n = \frac{1}{2\pi i} \int_C f(t)O_n(t) \, dt,$$

   where $z \mapsto f(z)$ is holomorphic in a disk $|z| < r$, and $C$ is the circle $|t| = \rho < r$. Verify.

6. Otto Schlömilch in 1857 studied expansions in terms of Bessel functions, of which the following is typical:

$$\sum_0^{\infty} a_n J_0(nx), \qquad 0 < x < \pi.$$

   Show that the series converges if $|a_n|$ is bounded and, further, the series $\Sigma_1^{\infty} n^{-1/2}|a_n - a_{n+1}|$ converges. When does the series converge for $x = 0$?

7. Given (6.3.25), where the remainder is $< C(a)x^{-3/2}$, find intervals on the positive $x$-axis in each of which $J_n(x)$ has a single zero and outside the union of which there are at most a finite number of zeros.

## 6.4. LAPLACE'S EQUATION

Pierre Simon Laplace (1749–1827) in his treatise *Théorie analytique des probabilitées* of 1817 considered a DE which carries his name and may be written as

$$(a_2 + b_2 x)\frac{d^2 w}{dx^2} + (a_1 + b_1 x)\frac{dw}{dx} + (a_0 + b_0 x)w = 0, \qquad (6.4.1)$$

where the $a$'s and $b$'s are given numbers. We set

$$a_2 + b_2 x = z \qquad (6.4.2)$$

and obtain

$$D_z(w) = zw''(z) + (c_0 + c_1 z)w'(z) + (d_0 + d_1 z)w(z) = 0. \qquad (6.4.3)$$

This equation has $z = 0$ and $z = \infty$ as singular points, the former regular singular, the latter irregular singular. The indicial equation at the origin is

$$r(r - 1) + c_0 r = 0, \qquad r_1 = 0, \qquad r_2 = 1 - c_0. \qquad (6.4.4)$$

If $c_0$ is not an integer, one solution is an entire function of $z$, while the other is $z^{1-c_0}$ times an entire function. Omitting some trivial cases, we find that the point at infinity is irregular singular with grade

$$g = 1 + \max(0, 0) = 1. \qquad (6.4.5)$$

It is desired to study this DE for large values of $z$. More precisely, we want asymptotic representations of the solutions in sectors of the plane, and the method developed in Section 5.6 is not satisfactory in this fringe case. Laplace was faced with this problem, and he invented what was later known as the *Laplace transform* to cope with the difficulty. Actually, a number of integrals go under this name. They are of the form

$$\int_C e^{zs} F(s)\, ds, \qquad (6.4.6)$$

where the weight function $F(s)$ and the path of integration $C$ have to be chosen so that (6.4.6) becomes a solution of (6.4.2). Assuming the right to apply the differential operator $D_z$ under the sign of integration, we get

$$D_z\left[\int_C e^{zs} F(s)\, ds\right] = \int_C D_z(e^{zs})F(s)\, ds, \qquad (6.4.7)$$

where

$$D_z(e^{zs}) = (zs^2 + c_0 s + d_0 + d_1 z)e^{zs}$$

$$= (s^2 + c_1 s + d_1)\frac{d}{ds}(e^{zs}) + (d_0 + c_0 s)e^{zs}.$$

Substitution in (6.4.7) and integration by parts leads to

$$[(s^2 + c_1 s + d_1)F(s)e^{zs}]_C - \int_C \{(s^2 + c_1 s + d_1)F'(s)$$

$$- [d_0 - c_1 + (c_0 - 2)s]F(s)\}e^{zs}\, ds. \qquad (6.4.8)$$

This will reduce to zero, provided that two conditions are met:

1. F is chosen as a solution of

$$(s^2 + c_1 s + d_1)F'(s) = -[c_1 - d_0 + (2 - c_0)s]F(s). \qquad (6.4.9)$$

2. The path of integration $C$ can be chosen so that

$$[(s^2 + c_1 s + d_1)F(s)e^{zs}]_C = 0. \qquad (6.4.10)$$

The first condition is sufficiently clear, but it leads to two distinct cases according as the equation

$$s^2 + c_1 s + d_1 = 0 \qquad (6.4.11)$$

has (i) distinct or (ii) equal roots.

In case (i) suppose that the roots are $s_1$ and $s_2$. Then

$$F(s) = (s - s_1)^\alpha (s - s_2)^\beta \qquad (6.4.12)$$

for a definite choice of $\alpha$ and $\beta$.

In case (ii) suppose that the double root is $s_0 = -\frac{1}{2}c_1$. Then

$$F(s) = (s - s_0)^\alpha \exp\left(-\frac{\beta}{s - s_0}\right), \qquad (6.4.13)$$

again for some choice of the parameters.

Condition 2 leads to various possibilities. The integrated part should vanish in the limit. If the roots are distinct, three different types of paths are available:

1. *Arcs* joining two singular points.
2. *Simple loops* starting at one singularity, surrounding another singularity, and returning to the starting points.
3. A *double loop* surrounding $s_1$ and $s_2$.

Using type 1, we have three choices: (1:i) from $s = s_1$ to $s = s_2$, (1:ii) from $s_1$ to $\infty$, and (1:iii) from $s_2$ to $\infty$. Here (1:i) requires that $\text{Re}\,(\alpha) > -1$, $\text{Re}\,(\beta) > -1$. If this condition is satisfied, the integrated part,

$$G(s, z) \equiv (s - s_1)^{\alpha+1}(s - s_2)^{\beta+1}e^{zs}, \qquad (6.4.14)$$

vanishes at both ends of $C$, and we have the solution

$$w_0(z) = \int_{s_1}^{s_2} (s - s_1)^\alpha (s - s_2)^\beta e^{sz}\, ds, \qquad (6.5.15)$$

which is an entire function of $z$. To use choice (1:ii) we must still have $\text{Re}\,(\alpha) > -1$, and we must approach the point at infinity along a line such that $\text{Re}\,(zs) \to -\infty$. For a given $z \neq 0$ such a direction can always be found.

For such a choice, $G(s, z)$ will again vanish at the end points and a solution is obtained. The situation is similar for choice (1:iii).

A simple loop from infinity and surrounding $s = s_1$ is shown in Figure 6.1. It is supposed that the line $\arg(s - s_1) = \omega_1$ is permissible at infinity.

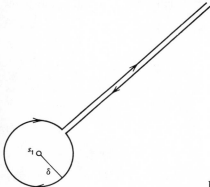

**Figure 6.1**

The loop starts at infinity, follows one edge of the line, makes a $\delta$-circuit in the positive sense about $s = s_1$, and retreats to infinity along the other edge. The asymptotic representation of this integral at infinity will be derived.

A double loop is shown in Figure 6.2. When $s$ describes $C$, it surrounds each of the points $s_1$ and $s_2$ twice but in the opposite direction, so that $G(s, z)$ returns to its original value and the integrated part vanishes.

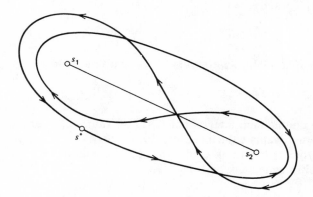

**Figure 6.2**

In case (ii) of condition 1 the weight function $F(s)$ is given by (6.4.13). There is a straight line through $s = s_0$ along which $\bar{\beta}(s - s_0)$ is purely imaginary, and the exponential function is bounded to one side of the line and unbounded to the other. There are essentially two different paths at our disposal, both shown in Figure 6.3. One starts and ends at $s = s_0$, where it has a cusp. It is a closed path which passes from the half-plane

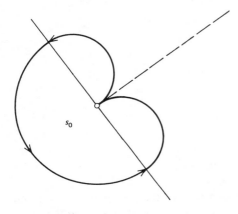

**Figure 6.3**

where the exponential function is bounded, through the half-plane where it is unbounded, and then back again. This solution is an entire function. The second choice starts at $s = s_0$ in the bounded half-plane and proceeds to infinity in a permissible direction. We shall not pay any further attention to case (ii).

Returning to case (i) and the double loop integral around $s_1$ and $s_2$, we set $s_2 - s_1 = j$ and $s = s_1 + jt$. Then

$$w_0(z) = j^{\alpha+\beta+1} e^{s_1 z} \int_{C_0} e^{jzt} t^{\alpha} (1 - t)^{\beta} \, dt, \qquad (6.4.16)$$

where $C_0$ is a double loop around zero and one. Hence

$$w_0(z) = j^{\alpha+\beta+1} e^{s_1 z} \sum_{k=0}^{\infty} \frac{(jz)^k}{k!} \int_{C_0} t^{\alpha+k} (1 - t)^{\beta} \, dt. \qquad (6.4.17)$$

Here the double loop integral exists for all values of $\alpha$ and $\beta$. It is then enough if we evaluate the integrals, say for Re $(\alpha) > -1$, Re $(\beta) > -1$, and use analytic continuation. Under the stated conditions the integrals may be explicitly evaluated, and if

$$J_k = \int_0^1 t^{\alpha+k} (1 - t)^{\beta} \, dt = \frac{\Gamma(\alpha + k + 1)\Gamma(\beta + 1)}{\Gamma(\alpha + \beta + k + 2)}, \qquad (6.4.18)$$

the corresponding double loop integral is

$$(1 - e^{-2\pi i\alpha})(1 - e^{-2\pi i\beta})J_k \equiv J_k^0. \tag{6.4.19}$$

To see this, suppose that we start the integration at $t = \delta$ with a value of the integrand which is real positive if $\alpha$ and $\beta$ are real positive. Here $0 < \delta < 1 - \delta < 1$. Proceed along the real axis to $t = 1 - \delta$, and there make a circuit in the negative sense around $t = 1$. The new determination of the integrand at $t = 1 - \delta$ equals $e^{-2\pi i\beta}$ times the old determination. Now proceed on the real axis to $t = \delta$, and there make a negative circuit around $t = 0$. Return to $t = \delta$ with $e^{-2\pi i(\alpha + \beta)}$ times the initial determination. Now proceed along the real axis to $t = 1 - \delta$, followed by a positive circuit about $t = 1$. The new determination at $t = 1 - \delta$ is $e^{-2\pi i\alpha}$ times the value encountered the first time when this point was reached. Proceed to $t = \delta$, and follow with a positive circuit around $t = 0$. This means returning to the starting point and the original value of the integrand. It follows that

$$J_k^0 = [1 - e^{-2\pi i\beta} + e^{-2\pi i(\alpha + \beta)} - e^{-2\pi i\alpha}] \int_\delta^{1-\delta} \cdots + o(\delta),$$

where the remainder represents the four circuit integrals. Letting $\delta \to 0$, we get (6.4.19). Draw a figure and follow the steps! It follows that

$$w_0(z) = (1 - e^{-2\pi i\alpha})(1 - e^{-2\pi i\beta})\Gamma(\beta + 1)j^{\alpha + \beta + 1}e^{s_1 z}$$
$$\times \sum_{k=0}^\infty \frac{\Gamma(\alpha + k + 1)}{\Gamma(\alpha + \beta + k + 2)} \frac{(jz)^k}{k!}. \tag{6.4.20}$$

At this stage we introduce the Pochhammer-Barnes notation,

$$_pF_q(a_1, a_2, \ldots, a_p; b_1, b_2, \ldots, b_q; z) = \sum_{k=0}^\infty \frac{(a_1)_k \ldots (a_p)_k}{k!(b_1)_k \ldots (b_q)_k} z^k. \tag{6.4.21}$$

With this notation we have

$$w_0(z) = j^{\alpha + \beta + 1}(1 - e^{-2\pi i\alpha})(1 - e^{-2\pi i\beta})\frac{\Gamma(\alpha + 1)\Gamma(\beta + 1)}{\Gamma(\alpha + \beta + 2)}$$
$$\times e^{s_1 z} {}_1F_1(\alpha + 1; \alpha + \beta + 2; jz), \tag{6.4.22}$$

where $j = s_2 - s_1$. If one of the quantities $\alpha$, $\beta$, $\alpha + \beta + 1$ is a negative integer, the right-hand side is to be replaced by its limit as the parameter approaches the critical value. Formula (6.4.22) holds also for (6.4.15) provided that the factor $(1 - e^{-2\pi i\alpha})(1 - e^{-2\pi i\beta})$ is omitted.

We shall find the asymptotic behavior of this solution after that of the simple loop integral from infinity to one of the points $s = s_1$ or $s = s_2$ has been found.

At this stage we need Hankel's integral (1864) for the reciprocal gamma function:

$$\frac{1}{\Gamma(a)} = \frac{1}{2\pi i} \int_C e^u u^{-a} \, du. \qquad (6.4.23)$$

Cut the $u$-plane along the negative real axis, and define $u^{-a} = e^{-a \log u}$, where the imaginary part of $\log u$ lies between $-\pi$ and $+\pi$. The contour of integration is a simple loop surrounding the negative real axis. It follows the lower edge of the cut from $-\infty$ to $-\delta$, $0 < \delta$, a positive $\delta$-circuit around the origin, and then along the upper edge of the cut back to $-\infty$. The integral exists for every finite value of $a$, and using, for instance, the theorem of Morera one proves that the integral defines an entire function of $a$. That this function is $1/\Gamma(a)$ is proved along lines now familiar to the reader.

Let us consider

$$w_1(z) = \int_L e^{zs} (s - s_1)^\alpha (s - s_2)^\beta \, ds, \qquad (6.4.24)$$

where $L$ is a simple loop to be defined. Mark a ray from $s = s_1$ to infinity such that the ray does not pass through $s = s_2$ and along the ray $\mathrm{Re}\,(zs) \to -\infty$ as $s \to \infty$. $L$ is now a simple closed loop surrounding the ray in the positive sense. Start at infinity with the value 0, and follow the appropriate edge of the ray until a point is reached at a distance $\delta$ from $s_1$, where $\delta < |j|$. Make a positive circuit around $s = s_1$, and return to infinity along the other edge of the cut. As above, set $s = s_1 + jt$. This maps the loop $L$ onto a loop $L_1$ and

$$w_1(z) = j^{\alpha+\beta+1} e^{s_1 z} \int_{L_1} e^{jzt} t^\alpha (1 - t)^\beta \, dt. \qquad (6.4.25)$$

The loop $L_1$ surrounds $t = 0$, it does not pass through $t = 1$, and the integral out to infinity exists. Expand $(1 - t)^\beta$ in binomial series

$$(1 - t)^\beta = \sum_{k=0}^\infty \frac{(-\beta)_k}{k!} t^k.$$

This series converges for $|t| < 1$, but for the time being we disregard this restriction and consider the formal series,

$$\sum_{k=0}^\infty \frac{(-\beta)_k}{k!} \int_{L_1} e^{jzt} t^{k+\alpha} \, dt.$$

Here we make the change of variable $jzt = u$. The resulting integral in the $k$th term of the formal series becomes

$$(jz)^{-\alpha-k-1} \int_C e^u u^{k+\alpha} \, du,$$

where $C$ is the loop occurring in the Hankel integral (6.4.23). The value of the integral then is

$$\frac{2\pi i}{\Gamma(-k-\alpha)} = (-1)^k 2i \sin(\pi\alpha)\Gamma(\alpha+k+1),$$

where we have used the symmetry formula for the gamma function,

$$\Gamma(x)\Gamma(1-x) = \frac{\pi}{\sin \pi x},$$

and the periodicity properties of $\sin \pi x$. Furthermore,

$$\Gamma(\alpha+1+k) = \Gamma(\alpha+1)(\alpha+1)_k.$$

This gives finally

$$w_1(z) = 2i \sin(\pi\alpha)\Gamma(\alpha+1)j^{\beta+1}z^{-\alpha-1}e^{s_1z} {}_2F_0\left(\alpha+1, -\beta; -\frac{1}{jz}\right). \quad (6.4.26)$$

The series naturally diverges, but it is an asymptotic representation in the sense of H. Poincaré (1886). He defined a formal power series to be *asymptotic* to $F(u)$:

$$F(u) \sim \sum_0^\infty a_n u^n \quad (6.4.27)$$

for approach to the origin along a ray or in a sector if, for all $n$,

$$\lim u^{-n}\left[F(u) - \sum_{k=0}^n a_k u^k\right] = 0. \quad (6.4.28)$$

To prove that (6.4.26) has such asymptotic properties it is necessary to go back to the point where divergent series entered the argument. We can always write

$$(1-t)^\beta = \sum_{k=0}^n \frac{(-\beta)_k}{k!} t^k + R_n(t). \quad (6.4.29)$$

From the obvious relation

$$(1-t)^\beta - 1 = -\beta \int_0^t (1-s)^{\beta-1} ds$$

we can obtain an expression for $R_n(t)$ by repeated integration by parts. The result may be written as

$$R_n(t) = \frac{(-\beta)_n}{n!} n \int_0^t (t-s)^n (1-s)^{\beta-n-1} ds$$

$$= \frac{(-\beta)_n}{n!} nt^{n+1} \int_0^1 (1-r)^n (1-rt)^{\beta-n-1} dr.$$

Now, if $t = R e^{i\theta}$, then $\theta \neq 0$ since $L_1$ does not go through $t = 1$. Furthermore,

$$\min |1 - rt| = |\sin \theta|,$$

and there exists an $M$ independent of $n$ such that

$$\left| \int_0^1 (1-r)^n (1-rt)^{\beta - n - 1} \, dr \right| < M |\text{cosec } \theta|^{n+1+|\beta|} \int_0^1 (1-r)^n \, dr.$$

Hence

$$R_n(t) = \frac{(-\beta)_n}{n!} t^{n+1} P_n(t) \qquad \text{and} \qquad |P_n(t)| \leqslant M |\text{cosec } \theta|^{n+1+|\beta|}. \qquad (6.4.30)$$

In the integral (6.5.25) we now replace $(1-t)^\beta$ by (6.4.29). Here the finite sum gives an expression involving a partial sum of the $_2F_0$-series, so we have to estimate the remainder using (6.4.30). Here

$$\left| \int_{L_1} e^{jzt} t^\alpha R_n(t) \, dt \right| \leqslant \frac{|(-\beta)_n|}{n!} \left| \int_{L_1} e^{jzt} t^{n+1+\alpha} P_n(t) \, dt \right|$$

$$\leqslant \frac{|(-\beta)_n|}{n!} M |\text{cosec } \theta|^{n+1+|\beta|} \left| \int_{L_1} e^{jzt} t^{n+1+\alpha} \, dt \right|.$$

Here we can choose $L_1$ so that $jzt$ becomes real negative. Thus $jzt = -|jz||t|$ and $|t^\alpha| < A |t|^{|\alpha|}$. The integral is then dominated by

$$2A \int_0^\infty \exp\left[-|jz||t|\right] |t|^{n+1+|\alpha|} |dt| = 2A \, \Gamma(n+2+|\alpha|) |jz|^{-n-2-|\alpha|}.$$

This gives

$$\left| \int_{L_1} e^{jzt} t^\alpha R_n(t) \, dt \right| < M_1 |\text{cosec } \theta|^{n+1+|\beta|} \frac{1}{n!} \Gamma(|\beta| + n)$$

$$\times \Gamma(|\alpha| + n + 2) |jz|^{-n-2-|\alpha|}. \qquad (6.4.31)$$

Set

$$2i \sin(\alpha\pi)\Gamma(\alpha + 1) j^\beta = K. \qquad (6.4.32)$$

Then

$$\left| K^{-1}(jz)^{-\alpha-1} e^{-s_1 z} w_1(z) - \sum_0^n \frac{1}{k!} (\alpha + 1)_k (-\beta)_k (-jz)^{-k} \right|$$

$$< M_1 |\text{cosec } \theta|^{n+1+|\beta|} \frac{1}{n!} \Gamma(|\beta| + n) \Gamma(|\alpha| + n + 2) |jz|^{-n-2-|\alpha|}. \qquad (6.4.33)$$

After multiplication by $|zj|^n$ the right-hand member still goes to zero as $z \to \infty$, so the $_2F_1$-series is indeed asymptotic to $K^{-1}(jz)^{-\alpha-1} e^{-s_1 z} w_1(z)$. It is seen that $w_1(z)$ goes to infinity in the half-plane $\text{Re}\,(s_1 z) > 0$ and to zero in $\text{Re}\,(s_1 z) < 0$ when $z \to \infty$.

Let $w_2(z)$ be the solution corresponding to a simple loop about $s = s_2$

from and to infinity. This solution has the same type of asymptotic expansion as (6.4.26), where now $s_1$ is replaced by $s_2$, $j$ by $-j$, and $\alpha$ and $\beta$ are interchanged. Thus

$$w_2(z) \sim 2i \sin(\beta\pi)\Gamma(\beta+1)(-j)^\alpha(-jz)^{-\beta-1}\,e^{\,s_2 z}$$
$$\times \,_2F_0\left(-\alpha, \beta+1; \frac{1}{jz}\right). \tag{6.4.34}$$

We can now also get the asymptotic representation of $w_0(z)$, defined by (6.4.16). Since the solutions are meromorphic functions of $\alpha$ and $\beta$, we may assume that $\mathrm{Re}\,(\alpha) > -1$, $\mathrm{Re}\,(\beta) > -1$, so that the loop integrals may be replaced by straight line integrals as in Figure 6.4. The three straight

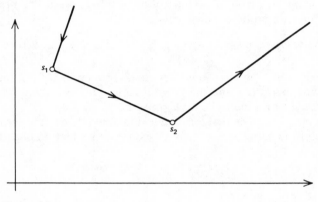

**Figure 6.4**

lines are closed by an arc of a circle of large radius. The integral along the closed contour is zero, and for suitably restricted values of $z$ the integral along the arc goes to zero as the radius becomes infinite. This implies that

$$\int_{s_1}^{s_2} = \int_{s_1}^{\infty} - \int_{s_2}^{\infty}. \tag{6.4.35}$$

If now the straight line integrals are expressed in terms of the loop integrals, we get the relation between $w_0(z)$, $w_1(z)$, and $w_2(z)$ and obtain the asymptotic formula for $w_0(z)$ from (6.4.26), (6.4.34), and (6.4.35). The unpleasantly difficult details are left to the diligent reader.

To illustrate we consider Bessel's equation

$$z^2 w'' + z w' + (z^2 - a^2)w = 0, \tag{6.4.36}$$

which is reduced to the form (6.4.2) by setting $w = z^a v$. Thus

$$zv'' + (2a + 1)v' + zv = 0, \tag{6.4.37}$$

so that $c_0 = 2a + 1$, $c_1 = 0$, $d_0 = 0$, $d_1 = 1$. The equation for $F(s)$ becomes

$$(s^2 + 1)F'(s) = (2a - 1)sF(s), \tag{6.4.38}$$

so that

$$F(s) = (s^2 + 1)^{a-1/2} \tag{6.4.39}$$

and

$$v(z) = \int_C e^{zs}(s^2 + 1)^{a-1/2} \, ds. \tag{6.4.40}$$

Here $C$ may be a loop integral or a straight line integral. The latter choice is admissible if Re $(a) > -\frac{1}{2}$. There are then three choices: an integral from $s = i$ to infinity or from $-i$ to infinity or from $-i$ to $+i$. The first choice leads to an asymptotic expansion,

$$(-2i)^{a+1/2}\Gamma(a + \tfrac{1}{2})z^{-a-1/2}e^{iz}{}_2F_0\left(a + \tfrac{1}{2}, -a + \tfrac{1}{2}; -\frac{i}{2z}\right). \tag{6.4.41}$$

Multiplication by $z^a$ gives the Bessel-Hankel function $H_a^{(1)}(z)$, the second choice leads to $H_a^{(2)}(z)$, and the integral from $-i$ to $+i$ gives $J_a(z)$, in all three cases up to a multiplicative constant.

Assuming $z = x$ to be real positive and taking only the first terms of the two asymptotic series, we obtain an expression which implies that

$$J_a(x) \sim \left(\frac{2}{\pi x}\right)^{1/2} \cos\left[x - \tfrac{1}{2}\pi(a + \tfrac{1}{2})\right] \tag{6.4.42}$$

in the first approximation.

The equations studied in this section go under the name of *confluent hypergeometric equations*. The hypergeometric equation

$$x(1 - x)\frac{d^2y}{dx^2} + [c - (a + b + 1)x]\frac{dy}{dx} + aby = 0 \tag{6.4.43}$$

has three singular points, 0, 1, $\infty$. If we now set $x = z/b$, a transformed equation is obtained with the singular points 0, $b$, $\infty$. Here we let $b \to \infty$, and find that the limiting equation may be written as

$$zw''(z) + (c - z)w'(z) + aw(z) = 0 \tag{6.4.44}$$

with the solution

$${}_1F_1(a; c; z) = \lim_{b \to \infty} {}_2F_1\left(a, b, c; \frac{z}{b}\right). \tag{6.4.45}$$

A second solution may be obtained by applying the same process to the solution $w_{02}(x)$ of (6.4.43). It is

$$z^{1-c}{}_1F_1(a - c + 1; 2 - c; z).  \qquad (6.4.46)$$

Equation (6.4.44) is the typical confluent hypergeometric equation.

It is clear that, if $a = -n$, a negative integer, then ${}_1F_1(a; c; z)$ is a polynomial of degree $n$ in $z$. We write

$$_1F_1(-n; 1 + \alpha; z) = \binom{n + \alpha}{n} L_n^{(\alpha)}(z)  \qquad (6.4.47)$$

and refer to $L_n^{(\alpha)}(z)$ as the $n$th (Abel-) *Laguerre polynomial* of order $\alpha$. If $\alpha = 0$, the superscript is usually dropped and $L_n(z)$ is the $n$th *Laguerre polynomial*, named for Edmond Laguerre (1834–1886), a brilliant Frenchman who enriched algebra, analysis, and complex geometry.

These polynomials have a number of important and interesting properties. We list only two, and ask the reader to verify them. The expansion

$$\sum_{n=0}^{\infty} L_n(x)t^n = (1 - t)^{-1} \exp\left(-\frac{zt}{1 - t}\right)  \qquad (6.4.48)$$

may be regarded as a *generating function* of the polynomials. Furthermore, we have the orthogonality property

$$\int_0^{\infty} e^{-x} L_m(x) L_n(x) \, dx = \delta_{mn}.  \qquad (6.4.49)$$

The set $\{e^{-x/2}L_n(x)\}$ is a complete orthonormal system for $L_2(0, \infty)$.

## EXERCISE 6.4

1. Determine the values of $\alpha$ and $\beta$ in (6.4.12).
2. Why is $w_0(z)$ of (6.4.15) an entire function of $z$?
3. Consider the equation $zw'' + w = 0$, which is a Laplace equation satisfied by Bessel functions. Evaluate the loop integral of Figure 6.3 for this case (calculus of residues or otherwise). Express it as a Bessel function.
4. Verify (6.4.20).
5. Verify (6.4.22).
6. Verify the Hankel integral (6.4.23).
7. Verify (6.4.34).
8. Find the asymptotic representation for $w_0(z)$.
9. Express the integral

$$\int_{-i}^{i} (1 + s^2)^{a-1/2} e^{zs} \, ds$$

   in terms of $J_a(z)$.
10. Explain how (6.4.42) is obtained.
11. Carry through the derivation of (6.4.46).
12. Prove that $e^z {}_1F_1(c - a, c; -z) = {}_1F_1(a, c; z)$.

13. Show that

$$n!L_n(z) = e^z \frac{d^n}{dz^n}(z^n e^{-z})$$

and obtain similar formulas for the $n$th polynomial of order $\alpha$.

14. Show that the zeros of Laguerre polynomials are real positive.

15. Try to prove (6.4.48). Note that the right member is holomorphic in $|t| < 1$ for any value of $z$ and may be expanded in a power series for $|t| < 1$ with coefficients which are functions of $z$. Discuss the coefficients.

16. Prove (6.4.49). [*Hint:* Try using (6.4.48).]

17. In Bessel's equation (6.4.36) set $t = -2iz$, $w = z^a e^{iz}u$. The resulting equation for $u$ as a function of $t$ is of form (6.4.44). Find the parameters.

### 6.5. THE LAPLACIAN: THE HERMITE-WEBER EQUATION; FUNCTIONS OF THE PARABOLIC CYLINDER

The DE's of classical mathematical physics usually arise from a boundary value problem involving one or an other of the partial DE's: Laplace's equation, Poisson's equation, the wave equation, or the heat equation. Frequently the problem suggests a change of variables which affects the Laplacian.

The two-dimensional Laplacian is

$$\frac{\partial^2 U}{\partial x^2} + \frac{\partial^2 U}{\partial y^2} \equiv \Delta_{xy}U. \tag{6.5.1}$$

The basic formula for the change of variables is given by

#### LEMMA 6.5.1

*Let* $U(x, y) = U(z)$ *be defined and twice continuously differentiable in a domain* $D$ *of the z-plane, and let* $D$ *be the one-to-one image of a domain* $D_0$ *under the conformal mapping*

$$z = f(u), \qquad u = s + it, \tag{6.5.2}$$

*where* $f(u)$ *is holomorphic in* $D_0$. *Then*

$$\Delta_{st}U = |f'(u)|^2 \Delta_{xy}U. \tag{6.5.3}$$

*Proof.* Straightforward computation shows that

$$\Delta_{st}U = \frac{\partial^2 U}{\partial s^2} + \frac{\partial^2 U}{\partial t^2} = \frac{\partial^2 U}{\partial x^2}\left[\left(\frac{\partial x}{\partial s}\right)^2 + \left(\frac{\partial x}{\partial t}\right)^2\right] + \frac{\partial^2 U}{\partial y^2}\left[\left(\frac{\partial y}{\partial s}\right)^2 + \left(\frac{\partial y}{\partial t}\right)^2\right]$$
$$+ \frac{\partial U}{\partial x}\left[\frac{\partial^2 x}{\partial s^2} + \frac{\partial^2 x}{\partial t^2}\right] + \frac{\partial U}{\partial y}\left[\frac{\partial^2 y}{\partial s^2} + \frac{\partial^2 y}{\partial t^2}\right]. \tag{6.5.4}$$

Here each of the last two brackets is zero since $x$ and $y$ are harmonic functions of $s$, $t$. By the Cauchy-Riemann equations each of the first two brackets equals $|f'(u)|^2$, so that (6.5.3) holds. ∎

In particular, if $U$ is a harmonic function of $x$, $y$, then $U[f(u)]$ is a harmonic function of $s$, $t$.

There are three problems in potential theory involving cylinders and axial symmetry. The cylinders are circular, parabolic, and elliptic, and they give rise to the *functions of the circular cylinder, the parabolic cylinder*, and *the elliptic cylinder*, respectively. The reader has already encountered the circular cylinder functions; they are the Bessel functions, but at this juncture it may be desirable to justify the name "circular cylinder functions."

It is required to find a solution of the Poisson equation [Siméon Denis Poisson, 1781–1840]:

$$\Delta_{xy} U + U = 0, \tag{6.5.5}$$

which is a function of $r$ and $\theta$ alone, $z = r\,e^{i\theta}$, twice continuously differentiable for $|z|^2 = x^2 + y^2 < R^2$, and which satisfies a given boundary condition on $|z| = R$; for instance, $U(r\,e^{i\theta})$ converges to a given function $f(\theta)$ in the $L_2(0, 2\pi)$-metric as $r \uparrow R$. The problem calls for polar coordinates, and then (6.5.5) becomes

$$\frac{\partial^2 U}{\partial r^2} + \frac{1}{r}\frac{\partial U}{\partial r} + \frac{1}{r^2}\frac{\partial^2 U}{\partial \theta^2} + U = 0. \tag{6.5.6}$$

Here it is possible to separate the variables. The equation has solutions of the form $U(r\,e^{i\theta}) = f(r)g(\theta)$. If we perform the indicated differentiations and divide the result by $f(r)g(\theta)$, it is seen that

$$r^2 f''(r) + r f'(r) + (r^2 - c)f(r) = 0, \tag{6.5.7}$$

$$g''(\theta) + cg(\theta) = 0, \tag{6.5.8}$$

where $c$ is a constant, but not an arbitrary constant, for $f(r)g(\theta)$ must satisfy (6.5.6) and be twice continuously differentiable. This requires $g$ to be periodic with period $2\pi$, so that $c$ must be the square of an integer, $c = n^2$. Then $f$ must satisfy

$$r^2 f''(r) + r f'(r) + (r^2 - n^2)f(r) = 0, \tag{6.5.9}$$

i.e., $f(r)$ is a Bessel function of integral order $n$. The continuity condition forces the solution to be a constant multiple of $J_n(z)$. Thus we see that

$$(c_n e^{ni\theta} + c_{-n} e^{-ni\theta})J_n(r) \tag{6.5.10}$$

is a solution of (6.5.6) for arbitrary integral values of $n$ and arbitrary constants $c_n$, $c_{-n}$. Since the problem is linear, the general solution is

$$c_0 J_0(r) + \sum_{n=1}^{\infty} J_n(r)(c_n e^{ni\theta} + c_{-n} e^{-ni\theta}). \qquad (6.5.11)$$

This should be $L_2$ with respect to $\theta$ uniformly in $r$ for $0 \le r \le R$, and since

$$J_n(r) = (\tfrac{1}{2}r)^n \frac{1}{n!}[1 + O(n^{-1})],$$

the $L_2$-condition is satisfied iff

$$\sum_{1}^{\infty} \left(\frac{1}{n!}\right)^2 (\tfrac{1}{2}R)^{2n}[|c_n|^2 + |c_{-n}|^2] \qquad (6.5.12)$$

is convergent. Here the boundary values must be given by the $L_2$-series,

$$f(\theta) \sim c_0 J_0(R) + \sum_{n=1}^{\infty} J_n(R)(c_n e^{ni\theta} + c_{-n} e^{-ni\theta}). \qquad (6.5.13)$$

On this note we leave the functions of the circular cylinder.

Our main task in this section is to study the corresponding problem for a domain bounded by a parabolic cylinder,

$$y^2 = 4R(R - x) \qquad (6.5.14)$$

as a right-hand boundary. The domain bounded by the parabola and cut along the negative real axis is mapped conformally on a vertical strip, $0 < s < (2R)^{1/2}$, by

$$z = f(s + it) = \tfrac{1}{2}u^2 = \tfrac{1}{2}(s^2 - t^2) + ist. \qquad (6.5.15)$$

We now take the partial DE in the form

$$\Delta_{xy} U - U = 0, \qquad (6.5.16)$$

so that the transformed equation is

$$\Delta_{st} U - (s^2 + t^2)U = 0. \qquad (6.5.17)$$

Here again the variables may be separated. The equation has solutions of the form $f(s)g(t)$ and

$$f''(s) - (c + s^2)f(s) = 0, \qquad g''(t) + (c - t^2)g(t) = 0, \qquad (6.5.18)$$

so the equations are almost identical. The solutions are *the functions of the parabolic cylinder.*

We take the equation in the normal form

$$w''(z) + (c - z^2)w(z) = 0, \qquad (6.5.19)$$

where $z$ is a complex variable and $c$ an arbitrary constant. This DE is also known as the *Hermite-Weber equation*. Charles Hermite (1822–1901) considered it in 1865; Heinrich Weber (1842–1913), in 1869. It may be reduced to a confluent hypergeometric equation by the substitution

$$t = z^2, \qquad u = \exp\left(\tfrac{1}{2}z^2\right) w, \qquad (6.5.20)$$

which leads to

$$t\frac{d^2u}{dt^2} + \left(\tfrac{1}{2} - t\right)\frac{du}{dt} + \tfrac{1}{2}(c - 1) u = 0 \qquad (6.5.21)$$

with solutions

$$_1F_1[\tfrac{1}{2}(1 - c), \tfrac{1}{2}; t] \qquad \text{and} \qquad t^{1/2} \,_1F_1(1 - \tfrac{1}{2}c, \tfrac{3}{2}; t). \qquad (6.5.22)$$

The case in which $c$ is a positive odd integer is particularly important. If $c = 4m + 1$, then

$$_1F_1(-m, \tfrac{1}{2}; t)$$

is a polynomial in $t$ of degree $m$. If $c = 4m + 3$, then

$$t^{1/2} \,_1F_1(-m, \tfrac{3}{2}; t)$$

is $t^{1/2}$ times a polynomial in $t$ of degree $m$. This means that, if $c = 2n + 1$, then (6.5.19) has a solution of the form

$$E_n(z) = \exp\left(-\tfrac{1}{2}z^2\right) H_n(z), \qquad (6.5.23)$$

where $H_n(z)$ is the $n$th *Hermite polynomial*, which is of degree $n$ and is an even or odd function of $z$ according as $n$ is an even or odd integer. We have the *generating function*

$$\exp(2zt - t^2) = \sum_{n=0}^{\infty} H_n(z) \frac{t^n}{n!} \qquad (6.5.24)$$

and the representation

$$H_n(z) = (-1)^n \exp(z^2) \frac{d^n}{dz^n} [\exp(-z^2)]. \qquad (6.5.25)$$

Not all authors use the same notation for the Hermite polynomials; the main variation is that $z^2$ is replaced by $\tfrac{1}{2}z^2$, as in E. T. Whittaker and G. N. Watson, *A Course in Modern Analysis* (1952).

The set $\{E_n(z)\}$ forms an orthogonal system for the interval $(-\infty, \infty)$. This follows from the DE. If $m$ and $n$ are distinct integers, then

$$E_m''(x) + (2m + 1 - x^2)E_m(x) = 0,$$
$$E_n''(x) + (2n + 1 - x^2)E_n(x) = 0.$$

Here we multiply the first equation by $E_n(x)$, the second by $E_m(x)$, and subtract to obtain

$$\frac{d}{dx}[E_m(x)E_n'(x) - E_n(x)E_m'(x)] = 2(m - n)E_m(x)E_n(x).$$

This is integrated from $-\infty$ to $+\infty$. It follows from (6.5.25) that the left member is integrable and the integral is zero:

$$\int_{-\infty}^{\infty} E_m(x)E_n(x)\,ds = \int_{-\infty}^{\infty} \exp(-x^2)H_m(x)H_n(x)\,dx = 0 \qquad (6.5.26)$$

if $m \neq n$. It may be shown that

$$\int_{-\infty}^{\infty} \exp(-x^2)[H_n(x)]^2\,dx = \pi^{1/2}\,2^n\,n!. \qquad (6.5.27)$$

The same technique gives

$$E_n(x)\,E_n'(x) + \int_x^{\infty} [E_n'(s)]^2\,ds + \int_x^{\infty} [s^2 - 2n - 1][E_n(s)]^2\,ds = 0. \qquad (6.5.28)$$

The identity implies that

$$E_n(x)\,E_n'(x) < 0 \qquad \text{for } x^2 \geq 2n + 1. \qquad (6.5.29)$$

In particular, the product cannot be zero for such values of $x$, and this extends to the interval $(-\infty, -(2n+1)^{1/2}]$ since the product is an odd function of $x$. Thus all real zeros of $H_n(x)$ belong to the interval $(-(2n+1)^{1/2}, (2n+1)^{1/2})$. Formula (6.5.25) implies that

$$H_{n+1}(z) - 2zH_n(z) + 2nH_{n-1}(z) = 0. \qquad (6.5.30)$$

This, in turn, implies that the set $\{H_n(x)\}$ forms a Sturmian chain. Hence, if $H_n(x_0) = 0$, then $H_{n+1}(x_0)H_{n-1}(x_0) < 0$. Since $H_0(x) \equiv 1$ and $H_{2k+1}(0) = 0$ for all $k$, we conclude that sgn $H_{2k}(0) = (-1)^k$ for all $k$. Furthermore, it is known that $H_n[(2n+1)^{1/2}] > 0$ for all $n$. From these facts we conclude step by step or by induction that all the zeros of $H_n(z)$ are real and that the zeros of $H_{n+1}(x)$ alternate with those of $H_n(x)$. The verification is left to the reader.

The asymptotic behavior of the solutions of (6.5.19) for large values of $z$ may be concluded from the results for confluent hypergeometric equations via (6.5.21) or by the method of asymptotic integration of Section 5.6. The results will be reviewed briefly. We have

$$Z(z) = \int_0^z (c - s^2)^{1/2}\,ds = \tfrac{1}{2}iz^2 + \tfrac{1}{2}ic \log z + O(z^{-2}). \qquad (6.5.31)$$

The lines $y = \pm x$ divide the plane into four sectors, in each of which there is a subdominant solution. For the sector $|\arg z| < \tfrac{1}{4}\pi$ this takes the form

$$E_1^+(z) = z^{(c-1)/2} \exp(-\tfrac{1}{2}z^2)[1 + o(1)], \qquad (6.5.32)$$

and the same formula represents $E_3^+(z)$ in the sector $\tfrac{3}{4}\pi < \arg z < \tfrac{5}{4}\pi$. Here $E_1^+(z)$ has no distant zeros in $|\arg z| < \tfrac{1}{4}\pi$, while $E_3^+(z)$ has no distant zeros in $|\arg(-z)| < \tfrac{1}{4}\pi$. For these two solutions, which are entire functions, the defect of the value 0 is normally $\frac{1}{2}$. The zeros of $E_1^+(z)$ satisfy $\lim \arg z_n \equiv \frac{3}{4}\pi$ or $-\frac{3}{4}\pi$, and $N(r, 0; E_1^+) = (1/\pi)r^2 + O(r)$. The value $c = 2n + 1$ is exceptional. Here $E_1^+(z)$ and $E_3^+(z)$ coincide and become multiples of $E_n(z)$, which has only $n$ zeros, so the defect of zero is one and zero is a Picard value. A similar situation holds when $c$ is a negative odd integer, for then

$$H_n(iz) \exp(\tfrac{1}{2}z^2) \tag{6.5.33}$$

is a solution for $c = -2n - 1$ and the subdominants $E_2^+(z)$ and $E_4^+(z)$ are linearly dependent. No other values of the parameter $c$ can give rise to such phenomena. We note that the set $\{E_n(z)\}$ corresponds to a Titchmarsh problem for (6.5.19).

## EXERCISE 6.5

1. Show that a solution of the problem

$$\Delta_{st} U - (s^2 + t^2)U = 0, \qquad 0 < s < (2R)^{1/2}, \qquad \lim U(s, t) = F(t),$$

where $F(t) \in L_2(-\infty, \infty)$ and convergence is in the $L_2$-metric, is of the form

$$\sum_{n=1}^{\infty} c_n \exp[\tfrac{1}{2}(t^2 - s^2)]H_n(s)H_n(it).$$

2. If $w(z, a)$ is a solution of (6.5.21), show that $w(iz, -a)$ is also a solution.

3. If $c = -1$, show that $w(z) = \exp(\tfrac{1}{2}z^2) \int_z^{\infty} \exp(-s^2)\, ds$ is a subdominant in the sector $|\arg z| < \tfrac{1}{4}\pi$, and find subdominants for the other sectors.

4. Find the DE satisfied by $H_n(z)$, and show that it is confluent hypergeometric.

5. Prove (6.5.27).

6. Verify (6.5.29).

7. Find $H_{2k}(0)$ and $H'_{2k+1}(0)$, and show that they are $O(k^{-1/2})$. (*Hint:* The formula for $\pi$ as the limit of factorials due to John Wallis may come in handy.)

8. Prove the statements in the text regarding the zeros of $H_n(x)$.

9. Since there are $n$ zeros of $H_n(x)$ in the interval $(-A, A)$, where $A = (2n + 1)^{1/2}$, the average distance between consecutive zeros is $\sim 2^{3/2} n^{-1/2}$. Show that the minimum distance is about $\pi(2n)^{-1/2}$.

10. Define two quadratic differential forms by

$$F_n(x) = [E'_n(x)]^2 + (2n + 1 - x^2)[E_n(x)]^2,$$

$$G_n(x) = \frac{[E'_n(x)]^2}{2n + 1 - x^2} + [E_n(x)]^2.$$

Show that the ordinates $|E_n(x_k)|$ at the consecutive extrema of $E_n(x)$ form an increasing sequence in $[0, (2n + 1)^{1/2}]$ and that $(2n + 1 - x_k^2)^{1/2}|E_n(x_k)|$ is a decreasing sequence.

11. Use the results of Problems 7 and 10 to show that on any fixed interval $(-b, b)$ the set $n^{1/2}|E_n(x)|$ is uniformly bounded.

### 6.6.  THE EQUATION OF MATHIEU; FUNCTIONS OF THE ELLIPTIC CYLINDER

We have left for consideration the elliptic cylinder. Let the axis of the cylinder be perpendicular to the $z$-plane at the origin, and let the intersection of the cylinder with the $z$-plane be the ellipse

$$E: \frac{x^2}{R^2+1} + \frac{y^2}{R^2} = 1. \tag{6.6.1}$$

The foci of $E$ are at the points $z = -1$ and $z = +1$. We cut the domain bounded by $E$ along the real axis from the left end point of the major axis to the left focus and from the right focus to the right end point of the major axis. The cut domain is the conformal image under the mapping

$$z = \sin u, \qquad u \equiv s + it \tag{6.6.2}$$

of the rectangle

$$-\tfrac{1}{2}\pi < s < \tfrac{1}{2}\pi, \qquad -c < t < c, \qquad c = \log[R + (R^2+1)^{1/2}]. \tag{6.6.3}$$

We have

$$z = x + iy, \qquad x = \sin s \cosh t, \qquad y = \cos s \sinh t$$

and

$$|f'(u)|^2 = \cos^2 s + \sinh^2 t.$$

Consider now the partial DE,

$$\Delta_{xy} U + KU = 0, \tag{6.6.4}$$

where $K$ is a positive parameter. The transformed equation is

$$\Delta_{st} U + K(\cos^2 s + \sinh^2 t)U = 0. \tag{6.6.5}$$

Here again the variables may be separated, and the equation admits of solutions of the form $f(s)g(t)$, which satisfy the equations

$$f''(s) + (C + K\cos^2 s)f(s) = 0, \tag{6.6.6}$$

$$g''(t) + (-C + K\sinh^2 t)g(t) = 0. \tag{6.6.7}$$

We shall consider the first equation in the normal form

$$w''(z) + (a + b\cos 2z)w(z) = 0. \tag{6.6.8}$$

This is known as *Mathieu's equation* after Émile Léonard Mathieu (1835–1890), who in 1868 studied the vibration of an elliptic membrane. It leads to another equation of type (6.6.4) after the time coordinate has been eliminated from the wave equation. Mathieu's name is also attached to a class of finite groups investigated by him.

Equation (6.6.8) contains two parameters, $a$ and $b$. In the physical problem just mentioned, $a$ and $b$ must be real, and it is also seen that the solution must be periodic with $\pi$ as a period or a half-period to be physically significant. This raises the question of the existence of periodic solutions of a DE with periodic coefficients. We shall have something to say about this general problem in the next section; here we shall present some aspects of the problem as posed by Mathieu's equation.

Equation (6.6.8) does not change if $z$ is replaced by $z + k\pi$, where $k$ is an integer. This means that if $w(z)$ is a solution so is $w(z + k\pi)$. The problem facing us is to decide under what conditions on the parameters $a$ and $b$ there exists a solution $w(z)$ with period $k\pi$,

$$w(z + k\pi) = w(z). \tag{6.6.9}$$

For the physical problem it is enough to treat the cases $k = 1$ and $2$.

In passing, we note that if $w(z)$ is a solution so is $w(-z)$ and

$$\tfrac{1}{2}[w(z) + w(-z)] \qquad \text{and} \qquad \tfrac{1}{2}[w(z) - w(-z)].$$

The first is an even function of $z$; the second, an odd function. This means that the search for periodic solutions may be restricted to the subsets of even and of odd solutions. A further simplification arises from the fact that the existence of even or odd periodic solutions is equivalent to the fact that certain boundary value problems have solutions. Now the existence of such solutions follows either from the general theory of such problems or from function-theoretical considerations. The second alternative is appropriate in a treatise on DE's in the complex plane and will be used here.

Suppose that $w(z)$ is an even solution of primitive ($=$ least) period $2\pi$. Then $w'(z)$ is odd and of period $2\pi$. We have $w'(0) = 0$ and $w'(-\pi) = w'(\pi)$ by the periodicity, while $w'(-\pi) = -w'(\pi)$ since the function is odd. Thus $w'(\pi) = 0$. Hence our solution must satisfy the two boundary conditions,

$$w'(0) = 0, \qquad w'(\pi) = 0. \tag{6.6.10}$$

Suppose conversely that a solution of (6.6.8) satisfies (6.6.10). In the DE (6.6.8) the factor $(a + b \cos 2z)$ is even, so a solution with $w(0) = c \neq 0$, $w'(0) = 0$ is an even function of $z$. By the same token $w(z)$ is an even function of $z - \pi$. Now $w(\pi - s) = w(\pi + s)$ implies successively that

$$w(2\pi + z) = w(-z) = w(z) \qquad \text{or} \qquad w(z + 2\pi) = w(z),$$

so that our solution does have the period $2\pi$.

Suppose instead that $w(z)$ is an odd solution of primitive period $2\pi$. Then $w(0) = 0$ and $w(-\pi) = -w(\pi)$, since the solution is odd, while, on the other hand, $w(-\pi) = w(\pi)$ by the periodicity. Hence $w(\pi) = 0$, and our solution satisfies the conditions

$$w(0) = 0, \qquad w(\pi) = 0. \qquad (6.6.11)$$

Conversely, if $w(z)$ is a solution satisfying these conditions, then the argument given above shows that $w(z)$ is an odd function of $z$ as well as of $z - \pi$. From $w(\pi - s) = -w(\pi + s)$ we infer that with $s = \pi - z$ we have $w(z) = -w(2\pi - z) = w(z - 2\pi)$ and, finally, $w(z + 2\pi) = w(z)$. Thus (6.6.11) implies that $w(z)$ is odd and has the period $2\pi$.

Next we consider the case in which $w(z)$ is odd and has the period $\pi$. This implies, in the first place, that $w(0) = 0$. Since $w(-z) = -w(z)$ and $w(z + \pi) = w(z)$, this gives $w(-\tfrac{1}{2}\pi) = w(\tfrac{1}{2}\pi)$ by the periodicity and $w(-\tfrac{1}{2}\pi) = -w(\tfrac{1}{2}\pi)$ by the oddness, so that $w(\tfrac{1}{2}\pi) = 0$. Thus

$$w(0) = 0, \qquad w(\tfrac{1}{2}\pi) = 0. \qquad (6.6.12)$$

Conversely, suppose that a solution satisfies these conditions. Then we see that $w(z)$ is an odd function of $z$ as well as of $z - \tfrac{1}{2}\pi$. Now from $w(\tfrac{1}{2}\pi - s) = -w(\tfrac{1}{2}\pi + s)$ we infer that $w(z + \pi) = -w(-z) = w(z)$. Thus (6.6.12) implies that $w(z)$ is odd and has period $\pi$.

Finally, if $w(z)$ is even and has period $\pi$, we show that

$$w'(0) = 0, \qquad w'(\tfrac{1}{2}\pi) = 0; \qquad (6.6.13)$$

and conversely, if these equations hold, then $w(z)$ is even and has period $\pi$.

If solutions of these four boundary value problems exist, they may be expanded in Fourier series, and corresponding to the four conditions we have the expansions

(i) $\displaystyle\sum_{n=0}^{\infty} c_{1,n}^{k} \cos(2n+1)z,$      (ii) $\displaystyle\sum_{n=0}^{\infty} c_{2,n}^{k} \sin(2n+1)z,$

$$(6.6.14)$$

(iii) $\displaystyle\sum_{n=0}^{\infty} c_{3,n}^{k} \sin 2nz,$      (iv) $\displaystyle\sum_{n=0}^{\infty} c_{4,n}^{k} \cos 2nz.$

These four types of periodic solutions are denoted by

$$\begin{array}{ll} \text{(i)} \quad \mathrm{ce}_{2k+1}(z), & \text{(ii)} \quad \mathrm{se}_{2k+1}(z), \\ \text{(iii)} \quad \mathrm{se}_{2k}(z), & \text{(iv)} \quad \mathrm{ce}_{2k}(z), \end{array} \qquad (6.6.15)$$

and the notation is such that

$$\lim \mathrm{ce}_m(z) = \cos mz, \qquad \lim \mathrm{se}_m(z) = \sin mz$$

when $b \to 0$.

In the applications $b$ is given and $a$ has to be determined so that one or another of the four sets of boundary conditions is satisfied.

We shall indicate a function-theoretical argument to prove the existence of infinitely many values of $a$ such that for a given $b$ one of the boundary value problems, e.g., (6.6.11), has a solution. The existence of a solution of (6.6.9) such that

$$w(0) = 0, \qquad w'(0) = 1,$$

is obvious and this holds for all values of $a$ and $b$. Let this solution be $w(z; a, b)$; it satisfies the Volterra integral equation

$$w(z) = z - \int_0^z (z - s)(a + b \cos 2s)w(s)\, ds. \qquad (6.6.16)$$

The problem is now to show that for a given value of $b \neq 0$ it is possible to find values of $a$ such that $w(\pi; a, b) = 0$. To do this we need, in the first place, an estimate of the rate of growth of $w(\pi; a, b)$ as a function of $a$ for fixed $b$. It is clearly an entire function of $a$. We take $z = x$ as real positive, and set $|w(x)| = u(x)$. This gives

$$u(x) \leqslant x + (|a| + |b|) \int_0^x (x - s)u(s)\, ds. \qquad (6.6.17)$$

From (6.6.17) we get, for example by successive substitutions,

$$u(x) \leqslant (|a| + |b|)^{-1/2} \sinh [(|a| + |b|)^{1/2}x]$$

and finally

$$|w(\pi; a, b)| \leqslant (|a| + |b|)^{-1/2} \sinh [(|a| + |b|)^{1/2}\pi]. \qquad (6.6.18)$$

It follows that $w(\pi; a, b)$ for fixed $b$ is an entire function of $a$ of order $\frac{1}{2}$ and normal type. An entire function of order at most $\frac{1}{2}$ either reduces to a polynomial and the order becomes 0, or has infinitely many zeros $\{a_n(b)\}$ with an exponent of convergence $\leqslant \frac{1}{2}$, i.e.,

$$\sum_{n=1}^{\infty} |a_n(b)|^{-\sigma} \qquad (6.6.19)$$

converges for any $\sigma > \frac{1}{2}$. The corresponding solutions $w(z; a_n(b), b)$ are the functions $se_{2n+1}(z)$ which satisfy (6.6.12). The probability of the function $w(\pi; a, b)$ reducing to a polynomial in $a$ would seem to be (and actually is) zero, but the author knows of no elementary argument that establishes this. With this reservation we have proved what we set out to do. We have, furthermore,

### THEOREM 6.6.1

*If b is real, the characteristic values of the boundary value problems (6.6.10)–(6.6.13) are real and the functions* $ce_{2n+1}(x)$, $se_{2n+1}(x)$, $se_{2n}(x)$, *and* $ce_{2n}(x)$ *form an orthogonal system for* $L_2(0, \pi)$.

*Proof.* Consider one of the functions $w_m(x)$ corresponding to $a = a_m$. Then, for $z = x$ real,

$$w''_m(x) = -(a_m + b \cos 2x)w_m(x). \tag{6.6.20}$$

Suppose that $a_m$ is not real, so that $w_m(x)$ is complex valued. Multiply both sides by the conjugate of $w_m(x)$, and integrate from 0 to $\pi$. Thus

$$\int_0^\pi \overline{w_m(x)} w''_m(x)\, ds = -\int_0^\pi (a_m + b \cos 2x)|w_m(x)|^2\, dx. \tag{6.6.21}$$

Integration by parts gives

$$[\overline{w_m(x)} w'_m(x)]_0^\pi = \int_0^\pi |w'_m(x)|^2\, dx - \int_0^\pi (a_m + b \cos 2x)|w_m(x)|^2\, dx. \tag{6.6.22}$$

Here the integrated part vanishes in both limits by virtue of the boundary conditions. Since the first integral is real positive, the second integral must be real, as is the case iff $a_m$ is real.

A glance at (6.6.14) shows that $\text{se}_m(x)$ and $\text{ce}_n(x)$ are orthogonal on $(0, \pi)$ regardless of the values of $m$ and $n$. It remains to prove the orthogonality of $\text{se}_m(x)$ and $\text{se}_n(x)$, $m \neq n$, and of $\text{ce}_m(x)$ and $\text{ce}_n(x)$.

Suppose that $w_m(x)$ with characteristic value $a_m$ satisfies (6.6.20), and let $w_n(x)$ with characteristic value $a_n$ satisfy

$$w''_n(x) = -(a_n + b \cos 2x)w_n(x), \qquad m \neq n. \tag{6.6.23}$$

Multiply (6.6.20) by $w_n(x)$ and (6.6.23) by $w_m(x)$, subtract, and integrate from 0 to $\pi$, obtaining

$$\int_0^\pi [w_n(x)w''_m(x) - w_m(x)w''_n(x)]\, dx = (a_n - a_m)\int_0^\pi w_m(x)w_n(x)\, ds.$$

The integral on the left can be carried out and gives

$$[w_n(x)w'_m(x) - w_m(x)w'_n(x)]_0^\pi,$$

which is zero by the boundary conditions. Since $a_m \neq a_n$ we see that, as asserted,

$$\int_0^\pi w_m(x)w_n(x)\, dx = 0. \quad \blacksquare$$

The periodic solutions of (6.6.8) and, in particular, the functions $\text{ce}_n(z)$ and $\text{se}_m(z)$ are known as *Mathieu functions of the first kind.* There are also *associated Mathieu functions* which are essentially the solutions of (6.6.7). Suppose that $a$ and $b$ are real. Then for any integer $k$ or 0 (6.6.8) is formally real on all the lines

$$z = \tfrac{1}{2}k\pi + iy, \qquad -\infty < y < +\infty, \tag{6.6.24}$$

the transformed equations being

$$\frac{d^2w}{dy^2} - [a + (-1)^k b \cosh 2y]w = 0. \qquad (6.6.25)$$

The solutions of these equations are the associated Mathieu functions. Their behavior depends on the sign of $b$ and the parity of $k$. If $b > 0$, $k$ even, the real solutions of (6.6.25) go hyperexponentially to infinity or, exceptionally, to zero. If $b > 0$, $k$ odd, the real solutions exhibit very rapid, strongly damped oscillations. If $b < 0$, the parity of $k$ should also be changed in this description.

To study the asymptotic behavior of these solutions or, equivalently, the behavior of $w(z)$ in the complex plane, we use the transformation of Liouville. Set

$$Z(z) = \int_{z_0}^{z} (a + b \cos 2s)^{1/2} \, ds, \qquad W(Z) = (a + b \cos 2z)^{1/4} w(z). \qquad (6.6.26)$$

The lines $x = (k + \frac{1}{2})\pi$, together with the real axis, divide the complex plane into infinitely many half-strips,

$$
\begin{aligned}
D_k^+ &: [z | z = x + iy, (k - \tfrac{1}{2})\pi < x < (k + \tfrac{1}{2})\pi, 0 < y], \\
D_k^- &: [z | z = x + iy, (k - \tfrac{1}{2})\pi < x < (k + \tfrac{1}{2})\pi, y < 0].
\end{aligned} \qquad (6.6.27)
$$

In each of these half-strips there is a unique normalized subdominant furnished by the discussion in Section 5.6. Let the subdominant in $D_0^+$ be denoted by $E(z)$. We know that its asymptotic form for $z$ restricted to $D_0^+$ is, for $b > 0$,

$$E(z) = \exp\left[-(\tfrac{1}{2}b)^{1/2} e^{-iz} + \tfrac{1}{2}iz\right][C_0 + O(e^{iz})], \qquad C_0 \neq 0. \qquad (6.6.28)$$

The symmetry properties of (6.6.8) show that $E(-z)$ is subdominant in $D_0^-$, and, more generally, the subdominants are given by

$$E(z - k\pi) \quad \text{in } D_k^+ \qquad \text{and} \qquad E(k\pi - z) \quad \text{in } D_k^-,$$

for $E(z - k\pi)$ is clearly subdominant in $D_k^+$ and, since the normalized subdominant is unique, the conclusion follows.

These properties have important consequences. $E(z)$ is an entire function defined in the whole finite plane. It goes to zero in $D_0^+$ and to infinity in the adjacent half-strips, $D_{-1}^+$ and $D_1^+$. This implies, among other things, that $E(z)$ and $E(z + \pi)$ are linearly independent; in particular, $E(z)$ cannot have the period $\pi$.

The solutions $E(z)$ and $E(-z)$ are linearly independent. To prove this, set

$$\tfrac{1}{2}[E(z) + E(-z)] = C(z), \qquad \tfrac{1}{2}[E(x) - E(-z)] = S(z).$$

Here $C(z)$ is even while $S(z)$ is odd. If $E(-z)$ were a constant multiple of

$E(z)$, then $S(z)$ would be a constant multiple of $C(z)$, and this is absurd. The possibility that $E(z)$ itself is even or odd is excluded by (6.6.32) below.

From these properties it follows that for no values of $a$ and $b$ can there exist two linearly independent solutions with the period $\pi$ or the period $2\pi$. To have two solutions with period $\pi$ is impossible, for this would require $E(z)$ to have the period $\pi$—an impossibility. If the solutions had period $2\pi$ but not $\pi$, one solution could be taken as $ce_{2n+1}(z)$ while the other was $se_{2n+1}(z)$. Then these solutions would have the property $f(z + \pi) = -f(z)$, and ultimately this would lead to $E(z)$ and $E(-z)$ being linearly dependent—another absurdity.

To complete the discussion we must justify (6.6.28), and this requires a discussion of the mapping $z \mapsto Z(z)$ defined by (6.6.26). The discussion can be restricted to the mapping of the half-strip $D_0^+$. See Figure 6.5 for

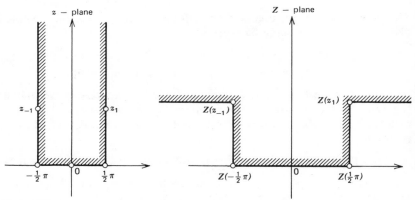

**Figure 6.5.**

the geometry of the mapping. We assume $a > b > 0$ and $z_0 = 0$. Then there are two branch points of $Z(z)$ on the vertical boundaries, given by

$$\pm \tfrac{1}{2}\pi + i\tfrac{1}{2} \log \frac{1}{b} [a + (a^2 - b^2)^{1/2}]. \qquad (6.6.29)$$

Then $Z = Z(z)$ maps the interior of $D_0^+$ on the interior of a region $U^+$ which contains an upper half-plane and is bounded by five line segments, two vertical and three horizontal. There are four vertices, two corresponding to the points $z = \pm \tfrac{1}{2}\pi$, and two with interior angles of $270°$ corresponding to the branch points.

Using the symmetry principle of H. A. Schwarz, we can continue $Z(z)$

analytically across each of the five boundary lines of $D_0^+$. Let us enter $D_1^+$ by crossing the right vertical boundary of $D_0^+$, above the branch point if $a > b > 0$. The image of $D_1^+$ under this determination of $Z(z)$ is a domain $U_1^-$, the mirror image of $U^+$ with respect to the horizontal line segment furthest to the right. Similarly we get an image $U_{-1}^-$ of $D_{-1}^+$ by reflecting $U^+$ in the horizontal line segment furthest to the left.

In the three domains $D_0^+$, $D_1^+$, and $D_{-1}^+$, the function $z \mapsto Z(z)$ has the same asymptotic character,

$$Z(z) = i(\tfrac{1}{2}b)^{1/2} e^{-iz} + O(1), \tag{6.6.30}$$

and the corresponding function $F(Z)$ is found to satisfy Hypothesis $F$ of Section 5.6, so the asymptotic integration theory applies. We have

$$F(Z) = -\tfrac{1}{4}Z^{-2} + O(Z^{-4}). \tag{6.6.31}$$

It follows that the subdominant solution $E(z)$ in $D_0^+$ is indeed given by (6.6.28), as asserted. The estimate for the remainder holds uniformly for $-\tfrac{3}{2}\pi + \epsilon \leqslant x \leqslant \tfrac{3}{2}\pi - \epsilon, 0 < y$. In particular, on the lines $x = \pm \pi$ we have

$$E(\pm \pi + iy) = \exp[(\tfrac{1}{2}b)^{1/2} e^y - \tfrac{1}{2}y][\pm iC_0 + O(e^{-y})]. \tag{6.6.32}$$

Finally, in $D_0^-$ we have

$$E(z) = \exp[(\tfrac{1}{2}b)^{1/2} e^{iz} - \tfrac{1}{2}iz][C_1 + O(e^{-iz})] \tag{6.6.33}$$

as $y \to -\infty$. Thus $E(z)$ is a dominant solution in $D_0^-$.

## EXERCISE 6.6

1. Are there any Mathieu functions having the primitive period $3\pi$?

2. Why do the boundary conditions (6.6.10)–(6.6.13) imply Fourier series of type (6.6.14)?

3. If $b$ is real, show that the Fourier coefficients are real.

4. It is desired to estimate the rate of decrease of the Fourier coefficients. Show that

$$\left| \int_0^\pi ce_{2n}(s) e^{2iks} \, ds \right| < M(t) e^{-2kt}, \qquad 0 < k,$$

where $M(t) = \max |ce_{2n}(z)|$ in the rectangle with vertices at $0$, $\pi$, $\pi + it$, $it$. [*Hint:* Integrate along the perimeter of the rectangle.]

5. A reasonable estimate of $M(t)$ is $C(n) \exp[(\tfrac{1}{2}b)^{1/2} e^t - \tfrac{1}{2}t]$. How should $t$ be chosen so that the estimate in Problem 4 will be as small as possible for a given $k > 0$, and what is the resulting estimate?

6. Using this estimate and Stirling's formula, show that the Fourier coefficients in (6.6.14) go to zero as

$$C(n)(\tfrac{1}{4}b)^{k+1/4}(2k + \tfrac{1}{2})^{-1/2}[\Gamma(2k + \tfrac{1}{2})]^{-1}.$$

7. Show that $\mathrm{ce}_m(z)$ and $\mathrm{se}_m(z)$ are either real or purely imaginary on each of the lines $x = (k + \frac{1}{2})\pi$.

8. Describe the map obtained when $z$ crosses one of the other three boundaries of $D_0^+$.

9. Find the explicit expression for $F(Z)$ in terms of $z$.

10. Why can an entire function of order $\frac{1}{2}$ have no Picard values?

11. Suppose that $a > b > 0$ and that $w(x)$ is a real solution of

$$w''(x) + (a + b \cos 2x)w(x) = 0.$$

Show that $w(x)$ has infinitely many real zeros, and try to estimate the number in the interval $[0, R]$.

## 6.7. SOME OTHER EQUATIONS

Mathieu's equation is a special case of a linear DE with simply periodic coefficients. We have seen that such an equation may have a periodic solution for a given value of $b$ and a suitable choice of $a$. What we have not seen is that the equation always has solutions of the form $e^{-cz}P(z)$, where $c$ is a constant and $P(z)$ is periodic—this regardless of the values of $a$ and $b$. This is a special case of a theorem discovered in 1883 by Gaston Floquet (1847–1920) for systems of linear DE's with periodic coefficients. We shall prove this theorem in matrix form. The proof requires more matrix theory than was developed in Section 1.4, however, so the first item on the agenda is matrix-valued analytic functions. In particular, we need the fact that a regular matrix has a logarithm.

If $\mathscr{A} \in \mathfrak{M}_n$, the algebra of $n$-by-$n$ matrices, then a matrix $\mathscr{B} \in \mathfrak{M}_n$ is called a *logarithm* of $\mathscr{A}$ if

$$\exp(\mathscr{B}) = \mathscr{A}. \tag{6.7.1}$$

Now $\exp(\mathscr{B})$ has an inverse, $\exp(-\mathscr{B})$, so we see that the existence of a logarithm of $\mathscr{A}$ requires that $\mathscr{A}$ have an inverse, i.e., be regular. This condition is also sufficient, but to show this fact a more elaborate argument is required. It is fairly simple if the characteristic values of $\mathscr{A}$ are simple, more complicated if there are multiple roots. Suppose that $s = s_i$ is a $\nu$-fold root of the *minimal equation* (this may be less than the multiplicity as a root of the characteristic equation); then there exist two matrices, $\mathscr{E}_j$ and $\mathscr{N}_j$. The former is an *idempotent*, and the latter a *nilpotent*. They commute and we have

$$(\mathscr{E}_j)^2 = \mathscr{E}_j, \qquad \mathscr{E}_j\mathscr{E}_k = \mathscr{E}_k\mathscr{E}_j = \mathbb{O}, \quad j \neq k, \tag{6.7.2}$$

$$\mathscr{E}_j\mathscr{N}_j = \mathscr{N}_j\mathscr{E}_j = \mathscr{N}_j, \qquad \mathscr{N}_j^{\nu-1} \neq \mathbb{O}, \quad \mathscr{N}_j^\nu = \mathbb{O}. \tag{6.7.3}$$

Suppose now that $f(s)$ is an analytic function which is holomorphic on the spectrum of $\mathscr{A}$, i.e., for $s = s_j$, $\forall\, j$. Then $f(\mathscr{A})$ can be defined, and we have

$$f(\mathscr{A}) = \sum_j [f(s_j)\mathscr{E}_j + f'(s_j)\mathscr{N}_j/1! + \cdots + f^{\nu-1}(s_j)\mathscr{N}_j^{\nu-1}/(\nu - 1)!]. \tag{6.7.4}$$

In particular, this formula is valid for $f(s) = \log s$ when $\mathscr{A}$ is regular. Thus a regular matrix has a logarithm. The reader who wants an exercise in matrix calculus is invited to prove that this definition of the logarithm satisfies condition (6.7.1). We can now state and prove the theorem of Floquet.

**THEOREM 6.7.1**

*Let $\mathscr{F}(z)$ be an n-by-n matrix of functions $f_{jk}(z)$ holomorphic in the strip S: $-\infty < x < \infty$, $-b < y < b$, and periodic with the real period $\omega$. Then the matrix DE*

$$\mathscr{W}'(z) = \mathscr{F}(z)\mathscr{W}(z) \tag{6.7.5}$$

*has a fundamental solution matrix of the form*

$$\mathscr{W}(z) = \mathscr{P}(z)\, exp\,(\mathscr{C}z), \tag{6.7.6}$$

*where $\mathscr{P}(z)$ is holomorphic in the strip S and of period $\omega$. If $\mathscr{A}$ is the transit matrix from $\mathscr{W}(z)$ to $\mathscr{W}(z + \omega)$, so that $\mathscr{W}(z + \omega) = \mathscr{W}(z)\mathscr{A}$, then $\mathscr{A}$ is nonsingular and $\mathscr{C} = (\log \mathscr{A})/\omega$.*

*Proof.* The matrix $\mathscr{W}(z)$ is a fundamental solution matrix if it is a solution matrix and nonsingular. If $\mathscr{W}(z)$ has this property, every other solution matrix is of the form $\mathscr{W}_0(z) = \mathscr{W}(z)\mathscr{M}$, where $\mathscr{M}$ is a constant matrix and $\mathscr{W}_0(z)$ is fundamental iff $\mathscr{M}$ is nonsingular. Since the DE is unchanged if $z$ is replaced by $z + \omega$, it follows that $\mathscr{W}(z + \omega)$ is fundamental iff $\mathscr{W}(z)$ is fundamental. In this case the matrix $\mathscr{A}$ is uniquely defined by $\mathscr{W}(z + \omega) = \mathscr{W}(z)\mathscr{A}$ and $\mathscr{A}$ is nonsingular. Now $\mathscr{A}$ has a logarithm. We can obviously assume that (6.7.5) has a solution of the form (6.7.6), but the problem here is to determine what properties $\mathscr{C}$ and $\mathscr{P}$ must have in view of the periodic character of $\mathscr{F}$.

Now $exp\,(\mathscr{M}_1 + \mathscr{M}_2) = exp\,(\mathscr{M}_1)\, exp\,(\mathscr{M}_2)$ iff $\mathscr{M}_1$ commutes with $\mathscr{M}_2$. We now set $\mathscr{C} = (1/\omega) \log \mathscr{A}$ and observe that

$$exp\,[\mathscr{C}(z + \omega)] = exp\,(\mathscr{C}z)\, exp\,(\mathscr{C}\omega) = exp\,(\mathscr{C}z)\mathscr{A}.$$

Furthermore,

$$\mathscr{W}(z + \omega) = \mathscr{P}(z + \omega)\, exp\,[\mathscr{C}(z + \omega)] = \mathscr{P}(z + \omega)\, exp\,(\mathscr{C}z)\mathscr{A}.$$

Comparing with $\mathscr{W}(z + \omega) = \mathscr{W}(z)\mathscr{A}$, we see that (6.7.6) is a solution if $\mathscr{P}(z + \omega) = \mathscr{P}(z)$, as asserted. ∎

Mathieu's equation is a special case of what is known as *Hill's equation,*

$$w''(z) + \left(\sum_{k=0}^{\infty} a_k \cos 2kz\right) w(z) = 0, \qquad (6.7.7)$$

named after the American astronomer G. W. Hill (1838–1914), who encountered such equations in lunar theory in 1877. In his study he was led to the use of infinite determinants, a field in which he was the pioneer. Hill's heuristic considerations were put on a firm basis by Henri Poincaré in his revision of celestial mechanics in the 1890's. G. Mittag-Leffler succeeded in having Hill's paper reproduced in *Acta Mathematica*, Vol. 8 (1886), and he suggested to one of his pupils, Helge von Koch, that the latter work in this new field. It was a labor of love, and in the 35 years of his scientific life von Koch produced general theories of infinite determinants and numerous applications to linear equations in infinitely many unknowns and to the theory of linear DE's.

In the early papers von Koch considered determinants of the form

$$D = \det\left(\delta_{jk} + a_{jk}\right), \qquad (6.7.8)$$

where

$$\sum \sum |a_{jk}| \qquad (6.7.9)$$

is convergent. Such determinants are absolutely convergent in a sense familiar from the theory of infinite products. The minors formed by the elements with subscripts $\leq n$ form a convergent sequence the limit of which is by definition the value of the determinant. All the elements of this sequence are dominated by

$$\prod_j \left(1 + \sum_k |a_{jk}|\right), \qquad (6.7.10)$$

which is then a bound for $D$. It is also a bound for all minors of $D$. The linear system

$$\sum_k (\delta_{jk} + a_{jk}) x_k = y_j, \qquad j = 0, 1, 2, \ldots, \qquad (6.7.11)$$

has a unique solution in the space of bounded sequences iff $D \neq 0$, while the homogeneous system is solvable if $D = 0$.

Several problems posed by Mathieu's equation and the more general Hill's equation lead to linear equations in infinitely many unknowns and hence to infinite determinants. Thus by the Floquet theory Hill's equation ought to have solutions of the form

$$e^{cz} \sum_{\infty}^{\infty} c_n e^{2niz}, \qquad (6.7.12)$$

and here we have the problem of determining $c$ and the $c_n$'s. If the series is substituted in (6.7.7) and the absolutely convergent series is multiplied out, one gets a system of homogeneous equations of the form

$$(c + 2ni)^2 c_n + \sum_{m=-\infty}^{\infty} a_m c_{n-m} = 0, \qquad n = \ldots, -2, -1, 0, 1, 2, \ldots. \quad (6.7.13)$$

If here the $n$th equation is divided by $[(c + 2ni)^2 + a_0]$, the resulting system has an absolutely convergent determinant, $D(ic)$ say. It follows that $c$ has to be determined from the equation

$$D(ic) = 0. \quad (6.7.14)$$

In this case it turns out that $D(ic)$ is an elementary function of $c$, and $c$ has to satisfy

$$\sinh^2 \tfrac{1}{2}\pi c = C,$$

where $C$ is a specific constant. Once $c$ is determined, the $c_n$'s are uniquely determined in terms of $c_0$ and the cofactors of $D(ic)$.

Another problem that leads to similar considerations is to determine the values of $a$ in terms of $b$ if, say, $\text{se}_n(z)$ is to be a solution of Mathieu's equation and the corresponding Fourier coefficients. We refrain from further details.

Linear DE's with doubly periodic coefficients have also been studied. There does not seem to exist a strict analogue of the theorem of Floquet, the difficulty being that the equation now has infinitely many fixed singularities at the poles of the coefficients, and these may be irregular singular or, if regular singular, may lead to branch points possibly involving logarithms.

The linear second order case leads to *Lamé's equation*, named after Gabriel Lamé (1795–1870). Important Lamé equations arise in Newtonian potential theory when elliptical coordinates are introduced in the Laplacian. Consider a family of confocal central quadrics:

$$\frac{x^2}{a^2 + s} + \frac{y^2}{b^2 + s} + \frac{z^2}{c^2 + s} = 1. \quad (6.7.15)$$

Through a given point $(x, y, z)$ in 3-space pass three quadrics of the system corresponding to the three roots $r, s, t$ of (6.7.15), regarded as a cubic equation in $s$. The roots are real; if $0 < a < b < c$, then $-c^2 < r < -b^2 < s < -a^2 < t < \infty$. The three roots are the *confocal coordinates* of $(x, y, z)$. The transformed equation has solutions of the form $f(r)g(s)h(t)$. The factors satisfy a linear second order DE of the Fuchsian class. These equations can be written in several different forms, algebraic or elliptic. One algebraic form is

$$\frac{d^2 u}{ds^2} + \frac{1}{2}\left(\frac{1}{a^2 + s} + \frac{1}{b^2 + s} + \frac{1}{c^2 + s}\right)\frac{du}{ds} = \frac{n(n+1)s + C}{4(a^2 + s)(b^2 + s)(c^2 + s)} u, \quad (6.7.16)$$

where $n$ is the degree of the ellipsoidal harmonic in $(x, y, z)$ and $C$ is a constant. This equation has four singular points, $-a^2, -b^2, -c^2$, and $\infty$, all regular singular. The indicial equations have roots $0, \frac{1}{2}$ at the first three points and $-\frac{1}{2}n, \frac{1}{2}(n + 1)$ at infinity. A simple linear translation will shift the finite singularities into new positions, $e_1, e_2, e_3$, with sum 0. In fact,

$$e_1 = -a^2 + \tfrac{1}{3}(a^2 + b^2 + c^2), \ldots, e_3 = -c^2 + \tfrac{1}{3}(a^2 + b^2 + c^2).$$

We can now introduce the Weierstrassian $\mathfrak{p}$-function corresponding to the equation $(v')^2 = 4(v - e_1)(v - e_2)(v - e_3)$ as a new independent variable and obtain the elliptic form of Lamé's equation:

$$w''(z) = [n(n + 1)\mathfrak{p}(z) + B]w(z), \tag{6.7.17}$$

where $B$ is a parameter. Here the singular points are the double poles of the $\mathfrak{p}$-function, points congruent to zero modulo period. The roots of the indicial equation at such a point are $n + 1$ and $-n$. Thus at every singularity there exists a holomorphic solution corresponding to the root $n + 1$. Since the difference of the indicial roots is an odd integer, we may expect the second solution in the local fundamental system to involve a logarithm. Actually, however, this does not happen. Suppose that the holomorphic solution is denoted by $w_1(z)$; then a linearly independent solution $w_2(z)$ is given by

$$w_1(z) \int^z [w_1(u)]^{-2} \, du. \tag{6.7.18}$$

In the neighborhood of the singularity, say at $u = 0$, the integrand is an even function of $u$, so no logarithms can arise from the integration of the negative powers in the Laurent expansion. This means that the solutions are all locally meromorphic and can be continued analytically along any path in the plane which avoids the singular points.

The solution is normally not a doubly periodic function, but for particular values of $B$ and $n$ one solution may be doubly periodic. A case in point is furnished by the equation

$$w'' = [e_1 + 2\mathfrak{p}(z)]w, \tag{6.7.19}$$

where the solutions are as follows:

$$[\mathfrak{p}(z) - e_1]^{1/2} \quad \text{and} \quad [\mathfrak{p}(z) - e_1]^{1/2}[\zeta(u + \omega_1) + e_1 z], \tag{6.7.20}$$

the first of which is doubly periodic whereas the second is not. Here the Weierstrass zeta function is $-\int \mathfrak{p}(t) \, dt$.

## EXERCISE 6.7

1. A possible way of defining $\log \mathscr{A}$ is by the logarithmic series

$$-\sum_1^\infty \frac{1}{n}(\mathscr{E} - \mathscr{A})^n.$$

   For what values of $\mathscr{A}$ does the series converge?

2. Give examples of idempotents and nilpotents in the matrix algebra $\mathfrak{M}_3$.

3. Suppose that $\mathscr{W}(z)$ is a fundamental solution of (6.7.5). Show that multiplication on the right of $\mathscr{W}(z)$ by a constant matrix yields a solution which is fundamental iff the multiplier is nonsingular.

4. Verify that $\exp(\mathscr{A} + \mathscr{B}) = \exp(\mathscr{A})\exp(\mathscr{B})$ if the matrices $\mathscr{A}$ and $\mathscr{B}$ commute. Try to find two noncommuting matrices in $\mathfrak{M}_2$ for which the addition formula breaks down.

5. Suppose that $D = \det(\delta_{jk} + a_{jk})$ is an infinite determinant with nonnegative elements satisfying the convergency condition (6.7.9). Form the sequence of principal minors $\{M_n\}$, where only elements with subscripts $\leq n$ occur in $M_n$. Show the convergence of the sequence, using the bound (6.7.10).

6. Verify that the roots of the indicial equations for (6.7.16) and (6.7.17) have the stated values.

7. Verify that (6.7.18) defines a solution of (6.7.17). If $w_1(z)$ is holomorphic at $z = 0$, verify that $w_2(z)$ is single valued at $z = 0$.
   The equation

$$(1 + z^2)^2 w''(z) + aw(z) = 0$$

   is known as *Halm's equation* [J. Halm, *Trans. Roy. Soc. Edinburgh* **41** (1906), 651–676].

8. Show that $z = \pm i$ are regular-singular points, and find the roots of the indicial equation. What is the nature of the point at $\infty$?

9. Find the general solution by the substitution

$$t = \arctan z, \qquad v = (1 + z^2)^{-1/2} w.$$

10. Let $z = x$ be real. Find the solution that tends to one as (i) $x \to -\infty$, (ii) $x \to +\infty$. When are these solutions identical?

## LITERATURE

This chapter corresponds to Chapter 7 of the author's LODE, which has an extensive bibliography. Special reference should be made to:

Whittaker, E. T. and G. N. Watson. *A Course in Modern Analysis.* 4th ed. Cambridge University Press, London, 1952.

The general theory of second order linear DE's starts in Chapter X, where Section 10.6 is devoted to the DE's of mathematical physics and where, following investigations of Felix Klein and Maxime Bôcher (1894), a certain second order DE with five singular points is singled out as the common source of the six equations of Bessel, Hermite, Lamé, Legendre, Mathieu, and Stokes, through confluence and choice of parameters. The program is carried out in Chapter XVIII. Chapters XIV–XVII deal with the hypergeometric case, Legendre's equation, the confluent hypergeometric case, and Bessel functions. There are a large number of exercises.

Much information is to be found in the following books:

Ince, E. L. *Ordinary Differential Equations*. Dover, New York, 1944.

Kamke, E. *Differentialgleichungen: Lösungen und Lösungemethoden*. Vol. 1, 3rd ed. Chelsea, New York, 1948.

See also the Bateman papers and:

Erdelyi, A., W. Magnus, F. Oberhettinger, and F. Tricomi, *Higher Transcendental Functions*. 3 vols. McGraw-Hill, New York, 1953–1955.

The standard treatise on Bessel functions is:

Watson, G. N. *Theory of Bessel functions*. 2nd ed. Cambridge University Press, London, 1962.

There is a large literature on confluent hypergeometric equations. In addition to Chapter XVI of Whittaker and Watson (*loc. cit.*), there are several books by Tricomi. See in particular:

Tricomi, F. G. *Fonctions hypergéometriques confluentes*. Gauthier-Villars, Paris, 1960.

The discussion in Sections 6.5 and 6.6 is based partly on papers by the author in *Ark. Mat., Astr., Fys.*, **18**, 26 (1924) and *Proc. London Math. Soc.* (2), **23**, (1924).

For the Floquet theory see the author's paper:

Floquet, G. Sur les equations différentielles linéaires à coefficients périodiques. *Ann. Sci. École norm.*, Sup. (2), **12** (1883), 47–89.

For the theory of infinite determinants see:

Hill, G. W. On the part of the motion of the lunar perigee which is a function of the mean motions of the sun and the moon. *Acta Math.*, **8** (1886), 1–36.

Koch, H. von. Sur les déterminants infinis et les équations différentielles linéaires. *Acta Math.*, **16** (1892), 217–295.

# 7
# REPRESENTATION THEOREMS

In this chapter we shall be concerned with the question of analytic representations of solutions of DE's not necessarily linear of the second order. Here the classical expansions in Taylor or Laurent series, with or without a multiplicative power and/or a logarithm, come to mind and are fairly obvious. Some problems lead to expansions in fractional powers for algebraic or algebroid solutions: these are perhaps less trivial. But for nonlinear equations we have also to allow expansions of one of the types

$$\sum_{n=0}^{\infty} a_n z^{\lambda_n}, \qquad \sum_{m=0}^{\infty} \sum_{n=0}^{\infty} a_{mn} z^{m+\lambda_n}, \qquad \sum_{m=0}^{\infty} \sum_{n=0}^{\infty} a_{mn} z^{\lambda_n} (\log z)^n.$$

Such series will be called *psi series* in the following discussion; the third type, in particular, is a *logarithmic psi series*.

Integral representations of the form

$$\int_C K(z, s) f(s) \, ds$$

are particularly important for the linear theory. The kernel $K(z, s) = (s - z)^{-a}$ is associated with the name of Euler; $e^{sz}$, with Laplace. Double and multiple integrals have also been pressed into service.

## 7.1. PSI SERIES

In the theory of nonlinear first-order DE's we may encounter equations with solutions having infinitely many algebraic branch points, all but a finite number of which are of the same order. The trivial equation

$$w' = w(w^2 - 1) \qquad \text{with } w(z) = C(C^2 - e^{2z})^{-1/2} \qquad (7.1.1)$$

is a case in point. There are branch points, all of order 1, at the points

249

$z_n = \log C + n\pi i$, where $n$ is any integer. For the class

$$w' = R(z, w),  \tag{7.1.2}$$

where $R$ is a rational function of $z$ and $w$, the numerator a polynomial in $w$ of degree $p$ and the denominator one of degree $q$, the typical branch point is of the type

$$c(z - a)^{-1/(p-q-1)},  \tag{7.1.3}$$

with at most a finite number of exceptions.

For more spectacular singularities we have to consider second order nonlinear equations. The deceptively simple special Emden equation,

$$zw'' = w^2,  \tag{7.1.4}$$

has a whole gallery of peculiar solutions and series expansions. Here $z = 0$ and $\infty$ are the fixed singularities. The function $2/z$ is a solution with a simple pole at the origin and a simple zero at infinity. This elementary solution is embedded in two one-parameter families of solutions of the form

$$\frac{1}{z}\left\{2 + \sum_{n=1}^{\infty} c_n a^n z^{-n\sigma}\right\} \quad \text{and} \quad \frac{1}{z}\left\{2 + \sum_{n=1}^{\infty} d_n b^n z^{n\tau}\right\},  \tag{7.1.5}$$

where

$$\left.\begin{array}{c} \sigma \\ \tau \end{array}\right\} = \tfrac{1}{2}(\sqrt{17} \pm 3).$$

The first series converges for large values of $|z|$; the second, for small values. Here $a$ and $b$ are the arbitrary parameters. These are rather simple types of psi series.

At the origin we also encounter logarithmic psi series. There are expansions of the form

$$\sum_{n=0}^{\infty} P_n\left(\log \frac{1}{z}\right) z^n,  \tag{7.1.6}$$

where $P_n(s)$ is a polynomial in $s$ of degree $[\tfrac{1}{2}(n + 1)]$; furthermore, $P_0(0)$ and $P_1(0)$ are arbitrary constants.

All this happens at the fixed singularities. Naturally the mobile ones are no better. Suppose that $z = a$ is such a point, say $a$ real positive. The test-ratio test shows that $w(z)$ becomes infinite as $6a(z - a)^{-2}[1 + o(1)]$, but the singularity is not a pole of order 2. Lacking a better term, we may call it a *pseudopole* of (apparent) order 2. At such a point we have logarithmic psi series of the form

$$\sum_{n=0}^{\infty} P_n[\log(z - a)](z - a)^{n-2},  \tag{7.1.7}$$

where now $P_n(s)$ is a polynomial of degree $[\frac{1}{6}n]$ and $P_6(0)$ is arbitrary.

This survey indicates that psi series are indispensable for the theory of nonlinear DE's. Such series are known from the theory of systems of the type

$$\frac{dx_1}{X_1} = \frac{dx_2}{X_2} = \cdots = \frac{dx_k}{X_k} = dt, \qquad (7.1.8)$$

where the $X_j$'s are functions of $x_1, x_2, \ldots, x_k, t$ holomorphic for small values of the variables and vanishing for $x_1 = x_2 = \cdots = x_k = 0$. See the monograph of H. Dulac (1934) and pp. 315–347 of J. Horn (1905).

Psi series give rise to a number of general questions in the mind of an analyst. Here are some:

1.    What analytic functions may be represented by a psi series of a given type?
2.    Given a psi series, what is its domain of absolute (nonabsolute, uniform) convergence?
3.    Does the function represented by the series necessarily have a singularity on the boundary of the domain of convergence?

It is easy to ask questions, but answering them is another matter. To the basic question 1 an obvious and trivial answer is: certain classes of solutions of selected nonlinear DE's. We obviously bypass the heart of the question, namely, What analytic properties of the solutions demand representations of this type? The author does not know the answer.

In regard to question 2, a little more can be said. Given a series

$$\sum_0^\infty a_n z^{\lambda_n}, \qquad (7.1.9)$$

where the $\lambda_n$'s form a strictly increasing sequence of nonnegative real numbers such that

$$\frac{\lambda_n}{\log n} \to \infty \qquad (7.1.10)$$

with $n$, an obvious extension of the Cauchy-Hadamard criterion shows that if

$$\limsup_{n \to \infty} |a_n|^{1/\lambda_n} = \mu, \qquad (7.1.11)$$

the series (7.1.9) converges absolutely for

$$|z| < \frac{1}{\mu}, \qquad (7.1.12)$$

and the terms are unbounded for $|z| > 1/\mu$. Nonabsolute convergence may possibly take place for a $z$ with $|z| = 1/\mu$. The function represented by the series is normally infinitely many valued but becomes single valued on the Riemann surface of the logarithm of $z$.

If $z \mapsto f(z)$ admits of a representation of type (7.1.9), it is unique, i.e., there is no nontrivial representation of zero by a series of type (7.1.9). This is proved in the same manner as for an ordinary power series by showing successively that all coefficients must be zero.

The same type of argument applies to series of the form

$$\sum_{m=0}^{\infty} \sum_{n=0}^{\infty} a_{mn} z^{m+\lambda_n}, \tag{7.1.13}$$

where the $\lambda_n$'s satisfy the conditions listed above. If the series has bounded terms for some value of $z \neq 0$, say $z = a$, it is absolutely convergent for $|z| < |a|$; hence the domain of absolute convergence is again of the form $|z| < R$, $-\infty < \arg z < +\infty$, for if

$$|a_{mn}| |a|^{m+\lambda_n} < M \qquad \text{and} \qquad |z| = r|a|, \tag{7.1.14}$$

where $0 < r < 1$, then

$$|a_{mn}| |z|^{m+\lambda_n} < M r^{m+\lambda_n} \tag{7.1.15}$$

and

$$\sum_{m=0}^{\infty} \sum_{n=0}^{\infty} r^{m+\lambda_n} = (1-r)^{-1} \sum_{n=0}^{\infty} r^{\lambda_n}. \tag{7.1.16}$$

Here the $\lambda$-series is convergent by the $\lambda$-root test by virtue of (7.1.10).

Again the representation of a function by a series of type (7.1.13), if at all possible, is unique, at least if no $\lambda_n$ is an integer, for if now $f(z)$ is identically zero and is represented by a series of type (7.1.13), each of the series $\sum_{n=1}^{\infty} a_{mn} z^{\lambda_n}$ must be a null series and this requires that all coefficients $a_{mn}$ be zero. If the $\lambda_n$'s can be integers, the representation is not necessarily unique.

We come finally to series of the form

$$\sum_{0}^{\infty} F_n\left(\log \frac{1}{z}\right) z^n, \tag{7.1.17}$$

where the functions $F_n(s)$ are entire functions of order $\beta < 1$. More precisely, there should exist positive numbers $\alpha$, $\beta$, $A$, and $B$, independent of $n$, such that

$$|F_n(r e^{i\theta})| < (A+r)^{\alpha n} \exp(nBr^\beta), \qquad 0 < \beta < 1, \tag{7.1.18}$$

for all $n$. Such a condition is obviously satisfied if the $F_n$'s are polynomials in $s$ of a degree which is dominated by a linear function of $n$. To discuss such a series we make the change of variables

$$s = \log \frac{1}{z} \quad \text{or} \quad z = e^{-s}.$$

The series then becomes

$$\sum_{n=0}^{\infty} F_n(s) e^{-ns}. \tag{7.1.19}$$

Set $z = x + iy$, $s = u + iv$, and note that the interval $0 < x < 1$ corresponds to $0 < u < \infty$. The series (7.1.19) will be absolutely convergent if

$$(A + r)^{\alpha} \exp(Br^{\beta} - r \cos \theta) < 1, \quad \theta = \arg s,$$

or

$$\alpha \log(A + r) + Br^{\beta} < r \cos \theta. \tag{7.1.20}$$

This inequality determines a domain $D$ in the $s$-plane. Any ray $\arg s = \theta$ with $-\frac{1}{2}\pi < \theta < +\frac{1}{2}\pi$ has the property that its distant part belongs to $D$. The transformation $z = e^{-s}$ maps $D$ conformally on a domain $S$ spread out on the Riemann surface of $\log(1/z)$. Each sheet of the surface contains a disk $|z| < r_n$ which belongs to $S$. Here $r_n$ goes to zero with $1/|n|$.

If a function $f(z)$ admits a representation by a series of type (7.1.18), it is unique, for if the sum of the series is identically zero, then each function $F_n(s)$ must be identically zero; this requires that the Taylor coefficients of each $F_n(s)$ be zero so only a trivial null series can exist.

Question 3 remains. For series of type (7.1.9) condition (7.1.10) is necessary for the existence of singularities on the boundary of the domain of convergence; however, it is not clear whether this condition is sufficient, and for the two other types of psi series no information appears to be available.

## EXERCISE 7.1

1. Try to derive (7.1.6). Set $\log(1/z) = s$.
2. Verify the claims of uniqueness for the series (7.1.9), (7.1.13), and (7.1.17).

## 7.2. INTEGRAL REPRESENTATIONS

We encountered several integral representations in Chapter 6. Here a general theory will be given, and as preparation some definitions and identities are needed. We shall be concerned with linear $n$th order DE's. The variables are complex, though with suitable restrictions the notions are meaningful also for real variables and are particularly useful for boundary value problems. Let the DE be

$$L(w) = \sum_{k=0}^{n} P_k(z) w^{(k)} = 0, \tag{7.2.1}$$

where the coefficients are locally analytic functions. The expression

$$L^*(w) = \sum_{k=0}^{n} (-1)^k [P_k(z) w]^{(k)} \tag{7.2.2}$$

will be called the *adjoint* of $L(w)$. We observe in passing that in the theory of boundary values the coefficients $P_k$ are normally replaced by their conjugates in defining the adjoint. We shall not do so here. $L(w)$ is *self-adjoint* if $L(w) = L^*(w)$, and *anti-self-adjoint* if $L^*(w) = -L(w)$.

We have by partial integration

$$\int U^{(p)} V \, ds = (-1)^p \int U V^{(p)} \, ds + \sum_{j+k=p-1} (-1)^j U^{(k)} V^{(j)}. \tag{7.2.3}$$

This gives

$$\int v L(u) \, ds = \int \left[ \sum P_k v u^{(k)} \right] ds = \sum \int u^{(k)} (P_k v) \, ds$$

$$= \sum u^{(k-1)} (P_k v) - \sum \int u^{(k-1)} (P_k v)' \, ds$$

$$= \sum u^{(k-1)} (P_k v) - \sum u^{(k-2)} (P_k v)'$$

$$+ \sum \int u^{(k-2)} (P_k v)'' \, ds. \tag{7.2.4}$$

After $n$ integrations by part we have

$$\int v L(u) \, ds = \int u L^*(v) \, ds + L^*(u, v), \tag{7.2.5}$$

where $L^*(u, v)$ is a bilinear form in the $u$'s, the $v$'s, and their derivatives. We have

$$L^*(u, v) = \sum_{m=1}^{n} \sum_{j+k=m-1} (-1)^j (P_m v)^{(j)} u^{(k)}. \tag{7.2.6}$$

The corresponding relation

$$v L(u) - u L^*(v) = \frac{d}{dz} L^*(u, v) \tag{7.2.7}$$

is known as *Lagrange's identity* after J. L. Lagrange (1736–1813), to whom several other identities are also accredited. The expression

$$\int_a^b [vL(u) - uL^*(v)] \, ds = [L^*(u, v)]_a^b \qquad (7.2.8)$$

is known as *Green's formula* after Cambridge don George Green (1793–1841), whose important contributions to potential theory, electricity, and magnetism were overlooked during his lifetime.

We now take up the problem of solving a linear DE (7.2.1) by means of a suitable integral of the form

$$\int_C K(z, t)F(t) \, dt, \qquad (7.2.9)$$

where the kernel $K(z, t)$ is given, subject to certain conditions to be stated later, whereas $F(t)$ and the path of integration $C$ have to be determined. It is assumed that $K(z, t)$ is locally holomorphic and that the differential operator $L = L_z$ may be taken under the integral sign so that

$$L_z\left[\int_C K(z, t)F(t) \, dt\right] = \int_C L_z[K(z, t)]F(t) \, dt.$$

It is further assumed that a differential operator $M = M_t$ exists such that

$$M_t[u] = \sum_{k=0}^{n} g_k(t)u^{(k)}(t) \qquad (7.2.10)$$

and has the following properties: (i) the functions $t \mapsto g_k(t)$ are holomorphic in some neighborhood $N(C)$ of the path of integration $C$ in the $t$-plane, except possibly at the end points of $C$ if $C$ is not a closed arc; and (ii) there exists a kernel $K_1(z, t)$ which is locally holomorphic in the two variables and such that

$$L_z[K(z, t)] = M_t[K_1(z, t)], \qquad (7.2.11)$$

and this gives

$$\int_C L_z[K(z, t)]F(t) \, dt = \int_C M_t[K_1(z, t)]F(t) \, dt. \qquad (7.2.12)$$

Here we can use Lagrange's identity. If $M_t^*(u)$ is the adjoint of $M_t(u)$ and $M_t^*(u, v)$ is the corresponding bilinear form, then

$$L_z(w) = \int_C K_1(z, t)M_t^*[F(t)] \, dt + \int_C \frac{d}{dt}[M_t^*(K_1, F)] \, dt. \qquad (7.2.13)$$

If now $F(t)$ is a solution, not identically zero, of the DE

$$M^*(u) = 0, \qquad (7.2.14)$$

and if the second integral in (7.2.13) vanishes for a suitable choice of $C$, then the function $w(z)$ of (2.7.9) is a solution of the DE (7.2.9).

Among the special kernels $K(z, t)$ used in this connection, the following are the most important:

$(z - t)^{\varepsilon - 1}$      Euler,

$e^{zt}$              Laplace,

$K(zt)$           Mellin,

$(-z)^t$           Barnes.

Various observations are in order concerning the choice of the path of integration $C$. Here $C$ may be an open arc or a closed curve. Suppose that $C$ is open and goes from $t = a$ to $t = b$, where one or both points may be infinity. If $M_1^*(K_1, F)|_{t=a}$ is to be zero, this requires that $t = a$ be a singular point of (7.2.14), for

$$M_1^*(K_1, F) = \sum_{m=0}^{n} \frac{\partial^m}{\partial t^m} [K_1(z, t)] \sum_{p=0}^{n-m-1} (-1)^p (g_{p+m+1}F)^{(p)},$$

and if the partials of $K_1$ with respect to $t$ actually depend on $z$, then $M^*(K_1, F)|_{t=a}$ can be identically zero iff

$$\sum_{p=0}^{n-m=1} (-1)^p (g_{p+m+1}F)^{(p)} = 0 \qquad (7.2.15)$$

holds for $t = a$ and $m = 0, 1, 2, \ldots, n - 1$. If $g_n(a) \neq 0$, this requires that $F(a) = F'(a) = \cdots = F^{(n-1)}(a) = 0$ and $F(t) \equiv 0$ contrary to the assumption. If the path $C$ extends to infinity, so that $b = \infty$, a similar argument shows that $t = \infty$ is a singularity of (7.2.14) if it is required that $M_1^*(K_1, F)$ vanish as $t \to \infty$.

If $C$ is a closed contour instead, $C$ must surround a singular point of $K$ or of $F$, for if the integrand in (7.2.9) is single valued and holomorphic inside $C$, the integral is identically zero and is of no interest for our problem. If the integrand is not single valued, it is essential that the integrand return to its original value after $t$ has described $C$. Here double loops may be useful (see Figure 6.2).

Double integrals have also been pressed into service for the representation problem. Here naturally the associated DE is a partial DE, say,

$$A(s, t) \frac{\partial^2 u}{\partial s\, \partial t} + B(s, t) \frac{\partial u}{\partial s} + C(s, t) \frac{\partial u}{\partial t} + D(s, t)u = 0. \qquad (7.2.16)$$

We denote the differential operator by $M_{s,t}$. Suppose that two kernels, $K(z, s, t)$ and $K^*(z, s, t)$, exist such that

$$L_z[K(z, s, t)] = M_{s,t}[K^*(z, s, t)]. \qquad (7.2.17)$$

Furthermore, let $F(s, t)$ be a solution of the partial DE

$$M^*_{s,t}(v) = \frac{\partial^2}{\partial s\, \partial t}(Av) - \frac{\partial}{\partial s}(Bv) - \frac{\partial}{\partial t}(Cv) + Dv = 0. \qquad (7.2.18)$$

Then

$$w(z) = \int_{C_1} \int_{C_2} K^*(z, s, t) F(s, t)\, ds\, dt \qquad (7.2.19)$$

is a solution of (7.2.1) provided that the paths $C_1$ and $C_2$ can be chosen so that $M^*(K^*, F)$ vanishes in the limits. Details are left to the reader.

Finally, in connection with the Barnes integral,

$$\int_C F(t)(-z)^t\, dt, \qquad (7.2.20)$$

it should be observed that the functional equation to be satisfied by $F(t)$ is normally not a DE but a *linear difference equation* instead. Conversely, certain classes of linear difference equations with polynomial coefficients are Mellin transforms of functions which satisfy a linear DE. See Section 7.6.

## EXERCISE 7.2

1. Verify (7.2.6).
2. Verify (7.2.13).
3. Find when (7.2.19) is a solution of (7.2.1).
4. In forming a solution of type (7.2.9) it is not necessary that $F(t)$ be the general solution of (7.2.14)—any particular solution will do. On the other hand, it may be advantageous to consider expressions like

$$\sum_k A_k \int_{C_k} K(z, t) F_k(t)\, dt,$$

where the $A$'s are constants, the $F$'s are solutions of (7.2.14), and the path $C_k$ is adjusted to the solution $F_k$. Under what conditions will this expression be a solution of (7.2.1)?

5. Consider the special hypergeometric equation

$$(1 - z^2)w'' - (a + b + 1)zw' - abw = 0,$$

and form the double integral

$$\int_0^\infty \int_0^\infty s^{a-1} t^{b-1} \exp\left(zst - \tfrac{1}{2}s^2 - \tfrac{1}{2}t^2\right) ds\, dt.$$

Show that this converges for $\mathrm{Re}\,(a) > 0$, $\mathrm{Re}\,(b) > 0$, $\mathrm{Re}\,(z) < 1$ and defines a solution of the equation in this half-plane. Show that the solution is a constant multiple of $w_{01}(z)$. Reconstruct the argument which led to this integral.

## 7.3.  THE EULER TRANSFORM

Quite an extensive theory is built around the Euler transform; Ludwig
Schlesinger (1864–1933) devotes 120 pages to this topic in his monumental
*Handbuch der Theorie der linearen Differentialgleichungen*, Vol. 2, 1897.
Incidentally, he coined the term *Euler transform* in imitation of "Laplace
transform," introduced by Poincaré. Here we are forced to squeeze an
abstract of the theory and some of its applications into a fraction of the
space that Schlesinger had at his disposal.

We write the transform

$$\int_C (t - z)^{\xi-1} F(t)\, dt \equiv w(z), \tag{7.3.1}$$

where $\xi$ is an arbitrary parameter at our disposal, and the problem is how
to choose the path $C$, the function $F$, and $\xi$ so that $w(z)$ becomes a
solution of the linear DE:

$$L(w) = \sum_{k=0}^{n} P_k(z) w^{(k)} = 0. \tag{7.3.2}$$

The $P_k$'s are polynomials in $z$ of a degree at most equal to the integer $p$.
Further restrictions will be introduced later.

We start by forming

$$L_z[(t - z)^{\xi-1}] = \sum_{k=0}^{n} (-1)^k k\,! P_k(z) C_{\xi-1,k} (t - z)^{\xi-k-1}, \tag{7.3.3}$$

where $C_{\xi-1,k}$ is the $k$th binomial coefficient of $\xi - 1$. Here we expand $P_k(z)$
in powers of $t - z$:

$$P_k(z) = \sum_{m=0}^{p} \frac{p_k^{(m)}(t)}{m!} (-1)^m (t - z)^m.$$

This gives

$$L_z[(t - z)^{\xi-1}] = \sum_{k=0}^{n} \sum_{m=0}^{p} (-1)^{k+m} C_{\xi-1,k} \frac{p_k^{(m)}(t)}{m!} (t - z)^{\xi-1-k+m}.$$

We set

$$f_j(t) = (-1)^{p-j} \sum_{k=0}^{n} C(\xi, j, k, p) \frac{p_k^{(p+k-j)}(t)}{(p + k - j)!} \tag{7.3.4}$$

with

$$C(\xi, j, k, p) = \frac{(\xi - 1)(\xi - 2) \cdots (\xi - k)}{(\xi + p)(\xi + p - 1) \cdots (\xi + p - j)}$$

and obtain

$$L_z[(t-z)^{\varepsilon-1}] = \sum_{j=0}^{n+p} f_j(t) \frac{d^j}{dt^j} [(t-z)^{\varepsilon+p-1}].$$ (7.3.5)

We then set

$$M_t(v) = \sum_{j=0}^{n+p} f_j(t) v^{(j)}(t).$$ (7.3.6)

This gives

$$L_z[(t-z)^{\varepsilon-1}] = M_t[(t-z)^{\varepsilon+p-1}].$$ (7.3.7)

Lifting a term from the theory of Abelian integrals, Schlesinger refers to this formula as the *theorem on interchange of argument and parameter,* for on the left side $t$ is the argument and $z$ the parameter, while on the right side $z$ is the argument and $t$ the parameter.

It follows that the associated function $F$ of (7.3.1) should be a solution of the linear DE

$$M_t^*(v) = 0,$$ (7.3.8)

which is of order $n + p$.

The singular points of (7.3.1) are the zeros of the polynomial $P_n(z)$, while those of (7.3.6) and hence also of (7.3.8) are the zeros of $f_{n+p}(t)$. Now (7.3.4) shows that $f_{p+n}(t)$ is a constant multiple of $P_n(t)$, so the two equations have the same singular points.

Suppose then that $F(t)$ is a solution, not identically zero, of (7.3.8). It follows that $w(z)$, defined by (7.3.1), is a solution of (7.3.2) provided that the path $C$ can be chosen so that

$$\int_C \frac{d}{dt} M_t^*[(t-z)^{\varepsilon+p-1}, F(t)] \, dt = 0,$$ (7.3.9)

for in that case we have

$$L_z\left[\int_C (t-z)^{\varepsilon-1} F(t) \, dt\right] = \int_C (t-z)^{\varepsilon+p-1} M_t^*[F(t)] \, dt = 0.$$

Various types of paths of integration $C$ are at our disposal. They all involve the singular points of the DE's, i.e., the zeros of $P_n(z)$ and the point at infinity. Let the distinct zeros of $P_n(z)$ be $a_1, a_2, \ldots, a_\nu$. Mark these points and the point $z$ in the complex plane. Denote by $[x, y]$ an integral from $t = x$ to $t = y$, denote by $[x, y)$ a simple loop starting at $t = x$ and surrounding $t = y$ in the positive sense, and, finally, let $(x, y)$ stand for a double loop surrounding $x$ and $y$. Here $x$ and $y$ take on values from the set $S$: $a_1, a_2, \ldots, a_\nu, z, \infty$. Double loop integrals always exist. The existence of other types will depend on the nature of the singularities, whether regular or irregular, and in the first case the roots of the corresponding indicial equation. In all cases care must be taken to ensure that $M^*[(t-z)^{\xi+p-1}, F(t)]$ returns to its original value after the path has been described.

It should be noted that in order to form the Euler transform (7.3.1) we do not need the general solution of (7.3.8). Any particular solution will do, since the variety of choices for $C$ provides enough freedom to enable one to find fundamental systems of solutions of (7.3.1). Furthermore, the parameter $\xi$ is at our disposal and can occasionally be used to simplify the work.

Further simplifications are available if the equation is of the Fuchsian type or, merely, the point at infinity is regular or regular singular. As will be seen in Section 9.6, the point at infinity will have the desired property iff the degrees of the coefficients

$$P_n, P_{n-1}, \ldots, P_0$$

are dominated by a strictly decreasing sequence. Hence, if the degree of $P_n$ is exactly $p$, the degree of $P_k$ is at most $p - n + k$. But then for $j < n$

$$p_k^{(p+k-j)}(t) \equiv 0.$$

By (7.3.4) this implies that the functions

$$f_0(t), f_1(t), \ldots, f_{n-1}(t)$$

are all identically zero, and the DE (7.3.6) takes the form

$$M_t(u) = \sum_{j=0}^{p} f_{n+j}(t) \frac{d^j}{dt^j} [u^{(n)}(t)], \tag{7.3.10}$$

so the equation is of order $p$ with respect to the $n$th derivative of $u$.

Further simplifications will take place if the equation is of the Fuchsian type, but we do not intend to pursue the general theory any further. In the rest of this section applications to Legendre's equation are given as illustrations. The general hypergeometric case will be discussed in the next section in some detail.

Given the Legendre equation,

$$L_z(w) = (1-z^2)w'' - 2zw' + a(a+1)w = 0, \tag{7.3.11}$$

which is of the Fuchsian class, it is desired to find solutions of the form (7.3.1) for a suitable choice of the parameter $\xi$. Here it is found that $M_t(u)$ is simply

$$M_t(u) = (1-t^2)u'' + 2(\xi-1)tu' + [a(a+1)-\xi(\xi-1)]u = 0.$$

If we choose $\xi$ so that the multiplier of $u$ is 0, which means either $\xi = a + 1$ or $-a$, we get for $\xi = -a$

$$M_t(u) = (1 - t^2)u'' - 2(a + 1)tu' = 0, \tag{7.3.12}$$

and the adjoint equation becomes

$$M_t^*(v) = (1 - t^2)v'' + 2(a - 1)tv' + 2av = 0. \tag{7.3.13}$$

This expression is an exact derivative and

$$M_t^*(w) = [(1 - t^2)v' + 2atv]' = 0. \tag{7.3.14}$$

a particular solution of which is our $F(t)$:

$$F(t) = (1 - t^2)^a. \tag{7.3.15}$$

Thus, if the path $C$ satisfies the appropriate conditions,

$$w(z) = \int_C (t - z)^{-a-1}(1 - t^2)^a \, dt \tag{7.3.16}$$

is a solution of Legendre's equation. If $\operatorname{Re}(a) > -1$, we can take $[-1, 1]$ for $C$. If $a$ is not an integer, we cut the plane along the real axis from $-1$ to $+1$; the integral is taken along the upper edge of the cut with $(1 - t^2)^a = 1$ for $t = 0$ and $(t - z)^{-a-1}$ defined in an obvious manner. Now

$$(t - z)^{-a-1} = (-z)^{-a-1}\left(1 - \frac{t}{z}\right)^{-a-1}, \qquad |z| > 1,$$

and here the second factor may be expanded in a binomial series. Multiplication by $(1 - t^2)^a$ and term-by-term integration are permitted. The result is of the form $(-z)^{-a-1}$ times a power series in $z^{-2}$, so the integral is a constant multiple of $Q_a(z)$. To get a representation of this function valid for all values of $a$, one may use a double loop $(-1, 1)$, but actually it is sufficient to use a lemniscate-shaped contour surrounding $t = +1$ in the positive sense and $t = -1$ in the negative. The right half of the circuit multiplies the integrand by $e^{2\pi i a}$, and the left half by $e^{-2\pi i a}$, so that we return to the starting point, say $t = 0$, with the original value of the integrand.

If $a = n$, a nonnegative integer, one may take for $C$ the circle $|t| = r > 1$. For $z$ inside the contour of integration, Cauchy's formula for the $n$th derivative, together with Problem 6.2:1, shows that

$$(-2)^{-n} \frac{1}{2\pi i} \int_C \frac{(1 - t^2)^n \, dt}{(t - z)^{n+1}} = P_n(z). \tag{7.3.17}$$

The case $a = -\frac{1}{2}$ is important on account of its connection with the theory of elliptic functions and modular functions. It is convenient to make a change of variables, so that the integral is taken to be

$$\int_C [t(1-t)(t-z)]^{-1/2}\, dt. \tag{7.3.18}$$

Here all paths of integration are permissible. If $s$ is an arbitrary point in the $z$-plane, then $[1, s]$ is an *elliptic integral* of $s$ *of the first kind* in the terminology of Legendre, who, however, spoke of elliptic *functions* in his *Traité des fonctions elliptiques* of 1825. Two years later N. H. Abel and C. G. J. Jacobi (1804–1851) showed independently of each other that, while the integral

$$\int_0^s [t(1-t)(t-z)]^{-1/2}\, dt = u \tag{7.3.19}$$

is an infinitely many-valued function of $s$, the inverse function $s = s(u)$ is a single-valued meromorphic and doubly periodic function of $u$. For this class of functions the name *elliptic function* is commonly accepted, while (7.3.19) has become known as an *elliptic integral*. It is of the first kind in the sense that it is everywhere bounded. Integrals of the second kind are obtained, for instance, by inserting a factor $t$ in the integrand. Such integrals become infinite with $s$. The inverse function is single valued; it is not doubly periodic but has additive periods. The third kind has logarithmic singularities.

The periods of (7.3.19) are of the form $4mK + 2niK'$, where

$$2K = \int_1^\infty [t(t-1)(t-z)]^{-1/2}\, dt, \tag{7.3.20}$$

$$2iK' = \int_z^1 [t(t-1)(t-z)]^{-1/2}\, dt. \tag{7.3.21}$$

Here $K = K(z)$ and $K' = K'(z)$ are holomorphic functions of $z$, real positive for $z$ real between zero and one. More precisely,

$$K(z) = \tfrac{1}{2}\pi F(\tfrac{1}{2}, \tfrac{1}{2}, 1; z), \tag{7.3.22}$$

$$K'(z) = K(1-z) = \tfrac{1}{2}\pi F(\tfrac{1}{2}, \tfrac{1}{2}, 1; 1-z). \tag{7.3.23}$$

That the periodicity modules satisfy a Legendre equation for $a = -\tfrac{1}{2}$ was first proved by Legendre.

The quantity

$$\tau = \frac{K'}{K}\, i \tag{7.3.24}$$

has a positive imaginary part, for any $z \neq 0, 1, \infty$. This is obvious for the case here considered, where $z$ is real and between zero and one, but it is also generally true. Set

$$q = e^{\pi i \tau}. \tag{7.3.25}$$

Since Im $(\tau) > 0$, it is seen that $|q| < 1$, so that series expansions in powers of $q$ converge rapidly. Such expansions are basic for the theory of *theta functions*, introduced by Jacobi in 1829. A typical theta series is

$$\vartheta_3(z, q) = 1 + 2 \sum_{n=1}^{\infty} q^{n^2} \cos 2nz, \tag{7.3.26}$$

which is an entire function of $z$. All the theta functions have a bearing on the theory of heat conduction—$\vartheta_3$, in particular, with the so-called *ring of Fourier*. Jacobi proved further that $z^{1/4}$ is the quotient of the *theta null series*,

$$z^{1/4} = \frac{\vartheta_2(0, q)}{\vartheta_3(0, q)}. \tag{7.3.27}$$

In the Jacobi theory $z^{1/2}$ is known as the *modulus* of the corresponding elliptic functions, usually denoted by $k$. The function defined by (7.3.27), known as the *elliptic modular function*, is a single-valued function of $\tau$ defined and holomorphic in the upper $\tau$-half-plane. It is obtained by inversion of the quotient of two fundamental solutions of Legendre's equation. It is invariant under the so-called *modular group*, a group of Möbius transformations of the upper half-plane into itself. Fuchs called attention to the fact that similar considerations apply to more general equations of the Fuchsian class. Poincaré elaborated this question into an extensive theory of *automorphic functions*: single-valued functions, invariant under a group of Möbius transformations, and normally with a natural boundary invariant under the group. Poincaré, always generous in naming the problems he considered after the author whose work had inspired him to delve deeper into the matter, termed these functions *fonctions Fuchsiennes*. Felix Klein (1849–1925) protested against that name, claiming that Fuchs had not considered such functions whereas he (Klein) had. Poincaré responded by calling the next class of automorphic functions which he discovered *fonctions kleiniennes*, since Klein had not considered them!

Incidentally, Picard built the proof of his theorem on the properties of the elliptic modular function (7.3.27): the existence of an entire function omitting the values 0, 1, $\infty$ would imply the existence of an entire function whose values have a positive imaginary part—an obvious impossibility.

All this goes to show the importance of Legendre's equation and of the Euler transform. The latter also adds to the understanding of hyperelliptic and more general Abelian integrals. Thus, if the integer $p > 1$ and if the

$a$'s are $2p$ distinct complex numbers, then

$$\int^s [(t - a_1)(t - a_2) \cdots (t - a_{2p-1})(t - a_{2p})(t - z)]^{-1/2}\, dt \qquad (7.3.28)$$

is a hyperelliptic integral of $s$ of the first kind and of genus $p$. It is infinitely many valued, and so is the inverse. The periodicity modules are integrals of the form $[a_j, a_{j+1}]$, $j = 0, 1, \ldots, 2p - 1$, where $a_0 = z$. They satisfy a linear DE of order $2p$ and of the Fuchsian class. The interested reader is referred to Schlesinger's Handbuch, Vol. 2, pp. 489–524, for further details.

### EXERCISE 7.3

1. Verify (7.3.3)–(7.3.5).
2. Verify (7.3.10) and preceding matters.
3. Verify (7.3.12)–(7.3.15).
4. Verify (7.3.22) and (7.3.23).
5. What is the DE satisfied by (7.3.18)?
6. Verify that

$$\int_{-1}^{1} (t - z)^{-a-1}(1 - t^2)^a\, dt$$

   is a function of $a$ times $Q_a(z)$. Find the multiplier.
7. Consider (7.3.16) for $a$ not an integer and $a > -\frac{1}{2}$. Let $C$ be a simple loop starting and ending at $t = -1$ and surrounding $t = 1$ and $t = z$ once in the positive sense. Assuming that $|z - 1| < 2$, expand the integral in powers of $1 - z$. Show that it is a constant multiple of $P_a(z)$, and find the multiplier.
8. Show that $F(\frac{1}{2}, \frac{1}{2}, 1; z) \neq 0$ for $z$ real, $0 < z < 1$.
9. Consider the quotient

$$\tau(z) = i\frac{K(1 - z)}{K(z)}.$$

   Show that it maps the upper $z$-half-plane conformally upon a curvilinear degenerate triangle above the circle $|\tau - \frac{1}{2}| = \frac{1}{2}$ and between the lines $\mathrm{Re}\,(\tau) = 0$ and $\mathrm{Re}\,(\tau) = 1$. There are three horn angles.
10. The theta function $\vartheta_3(z, q)$ is defined by (7.3.26). The other Jacobi thetas are

$$\vartheta_1(z, q) = 2\sum_{n=0}^{\infty} (-1)^n\, q^{(2n+1)^2/4} \sin(2n + 1)z,$$

$$\vartheta_2(z, q) = 2\sum_{n=0}^{\infty} q^{(2n+1)^2/4} \cos(2n + 1)z,$$

$$\vartheta_4(z, q) = 1 + 2\sum_{n=1}^{\infty} (-1)^n\, q^{n^2} \cos 2nz.$$

Show that all four functions are entire functions of $z$ of order 2 for fixed $q$, $|q| < 1$.

11. Estimate $T(r, F)$ for $\vartheta_3(z, q)$ if $q$ is real, $0 < q < 1$.

12. Show that the thetas have period $\pi$ or $2\pi$. Show that $\vartheta_3(z + \pi\tau, q) = q^{-1} e^{-2iz} \vartheta_3(z, q)$, and find similar relations for the other thetas.

13. Show that the quotient of two thetas is an elliptic function.

14. The theta null series are obtained by setting $z = 0$ in the expression for $\vartheta_j(z, q)$, $j = 2, 3, 4$. They are analytic functions of $\tau$, defined in the upper $\tau$-half-plane. If $\tau$ is purely imaginary, try to estimate the rate of growth of $\vartheta_3(0, q)$ as $q \to 1$.

## 7.4. HYPERGEOMETRIC EULER TRANSFORMS

Let us consider the hypergeometric DE

$$z(1 - z)w'' + [c - (a + b + 1)z]w' - abw = 0. \tag{7.4.1}$$

Here we know that the solutions are representable as Euler transforms. The object of the present section is threefold: (i) to derive the integrals by a more elementary method, (ii) to identify the various integrals with the known series solutions, and (iii) to use the integrals to obtain some of the transit formulas.

For these problems the Euler beta function is essential. For $\mathrm{Re}\,(u) > 0$, $\mathrm{Re}\,(v) > 0$ we have

$$\int_0^1 s^{u-1}(1 - s)^{v-1}\, ds = \frac{\Gamma(u)\Gamma(v)}{\Gamma(u + v)}. \tag{7.4.2}$$

If instead of an integral $[0, 1]$ we use a double loop $(0, 1)$, we get

$$\int_{(0,\,1)} s^{u-1}(1 - s)^{v-1}\, ds = (1 - e^{-2\pi i u})(1 - e^{-2\pi i v})\frac{\Gamma(u)\Gamma(v)}{\Gamma(u + v)}. \tag{7.4.3}$$

This becomes indeterminate if $u$ or $v$ or both are negative integers or zero. We get the correct value by replacing the right-hand member by its limit as $u$ or $v$ approaches the value in question. In either case it is finite, and it is zero if both $u$ and $v$ approach nonpositive integers.

As in the proof of Lemma 6.1.1, we start by finding an integral representation for the hypergeometric series $F(a, b, c; z)$ when $|z| < 1$ and $c$ is not a negative integer or zero. We note that for $|z_s| < 1$

$$(1 - zs)^{-a} = \sum_{n=0}^{\infty} \frac{(a)_n}{(1)_n}(zs)^n.$$

For a fixed $z$, $|z| < 1$, this converges uniformly with respect to $s$ on the interval $[0, 1]$, and the series may be multiplied by an integrable function of $s$ and integrated term by term between zero and one. We choose this function as $s^{u-1}(1-s)^{v-1}$, where $u$ and $v$ are to be disposed of later and for the moment only satisfy $\mathrm{Re}\,(u) > 0$, $\mathrm{Re}\,(v) > 0$. This gives successively

$$\int_0^1 s^{u-1}(1-s)^{v-1}(1-zs)^{-a}\,ds = \sum_{n=0}^\infty \frac{(a)_n}{(1)_n} z^n \int_0^1 s^{u+n-1}(1-s)^{v-1}\,ds$$

$$= \sum_{n=0}^\infty \frac{(a)_n}{(1)_n} \frac{\Gamma(u+n)\Gamma(v)}{\Gamma(u+v+n)} z^n$$

$$= \frac{\Gamma(u)\Gamma(v)}{\Gamma(u+v)} \sum_{n=0}^\infty \frac{(a)_n(u)_n}{(1)_n(u+v)_n} z^n.$$

Here we choose $u = b$, $v = c - b$ to get

$$\int_0^1 s^{b-1}(1-s)^{c-b-1}(1-zs)^{-a}\,ds = \frac{\Gamma(c-b)\Gamma(b)}{\Gamma(c)} F(a, b, c\,;z), \quad (7.4.4)$$

where we have to assume that $\mathrm{Re}\,(b) > 0$, $\mathrm{Re}\,(c - b) > 0$. If these conditions are not satisfied, the integral $[0, 1]$ must be replaced by a double loop $(0, 1)$ with corresponding modification of the multiplier.

The result is not yet of the Euler transform type but becomes of the canonical form by the change of variable $t = 1/s$. This gives

$$\int_1^\infty t^{a-c}(t-1)^{c-b-1}(t-z)^{-a}\,dt = \frac{\Gamma(c-b)\Gamma(b)}{\Gamma(c)} F(a, b, c\,;z). \quad (7.4.5)$$

The conditions on the parameters are unchanged. Now this representation is valid in the whole $z$-plane cut along the positive real axis from one to plus infinity.

We can now consider the integrals

$$[x, y] = \int_x^y t^{a-c}(t-1)^{c-b-1}(t-z)^{-a}\,dt. \quad (7.4.6)$$

Here we have six possible choices for $[x, y]$, valid under different conditions on the parameters. The list below gives the different choices of path and the corresponding conditions on the parameters:

| | |
|---|---|
| $[0, 1]$ | $\mathrm{Re}\,(a - c) > -1, \mathrm{Re}\,(c - b) > 0$; |
| $[0, z]$ | $\mathrm{Re}\,(a - c) > -1, \mathrm{Re}\,(a) < 1$; |
| $[-\infty, 0]$ | $\mathrm{Re}\,(a - c) > -1, \mathrm{Re}\,(b) > 0$; |
| $[1, \infty]$ | $\mathrm{Re}\,(c) > \mathrm{Re}\,(b) > 0$; |
| $[1, z]$ | $\mathrm{Re}\,(a) < 1, \mathrm{Re}\,(b) < \mathrm{Re}\,(c)$; |
| $[z, \infty]$ | $\mathrm{Re}\,(a) < 1, \mathrm{Re}\,(b) > 0$. |

We start our survey with a brief look at the integral $[0, 1]$. For large values of $z$ it is clearly $O(z^{-a})$, which suggests that it is a constant multiple of $w_{\infty 1}(z)$. It is easy to confirm this by series expansion, and the multiplier is obtained by multiplying the integral by $z^a$ and passing to the limit with $z$, thus obtaining

$$[0, 1] = e^{\pi i(c-a-b-1)} \frac{\Gamma(a - c + 1)\Gamma(c - b)}{\Gamma(a - b + 1)} w_{\infty,1}(z). \tag{7.4.7}$$

The DE is not altered if $a$ and $b$ are interchanged, and this shows that an interchange of $a$ and $b$ in an analytic expression for a solution preserves the property of being a solution. This procedure does not always give a new solution. In the present case, however, it does, and we have

$$\int_0^1 t^{b-c}(1-t)^{c-a-1}(t - z)^{-b} \, dt$$

$$= e^{\pi i(c-a-b-1)} \frac{\Gamma(b - c + 1)\Gamma(c - a)}{\Gamma(b - a + 1)} w_{\infty,2}(z). \tag{7.4.8}$$

Our next item is the integral $[0, z]$. Here a change of the variable of integration, setting $t = zs$, gives

$$[0, z] = z^{1-c} \int_0^1 s^{a-c}(sz - 1)^{c-b-1}(s - 1)^{-a} \, ds,$$

and the integral is a holomorphic function of $z$ at the origin. This shows that $[0, z]$ is a constant multiple of $w_{0,2}(z)$. Letting $z \to 0$ in the integral gives the multiplier and shows that

$$[0, z] = e^{\pi i(c-a-b-1)} \frac{\Gamma(a - c + 1)\Gamma(1 - a)}{\Gamma(2 - a)} w_{0,2}(z). \tag{7.4.9}$$

The fact that the integral $[-\infty, 0]$ tends to a finite limit as $z \to 1$ and appears to be holomorphic there would indicate that $[-\infty, 0]$ is a constant multiple of $w_{11}(z)$. This is indeed the case, and

$$[-\infty, 0] = e^{\pi i(b-1)} \frac{\Gamma(a - c + 1)\Gamma(b)}{\Gamma(a + b - c + 1)} w_{1,1}(z). \tag{7.4.10}$$

The case $[1, \infty]$ served as our starting point, so we can pass to the integral $[1, z]$. Here the substitution $t - 1 = s(z - 1)$ gives $(1 - z)^{c-a-b}$ times a function holomorphic and different from zero at $z = 1$. The usual steps give

$$[1, z] = a^{\pi i(c-b-1)} \frac{\Gamma(c - b)\Gamma(1 - a)}{\Gamma(c - a - b + 1)} w_{1,2}(z). \tag{7.4.11}$$

Our final item is $[z, \infty]$. Here the transformation $t = z/s$ shows that $[z, \infty]$ is a multiple of $w_{\infty,2}(z)$. Thus

$$[z, \infty] = e^{\pi i (c - a - b - 1)} \frac{\Gamma(b)\Gamma(1 - a)}{\Gamma(b - a + 1)} w_{\infty,2}(z). \qquad (7.4.12)$$

Again, by interchanging $a$ and $b$ in the integral, we get an alternative representation of $w_{\infty,1}(z)$. This finishes the discussion of the six (eight) individual integrals.

We can also obtain transit formulas from the integral representations. For this see Figure 7.1. There are six paths of integration for $z$ nonreal. They divide the $t$-plane into four triangular regions, of which three are degenerate. If $\mathrm{Re}\,(b) > 0$, integrals along arcs of the circle $|t| = R$ go to zero as $R \to \infty$. By the usual argument in complex integration it follows that the sum of the integrals along each of the four triangular boundaries is zero. This gives four sets of homogeneous linear equations between six unknowns, each involving three terms. We can write the set as follows:

$$
\begin{aligned}
& & I_1 &= [0, 1], \\
I_1 + 0 + I_3 + I_4 + 0 + 0 &= 0, & I_2 &= [0, z], \\
I_1 - I_2 + 0 + 0 + I_5 + 0 &= 0, & I_3 &= [-\infty, 0], \\
0 + I_2 + I_3 + 0 + 0 + I_6 &= 0, & I_4 &= [1, \infty], \\
0 + 0 + 0 + I_4 - I_5 + I_6 &= 0. & I_5 &= [1, z], \\
& & I_6 &= [z, \infty].
\end{aligned}
\qquad (7.4.13)
$$

In the 4-by-6 matrix all 4-by-4 minors are zero, but there are 3-by-3 minors which are not zero. There are obviously only three independent equations. This means that we get relations between three solutions, each of a different fundamental system. These are not transit formulas in the ordinary sense, nor do they readily lend themselves to the obtaining of ordinary transit formulas.

Before leaving the applications of the Euler transformation to the hypergeometric equation, let us note that (7.4.5) yields the value of $f(a, b, c)$ when $\mathrm{Re}\,(c - a - b) > 0$. We can then let $z \to 1$ to obtain

$$\int_1^\infty t^{a-c}(t - 1)^{c-a-b-1}\, dt = \frac{\Gamma(c - b)\Gamma(b)}{\Gamma(c)} f(a, b, c).$$

Here the left member equals

$$\frac{\Gamma(b)\Gamma(c - a - b)}{\Gamma(c - a)},$$

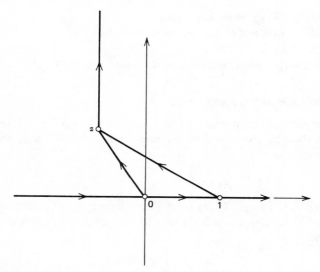

**Figure 7.1**

so that

$$f(a, b, c) = \frac{\Gamma(c)\Gamma(c - a - b)}{\Gamma(c - a)\Gamma(c - b)},$$ (7.4.14)

in agreement with (6.1.29).

For a representation of hypergeometric functions based on uniformization by theta functions, see Section 10.5.

## EXERCISE 7.4

1. Assuming (7.4.2), prove (7.4.3) when both are meaningful and use analytic continuation.

2. If (7.4.4) is replaced by a double loop (0, 1), how should the right member be changed?

3. The integral [1, ∞] of (7.4.5) satisfies the hypergeometric DE by construction. Why do the other integrals [x, y] satisfy? The three integrals involving z in one of the limits may require special consideration.

4. Verify the limits on the parameters needed for the existence of the integrals [x, y].

5. Discuss (7.4.7).

6. Try to find some justification for the interchange principle used in deriving (7.4.8). Permanency of functional equations?

7. Obtain (7.4.9) by series expansion.
8. Obtain (7.4.10) and (7.4.11) by series expansion.
9. Verify (7.4.12).
10. Find a 3-by-3 minor, not zero, of the 4-by-6 matrix (7.4.13).

## 7.5.   THE LAPLACE TRANSFORM

This transform was encountered in Section 6.4 in the second order case. Here we shall be concerned with a linear $n$th order DE with polynomial coefficients

$$L_z(w) = \sum_{k=0}^{n} P_k(z) w^{(k)}, \qquad P_k(z) = \sum_{p=0}^{m} C_{pk} z^p, \qquad (7.5.1)$$

where $C_{mn} = 1$, so that $P_n(z)$ is of exact degree $m$ while the other $P_k$'s may be of lower degree. It is desired to find solutions of this equation of the form

$$w(z) = \int_C e^{zt} F(t)\, dt \qquad (7.5.2)$$

for a suitable choice of $F$ and of the path of integration $C$. Here

$$L_z(e^{zt}) = \sum_{k=0}^{n} P_k(z) t^k e^{zt}. \qquad (7.5.3)$$

This equals

$$L_z(e^{zt}) = \sum_{k=0}^{n} \sum_{p=0}^{m} C_{pk} t^k z^p e^{zt}$$

$$= \sum_{p=0}^{m} \frac{d^p}{dt^p} (e^{zt}) \sum_{k=0}^{n} C_{pk} t^k = M_t(e^{zt}),$$

where

$$M_t(u) = \sum_{p=0}^{m} Q_p(t) u^{(p)} \qquad \text{with } Q_p(t) = \sum_{k=0}^{n} C_{pk} t^k. \qquad (7.5.4)$$

The adjoint is

$$M_t^*(u) = \sum_{j=0}^{m} (-1)^j \frac{d^j}{dt^j} [Q_j(t) u] \qquad (7.5.5)$$

with the associated bilinear differential form

$$M_t^*[f, g] = \sum_{j=0}^{m} \sum_{i=0}^{j-1} (-1)^i \frac{d^i}{dt^i} [Q_j(t) g] \frac{d^{j-i-1}}{dt^{j-i-1}} (f). \qquad (7.5.6)$$

Here we set

$$f = e^{zt}, \qquad g = F$$

and apply Lagrange's identity,

$$FL_z(e^{zt}) - e^{zt}M_t^*(F) = \frac{d}{dt} M_t^*(e^{zt}, F),$$

to obtain

$$L_z\left\{\int_C e^{zt}F(t)\,dt\right\} = \int_C e^{zt}M_t^*[F]\,dt + \int_C \frac{d}{dt}[M_t^*(e^{zt}, F)]\,dt.$$

If now $F$ is chosen as a solution of

$$M_t^*(u) = 0, \tag{7.5.7}$$

and if $C$ can be chosen so that

$$\int_C \frac{d}{dt}[M_t^*(e^{zt}, F(t)]\,dt = 0, \tag{7.5.8}$$

then (7.5.2) will furnish a solution of (7.5.1) of the desired type. We note that it is not necessary to find the general solution of (7.5.7); any solution, the simpler the better, will do as long as it is not identically zero.

The path $C$ has to be adjusted to the singularities of the adjoint equation. These are the zeros of

$$Q_m(t) = \sum_{p=0}^{n} C_{mp}t^p \tag{7.5.9}$$

plus the point at infinity. The use of the Laplace transform is normally not advisable when $m \geq n$; only when the order of $M^*$ is less than that of $L$ can one expect a simplification of the integration problem. If $m < n$, the point at infinity cannot be regular singular; in fact, we may expect some transcendental entire solutions.

The case considered by Laplace in 1812 involved $m = 1$, so that the adjoint equation is solvable by quadratures. Let the DE of Laplace be

$$\sum_{k=0}^{n} (C_{1k}z + C_{0k})w^{(k)} = 0. \tag{7.5.10}$$

Here

$$Q_1(t) = \sum_{p=0}^{n} C_{1p}t^p, \qquad Q_0(t) = \sum_{p=0}^{n} C_{0p}t^p, \tag{7.5.11}$$

so that

$$u(t) = \frac{1}{Q_1(t)} \exp\left[\int \frac{Q_0(s)}{Q_1(s)}\,ds\right]. \tag{7.5.12}$$

Here $Q_0/Q_1$ should be written as the sum of partial fractions, and there the relation between the degrees of $Q_0$ and $Q_1$ becomes important. Let the degrees of $Q_0$ and $Q_1$ be $q_0$ and $q_1$, respectively. If $q_0 - q_1 = q$, the partial fraction expansion will be of the form

$$\frac{Q_0(t)}{Q_1(t)} = \sum_{j=0}^{q} c_j t^j + \sum_{j=0}^{k} \sum_{p=1}^{\mu_j} b_{jp}(t - a_j)^{-p}, \qquad q \geqslant 0, \qquad (7.5.13)$$

where $\Pi(t - a_j)^{\mu_j} = Q_1(t)$. This gives

$$\int \frac{Q_0(s)}{Q_1(s)} ds = C + \sum_{j=0}^{q} \frac{c_j}{j+1} t^{j+1} + \sum_{j=1}^{k} b_{j1} \log(t - a_j)$$
$$+ \sum_{j=1}^{k} \sum_{p=2}^{\mu_j} b_{jp}(p - 1)^{-1}(t - a_j)^{1-p},$$

whence

$$u(t) = \prod_{j=1}^{k} (t - a_j)^{b_{j1} - \nu_j}$$
$$\times \exp\left[ \sum_{0}^{q} c_j(j + 1)^{-1} t^{j+1} + \sum_{j=1}^{k} \sum_{p=2}^{k_j} b_{jp}(p - 1)^{-1}(t - a_j)^{1-p} \right]. \qquad (7.5.14)$$

Thus, if either $q \geqslant 0$ or $Q_1 = 0$ has multiple roots, $u(t)$ will have (isolated) essential singularities at infinity and/or at the multiple roots.

Here the simplest case is obviously $q = 0$, and $Q_1 = 0$ has only simple roots. In this case we have

$$U(t) = \prod_{j=1}^{n} (t - a_j)^{b_j - 1}, \qquad (7.5.15)$$

where $b_j$ is the residue of $Q_0/Q_1$ at $t = a_j$. With this choice of $F(t)$ we can take for $C$ one of the line segments $[a_j, a_{j+1}]$ or the corresponding double loop $(a_j, a_{j+1})$. The first choice is permissible if both residues, $b_j$ and $b_{j+1}$, have positive real parts. The double loop is always permissible. We can also consider simple loops around the points $a_j$, where the loop begins and ends at $t = \infty$ and the path approaches infinity in such a manner that $\mathrm{Re}\,(zt) \to -\infty$. This gives a fairly large choice of paths of integration.

We note that (7.5.1) has only one finite singularity, $z = -C_{0n}/C_{1n}$, and this point is obviously a regular singularity. The corresponding indicial equation has the roots

$$0, 1, 2, \ldots, n - 2$$

and an $n$th root which is ordinarily not an integer. To the $n - 1$ integral roots correspond $n - 1$ linearly independent transcendental entire functions, while to the $n$th root corresponds a power of $z$ times an entire function. All these entire functions are of order 1. For exceptional values of the parameters $C_{pk}$ polynomial solutions may occur.

The Laplace transformation is effective and useful when the order of the irregular singularities is at most unity. Here the order is measured by the maximal rate of growth of the maximum modulus for approach to the singularity. If the order exceeds one, the Laplace transform can no longer cope with the problem. Here George David Birkhoff (1884–1944) showed the way to surmount the difficulties. In a number of papers starting in 1909, he proposed the generalized Laplace transform

$$\int_C \exp(z^p s) \left[ \sum_{j=0}^{p-1} z^j u_j(s) \right] ds = w(z), \qquad (7.5.16)$$

which he studied in some detail. Here the expected order of the singularities exceeds $p - 1$ but is at most $p$. The $p$ functions $u_j(s)$ and the path $C$ are to be determined.

Let the DE be of order $p$:

$$L_z(w) \equiv \sum_{j=0}^{p} P_j(z) w^{(j)} = 0, \qquad (7.5.17)$$

where the coefficients are polynomials in $z$ and

$$\deg P_j = k_j \leq (j-1)p \qquad (7.5.18)$$

with equality for at least one value of $j$.

Substitute (7.5.16) in (7.5.17). The result is a conglomerate of terms of the form

$$\int_C \exp(z^p s) s^j z^{j(p-1)+m+k} u_m(s) \, ds. \qquad (7.5.19)$$

Here $j$ goes from 0 to $p - 1$, and so does $m$ while $k \leq k_j$. Each such term is now subjected to integration by parts at most $p - 1$ times in order to reduce the exponent $j(p - 1) + m + k$ to at most $p - 1$. The idea is that

$$\int_C U(s, z) \, d_s \exp(z^p s) = [U(s, z) \exp(z^p s)]_C$$
$$- \int_C \exp(z^p s) \frac{\partial U(s, z)}{\partial s} \, ds.$$

Here $\partial U(s, z)/\partial s$ as a polynomial in $z$ is of degree $p$ units less than $j(p - 1) + m + k$, and because of the assumptions on $k_j$ not more than $p - 1$ integrations by parts are needed to lower the exponent of $z$ to at most $p - 1$. The various integrated parts are supposed to be eliminated by the choice of $C$. Each time we integrate, $u_m(s)$ will be differentiated. The net result of the reduction process is of the form

$$\int_C \exp(z^p s) \left[ \sum_{j=0}^{p-1} z^j L_j(u_0, u_1, \ldots, u_{p-1}; s) \right] ds, \qquad (7.5.20)$$

where each $L_j$ is a linear homogeneous differential expression in the $u$'s and their derivatives up to at most order $p - 1$. The coefficients are polynomials in $s$. To get a solution we solve the simultaneous DE's

$$L_j(u_0, u_1, \ldots, u_{p-1}; s) = 0, \qquad j = 0, 1, 2, \ldots, p - 1, \qquad (7.5.21)$$

and choose $C$ so that the integrated parts vanish in the limit. Rectilinear paths may be permissible, as well as loops and double loops. One can integrate to infinity, provided that the line of integration is such that $\mathrm{Re}\,(z^p s) \to -\infty$ along the path.

Birkhoff works with matrix equations of the form

$$z \mathcal{Y}'(z) = \mathcal{A}(z)\mathcal{Y}(z). \qquad (7.5.22)$$

Here $\mathcal{A}(z) = (a_{jk}(z))$ is a $p$-by-$p$ matrix where the diagonal elements are polynomials of degree $p$ and the other elements are of degree $p - 1$ at most. Birkhoff claimed that a DE having an irregular singularity of rank $p$ at infinity could be reduced to this normal form by suitable analytic transformations. Gaps in Birkhoff's proof were later filled in by H. L. Turrettin in 1955 and 1963.

**EXERCISE 7.5**

1. Verify (7.5.4)–(7.5.6).
2. In (7.5.15) let $u(t) = (t^3 - 1)^{1/2}\, dt$ and consider

$$\int_C e^{zt}(t^3 - 1)^{1/2}\, dt.$$

   Discuss the various choices for $C$. Here is the solution; where is the DE?
3. Supply additional details in the discussion of (7.5.19).
4. Apply Birkhoff's method to the equation $w'' - 4z^2 w = 0$, which is integrable in terms of Bessel functions.
5. Find sufficient conditions so that the DE $L_j = 0$ involves only $u_j$.
6. If $L_j = 0$ involves only $u_j$, when are the singularities of the equation regular singular?

## 7.6. MELLIN AND MELLIN–BARNES TRANSFORMS

In the 1890's the Finnish mathematician Hjalmar Mellin (1854–1933) devoted much attention to linear DE's involving the differential operator

$$\vartheta_z = z \frac{d}{dz}, \qquad (7.6.1)$$

which possesses the property of having the powers of $z$ as *characteristic functions* in the sense that

$$\vartheta_z z^a = a z^a \tag{7.6.2}$$

for any constant $a$. The powers of the operator are defined recursively by

$$\vartheta_z^n = \vartheta_z \vartheta_z^{n-1}. \tag{7.6.3}$$

Thus $\vartheta_z^n(z^a) = a^n z^a$.

Now let $F(u)$ and $G(u)$ be polynomials in $u$ with constant coefficients, let $n$ be a positive integer, and consider the DE

$$L_z(w) = z^n F(\vartheta_z) w + G(\vartheta_z) w = 0. \tag{7.6.4}$$

Let $H(u)$ be any polynomial and $K(z)$ any solution of the DE

$$z^n F(\vartheta_z) u - H(\vartheta_z) u = 0; \tag{7.6.5}$$

then $K(zt)$ satisfies the partial DE

$$\{z^n [F(\vartheta_z)] + G(\vartheta_z)\} U = \{G(\vartheta_t) + t^{-n} H(\vartheta_t)\} U, \tag{7.6.6}$$

so that

$$L_z[K(zt)] = M_t[K(zt)]$$

with obvious notation. Hence the integral

$$w(z) = \int_C K(zt) v(t)\, dt \tag{7.6.7}$$

is a solution of (7.6.4), provided that $v(t)$ is a solution of

$$M_t^*(v) = 0, \tag{7.6.8}$$

where $M_t^*$ is the adjoint operator of $M_t$, and provided that the corresponding associated bilinear form can be nullified. We refrain from further details.

Let $t \mapsto f(t)$ be defined and continuous for $0 \le t \le \infty$, and suppose that $t^\rho f(t) \in L(0, \infty)$ for $-1 < \rho < \omega < \infty$. Then the Mellin transform

$$\mathfrak{M}[f](s) = \int_0^\infty t^{s-1} f(t)\, dt \equiv v(s) \tag{7.6.9}$$

exists for $0 < \mathrm{Re}\,(s) < \omega + 1$ and is a holomorphic function of $s$ in this strip and also bounded on vertical lines. The inverse transformation is

$$\mathfrak{M}^{-1}[v](t) = \frac{1}{2\pi i} \int_{c-i\infty}^{c+i\infty} v(s) t^{-s}\, ds, \qquad 0 < c < \omega + 1. \tag{7.6.10}$$

Here the integral may have to be taken as a Cauchy principal value or, possibly, in the $(C, 1)$-sense. The Mellin transform has a number of interesting properties, some of which are presented as problems below.

The most elegant application of the Mellin transform to DE's was made in 1908 by the Cambridge don Ernest William Barnes (1874–1955), later the Right Reverend the Lord Bishop of Birmingham, who studied the hypergeometric DE by this method. By Problem 6.1:24 the equation may be written as

$$z(\vartheta_z + a)(\vartheta_z + b)w = \vartheta_z(\vartheta_z + c - 1)w. \qquad (7.6.11)$$

This is now to be satisfied by an integral of the form

$$w(z) = \frac{1}{2\pi i} \int_C v(s)(-z)^s \, ds, \qquad (7.6.12)$$

where $C$ is a vertical line (with possible deviations to avoid singularities). Here $v(s)$ has to be determined so that $w(z)$ satisfies (7.6.11). Assuming the right to apply the operator $\vartheta_z$ and its powers under the sign of integration, we obtain

$$-\int_C (s + a)(s + b)v(s)(-z)^{s+1} \, ds = \int_C s(s + c - 1)v(s)(-z)^s \, ds.$$

In the right member we replace $s$ by $t + 1$, and write again $s$ instead of $t$. This gives

$$\int_C [(s + a)(s + b)v(s) + (s + 1)(s + c)v(s + 1)](-z)^{s+1} \, ds = 0. \quad (7.6.13)$$

This is legitimate, and the shift of the line of integration back to $C$ is allowed, provided that for the appropriate $v(s)$ we have

$$\lim s^2 v(s)(-z)^s = 0. \qquad (7.6.14)$$

This replaces the bilinear form condition for the previously examined transforms. Equation (7.6.14) should hold uniformly in finite vertical strips as $|\tau| \to \infty$, $s = \sigma + i\tau$, for fixed $z$.

Thus (7.6.12) will be a solution of (7.6.11) if (i) $v(s)$ is a solution of the difference equation

$$v(s + 1) = -\frac{(s + a)(s + b)}{(s + 1)(s + c)} v(s), \qquad (7.6.15)$$

and (ii) condition (7.6.14) is satisfied for the solution of (7.6.15) that has been selected.

The difference equation (7.6.15) evidently is solvable in terms of gamma functions. A solution of

$$u(s + 1) = (s + d)u(s) \quad \text{is } \Gamma(s + d). \qquad (7.6.16)$$

The equation

$$u(s+1) = -\frac{u(s)}{s+1} \qquad (7.6.17)$$

has various solutions, but $\Gamma(-s)$ is the most suitable for our purposes. Combining, we see that

$$v(s) = \frac{\Gamma(s+a)\Gamma(s+b)}{\Gamma(s+c)}\Gamma(-s) \qquad (7.6.18)$$

is a solution of (7.6.15).

We have now to verify condition (7.6.14). Here we need to use various properties of the gamma function. We have

$$\Gamma(t)\Gamma(1-t) = \frac{\pi}{\sin \pi t},$$

which shows that

$$\left|\Gamma(\tfrac{1}{2}+i\tau)\right| = \left(\frac{\pi}{\cosh \pi\tau}\right)^{1/2} < (2\pi)^{1/2}\exp\left(-\tfrac{1}{2}\pi|\tau|\right).$$

Furthermore,

$$\lim \frac{\Gamma(s+d)}{s^d\Gamma(s)} = 1$$

for $s$ tending to infinity in such a manner that its distance from the negative real axis becomes infinite. If $-z = re^{i\theta}, |\theta| < \pi$, and $s = \sigma + i\tau$, then

$$\left|(-z)^s\right| = r^\sigma e^{-\theta\tau}.$$

Combining, we see that there are numbers $m$ and $M$ depending on $a, b, c$, and $x = \sup|\sigma|$ so that

$$\left|s^2 v(s)(-z)^s\right| \leqslant M(|\tau|+1)^m r^x \exp\left[-(\pi - |\theta|)|\tau|\right]; \qquad (7.6.19)$$

and if $|\theta| < \pi - \epsilon, 0 < \epsilon$, and $r \leqslant R < \infty$, the right member of (7.6.19) goes to zero as $|\tau| \to \infty$, uniformly with respect to $r$, $\theta$, and $\sigma$. Thus (7.6.14) is satisfied, and a solution of (7.6.11) is given by

$$w(z) \equiv \frac{1}{2\pi i}\int_{-1/2-i\infty}^{-1/2+i\infty} \frac{\Gamma(s+a)\Gamma(s+b)}{\Gamma(s+c)}\Gamma(-s)(-z)^s \, ds. \qquad (7.6.20)$$

Here it is assumed that the points $-a$ and $-b$ lie to the left of the path of integration. If this is not so at the outset, the line may be deformed locally to achieve it.

Here $w(z)$ is a multiple of $w_{01}(z)$,

$$w(z) = \frac{\Gamma(a)\Gamma(b)}{\Gamma(c)} F(a, b, c; z). \tag{7.6.21}$$

This can be verified for $|z| < 1$ by contour integration. If in (7.6.21) the integral is taken along a rectangle with vertices at $s = -\frac{1}{2} \pm iT$ and $-\frac{1}{2} + N \pm iT$, the integral is the sum of the residues inside, i.e.,

$$\sum_{n=0}^{N} \frac{\Gamma(a+n)\Gamma(b+n)}{\Gamma(1+n)\Gamma(c+n)} z^n.$$

Here we can let first $T$ and then $N$ go to infinity. For $|z| < 1$ the sum of the residues converges to (7.6.21), and the integrals along $\tau = \pm T$ and $\sigma = -\frac{1}{2} + N$ go to zero as $T$ and $N$ become infinite. The details are left to the reader.

From (7.6.20) we can also find the expression of $w_{01}(z)$ in terms of $w_{\infty 1}(z)$ and $w_{\infty 2}(z)$, assuming $|z| > 1$ and $a - b$ not congruent to zero (modulo 1). This requires another contour integration argument. This time we integrate along the boundary of a rectangle with vertices at $s = -\frac{1}{2} \pm iT$ and $s = -\frac{1}{2} - N \pm iT$. Here it assumed that the poles of $\Gamma(a + s)$ and $\Gamma(b + s)$ are avoided by the contour, so that all poles are ultimately inside the contour. The integral is the sum of the residues. At $s = -a - n$ the residue is

$$(-1)^n \frac{\Gamma(b-a-n)\Gamma(a+n)}{\Gamma(c-a-n)\Gamma(n+1)} (-z)^{-a-n}. \tag{7.6.22}$$

Interchanging $a$ and $b$, we get the residue at $s = -b - n$. We let $T$ and $N$ tend to infinity; this makes the integrals along $\tau = \pm T$ and $\sigma = -\frac{1}{2} - N$ go to zero, while the sum of the residues converges. The various gamma functions of the type $\Gamma(u - n)$ are reduced to

$$\Gamma(u - n) = (-1)^{n+1} \frac{\pi}{\sin u\pi} \frac{1}{\Gamma(u+1+n)}.$$

For $n = 0$ the same formula is used to eliminate the sine functions in the final result. After much simplification we emerge with the final result:

$$\frac{\Gamma(a)\Gamma(b)}{\Gamma(c)} F(a, b, c; z) = \frac{\Gamma(a)\Gamma(a-b)}{\Gamma(a-c)} (-z)^{-a} F(a, a-c+1, a-b+1; z^{-1})$$

$$+ \frac{\Gamma(b)\Gamma(b-a)}{\Gamma(b-c)} (-z)^{-b} F(b, b-c+1, b-a+1; z^{-1}). \tag{7.6.23}$$

Barnes could also obtain the transit formula expressing $w_{01}(z)$ in terms of $w_{11}(z)$ and $w_{12}(z)$, but for this the reader is referred to the presentation in Whittaker and Watson, *A Course in Modern Analysis*, Chapter XIV.

The Barnes integral representations for hypergeometric functions have been extended to the abstract case of (6.1.8) by Hans Wilhelm Burmann (1971), who also derived the transit formulas which the present author was unable to find in his discussion in 1968.

The use of the Mellin–Barnes transform is not limited to the hypergeometric equation. In fact, any equation of the form (7.6.4) can be handled in this manner, though the corresponding difference equation for $v(s)$ may become puzzling. We shall illustrate both facts by Bessel's equation. This may be written as

$$(\vartheta^2 - a^2)w + z^2w = 0, \qquad (7.6.24)$$

and the corresponding difference equation is

$$v(s+2) = -\frac{v(s)}{(s+2+a)(s+2-a)}. \qquad (7.6.25)$$

Here the span is two, but it may be reduced to one by setting $2s = t$, $v(2s) = u(t)$; this gives

$$u(t+1) = \frac{2^{-2}u(t)}{(t+1+\tfrac{1}{2}a)(\tfrac{1}{2}a - t - 1)}, \qquad (7.6.26)$$

a solution of which is

$$u(t) = 2^{-2t}\frac{\Gamma(\tfrac{1}{2}a - t)}{\Gamma(t+1+\tfrac{1}{2}a)}.$$

Thus

$$v(s) = 2^{-s}\frac{\Gamma(\tfrac{1}{2}-\tfrac{1}{2}s)}{\Gamma(\tfrac{1}{2}s + \tfrac{1}{2}a + 1)}, \qquad (7.6.27)$$

and finally

$$w(z) = \frac{1}{2\pi i}\int_{-1/2-i\infty}^{-1/2+i\infty} 2^{-s}\frac{\Gamma(\tfrac{1}{2}a - \tfrac{1}{2}s)}{\Gamma(\tfrac{1}{2}s + \tfrac{1}{2}a + 1)}(-z)^s\, ds. \qquad (7.6.28)$$

Contour integration shows that this is $e^{a\pi i}J_a(z)$. The formula does not seem to add much to our knowledge of Bessel functions.

We have seen that the use of the Barnes transform in the theory of linear DE's leads to a difference equation for the weight function $v(s)$. Conversely, certain classes of linear difference equations can be solved by the use of the inverse transformation. This mode of attack was initiated and developed by Niels Erik Nörlund in a number of papers in the 1910's. We shall illustrate by considering the difference equation satisfied by the Legendre functions. See Problem 6.2:14. We write it as

$$(a+2)X(a+2) - (2a+3)zX(a+1) + (a+1)X(a) = 0, \qquad (7.6.29)$$

and try to solve it by an integral

$$X(a) = \int_C t^{a-1} Y(t)\, dt. \tag{7.6.30}$$

It is found that $Y(t)$ has to be a solution of

$$t(t^2 - 2zt + 1)Y'(t) + (zt - 1)Y(t) = 0. \tag{7.6.31}$$

We can take

$$Y(t) = t(t^2 - 2zt + 1)^{-1/2}. \tag{7.6.32}$$

We set

$$\xi = z - (z^2 - 1)^{1/2}, \qquad \frac{1}{\xi} = z + (z^2 - 1)^{1/2}, \tag{7.6.33}$$

where the square root is real positive for $z = x$ real and $> 1$. Let $C_1$ be a closed curve surrounding $t = \xi$ and $t = 1/\xi$ in the positive sense, leaving $t = 0$ on the outside. Let $C_2$ be the ray from $t = 0$ to $t = \xi$. Then for $\mathrm{Re}\,(a) > -1$

$$P_a(z) = -\frac{1}{2\pi i} \int_{C_1} \frac{t^a\, dt}{(t^2 - 2zt + 1)^{1/2}}, \tag{7.6.34}$$

$$Q_a(z) = \int_{C_2} \frac{t^a\, dt}{(t^2 - 2zt + 1)^{1/2}}. \tag{7.6.35}$$

## EXERCISE 7.6

1. Justify (7.6.6).
2. Why is (7.6.9) bounded on vertical lines in the strip?
3. Justify (7.6.10).
4. Prove (7.6.15), (7.6.18), and (7.6.19).
5. Give a detailed proof of (7.6.23).
6. Prove that for $\arg(-z) < \pi$

$$\Gamma(a)(1-z)^{-a} = \frac{1}{2\pi i} \int_{-1/2-i\infty}^{-1/2+i\infty} \Gamma(s+a)\Gamma(-s)(-z)^s\, ds.$$

7. Fill in the missing details in the discussion leading to (7.6.28).
8. Prove (7.6.34) and (7.6.35). Is it clear that the integrals define solutions of Legendre's equation (6.2.1) as functions of $z$? If so, why is the first integral holomorphic at $z = 1$ and the second integral goes to zero as $z$ becomes infinite? What is the limit of $\xi$?
9. According to (6.3.7), the Bessel function $J_a(z)$ satisfies the second order linear difference equation

$$z X(a+2) - 2a X(a+1) + z X(a) = 0.$$

Try to solve this equation by an integral of type (7.6.30). Determine $Y(s)$ and suitable paths of integration $C$.

## LITERATURE

For psi series see:

Dulac, M. *Points singuliers des équations différentielles.* Mémorial des Sci. Math, No. 61. Gauthier-Villars, Paris, 1934.

Hille, E. On a class of series expansions in the theory of Emden's equation. *Proc. Roy. Soc. Edinburgh,* (A) **71**:8 (1972/73), 95–110.

———. A note on quadratic systems. *Proc. Roy. Soc. Edinburgh,* (A) **72**:3 (1972/73), 17–37.

Horn, J. *Gewöhnliche Differentialgleichungen beliebiger Ordnung.* Sammlung Schubert, 50. Göschen, Leipzig, 1905.

Smith, R. Singularities of solutions of certain plane autonomous systems. *Proc. Roy. Soc. Edinburgh,* (A) **72**:26 (1973/74), 307–315.

General surveys of integral representations of solutions of linear DE's are found in Chapter VIII of:

Ince, E. L. *Ordinary Differential Equations.* Dover, New York, 1944;

and on pp. 90–100 of

Kamke, E. *Differentialgleichungen: Lösungsmethoden und Lösungen.* 3rd ed. Chelsea, New York, 1944.

Euler and Laplace transforms are well presented by Ince, *loc. cit.* A detailed account of Laplace transforms is given also on pp. 401–435 of Vol. 1 and of Euler transforms on pp. 405–534 of Vol. 2 of:

Schlesinger, L. *Handbuch der Theorie der linearen Differentialgleichungen.* Teubner, Leipzig, 1896–1897.

Further references for Sections 7.3–7.5 are:

Birkhoff, G. D. Singular points of ordinary differential equations. *Trans. Amer. Math. Soc.,* **10** (1909), 436–470.

———. Equivalent singular points of ordinary linear differential equations. *Math. Ann.,* **74** (1913), 134–139.

Laplace, P. S. *Théorie analytique des probabilitées.* Vol. 1, 1812.

Poincaré, H. Sur les intégrales irregulières des équations linéaires. *Acta Math.,* **8** (1885), 295–344.

Turrettin, H. L. Convergent solutions of ordinary linear homogeneous differential equations in the neighborhood of an irregular singular point. *Acta Math.,* **93** (1955), 27–66.

———. Reduction of ordinary differential equations to the Birkhoff canonical form. *Trans. Amer. Math. Soc.,* **107** (1963), 485–507.

See also pp. 198–209 and 266–272 of the author's LODE and the literature quoted therein. The discussion of the Birkhoff-Laplace transform there in Section 6.9 is based on:

Miller, J. B. Solution in Banach algebras of differential equations with irregular singular points. *Acta Math.,* **110** (1963), 209–231.

The Mellin transform is discussed in Ince, *loc. cit.* For the Barnes–
Mellin transform, see Chapter XIV of:

Whittaker, G. T. and G. N. Watson, *A Course in Modern Analysis.* 4th ed. Cambridge
University Press, London, 1952.

The original papers are:

Barnes, E. W. A new development of the theory of the hypergeometric functions. *Proc.
London Math. Soc.* (2), **6** (1908), 141–177.

Burmann, H. W. Die hypergeometrische Differentialgleichung über Banachalgebren.
Habilitationsschrift, Göttingen, 1971; *Math. Zeit.*, **125** (1972), 139–176.

Mellin, Hj. Über die fundamentale Wichtigkeit des Satzes von Cauchy für die Theorien der
Gamma- und der hypergeometrischen Funktionen. *Ann. Sci. Soc. Fenn.*, **20** (1895), No.
12, 115 pp.

Tables of Mellin transforms are given in:

Colombo, S. and J. Lavoine. *Transformations de Laplace et de Mellin.* Mémorial des Sci.
Math., No. 160. Gauthier-Villars, Paris, 1972.

The standard treatise on difference equations is:

Nörlund, N. E. *Vorlesungen über Differenzenrechnung.* Grundlehren, No. 13. Springer-
Verlag, Berlin, 1924.

# 8

# COMPLEX
# OSCILLATION THEORY

We shall be concerned with second order linear DE's and, more specifically, with the oscillatory behavior of solutions in the complex domain when the coefficients of the DE are analytic functions. This is essentially a question of the distribution in the plane of zeros and extrema of the solution. In a few cases boundary value problems make sense in the complex plane. This is in contrast to the behavior of real solutions on the real line, where since the days of Joseph Liouville (1809–1882) and Jacques Charles François Sturm (1803–1855), effectively from the 1830's, boundary value problems and oscillatory properties of the solutions have been the objects of much research.

That there could be interesting oscillation problems for the complex plane was not realized until much later. The pioneer was Adolf Hurwitz (1859–1919), who in 1889 examined the zeros of the Bessel function $J_a(z)$ for real values of $a$. This was followed by work on the hypergeometric equation by E. B. Van Vleck in 1902, by Hurwitz again in 1907, and by P. Schafheitlin in 1908. Their methods were essentially function theoretical. The Bessel functions or, rather, their logarithmic derivatives were considered by Pierre Boutroux in his profound investigation of the Painlevé transcendents in 1913–1914. The displacement of the poles (zeros of Bessel functions) with changing initial conditions served as prototype for the much more profound problem posed by the Painlevé functions.

From 1918 to 1924 the present author re-examined these problems, using a variety of methods (Sturmian, Green's transform, transformation of Liouville, asymptotic integration, variation of parameters), and applied these methods to Legendre's, the Hermite-Weber, and Mathieu's equations.

The Schwarzian derivative and univalence were brought to bear on the problem by Z. Nehari (1949) and by his pupils P. Beesack and B. Schwartz, while C-T. Taam used comparison theorems in 1952.

## 8.1. STURMIAN METHODS; GREEN'S TRANSFORM

We shall be concerned mainly with second order linear DE's of the form

$$w'' + G(z)w = 0, \tag{8.1.1}$$

where $G(z)$ is holomorphic except for poles in the finite plane. Our general object is to obtain information about the distribution of the zeros and extrema of solutions of such an equation. Here *extremum* denotes a zero of the derivative.

Our first observation is that complex oscillation theory is essentially nonoscillation theory: such a solution cannot have zeros on a certain curve or in a domain of the complex plane or in a neighborhood of a given point. But rarely do we obtain positive information regarding the existence of zeros. This negative character has been apparent from the very beginning of the theory. Thus Hurwitz proved that $J_a(z)$ can have no complex zeros if $a > -1$. There is an old result of A. R. Forsyth (1900), according to which a solution of (8.1.1) with a zero at a nonsingular point $z = a$ can have no other zero in the domain

$$|z - a|^2 |G(z)| < 4. \tag{8.1.2}$$

No information on the rate of growth of $|G(z)|$ or on $\arg G(z)$ can exclude solutions of the form

$$P(z) e^{Q(z)}, \tag{8.1.3}$$

where $P$ and $Q$ are polynomials in $z$. We can assert, however, that (8.1.1) cannot have two linearly independent solutions of this form unless $G(z)$ is a constant, for the Wronskian is a constant, and if $P e^Q$ and $R e^S$ are the solutions, then

$$e^{Q+S}[PR' - P'R + PR(S' - Q')] = C, \tag{8.1.4}$$

leads to the stated conclusion.

Here $Q + S$ and the expression between brackets are polynomials. The latter expression cannot be zero for any value of $z$. Thus it must be a constant. Now, if $e^{Q+S}$ is a constant, the exponent must also have this property. From $Q + S = K$ we get $S' = -Q'$, and this gives

$$RP' - PR' - 2Q'PR = C_1. \tag{8.1.5}$$

Here there are several possibilities. If both $P$ and $R$ are constants, there exists a constant $a \neq 0$ such that $Q' = a$ and the solution is

$$A e^{az} + B e^{-az}, \qquad G(z) = -a^2. \tag{8.1.6}$$

If $P$ is a constant but not $R$, then

$$P_0 R'(z) - 2Q'(z)P_0 R(z) = K.$$

If $Q'$ is not identically zero, this is impossible, for the degree of $R'$ is one unit less than the degree of $R$ and hence at least two units less than that of $Q'R$. If, on the other hand, $Q' \equiv 0$, there exists a constant $b$ such that $R'(z) \equiv b$ and the solution is

$$a + bz \qquad \text{with } G(z) \equiv 0. \qquad (8.1.7)$$

Finally, if neither $P$ nor $R$ is a constant, one sees that $Q' \not\equiv 0$ *is impossible. If $Q' \equiv 0$, then*

$$PR' - RP' = C. \qquad (8.1.8)$$

Since $P$ and $R$ are at least of degree 1, $PR'$ and $RP'$ must be of the same degree, and this requires that $P$ and $R$ be of the same degree, say $k$. From (8.1.8) it follows that the zeros of $P$ and $R$ are simple and they have no zeros in common. On the other hand, we have

$$-G(z) = \frac{P''}{P} = \frac{R''}{R}.$$

This shows that any zero of $P$ must be a singular point of the DE. Since there are no finite singularities, $P''$ must vanish at all the zeros of $P$. But $P''$ has only $k - 2$ zeros and cannot vanish at the $k$ zeros of $P$. Hence $P''$ is identically zero, so that $P(z)$ is linear. This shows also that $G(z) \equiv 0$. The assertion is now proved: $G(z)$ must be a constant.

In a general way we can say that the modulus of $G(z)$ controls the frequency of the oscillation, while the argument of $G(z)$ governs the orientation and twist of the zeros. This vague statement will be made more precise later.

Keeping in mind the negative character of the expected information, let us see how the venerable methods of Sturm can be adjusted to explorations in the complex plane. The prototype is the identity (5.5.7), i.e.,

$$[w(t)w'(t)]_{t_1}^{t_2} - \int_{t_1}^{t_2} [w'(t)]^2 \, dt - \int_{t_1}^{t_2} G(t)[w(t)]^2 \, dt = 0, \qquad (8.1.9)$$

which shows that for $G(t) > 0$ in the interval $(a, b)$ the equation

$$w'' - G(t)w = 0 \qquad (8.1.10)$$

can have no nontrivial real solution for which $w(t)w'(t)$ vanishes more than once in $(a, b)$. Here (8.1.9) is obtained from (8.1.10) by multiplying by $w(t)$, integrating from $t_1$ to $t_2$, and integrating once by parts.

Consider now (8.1.1). Here $[w(t)]^2$ and $[w'(t)]^2$ are normally not real, nonnegative, whereas $|w(z)|^2$ and $|w'(z)|^2$ are. This suggests multiplication of (8.1.1) by $\overline{w(z)}$. The interval $(a, b)$ is now to be replaced by an oriented rectifiable arc $C$ in the complex plane. It is assumed that $C$ has piecewise, continuously turning semitangents. Usually $C$ will be a polygonal line or simply a line segment. The next hurdle to overcome is the meaning of the derivative of the conjugate along $C$, the derivative taken in the direction of the local positive semitangent. Here we use the parametric representation of $C$ in terms of arc length, $t = t(\sigma)$, and the definition of the Riemann-Stieltjes integral to obtain

$$\frac{d}{dt} \overline{w(t)} = \overline{w'(t)t'(\sigma)},$$

and the integral may be written as

$$\int_{z_1}^{z_2} |w'(t)|^2 \, \overline{dt}.$$

Finally we have the identity

$$[\overline{w(z)}w'(z)]_{z_1}^{z_2} - \int_{z_1}^{z_2} |w'(t)|^2 \, \overline{dt} + \int_{z_1}^{z_2} G(t)|w(t)|^2 \, dt = 0. \quad (8.1.11)$$

known as *Green's transform*. Here $z_1$ and $z_2$ are any two points on the arc $C$ such that $z_1$ precedes $z_2$ in the positive orientation of $C$. We shall give a number of applications of this identity, starting with some simple cases.

First we need some conventions. A domain $D$ in the complex plane will be called *vertically convex* if every vertical line which has a nonvoid intersection with $D$ intersects $D$ in an open line segment. Similarly, $D$ is *horizontally convex* if horizontal lines, which do intersect $D$, intersect in an open line segment.

### THEOREM 8.1.1

*Suppose that* (i) G(z) *is holomorphic in a domain* D *symmetric with respect to the real axis and vertically convex,* (ii) *the real axis intersects* D *in the interval* (a, b) *and* G(z) *is real for* z = x, a < x < b, (iii) P(x, y) = Re [G(x + iy)] *is positive in* D, *and* (iv) w(z) *is a nontrivial solution of* (8.1.1) *which is real on* (a, b). *Then* w(z)w'(z) *has only real zeros in* D, *if any.*

*Proof.* We set $G(z) = P(x, y) + iQ(x, y)$. The complex zeros, if any, would occur in conjugate complex pairs. Suppose that $z_2 = x_1 + iy_1$ is a zero with $y_1 > 0$. Then $x_1 - iy_1$ would also be a zero of $w(z)w'(z)$. We take $z_1 = x_1 - iy_1$, $z_2 = x_1 + iy_1$ in (8.1.1.11) and integrate along the vertical line joining these two points. We have

$$i \int_{-y_1}^{y_1} |w'(x_1 + iy)|^2 \, dy + \int_{-y_1}^{y_1} [P(x_1, y) + iQ(x_1, y)] |w(x_1 + iy)|^2 \, dy = 0.$$

$$(8.1.12)$$

By assumption the real part of this is positive, and we have a contradiction. Since $G(x) = P(x)$ is positive in $(a, b)$, there may be real zeros. If by any chance $b - a > \pi [\min P(x, 0)]^{-1/2}$, there are certainly solutions with zeros in $(a, b)$. ■

The same method gives

**THEOREM 8.1.2**

*Let* $G(z)$ *be holomorphic in a vertically convex domain* D, *let* $P(x, y) = Re \, [G(x + iy)] > 0$ *in* D, *and let* $w(z)$ *be a nontrivial solution of* (8.1.1). *Then no two zeros of* $w(z)w'(z)$ *can lie on the same vertical line in* D. *The same result holds if* $Q(x, y) = Im \, [G(z)]$ *keeps a constant sign in* D.

Another vertical line result of some interest is

**THEOREM 8.1.3**

*Let* $G(z)$ *be holomorphic in a vertically convex domain* D, *symmetric to the real axis, on which* $G(z)$ *is supposed to be real. Let* $Q(x, y)$ *keep a constant sign in the upper half of* D, *the lower boundary of which is the interval* (a, b) *of the real axis. Let* $w(z)$ *be a nontrivial solution of* (8.1.1), *real on the interval* (a, b) *so that* $w(x)w'(x)$ *is real on* (a, b), *where it keeps a constant sign the same as that of* $Q(x, y)$ *above the axis. Then* $w(z)w'(z)$ *has no complex zeros in* D.

*Proof.* · If the theorem were false, there would be a zero $x_0 + iy_0$ with $y_0 > 0$, and the following expression must vanish:

$$w(x_0)w'(x_0) - i \int_0^{y_0} |w'(x_0 + iy)|^2 \, dy - i \int_0^{y_0} (P + iQ) |w(x_0 + iy)|^2 \, dy.$$

The assumptions imply that the real part of this cannot be zero, so the assumption of complex zeros evidently leads to a contradiction. ■

An important consequence of Theorem 8.1.2 is that in a vertically convex domain $D$ where either $P(x, y) > 0$ or $Q(x, y)$ keeps a constant sign the zeros of $w(z)w'(z)$ can be ordered in a finite or infinite sequence according to increasing values of the abscissa.

Similar results hold for horizontally convex domains $D$ in which either $P(x, y) < 0$ or $Q(x, y)$ keeps a constant sign. We shall denote as **THEOREM 8.1.4** the result obtained by replacing in the wording of

Theorem 8.1.2 the expressions "vertically convex," "$P(x, y) > 0$," and "vertical line" by "horizontally convex," "$P < 0$," and "horizontal line," respectively.

Analogous results hold for solutions real on a segment of the imaginary axis. Here is one such result, the proof of which is left to the reader.

**THEOREM 8.1.5**

*Let* $G(z)$ *be holomorphic in a horizontally convex domain* D *symmetric with respect to the imaginary axis on which* $G(z)$ *is real negative. Suppose that* $P(z) < 0$ *in* D. *Let* $w(z)$ *be a nontrivial solution of* (8.1.1) *which is real on that part of the imaginary axis between* ci *and* di *which lies in* D. *Then* $w(z)w'(z)$ *may have zeros between* ci *and* di *on the imaginary axis but nowhere else in* D.

Note that $w'(z)$ is purely imaginary on the imaginary axis.

We note further that in a horizontally convex domain where either $P(x, y) < 0$ or $Q(x, y)$ keeps a constant sign we may order the zeros of $w(z)w'(z)$ according to increasing ordinates, for under the stated assumptions there can be at most one zero on each horizontal line. Now, if the domain is convex and hence both horizontally and vertically convex, we have two orderings of the zeros, one by increasing abscissas and the other by increasing ordinates. Are these orders consistent? The answer is that either the two orders agree or one is the reverse of the other. This follows from

**THEOREM 8.1.6**

*Suppose that* $G(z)$ *is holomorphic in a convex domain* D *where* $\vartheta_1 < \arg G(z) < \vartheta_2$ *and* $\vartheta_2 - \vartheta_1 < \pi$. *Let* $z = a$ *and* $z = b$ *be two zeros in* D *of* $w(z)w'(z)$. *Then either* $\arg (b - a)$ *or* $\arg (a - b)$ *lies between* $-\frac{1}{2}\vartheta_2$ *and* $-\frac{1}{2}\vartheta_1$.

*Proof.* This follows from (8.1.11), where $z_1 = a$, $z_2 = b$, $t = a + s e^{i\theta}$, $\theta = \arg (b - a)$, and $r = |b - a|$. We have then that

$$e^{2i\theta} \int_0^r G(a + s e^{i\theta}) |w(a + s e^{i\theta})|^2 \, ds$$

is real positive. The argument of this expression, however, lies between $\vartheta_1 + 2\theta$ and $\vartheta_2 + 2\theta$, so that

$$\vartheta_1 + 2\theta < 0 < \vartheta_2 + 2\theta,$$

from which the assertion follows. ∎

We can now compare the two orderings of the zeros of $w(z)w'(z)$ in D.

If $Q(x, y) > 0$ in $D$, we have

$$\vartheta_1 = 0, \qquad \vartheta_2 = \pi, \qquad \text{whence } -\tfrac{1}{2}\pi < \arg(b - a) < 0.$$

If the zeros in $D$ have been ordered by increasing abscissas, and if

$$z_n = x_n + iy_n, \qquad z_{n+1} = x_{n+1} + iy_{n+1}$$

are "consecutive" zeros of $w(z)w'(z)$, then

$$x_n < x_{n+1}, \qquad y_n > y_{n+1} \qquad (8.1.13)$$

by the inequality proved for the argument of $z_{n+1} - z_n$. In this case, the order by increasing ordinates is opposite to the order by increasing abscissas. If instead $Q(x, y) < 0$, then $0 < \arg(b - a) < \tfrac{1}{2}\pi$ and

$$x_n < x_{n+1}, \qquad y_n < y_{n+1}, \qquad (8.1.14)$$

so the two orderings are consistent.

For the next theorem we need the notion of a curve being an *indicatrix*, a notion that will reappear again and again in the following. It is supposed that $G(z)$ has no other singularities than poles in the finite plane and that $z = z_0$ is not singular. Consider

$$Z(z) = \int_{z_0}^{z} [G(t)]^{1/2} \, dt = X(z) + iY(z). \qquad (8.1.15)$$

In general, this is an infinitely many-valued function. At the zeros of $G(z)$ the function $Z(z)$ has in general algebraic branch points, while the poles give algebraic and/or logarithmic branch points. The determinations of $Z(z)$ are of the form

$$\pm Z(z) + a,$$

where $a$ is a number depending on the path leading from $z_0$ to $z$. It will be zero if $Z(z)$ has at most one algebraic branch point and no logarithmic ones.

For the following it is essential to consider the curve net

$$X(z) = C_1, \qquad Y(z) = C_2. \qquad (8.1.16)$$

Curves of the first family will be referred to as *anticlinals*, and those of the second as *indicatrices* for, as we shall see, they indicate the direction to nearby zeros. These curves should be traced on the Riemann surface of $Z(z)$. Actually the net is the same on all sheets of the surface, but the labeling of individual curves may differ from one sheet to another. Through a point $z_1$, which is neither a zero nor a pole of $G(z)$, passes one and only one curve of each family. The situation is different at poles and zeros of $G(z)$.

For the discussion immediately following, the crucial property of the indicatrix is the value of its slope. At a point $z_1$, neither a zero nor a pole of $G(z)$, the local indicatrix

$$Y(z) = Y(z_1)$$

has a unique tangent and its slope is $\tan \theta_1$, where

$$\theta_1 \equiv -\tfrac{1}{2} \arg G(z_1) \qquad (\text{mod } \pi). \qquad (8.1.17)$$

With this background we can state a *convexity theorem*:

**THEOREM 8.1.7**

*Suppose that* D *is a convex domain where* G(z) *is holomorphic and* $\neq 0$. *In* D *suppose that* $\vartheta_1 < \arg$ G(z) $< \vartheta_2$, *where* $\vartheta_2 - \vartheta_1 < \pi$. *Furthermore, no indicatrix shall have a point of inflection in* D. *Let* w(z) *be a nontrivial solution of* (8.1.1) *with a finite number of zeros and extrema in* D. *Then these points form the vertices of a convex polygon.*

*Proof.* Without loss of generality we may assume that $\vartheta_1 = 0$ so that $Q(x, y) > 0$ in $D$. If this is not so at the outset, we can bring it about by a suitable change of independent variable: a rotation about the origin will preserve convexity, indicatrices, and the nonexistence of points of inflection.

Now the absence of points of inflection means that each individual indicatrix has a curvature of the same sign, and the continuity of $G(z)$ and the condition $G(z) \neq 0$ mean that the sign of the curvature is the same for all indicatrices in $D$. There are then only two possibilities:

1. The indicatrices are concave upward and lie above the local curve tangent.
2. The indicatrices are concave downward and lie below the curve tangent.

It is enough to discuss one of these cases, and we take the first.

Mark the zeros of $w(z)w'(z)$, say $z_1, z_2, \ldots, z_n$ with $z_k = x_k + iy_k$, where now

$$x_k < x_{k+1}, \qquad y_k > y_{k+1}, \qquad k = 1, 2, \ldots, n - 1.$$

Suppose that the indicatrices through the points $z_k$ are plotted as

$$I_k \colon Y(z) = Y(z_k),$$

and let $T_k$ be the tangent of $I_k$ at $z = z_k$. Then $T_k$ divides $D$ into two complementary parts: $D_k^+$ above $T_k$ and $D_k^-$ below and on $T_k$. It is claimed that no point in one of the regions $D_k^-$ can be a zero or extremum of our solution $w(z)$. If this were false, we could find an $r > 0$ and a $\theta$, $\theta_k - \pi \leqslant \theta \leqslant \theta_k$, $\theta_k = -\frac{1}{2} \arg G(z_k)$, such that

$$- \int_0^r |w'(z_k + s\,e^{i\theta})|^2 \, ds + e^{2i\theta} \int_0^r G(z_k + s\,e^{i\theta})|w(z_k + s\,e^{i\theta})|^2 \, ds = 0.$$

Here we take the imaginary part and obtain

$$\int_0^r \sin\,[2\theta + \gamma(s)]g(s)|w(z_k + s\,e^{i\theta})|^2 \, ds = 0, \qquad (8.1.18)$$

where

$$g(s) = |G(z_k + s\,e^{i\theta})|, \qquad \gamma(s) = \arg G(z_k + s\,e^{i\theta}).$$

Now (8.1.18) is clearly not true as long as $\sin\,[2\theta + \gamma(s)]$ keeps a constant sign. For it to change its sign, there must be a value $s$ where $2\theta + \gamma(s) \equiv 0$ (mod $\pi$). At such a point, according to (8.1.17), the line would have to be tangent to the local indicatrix, so a question arises: Is it possible to draw a tangent from $z = z_k$ to an indicatrix in $D_k^-$? Now all these indicatrices turn their concave side toward $z_k$ so no tangent can be drawn, and this means that (8.1.18) cannot hold for points in $D_k^-$. From this fact we conclude that no zeros or extrema of $w(z)$ can be located in $\cup \, D_k^-$ and all points $z_j$ are in $\cap \, D_k^+$. Now this implies convexity of the polygonal line $z_1, z_2, \ldots, z_n, z_1$ unless all $z_k$'s lie on a straight line so that the polygon degenerates. This can happen iff $G(z)$ is a constant. Excluding this fairly trivial case, we have a nondegenerate polygon and the angle between the sides joining $z_{k-1}$ with $z_k$ and $z_k$ with $z_{k+1}$ is less than $\pi$, so all interior angles at the vertices are $< \pi$. This proves the convexity. ∎

Now in the chain $z_1, z_2, \ldots, z_n$ some points are zeros, whereas others are extrema. Do they alternate? This question was posed by the author over 50 years ago and remains unanswered. In some cases, however, it is possible to answer the question in the affirmative by an argument based on variation of parameters. We shall encounter some such instances later.

In the statement of the convexity theorem the condition regarding no points of inflection is puzzling and leads naturally to the question of how one can verify such a condition. Now the system of indicatrices is the set of integral curves of a certain linear homogeneous first order DE, and it is well known that for such DE's one can determine various loci where this or that phenomenon occurs. The points of inflection present such a phenomenon, and for the family $Y(z) + C$ we get the *inflection locus*

$$\text{Im}\,\{G'(z)[G(z)]^{-3/2}\} = 0. \qquad (8.1.19)$$

This says that the derivative of $[G(z)]^{-1/2}$ is real along the locus. It is a part of the locus

$$\text{Im}\,\{[G'(z)]^2[G(z)]^{-3}\} = 0, \tag{8.1.20}$$

the other part of which is the inflection locus of the anticlinals.

**EXERCISE 8.1**

1. Verify the assertions concerning solutions of the form $P e^{Q}$. Find an expression for $G(z)$ when such a solution occurs.
2. A convex domain is horizontally and vertically convex. Is the converse true?
3. Use (8.1.11) to show that, if $G(z) = a^2$, a constant, and if $w(z)$ is a nontrivial solution with a zero at $z = z_0$, then all its zeros and extrema lie on a straight line through $z_0$. Find the line.
4. How should $\omega$ be chosen, $|\omega| = 1$, so that the rotation $z = \omega t$ leads to a DE with the properties used in the proof of Theorem 8.1.7?
5. If $G^{1/2} = U + iV$, show that the indicatrices satisfy the DE $U\,dy + V\,dx = 0$, while the anticlinals satisfy $U\,dx - V\,dy = 0$.
6. Use this fact to verify (8.1.19).
7. Why does (8.1.19) imply (8.1.20)? Find the inflection locus of the anticlinals.
8. If $G(z) = z - a$, sketch the indicatrices and find their inflection locus.
9. In Problem 8 take $a = 0$ and the solutions determined by (i) $w_1(0) = 0$, $w_1'(0) = 1$, (ii) $w_2(0) = 1$, $w_2'(0) = 0$. Find where their zeros are located.

## 8.2. ZERO-FREE REGIONS AND LINES OF INFLUENCE

Various *zero-free regions* have been encountered in Section 8.1. *A region $S(z_0)$ in the complex plane is a zero-free region with respect to a solution $w(z)$ satisfying a condition at $z = z_0$ if $w(z)w'(z) \neq 0$ in $S(z_0)$).* There are various ways of constructing such regions.

We may consider (8.1.1) as the equation of motion of a "particle" $w(z)$ when $z$ describes a path $C$ and as a consequence $w(z)$ describes an orbit $\Gamma$ in the $w$-plane. A zero of $w(z)$ encountered by $C$ corresponds to a passage of $\Gamma$ through the origin. A zero of $w'(z)$ is another story. If

$$w(z) = R(z)\, e^{i\Theta(z)}, \tag{8.2.1}$$

then

$$w'(z) = [R'(z) + iR(z)\Theta'(z)]\, e^{i\Theta(z)}. \tag{8.2.2}$$

If this is to be zero at $z = z_1$, we must have

$$R'(z_1) = \Theta'(z_1) = 0, \tag{8.2.3}$$

and normally these are simple zeros where $R'$ and $\Theta'$ change their signs.

The orbit $\Gamma$ has a *cusp* for this value of $z$, and both branches of $\Gamma$ lie to the same side of the cusp tangent.

Now it is clear that between consecutive passages through the origin of $\Gamma$ there must be a point where $R(z)$ has a maximum, but there is no reason why this should correspond to a cusp and a zero of $w'(z)$. Let us now note the equations of motion. We have

$$w''(z) = \{R''(z) - R(z)[\Theta'(z)]^2 + i[2R'(z)\Theta'(z) + R(z)\Theta''(z)]\} e^{i\Theta(z)}.$$
(8.2.4)

If $z = r e^{i\theta}$ and the differentiations are with respect to $r$ for fixed $\theta$, the DE takes the from

$$R''(r) - R(r)[\Theta'(r)]^2 + i[2R'(r)\Theta'(r) + R(r)\Theta''(r)]$$
$$+ g(r) e^{i\gamma(r)+2i\theta}R(r) = 0,$$
(8.2.5)

where

$$G(z) = g(r) e^{i\gamma(r)}.$$

Here we separate reals and imaginaries and obtain the real-valued equations of motion:

$$R''(r) + \{g(r) \cos [\gamma(r) + 2\theta] - [\Theta'(r)]^2\}R(r) = 0, \qquad (8.2.6)$$

$$\{\Theta'(r)[R(r)]^2\}' + g(r) \sin [\gamma(r) + 2\theta][R(r)]^2 = 0, \qquad (8.2.7)$$

where the second equation has been multiplied by $R(r)$ to bring it into self-adjoint form.

These equations are satisfied by

$$R(r) = |w(z_0 + r e^{i\theta})| \qquad \text{and} \qquad \Theta(z) = \arg w(z_0 + r e^{i\theta})$$

for small positive values of $r$. The functions are continuous together with their first and second order derivatives. As $r$ increases, various things may happen. We mark on the ray $\arg (z - z_0) = \theta$ the points where one of the following events takes place: the point is (i) a zero of $G(z)$, (ii) a singularity of $G(z)$, (iii) a zero of $w(z)$, (iv) a zero of $w'(z)$.

Let $z_1 = r_1 e^{i\theta}$ be the first such point. The first two categories may be considered as known, but disregarding these points, we need a lower bound for an $r_1$ corresponding to categories (iii) and (iv). We suppose $w(z)$ to be fixed by the conditions

$$w(z_0) = 0, \qquad w'(z_0) = 1.$$

Equation (8.2.6) leads to the observation that

$$g(r) \cos [\gamma(r) + 2\theta] - [\Theta'(r)]^2 \leq g(r),$$

and this, in turn, leads to the following question. Equation (8.2.6) is satisfied by $R(r)$ in the interval $(0, r_1)$ and $R(0) = R(r_1) = 0$. Consider the equation

$$v''(r) + g(r)v(r) = 0, \qquad v(0) = 0, \qquad v'(0) = 1. \tag{8.2.8}$$

Suppose that the least positive zero of $v(r)$ is $r_0$. If

$$g_1(r) = g(r) \cos \left[ \gamma(r) + 2\theta \right] - [\Theta'(r)]^2,$$

we have

$$R(r_0)v'(r_0) - R'(r_0)v(r_0) + \int_0^{r_0} [g(s) - g_1(s)]v(s)R(s) \, ds = 0.$$

Here $v(r_0) = 0$, $v'(r_0) < 0$, and the bracket is nonnegative. It follows that $R(r) > 0$ for $0 < r < r_0$. This gives

**THEOREM 8.2.1**

*Let* $v(r) = v(r, \theta)$ *be the solution of*

$$v''(r) + |G(z_0 + r\, e^{i\theta})|v(r) = 0, \qquad v(0) = 0, \qquad v'(0) = 1. \tag{8.2.9}$$

*Let the least positive zero of* $v(r, \theta)$ *be* $r_0(\theta)$. *Then the solution* $w(z)$ *of* (8.1.1) *with* $w(z_0) = 0$, $w'(z_0) = 1$, *has no zeros on the open interval* $(z_0, z_0 + r_0(\theta)\, e^{i\theta})$.

**COROLLARY**

*Let* $G(z)$ *be holomorphic in the disk* $|z - z_0| < R$ *and satisfy* $|G(z)| \leqslant \pi^2 R^{-2}$ *there; then the solution* $w(z)$ *with* $w(z_0) = 0$, $w'(z_0) = 1$ *has no other zeros in* $|z - z_0| < R$.

Here we have the comparison equation

$$v''(r) + \left(\frac{\pi}{R}\right)^2 v(r) = 0 \tag{8.2.10}$$

with

$$v(r) = \sin \left(\frac{\pi}{R} r\right)$$

and the only zero in $(-R, R)$ is $r = 0$.

An alternative formulation of the corollary is:

*If* $G(z)$ *is holomorphic in* $|z - z_0| < R$ *and if* $|G(z)| \leqslant M$ *there, then* $w(z)$ *has no zeros in the punctured disk,*

$$0 < |z - z_0| < \min (R, \pi M^{-1/2}). \tag{8.2.11}$$

This should be compared with Forsyth's bound, given in (8.1.3).

We obtain similar results when $w(z)$ has an extremum at $z = z_0$, but here the zero-free region is smaller.

**THEOREM 8.2.2**

*If* $G(z)$ *is holomorphic in* $|a - z_0| < R$ *and if* $|G(z)| \leq M$ *there, then the solution of* (8.1.1) *with* $w(z_0) = 1$, $w'(z_0) = 0$ *has no zeros in*

$$|z - z_0| < min\ (R, \tfrac{1}{2}\pi M^{-1/2}).  \qquad (8.2.12)$$

The comparison equation is still (8.2.10), but now the solution is

$$\cos\left(\frac{\pi}{R}\,r\right),$$

and the least positive zero is $\tfrac{1}{2}\pi R$. Theorems 8.2.1 and 8.2.2 are due to Choy-Tak Taam (1952).

So far the zero-free regions have been obtained using comparison theorems, a classical Sturmian device.

The conclusion to be drawn from (8.2.7) is of a different nature, as we have already seen in the proof of Theorem 8.1.7. We consider a solution for which $w(z_0)w'(z_0) = 0$. This implies $\Theta'(0)R(0) = 0$. We now integrate (8.2.7) with respect to $r$, keeping $\theta$ fixed:

$$\Theta'(r)[R(r)]^2 + \int_0^r g(s) \sin\left[\gamma(s) + 2\theta\right][R(s)]^2\ ds = 0.  \qquad (8.2.13)$$

This shows that we cannot have $\Theta'(r)[R(r)]^2 = 0$ as long as

$$\sin\left[\gamma(s) + 2\theta\right]$$

has a constant sign in $0 < s < r$. Let us now mark the first point, $z_1 = z_1(r, \theta) \neq z_0$, where either $G(z)$ is singular or

$$\arg G(z_0 + r\,e^{i\theta}) + 2\theta \equiv 0 \qquad (\text{mod } \pi).  \qquad (8.2.14)$$

We have then

**THEOREM 8.2.3**

*A solution of* (8.1.1) *for which* $w(z_0)w'(z_0) = 0$, *where* $G(z)$ *is holomorphic at* $z_0$, *admits the set*

$$S_0(z_0) = \{z; z = z_0 + \alpha[z_1(r, \theta) - z_0], 0 < \alpha \leq 1, 0 \leq \theta < 2\pi\}  \qquad (8.2.15)$$

*as a zero-free star in which* $w(z)w'(z) \neq 0$.

Condition (8.2.14) is familiar to us from the proof of Theorem 8.1.7. It asserts that at the point $z_1$ in question the ray $\arg(z - z_0) = \theta$ is tangent to the local indicatrix $Y(z) = Y(z_1)$. If $z_0$ is not a zero of $G(z)$, then there is one and only one indicatrix through $z_0$; and if in addition $z_0$ is not on the indicatrix inflection locus (8.1.19), then in some partial neighborhood of $z = z_0$ the indicatrices turn their concave side toward $z_0$. Let $T_0$ be the tangent of the indicatrix $Y(z) = Y(z_0)$. Then one side of $T_0$ belongs to $S_0(z_0)$, whereas the other does not (compare the situation encountered in the proof of Theorem 8.1.7). Let us call the two sides the "on side" and the "off side," respectively. On the off side of $T_0$ we can draw tangents to the indicatrices from $z_0$. From this we get the following

## COROLLARY

*The boundary of $S_0(z_0)$ is the locus of the nearest points of tangency between the pencil of straight lines through $z = z_0$ and the system of indicatrices plus part of the tangent $T_0$, and possibly also parts of lines beyond the singular points.*

As observed above, this picture will have to be modified if $z = z_0$ is a zero of $G(z)$ or belongs to the indicatrix inflection locus.

We turn to the construction of zero-free regions based on the use of the corresponding Green's transform (8.1.7). We start by finding a zero-free star. Suppose that $w(z)$ is a solution of (8.1.1) such that $w(z_0)w'(z_0) = 0$, where, as usual, $z_0$ is not a singular point of the DE. We integrate along the ray $\arg(z - z_0) = \theta$ to obtain

$$e^{i\theta}\overline{w(z)}w'(z) = \int_0^r |w'(t)|^2 \, ds - \int_0^r g(s) \, e^{i\gamma(s)+2i\theta}|w(t)|^2 \, ds. \qquad (8.2.16)$$

Here

$$t = z_0 + s \, e^{i\theta}, \qquad z = z_0 + r \, e^{i\theta},$$
$$g(s) = |G(t)|, \qquad \gamma(s) = \arg G(t).$$

The first integral on the right is positive; the second one has an argument which lies between

$$\min \gamma(s) + 2\theta \qquad \text{and} \qquad \max \gamma(s) + 2\theta,$$

provided the difference between the maximum and the minimum is $< \pi$. Let us mark on the ray $\arg(z - z_0) = \theta$ the first point where (i) $G(z)$ is singular or (ii) $\arg G(z) + 2\theta \equiv 0 \,(\text{mod } \pi)$ or (iii) $\max \arg G(z) - \min \arg G(z) = \pi$. Here the maximum and the minimum refer to the points on the ray from $z = z_0$ to the point in question. Let the marked point correspond to $s = r(\theta)$, and define

$$S(z_0) = [z; z = z_0 + s \, e^{i\theta}, 0 < s \leqslant r(\theta), 0 \leqslant \theta < 2\pi]. \qquad (8.2.17)$$

This is a star containing the star $S_0(z_0)$, and the inclusion may be proper. We have

**THEOREM 8.2.4**

*If* $w(z)$ *is a solution of* (8.1.1) *such that* $w(z_0)w'(z_0) = 0$, *where* $z_0$ *is nonsingular, then* $w(z)w'(z) \neq 0$ *in* $S(z_0)$.

The proviso that $z_0$ be nonsingular may be disregarded if $w(z)$ is a holomorphic solution at $z = z_0$ or $w(z)w'(z)$ tends to the limit 0 as $z \to z_0$ and the integrals in (8.2.16) exist for the lower limit 0.

This is only one of the ways in which the restrictions involving the use of Green's transforms may be relaxed. We are not restricted in the discussion to the use of rays; any oriented rectifiable arc $C$ will do if it leads to a *definite vector field* in a sense to be clarified. We start by taking a solution $w(z)$ defined at the nonsingular point $z_0$ by the *initial parameter*

$$\frac{w'(z_0)}{w(z_0)} = \lambda. \tag{8.2.18}$$

This may be zero, infinity, or a complex number. In the third case set $\omega = \arg \lambda$. Consider the identity

$$\overline{w(z)}w'(z) = \overline{w(z_0)}w'(z_0) + \int_C |w'(t)|^2 \, \overline{dt} - \int_C G(t)|w(t)|^2 \, dt. \tag{8.2.19}$$

The first term on the right either is zero or is a nonzero vector whose argument is $\omega$. It is assumed that $C$ is piecewise smooth and has a continuously turning positive semitangent except for a finite number of points. Let the slope angle of the semitangent be $\theta(z)$. With the first integral is then associated the set of vectors

$$\{e^{-i\theta(z)}; z \in C\}, \tag{8.2.20}$$

while

$$\{\exp i \, [\arg G(z) - \pi + \theta(z)]; z \in C\} \tag{8.2.21}$$

goes with the second integral. The union of these two sets and the vector $e^{i\omega}$ is *the vector field of $C$ with respect to* $w(z)$. The vector field is *definite* if all three components are confined to an angle of opening $< \pi$, and in this case $C$ is called a *line of influence* of the solution $w(z)$ with respect to $z = z_0$.

Thus $C$ is an *li* ($-$line of influence) if there exists a $\vartheta$ such that (i) $\vartheta < \omega < \vartheta + \pi$, (ii) $\vartheta < -\theta(z) < \vartheta + \pi, z \in C$, and further (iii) $\vartheta < \arg G(z) - \pi + \theta(z) < \vartheta + \pi, z \in C$.

If the vector field is definite, the sum of the three vectors in the right-hand member of (8.2.19) cannot be zero, so $w(z)w'(z)$ cannot vanish at the upper end point of $C$. In the three inequalities we can replace "$<$" by "$\leqslant$" provided "$<$" holds either in (i) or in a subset of $C$ of positive measure in (ii) and (iii).

The totality of points $z$ which can be reached by an li from $z = z_0$ with respect to the given solution $w(z)$ is defined as the *domain of influence* DI $(w, z_0)$ of $w(z)$ with respect to $z_0$. It is clearly a zero-free region.

**THEOREM 8.2.5**

*The solution* w(z) *specified by condition* (8.2.18) *satisfies* w(z)w′(z) $\neq$ 0 *in* DI (w, z₀).

The search for lines of influence can be standardized to some extent. Various curve nets present themselves as possibilities, and one can always try polygonal lines, where the sides are parallel to the coordinate axes.

As an example take the equation

$$w'' + zw = 0, \qquad z_0 = a > 0, \qquad \lambda \text{ real.} \tag{8.2.22}$$

Let $C$ be the line from $a$ to $b$, $b$ real $> a$, followed by the vertical from $z = b$ to $z = b + iy$, $0 < y$. This is an li. For $\omega = 0$ or $\pi$, the set (8.2.20) consists of two vectors, one of argument $-\frac{1}{2}\pi$ and the other 0, while the set (8.2.21) has one vector along the negative real axis and the others in the fourth quadrant. The vector field is then definite, so the solution can have neither zeros nor extrema $x_1 + iy_1$ with $a \leqslant x_1, 0 < y_1$. Note that the segment $(a, b)$ is not an li. It is recommended that the reader plot the vector field.

Another suitable net is formed by the curves

$$\mathbf{G}(z) = \int_a^z G(t)\, dt = \mathbf{G}_1(z) + i\mathbf{G}_2(t), \tag{8.2.23}$$

$$\mathbf{G}_1(z) = C_1, \qquad \mathbf{G}_2(z) = C_2. \tag{8.2.24}$$

This is the G-net; the two sets are orthogonal trajectories of each other, and through a point $z$ which is neither a zero nor a pole of $G(z)$ passes one and only one curve of each family. If we integrate along a member of the $\mathbf{G}_1$-family, then

$$-\int_C G(t)|w(t)|^2\, dt = -i \int_C |w(t)|^2\, d\mathbf{G}_2 \tag{8.2.25}$$

is purely imaginary, whereas the integral along a $\mathbf{G}_2$-curve gives

$$-\int_C |w(t)|^2\, d\mathbf{G}_1, \tag{8.2.26}$$

which is real. These are useful paths of integration as long as the vectors $\{e^{-\theta(z)}; z \in C\}$ give a definite vector field in conjunction with (8.2.25) or (8.2.26), as the case may be.

As an illustration of the use of this net let us take the Hermite-Weber equation in the form

$$w''(z) + (c^2 - z^2)w(z) = 0, \qquad 0 < c. \qquad (8.2.27)$$

Here

$$\begin{aligned}
\mathbf{G}_1(x, y): c^2 x - \tfrac{1}{3}x^3 + xy^2 = C_1, \\
\mathbf{G}_2(x, y): c^2 y - x^2 y + \tfrac{1}{3}y^3 = C_2.
\end{aligned} \qquad (8.2.28)$$

Let us consider a solution of (8.2.27) which on the real axis is real and on the imaginary axis satisfies a restriction on the initial parameter,

$$\lambda(iy_0) = \frac{w'(iy_0)}{w(iy_0)}, \qquad (8.2.29)$$

to the effect that its argument $\omega = \omega(iy_0)$ will not fall in the first quadrant for $y_0 > 0$ or in the fourth quadrant for $y_0 < 0$. If these conditions can be satisfied for all $y_0$, there exist two domains, $D$ and $\bar{D}$, in which $w(z)w'(z) \neq 0$. Here $D$ is bounded by the hyperbola $x^2 - \tfrac{1}{3}y^2 = c^2, x < -c$, $0 < y$, the segment $(-c, 0)$ of the real axis and the positive imaginary axis. $\bar{D}$ is the conjugate domain.

We can integrate along a $\mathbf{G}_2$-curve from $t = iy_0$ to $t = z \in D$, where $y_0 > 0$. We have then

$$\overline{w(z)}w'(z) = \lambda(iy_0)|w(iy_0)|^2 + \int_{iy_0}^{z} |w'(t)|^2 \, \overline{dt}$$
$$- \int_{iy_0}^{z} |w(t)|^2 \, d\mathbf{G}_1(t). \qquad (8.2.30)$$

Here $\lambda(iy_0)$ does not lie in the first quadrant, the first integral lies in the third quadrant, and the third term is real negative. The resultant of the three vectors cannot be the zero vector, no matter what value $\lambda(iy_0)$ has, as long as it is not in the first quadrant. This proves the assertion for $D$. The same type of argument works if $y_0 < 0$.

### EXERCISE 8.2

1. Verify (8.2.2)–(8.2.7).
2. Why does a zero of $w'(z)$ correspond to a cusp of the orbit, and why do the two branches lie to the same side of the cusp tangent?
3. At a zero $r_0$ of $R(r)$ this function changes its sign. Does this mean that the absolute value of $w(z)$ becomes negative? At such a point $\Theta(r)$ is discontinu-

ous. Show that there are finite left- and right-hand limits which differ by an odd multiple of $\pi$. Show that $\Theta'(r_0 + 0) = \Theta'(r_0 - 0) = 0$.

4.  Verify Theorem 8.2.3.

5.  Let $D$ be a convex domain in the $z$-plane, where $G(z)$ is holomorphic and $\neq 0$. Take $z_0 \in D$, and consider the pencil of straight lines through $z_0$. Find the locus of tangency of the lines with the indicatrices in $D$. Show that this locus is a part of $\text{Im}\,[(z - z_0)^2 G(z)] = 0$, and show that the rest is the tangent locus of the anticlinals.

6.  For the equation $zw'' - w = 0$ and $z_0 = 1 + i$, determine the star $S(z_0)$ with respect to a solution for which $w(z_0)w'(z_0) = 0$.

7.  The equation in Problem 6 has a regular-singular point at the origin where there is a solution which is an entire function of $z$. Show that it is possible to define the star $S(0)$ for this solution, and determine it.

8.  For the same equation take a solution real on the positive real axis and a point $z = a$, $0 < a$, and $w(a) > 0$, $w'(a) < 0$. Take the path $z = a + re^{i\theta}$, $0 < r$, $0 \le \theta < 2\pi$. Determine the vector field of the path. Are there any limitations on $r$?

9.  In (8.2.30) show that the first integral is a vector in the third quadrant, as asserted.

10. Verify that the condition imposed regarding $\lambda(iy_0)$ in the discussion of (8.2.30) is satisfied by $E_a(z)$, where $c^2 = 2a + 1$.

## 8.3. OTHER COMPARISON THEOREMS

The results to be proved or stated in this section are due to Z. Nehari and his school, P. R. Beesack and B. Schwarz, as well as V. V. Pokarnyi and C. T. Taam. They are mostly nonoscillation theorems with a bearing on the theory of *univalence* of analytic functions (cf. Chapter 10). We start with a result obtained by Taam (1953), for which we give a proof due to Beesack (1956) involving a generalization of Green's transform and another quadratic integral identity.

Consider the equation

$$w'' + G(z)w = 0 \tag{8.3.1}$$

and the corresponding Green's transform,

$$[\overline{w(z)}w'(z)]_a^b - \int_a^b |w'(t)|^2\,dt + \int_a^b G(t)|w(t)|^2\,dt = 0. \tag{8.3.2}$$

Here $a$ and $b$ are two zeros of $w(z)w'(z)$ in a convex domain $D$ where $G(z)$ is holomorphic, $t = a + re^{i\theta}$, and $0 < r \le r_1 = |b - a|$. Hence

$$\int_0^{r_1} |w'|^2\,dr = \int_0^{r_1} e^{2i\theta} G|w|^2\,dr \equiv \int_0^{r_1} (g_1 + ig_2)|w|^2\,dr.$$

This gives

$$\int_0^{r_1} |w'|^2 \, dr = \int_0^{r_1} g_1 |w|^2 \, dr, \qquad \int_0^{r_1} g_2 |w|^2 \, dr = 0.$$

Next let $\lambda$ and $\mu$ be real numbers, $\lambda^2 + \mu^2 \neq 0$. Multiply the first equation by $\lambda$ and the second by $\mu$, and add to obtain

$$\int_0^{r_1} (\lambda g_1 + \mu g_2) |w|^2 \, dr = \lambda \int_0^{r_1} |w'|^2 \, dr. \tag{8.3.3}$$

This is Beesack's modification of Green's identity. The second required identity we state as a lemma:

**LEMMA 8.3.1**

*Let* $F(r)$ *be real and continuous in the interval* $(a, b)$, *and let* $(t-a)^2 \times (t-b)^2 F(t)$ *be bounded in* $(a, b)$. *Suppose that the Riccati equation*

$$Y'(t) = F(t) + [Y(t)]^2 \tag{8.3.4}$$

*has a solution* $Y(t)$ *continuous together with its first derivative and such that* $(t-a)(b-t)Y(t)$ *is bounded in* $(a, b)$. *If* $y(t)$ *is a piecewise smooth function and* $y(a) = y(b) = 0$, *then*

$$\int_a^b \{[y'(t)]^2 - F(t)[y(t)]^2\} \, dt = \int_a^b [y'(t) + Y(t)y(t)]^2 \, dt. \tag{8.3.5}$$

*Proof (sketch).* The integral in the left member exists since the assumptions on $y(t)$ imply that $|y(t)| \leq K(b-t)(t-a)$, so that $F(t)[y(t)]^2$ is bounded and continuous; further left- and right-hand derivatives exist everywhere and are bounded and equal with a finite number of exceptions. If the right-hand integral exists, it equals

$$\int_a^b [y'(t)]^2 \, dt + 2 \int_a^b y(t)y'(t)Y(t) \, dt + \int_a^b [Y(t)y(t)]^2 \, dt. \tag{8.3.6}$$

The first integral exists and is also present in the left member of (8.3.5). An integration by parts, using the vanishing of $Y(t)[y(t)]^2$ in the limits, reduces the second integral to

$$-\int_a^b [y(t)]^2 Y'(t) \, dt = -\int_a^b [y(t)]^2 F(t) \, dt - \int_a^b [y(t)Y(t)]^2 \, dt.$$

Here the first integral on the right exists, and the second integral cancels the third integral in (8.3.6). Hence the right member of (8.3.5) exists and equals the left member, so the lemma is proved. The important thing here is that the right member is nonnegative and can be zero iff the integrand is identically zero. ■

**THEOREM 8.3.1**

Let $G(z)$ be holomorphic in a domain $D$ containing the line segment $z = z_0 + r\,e^{i\theta}$, $0 \le r \le R$. Set

$$e^{2i\theta} G(z_0 + r\,e^{i\theta}) = g_1(r, \theta) + ig_2(r, \theta). \qquad (8.3.7)$$

Let $Q(r)$ be continuous in $(0, R)$, and suppose that the DE

$$v''(r) + Q(r)v(r) = 0 \qquad (8.3.8)$$

has a solution with a zero at $r = 0$ and no zeros in $(0, R)$. Suppose that real numbers $\lambda, \mu$ can be found, $\lambda \ge 0$, $\lambda^2 + \mu^2 > 0$, such that

$$\lambda g_1(r, \theta) + \mu g_2(r, \theta) \le \lambda Q(r), \qquad 0 \le r \le R. \qquad (8.3.9)$$

Then any nontrivial solution of (8.3.1) with $w(z_0) = 0$ is unequal to zero on the segment $(z_0, z_0 + R\,e^{i\theta})$ unless $g_2(r, \theta) \equiv 0$. Even if $g_2(r, \theta)$ should be $\equiv 0$, the statement holds, provided that $\lambda > 0$. Moreover, if strict inequality holds in (8.3.9) at a single point, then $w(z)$ has no zeros in $(z_0, z_0 + R\,e^{i\theta})$.

*Proof.* Suppose that $w(z_1) = 0$, $z_1 = z_0 + r_1\,e^{i\theta}$, $0 < r_1 < R$. Now (8.3.3) with (8.3.9) gives

$$\lambda \int_a^b |w'|^2\, dr \le \lambda \int_a^b Q(r)|w|^2\, dr.$$

Since $\lambda$ may be zero, this case must be settled before we can proceed. If $\lambda = 0$, then $\mu \ne 0$, and (8.3.9) shows that either $g_2(r, \theta)$ has a constant sign opposite to that of $\mu$, or else $g_2(r, \theta) \equiv 0$. In the first case, (8.3.3) has a negative left member while the right one is zero—a clear contradiction. From this it follows that $w(z)$ can have no zero in $(z_0, z_0 + R\,e^{i\theta})$. The second alternative, $g_2(r, \theta) \equiv 0$, cannot arise if strict inequality holds in (8.3.9) for at least one value of $r$ and hence in some interval.

Suppose now that strict inequality holds nowhere and $g_2(r, \theta) \equiv 0$; then for $\lambda > 0$ (8.3.9) shows that $g_1(r, \theta) = Q(r)$ and the desired conclusion is immediate. Suppose now that $g_2(r, \theta) \not\equiv 0$ and that $\lambda > 0$. Note that

$$\int_0^{r_1} |w'|^2\, dt \le \int_0^{r_1} Q(r)|w|^2\, dr,$$

and set

$$|w(z_0 + r\,e^{i\theta})| = W(r).$$

We observe that $W'(r)$ exists for $0 < r < r_1$. Moreover, by (8.2.1) and (8.2.2), we see that $W'(r) = R'(r)$, so that

$$|W'(r)| \le |w'(z)|, \qquad z = z_0 + r\,e^{i\theta}$$

and

$$\int_0^{r_1} [W'(r)]^2 \, dr \le \int_0^{r_1} Q(r)[W(r)]^2 \, dr. \qquad (8.3.10)$$

We can now use Lemma 8.3.1 to obtain the opposite inequality and a contradiction. Take

$$a = 0, \qquad b = r_1, \qquad y(t) = W(t), \qquad F(t) = Q(t), \qquad Y(t) = -\frac{v'(t)}{v(t)}.$$

Since $v(t)$ has no zeros in $(0, R)$, $Y(t)$ satisfies the Riccati equation

$$Y'(t) = Q(t) + [Y(t)]^2$$

in $(0, R)$, and $t(R - t)Y(t)$ is bounded. Furthermore, $W(t)$ is smooth and $W(0) = W(r_1) = 0$. Since the assumptions of the lemma are satisfied, we get

$$\begin{aligned}
\int_0^{r_1} [W'(t)]^2 \, dt &= \int_0^{r_1} Q(t)[W(t)]^2 \, dt \\
&\quad + \int_0^{r_1} [W'(t) + Y(t)W(t)]^2 \, dt \\
&\ge \int_0^{r_1} Q(t)[W(t)]^2 \, dt, \qquad (8.3.11)
\end{aligned}$$

and here equality can hold iff

$$W'(t) + Y(t)W(t) = 0, \qquad W(t) \equiv Cv(t).$$

If this should happen, $W(t) \ne 0$ for $0 < t < R$. Finally, if $W(t)$ is not a constant multiple of $v(t)$, the resulting inequality (8.3.11) contradicts (8.3.10). It follows in either case that $w(z) \ne 0$ on $(z_0, z_0 + R e^{i\theta})$, as asserted. ∎

As special cases of this theorem we list the following:

1. $Q(r) = \frac{1}{4}\pi^2 R^{-2}$.
2. $\lambda = 1, \mu = 0, Q(r) = g_1(r, \theta)$.
3. $\lambda = 0, g_2(r, \theta) \ne 0$ on $(0, R), \mu = -\operatorname{sgn} g_2$.
4. $\lambda = 1, \mu = 0, Q(r) = g_1(r, \theta) < 0$.

Here case 1 is related to the Corollary of Theorem 8.2.1 and to Theorem 8.2.2. The result is due to Z. Nehari (1949). We state it here as

## THEOREM 8.3.2

Let $G(z)$ be holomorphic in $|z - z_0| < R$ and satisfy $|G(z)| \le \frac{1}{4}\pi^2 R^{-2}$. Then no solution of (8.3.1) can have more than one zero in the disk.

Since we have $g_1(r, \theta) \leq |e^{2i\theta}G(z_0 + r e^{i\theta})| \leq \frac{1}{4}\pi^2 R^{-2}$, we can take $\lambda = 1$, $\mu = 0$ and the majorant equation

$$v'' + \tfrac{1}{4}\pi^2 R^{-2}v = 0.$$

There are real solutions with equidistant real zeros, and the distance between consecutive zeros equals the diameter of the disk.

Case 2 follows from (8.2.6) since

$$g(r) \cos [\gamma(r) + 2\theta] - [\Theta'(r)]^2 \leq g_1(r, \theta).$$

This case is essentially covered by Theorems 8.2.1 and 8.2.3. Case 3 follows from (8.2.6) and the proof of Theorem 8.2.1, while case 4 is essentially Theorem 8.2.2.

Beesack has combined comparison theorems with fractional linear transformations to obtain

**THEOREM 8.3.3**

*Let* $G(z)$ *be holomorphic in the unit disk,* $|z| < 1$, *and let* $M(r)$ *be the maximum modulus of* $G(z)$, *i.e.,* $M(r) = max\ |G(r\ e^{i\theta})|$. *Suppose that* $(1 - r^2)^2 M(r)$ *is nonincreasing in* $(0, 1)$ *and that the DE*

$$v'' + M(r)v = 0 \tag{8.3.12}$$

*has a solution without zeros in the interval* $(-1, 1)$. *Then no solution of* (8.3.1) *can have more than one zero in* $|z| < 1$.

*Proof.* Suppose that there is a solution of (8.3.3) with two zeros, $a$ and $b$ in the unit disk. By a suitable Möbius transformation we can bring it about that a transformed DE has two zeros in the interval $(-1, 1)$. If $a$ and $b$ are already real, the next step is to be omitted. If not, we observe that there is a unique circle $\Gamma$ passing through $a$ and $b$ and orthogonal to the unit circle. We may assume this circle to be symmetric to the imaginary axis, for if this is not so at the outset we can use a rotation to achieve it. This means replacing $G(z)$ by $\omega^2 G(\omega z)$, where $|\omega| = 1$. It does not change the maximum modulus or any of the hypotheses.

Suppose now that $\Gamma$ intersects the imaginary axis at $z = \beta i$, $0 < \beta < 1$. The transformation

$$\zeta = \frac{z - \beta i}{1 + \beta i z} \tag{8.3.13}$$

will then map the arc of $\Gamma$ in the unit disk on the interval $-1 < \zeta < +1$, and the points $z = a$ and $z = b$ go into points on the interval $(-1, 1)$, say

$-1 < \xi_1 < \xi_2 < 1$. The Möbius transformation (8.3.13) takes (8.3.1) into

$$Y'' - \frac{2\beta i}{1 - i\beta\zeta} Y' + \frac{(1 - \beta^2)^2}{(1 - i\beta\zeta)^4} G\left(\frac{\zeta + i\beta}{1 - i\beta\zeta}\right) Y = 0, \qquad (8.3.14)$$

where $Y(\zeta) = w(z)$ and the primes denote differentiation with respect to $\zeta$. Here we can remove the first derivative by setting $u(\zeta) = (1 - i\beta\zeta) Y(\zeta)$, which leads to the DE

$$\frac{d^2u}{d\zeta^2} + \frac{(1 - \beta^2)^2}{(1 - i\beta\zeta)^4} G\left(\frac{\zeta + i\beta}{1 - i\beta\zeta}\right) u = 0. \qquad (8.3.15)$$

Here we are interested in real values of $\zeta$ between $-1$ and $+1$. We have then

$$\left|\frac{\zeta + i\beta}{1 - i\beta\zeta}\right| = \left(\frac{\beta^2 + \zeta^2}{1 + \beta^2\zeta^2}\right)^{1/2}.$$

Now (8.3.15) has a solution with two zeros in the interval $(-1, 1)$. We are thus led to the comparison equation

$$\frac{d^2v}{d\zeta^2} + \frac{(1 - \beta^2)^2}{(1 + \beta^2\zeta^2)^2} M\left[\left(\frac{r^2 + \beta^2}{1 + \beta^2 r^2}\right)^{1/2}\right] v(r) = 0 \qquad (8.3.16)$$

and an auxiliary hypothesis which will be eliminated afterward.

Hypothesis H. No solution of the DE (8.3.16) can have more than one zero in the interval $(-1, 1)$ for any value of $\beta$, $0 < \beta < 1$.

Our discussion has shown that the DE (8.3.15) has a solution with two zeros in $(-1, 1)$, and if Hypothesis H holds this contradicts Theorem 8.2.1, which allows a solution of (8.3.15) to have at most one zero in $(-1, 1)$.

To complete the proof of Theorem 8.3.3 we have to show that Hypothesis H is equivalent to the condition that $(1 - r^2)^2 M(r)$ is nonincreasing. We shall show that

$$\frac{(1 - \beta^2)^2}{(1 + \beta^2 r^2)} M\left[\left(\frac{\beta^2 + r^2}{1 + \beta^2 r^2}\right)^{1/2}\right] \leqslant M(r) \qquad (8.3.17)$$

for $0 \leqslant \beta < 1$, $0 \leqslant r < 1$. To prove the equivalence of the two conditions, assume first that $(1 - r^2)^2 M(r)$ is nonincreasing. Set $x^2 = (\beta^2 + r^2)(1 + \beta^2 r^2)^{-1}$. Then $r^2 \leqslant x^2 \leqslant 1$ and

$$(1 - x^2)^2 M(x) \leqslant (1 - r^2)^2 M(r),$$

and this reduces to (8.3.17). Conversely, suppose that (8.3.17) holds, and let $r$ and $x$ be such that $0 \leqslant r \leqslant x < 1$. Define $\beta^2$ so that $x^2 = (\beta^2 + r^2)(1 + \beta^2 r^2)^{-1}$; then $\beta, r$ satisfy condition (8.3.17). The inequality now reduces to

$$(1 - r^2)^2 M(r) \leqslant (1 - x^2) M(x), \qquad (8.3.18)$$

and since $r \leqslant x$ this proves the nonincreasing property of the product. This completes the proof of Theorem 8.3.3.  ∎

### COROLLARY

*No solution of* (8.3.1) *can have more than one zero in* $|z| < 1$ *if* $(1 - r^2)^2 M(r)$ *is nonincreasing in* $(0, 1)$ *and if the DE*

$$y'' + M(r)y = 0 \qquad (8.3.19)$$

*has a solution which does not vanish in* $-1 < r < 1$.

In other words, all the solutions of (8.3.1) are univalent in the unit disk. This result is due to Nehari (1954), who pointed out that it includes three sharp special cases: (i) $M(r) = \frac{1}{4}\pi^2$, (ii) $M(r) = 2(1 - r^2)^{-1}$, (iii) $M(r) = (1 - r^2)^{-2}$. The first and the third case had been proved earlier by Nehari (1949). The second case is contained in a theorem stated by V. V. Pokornyi (1951). Beesack showed that the entire theorem of Pokornyi is covered by Theorem 8.3.3.

To apply the theorem one must be able to refer to a majorant equation $y'' + M(r)y = 0$ with the desired properties. For case (i) we can refer to the corollary of Theorem 8.2.1. For case (iii) we note that the equation

$$w'' + (1 - z^2)^{-2} w = 0 \qquad (8.3.20)$$

has the general solution

$$w(z) = (1 - z^2)^{1/2} \left( C_1 + C_2 \log \frac{1 - z}{1 + z} \right), \qquad (8.3.21)$$

which has at most one zero in $(-1, 1)$ and no zeros if $C_1 = 1$, $C_2 = 0$. Here we cannot introduce a constant multiplier $a > 1$ before the factor $(1 - z^2)^{-2}$ without getting solutions with infinitely many zeros in $(-1, 1)$.

The general Beesack-Pokarnyi theorem involves

$$M(r) = \begin{cases} 2^{3-2\lambda} \pi^{2(1-\lambda)} (1 - r^2)^{-\lambda}, & 0 \leqslant \lambda < 1, \\ 2^{3-\lambda} (1 - r^2)^{-\lambda}, & 1 \leqslant \lambda \leqslant 2, \end{cases} \qquad (8.3.22)$$

for $0 \leqslant r < 1$. The majorant equations are fairly complicated, and we refer to the original article for details.

Beesack has some interesting results concerning zeros in a neighborhood of a regular-singular point, a problem broached by the present author in 1925, using the Liouville transformation and the Green transform. Beesack proves

**THEOREM 8.3.4**

Let $G(z)$ be holomorphic except for a double pole at $z = 0$. Suppose that

$$|G(z)| \leq \tfrac{1}{4}|z|^{-2} + \lambda |z|^{-1} - \lambda^2 \qquad (8.3.23)$$

for all $z$ and some fixed $\lambda \geq 0$. If $z = a \neq 0$ is a zero of $w(z)$, a solution of (8.3.1), then $w(z)$ can have other zeros neither in the half-plane $|a + r e^{i\theta}| \geq r$ nor in the disk $|z - a| < |a|$.

For details the reader is referred to the original paper (Beesack, 1956).

Beesack considers a number of other problems such as bounds for the distance between zeros, nonoscillation in sectors, and strips or circles. Some of these may be reducible to cases already settled by the use of conformal mapping. As a sample we consider a special result due to B. Schwarz (1955).

**THEOREM 8.3.5**

Let $G(z)$ be holomorphic in the right half-plane, where it satisfies

$$|G(x + iy)| \leq \tfrac{1}{4}x^{-2}. \qquad (8.3.24)$$

Then no solution of (8.3.1) can have more than one zero in the right half-plane.

*Proof.* The transformation

$$t = \frac{1 - z}{1 + z}, \qquad w(z) = (1 + t)^{-1} Y(t)$$

maps the right $z$-half-plane into the unit disk of the $t$-plane and carries (8.3.1) into

$$Y''(t) + 4(1 + t)^{-4} G\left(\frac{1 - t}{1 + t}\right) Y(t) = 0. \qquad (8.3.25)$$

Here the coefficient of $Y(t)$ does not exceed in absolute value

$$(1 - |t|^2)^{-2}, \qquad (8.3.26)$$

so we are back to the corollary of Theorem 8.3.3. ∎

We end this discussion with a result of the present author, which is stated without proof.

**THEOREM 8.3.6**

Let $G(z)$ be holomorphic in a sector $S: |arg\ z| \leq \alpha < \pi$. Let $G(x + iy)$ be *integrable out to infinity for fixed* y *and satisfy*

$$\left| z \int_z^\infty G(t)\, dt \right| \leq \rho < \tfrac{1}{4}, \qquad z \in S, \tag{8.3.27}$$

*where the path of integration is parallel to the positive real axis. Then* (8.3.1) *has a solution without zeros in* S.

For proof see Hille (1948), pp. 695–698.

In view of this result one might be inclined to believe that all solutions have only a small number of zeros in S. That this is not the case is shown by an example used by Beesack for a similar purpose:

$$w'' + e^{-2z} w = 0, \tag{8.3.28}$$

which is satisfied by any Bessel function of order 0 in the variable $e^{-z}$. Any sector S with $\alpha < \tfrac{1}{4}\pi$ will do. We choose a solution of (8.3.28) with a large positive zero $x_0$. This solution also has complex zeros $x_0 + k\pi i$, $k = 0, \pm 1, \pm 2, \ldots$ . The larger $x_0$, the more zeros are in S. There is no solution with infinitely many zeros in S, but there is obviously no finite upper bound on the number of zeros that a solution can have in S.

### EXERCISE 8.3

1. Verify the various statements about the circle in the proof of Theorem 8.3.3.
2. Verify (8.3.14)–(8.3.16).
3. Check the equivalence proof in Theorem 8.3.3.
4. Let $w'' + a(1 - z^2)^{-2} w = 0$ with $a$ real $> 1$. Show the existence of solutions with infinitely many zeros in $(-1, 1)$.
5. Verify (8.3.25) and (8.3.26).
6. Give an example of a function satisfying (8.3.24).
7. Answer Problem 6 for (8.3.27).
8. Determine the half-plane $|a + r e^{i\theta}| \geq r$.
9. Show that the function $Y(t) = \lambda - \tfrac{1}{2}t^{-1}$ satisfies $Y'(t) = [Y(t)]^2 + F(t)$ with $F(t) = \tfrac{1}{4}t^{-2} + \lambda t^{-1} - \lambda^2$ so that $F(t)$ and $Y(t)$ satisfy the hypotheses of Lemma 8.3.1. How does this lead to condition (8.3.23)?
10. Show that the inequality $|G(a + r e^{i\theta})| \leq \tfrac{1}{4}r^{-2} + \lambda r^{-1} - \lambda^2$ is satisfied if $|a + r e^{i\theta}| \geq r$. Conclusion? How does the condition $|z - a| < |a|$ lead to a zero-free region?

### 8.4.  APPLICATIONS TO SPECIAL EQUATIONS

The methods developed in the preceding sections will now be applied to some special equations. Some of the equations treated in Chapter 6 are considered again here from this point of view.

## The Hypergeometric Equation

The equation

$$z(1-z)u'' + [c - (a + b + 1)z]u' - abu = 0 \tag{8.4.1}$$

is not in a form suitable for our consideration and is replaced by the invariant form

$$w'' + \tfrac{1}{4}[(1 - \alpha_1^2)z^{-2} + (1 - \alpha_2^2)(z - 1)^{-2} + (\alpha_1^2 + \alpha_1^2 + \alpha_2^2 - \alpha_3^2 - 1)$$
$$\times z^{-1}(z - 1)^{-1}]w = 0. \tag{8.4.2}$$

Here

$$w(z) = z^\rho(1 - z)^\sigma u(z), \qquad \rho = \tfrac{1}{2}(1 - \alpha_1), \qquad \sigma = \tfrac{1}{2}(1 - \alpha_2),$$
$$a = \tfrac{1}{2}(1 - \alpha_1 - \alpha_2 + \alpha_3), \qquad b = \tfrac{1}{2}(1 - \alpha_1 - \alpha_2 - \alpha_3), \qquad c = 1 - \alpha_1. \tag{8.4.3}$$

In 1907 A. Hurwitz got results for $u_{01}(z) = F(a, b, c; z)$. Here we shall aim at $u_{\infty,1}(z) = (-z)^{-a}F(a, a - c + 1, a - b + 1; z^{-1})$ and large neighborhoods of infinity. For this choice of $u(z)$ we must assume that $w(z) \to 0$ as $z$ becomes infinite. If the coefficients are real, as we suppose in the following, we must have $a - b = \alpha_3 > 1$.

We can now use Green's transform and integration to infinity along vertical or horizontal lines. If the coefficient of $w$ in (8.4.2) is denoted by $G(z) = P(x, y) + iQ(x, y)$, $z = x + iy$, we have

$$[w(t)w'(t)]_z^\infty - \int_z^\infty |w'(t)|^2 \, \overline{dt} + \int_z^\infty (P + iQ)|w|^2 \, dt = 0.$$

Now $w(z)$ goes to zero as $z^{1-a+b}$ as $z$ becomes infinite. This implies that the integrated part goes to zero and the first integral exists. The second integrand is $o(z^{-2})$, so again the integral exists. If now the integral is taken along a vertical line in the upper half-plane, the sign of $Q(x, y)$ along the path of integration is decisive for the success of this mode of attack. We have

$$Q(x, y) = -2y\{Ax|z|^{-4} + B(x - 1)|z - 1|^{-4} + C(x - \tfrac{1}{2})|z|^{-2}|z - 1|^{-2}\},$$
$$A = \tfrac{1}{4}(1 - \alpha_1^2), \qquad B = \tfrac{1}{4}(1 - \alpha_2^2), \qquad C = \tfrac{1}{4}(\alpha_1^2 + \alpha_2^2 - \alpha_3^2 - 1). \tag{8.4.4}$$

To fix the ideas, suppose that all three coefficients, $A$, $B$, $C$, are of the same sign or are zero; then $Q(x, y)$ keeps a constant sign in the four quarter-planes

$$x \le 0, y \le 0; \qquad x \le 0, y \ge 0; \qquad x \ge 1, y \ge 0; \qquad x \ge 1, y \le 0.$$

These quadrants are thus zero-free regions for

$$(-z)^{-a}F(a, a - c + 1, a - b + 1; z^{-1})$$

under the stated conditions on $A$, $B$, $C$. The strip $0 < x < 1$ is in doubt. If, however, $A \geqslant B \geqslant 0$, $C \geqslant 0$, the right half of the strip is zero-free.

Here we get better results by integrating radially from a hypothetical zero to infinity. Now

$$-\int_z^\infty |w'|^2 \, dr + \int_z^\infty G(r \, e^{i\theta}) \, e^{2i\theta} |w|^2 \, dr = 0.$$

Then the imaginary part of $G(r \, e^{i\theta}) \, e^{2i\theta}$ is

$$\sin \theta \{ 2B |r - e^{-i\theta}|^{-4} (r + \cos \theta) + Cr^{-1}|r - e^{-i\theta}|^{-2} \}. \tag{8.4.5}$$

If $B$ and $C$ have the same sign and if $0 < \theta < \tfrac{1}{2}\pi$, the expression is not zero and we see that there can be no zeros in the open first and fourth quadrants. With these fragmentary results we leave the hypergeometric equation.

## Legendre's Equation

Here more satisfactory results are obtainable. We start by considering

$$P_a(z) = F[a + 1, - a, 1; \tfrac{1}{2}(1 - z)],$$

which satisfies the equation

$$\frac{d}{dz}\left[ (1 - z^2) \frac{dw}{dz} \right] + a(a + 1)w = 0. \tag{8.4.6}$$

It is assumed that either $a$ is real $\geqslant -\tfrac{1}{2}$ or $a = -\tfrac{1}{2} + mi$, $0 < m$. In either case $P_a(z)$ is real for $z$ real, $z > -1$. We know that $P_n(z)$ is a polynomial of degree $n$, all the zeros of which are real and located in the interval $(-1, 1)$. For a nonintegral value of $a$ the $z$-plane should be cut along the real axis from $-\infty$ to $-1$ in order to deal with a single-valued function. We shall show that in the cut plane $P_a(z)$ has only real zeros; for $a$ real $-\tfrac{1}{2} \leqslant a$ the zeros lie in the interval $(-1, 1)$, and for $a = -\tfrac{1}{2} + im$ they lie in $(1, \infty)$ and are infinitely many. In the real case the number of zeros is $(a]$, where this symbol denotes the least integer $> a$ but less than $a + 1$. To show the reality of the zeros we use Green's transform—more precisely, the zero-free star $S_0(1)$.

We have then, for $w = P_a(z)$, $z = 1 + r \, e^{i\theta}$, and supposing that $1 + R \, e^{i\theta}$ is a zero of $P_a P_a'$, we obtain

$$\int_0^R (2r \, e^{-i\theta} + r^2)|w'|^2 \, dr + a(a + 1) \int_0^R |w|^2 \, dr = 0, \tag{8.4.7}$$

the imaginary part of which is

$$2 \sin \theta \int_0^R r|w'|^2 \, dr.$$

This is zero iff $\theta$ is congruent to zero (modulo $\pi$). Thus the zeros under consideration are real. If $\theta = 0$ and $a \geq 0$, the left member of (8.4.7) is positive and no zeros can belong to $(1, \infty)$. If $\theta = \pi$, $R < 2$, $a(a + 1) < 0$, we have again a contradiction. If $a$ is real, there can be no zeros on the edges of the cut, for on the cut the imaginary part of $P_a(z)$ is $\pm \pi i P_a(-z) \neq 0$. One way of establishing the number of zeros in $(-1, 1)$ of $P_a(z)$, $-\frac{1}{2} < a$, is to find the change in the argument when $z$ describes a great semicircle in the upper half-plane. For the case $a = -\frac{1}{2} + mi$, the Liouville transformation maps (8.4.6) on a perturbed sine equation. Since

$$Z(z) = (\tfrac{1}{4} + m^2)^{1/2} \log [z + (z^2 - 1)^{1/2}], \qquad (8.4.8)$$

the number of zeros of $P_a(x)$, $a = -\frac{1}{2} + mi$, in the interval $(1, A)$ is approximately

$$\frac{1}{\pi} (\tfrac{1}{4} + m^2)^{1/2} \log 2A. \qquad (8.4.9)$$

We turn now to the function

$$Q_a(z) = \frac{\Gamma(\tfrac{1}{2})\Gamma(a + 1)}{\Gamma(a + \tfrac{3}{2})} (2z)^{-1-a} F(\tfrac{1}{2}a + \tfrac{1}{2}, \tfrac{1}{2}a + 1, a + \tfrac{3}{2}; z^{-2}) \qquad (8.4.10)$$

where $a$ is real, $-\frac{1}{2} < a$. To make the function single valued we cut the plane along the real axis from $-\infty$ to $+1$. We can restrict ourselves to the closed right half-plane since if $z_0$ is a zero so is $-z_0$. Here we use the Green's transform and integration along rays from the origin, starting at a hypothetical zero $z_0$ and going to $\infty$. From the behavior of $Q_a(z)$ for large values of $z$ we see that $Q_a(z)Q_a'(z)(1 - z^2) \to 0$ as $z \to \infty$. Furthermore, $|Q_a'(z)|^2(1 - z^2)$ and $|Q_a(z)|^2$ are integrable out to infinity. We thus get

$$-\int_{r_0}^{\infty} (e^{-2i\theta} - r^2)|w'|^2 \, dr + a(a + 1) \int_{r_0}^{\infty} |w|^2 \, dr = 0. \qquad (8.4.11)$$

and here the imaginary part is

$$-\sin 2\theta \int_{r_0}^{\infty} |w'|^2 \, dr,$$

which is zero iff $\sin 2\theta = 0$ or $\theta$ is a multiple of $\frac{1}{2}\pi$. For $\theta = \frac{1}{2}\pi$ and $a \geq 0$ this leads to a contradiction, as shown by (8.4.11). For $\theta = 0$, $r_0 > 1$, $a > 0$ we again get a contradiction. We conclude that for $a \geq 0$ the function $Q_a(z)$ can have no zeros in the cut $z$-plane. For $-\frac{1}{2} < a < 0$ we are in doubt concerning the existence of real zeros $> 1$. But for such values of $a$ the coefficients of $Q_a(z)$ in the representation (8.4.10) are all positive, so that there can be no zeros on the interval $(1, \infty)$. Thus we have shown that $Q_a(z)$ for $a > -\frac{1}{2}$ has no zeros in the cut plane.

## Bessel's Equation

We start by proving Hurwitz's theorem, stating that $J_a(z)$ for $a > -1$ has only real zeros. Suppose contrariwise that $z = c$ is a nonreal zero of $J_a(z)$. The function $J_a(ct)$ in its dependence on $t$ satisfies the DE

$$(tu')' + (c^2 t - a^2 t^{-1})u = 0. \qquad (8.4.12)$$

Here we multiply by $\bar{u}$ and integrate from $t = h > 0$ to $t = 1$. An integration by parts gives

$$[t\bar{u}u']_h^1 - \int_h^1 t|u'(t)|^2 \, dt + \int_h^1 (c^2 t - a^2 t^{-1})|u(t)|^2 \, dt = 0. \qquad (8.4.13)$$

Here $u(1) = J_a(c) = 0$ by assumption, so the equation reduces to

$$-h\overline{u(h)}u'(h) - \int_h^1 t|u'(t)|^2 \, dt + \int_h^1 (c^2 t - a^2 t^{-1})|u(t)|^2 \, dt = 0.$$

Now, if a complex number is zero, its conjugate is also zero and so is their difference. In forming the latter, the first integral drops out and in the second integral the term involving $a^2 t^{-1}$ also drops out; therefore we are left with

$$-h[\overline{u(h)}u'(h) - u(h)\overline{u'(h)}] + (c^2 - \bar{c}^2)\int_h^1 t|u(t)|^2 \, dt = 0. \qquad (8.4.14)$$

Here the first expression is

$$F(a)|c|^{2a}(c^2 - \bar{c}^2)h^{2a+2}[1 + O(h^2)], \qquad (8.4.15)$$

and this goes to zero with $h$ for any $a > -1$. Here $F(a)$ is an analytic real-valued function of $a$ of no importance for the following. Hence we are left with

$$(c^2 - \bar{c}^2)\int_0^1 t|u(t)|^2 \, dt = 0.$$

The integrand is $O(t^{2a+1})$, and hence the integral exists. Since the integral cannot be zero, the only way for the result to hold is that

$$c^2 = \bar{c}^2 \qquad \text{or either} \qquad c = \bar{c} \quad \text{or} \quad c = -\bar{c}.$$

In the first case the zero would be real; in the second, purely imaginary. The second alternative cannot hold, however, if $a > -1$, for in the expansion

$$J_a(iy) = \exp\left(\tfrac{1}{2}\pi i a\right) \sum_{k=0}^{\infty} \frac{1}{\Gamma(k+1)\Gamma(k+a+1)} \left(\frac{y}{2}\right)^{a+2k} \qquad (8.4.16)$$

all the coefficients are positive. This proves Hurwitz's theorem.

Hurwitz also discussed the case of $a < -1$, where in addition to the real zeros there are complex zeros. If $-n - 1 < a < -n$, there are $n$ pairs of conjugate complex zeros symmetrically located with respect to the origin. If $n$ is odd, there is one and only one pair of purely imaginary zeros.

Next we take up the complex zeros of a solution $C_a(z)$ of Bessel's equation such that $C_a(z)$ is not real on the real axis (or a complex number times a real solution). Suppose that $z = z_1 = x_1 + iy_1$ is a zero of $C_a(z)$ in the first quadrant. To study the implications of such a zero we consider the DE

$$w'' + [1 - (a^2 - \tfrac{1}{4})z^{-2}]w = 0, \tag{8.4.17}$$

which is satisfied by $z^{1/2}C_a(z)$. Here there are three cases according as $a^2 < \tfrac{1}{4}$, $= \tfrac{1}{4}$, or $> \tfrac{1}{4}$. The case $a^2 = \tfrac{1}{4}$ does not pose any problem, since here (8.4.17) is a pure sine equation and the solution which has a zero at $z = z_1$ is $C \sin(z - z_1)$ with the zeros $z_1 \pm n\pi$.

Let $a^2 \neq \tfrac{1}{4}$, multiply (8.4.18) by $\bar{w}$, and integrate along the line $y = y_1$. For $x \geq 0$ there can be no other zeros on the horizontal line through $z = z_1$, for if $x_2 = x_2 + iy_1$ were a second zero, then

$$-\int_{x_1}^{x_2} |w'|^2 \, dx + \int_{x_1}^{x_2} (P + iQ)|w|^2 \, dx = 0, \tag{8.4.18}$$

where

$$P(x, y) = 1 - (a^2 - \tfrac{1}{4})|z|^{-4}(x^2 - y^2), \qquad Q(x, y) = (a^2 - \tfrac{1}{4})|z|^{-4}2xy, \tag{8.4.19}$$

and $y = y_1$. The imaginary part of (8.4.18) is then

$$2y_1(a^2 - \tfrac{1}{4}) \int_{x_1}^{x_2} x|w(x + iy_1)|^2 \, dx \neq 0$$

for $x_2 \geq 0$. In the same manner it is proved that no two zeros in the first quadrant can have the same abscissas.

Suppose now that $z_2 = x_2 + iy_2$, $x_1 < x_2$, is a second zero of $C_a(z)$ in the first quadrant. Here $y_1 < y_2$ or $y_1 > y_2$ according as $a^2 - \tfrac{1}{4} < 0$ or $> 0$. This follows from Theorem 8.1.6, for the sign of $Q(x, y)$ is the same as that of $a^2 - \tfrac{1}{4}$. Thus

$$0 < \arg G < \pi, \qquad -\tfrac{1}{2}\pi < -\tfrac{1}{2}\arg G < 0 \qquad \text{if } a^2 > \tfrac{1}{4},$$
$$-\pi < \arg G < 0, \qquad 0 < -\tfrac{1}{2}\arg G < \tfrac{1}{2}\pi \qquad \text{if } a^2 < \tfrac{1}{4},$$

whence the result follows.

In any case, if the zeros $z_n = x_n + iy_n$ are ordered according to increasing abscissas, $x_n < x_{n+1}$, the ordinates form a bounded monotone sequence of positive numbers which tend to a positive limit as $n \to \infty$. Furthermore, $\lim(x_{n+1} - x_n) = \pi$. This follows from the discussion in

Section 5.6, since Hypothesis F is satisfied. If $w(z)$ is an oscillatory solution of (8.4.17), there is, by Theorem 5.6.1, an associated asymptotic sine function, $C \sin(z - z_0)$, such that $\lim(z_n - z_0 - n\pi) = 0$, and here $z_0$ cannot be real, for there is a one-to-one correspondence between solutions $C_a(z)$ and $z_0$, reduced modulo $\pi$, and real values of $z_0$ correspond to solutions real on the real axis.

It is natural to ask whether the convexity theorem, Theorem 8.1.7, applies in the present case to the zeros in the first quadrant. Now the argument of $G(z)$ lies between 0 and $\pi$ if $a^2 - \frac{1}{4} > 0$ and between $-\pi$ and 0 if $m = a^2 - \frac{1}{4} < 0$. This is satisfactory, but the inflection locus of the indicatrices places restrictions on what part of the quadrant is admissible. By (8.1.19) this curve is

$$\text{Im}\{G'(z)[G(z)]^{-3/2}\} = 0, \tag{8.4.20}$$

which forms a part of the combined inflection locus

$$\text{Im}\{[G'(z)]^2[G(z)]^{-3}\} = 0. \tag{8.4.21}$$

The sign of $m$ affects the system of indicatrices in various ways. Thus the origin, which is a topological singularity of the system, is a *center* (= vortex) of the system if $m > 0$ but a *star point* if $m < 0$. The curve (8.4.21) is a sextic which factors into $xy = 0$, i.e., the axes, times the quartic

$$(3x^2 - y^2)(\tfrac{1}{3}x^2 - y^2) - 2m(x^2 - y^2) + m^2 = 0. \tag{8.4.22}$$

The case $m > 0$ is sketched roughly in Figure 8.1. For $m < 0$ the graph is obtained by interchanging $x$ and $y$, i.e., a rotation of the figure through an angle of 90°. If $m > 0$, there are double points with real tangents at $z = m^{1/2}$ and $z = -m^{1/2}$. The lines $y = \pm 3^{1/2}x$ and $y = \pm 3^{-1/2}x$ are asymptotes. Now in the upper half-plane the two branches of (8.4.22) whose asymptotes make the angles 30° and 150° with the horizontal belong to the inflection locus of the indicatrices. They divide the upper half-plane into three curvilinear sectors in each of which the convexity theorem holds. In the middle sector the indicatrices are concave downward; in the other sectors, concave upward. It may be shown that a solution $C_a(z)$ which has infinitely many zeros in the first quadrant can have at most a finite number in the fourth quadrant.

We still have to consider the Bessel-Hankel functions which correspond to the solutions $e^{iz}$ and $e^{-iz}$ of the associated sine equation. It is enough to consider the first case, $H_a^{(1)}(z)$, which is $z^{-1/2}$ times a subdominant solution of (8.4.17). If $z_0 = x_0 + iy_0$ is a zero, then, using the Green's transform and integration along the vertical line from $z_0$ to

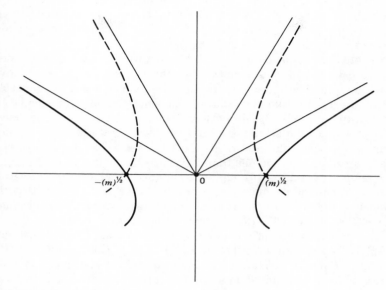

**Figure 8.1**

$z_0 + i\infty$, we see that

$$\int_{y_0}^{\infty} Q|w|^2 \, dy = 0, \qquad \int_{y_0}^{\infty} [|w'|^2 + P|w|^2] \, dy = 0.$$

In the upper half-plane the first of these identities holds only on the imaginary axis, while the second one cannot hold in the part of the plane, visible from $i\infty$, where $P \geq 0$. The curve $P = 0$ is the lemniscate

$$(x^2 + y^2)^2 = m(x^2 - y^2), \qquad m = a^2 - \tfrac{1}{4}.$$

The foci of this curve are at $z = \pm(\tfrac{1}{2}m)^{1/2}$ if $m > 0$ and at $\pm i(\tfrac{1}{2}|m|)^{1/2}$ if $m < 0$. For $m > 0$ this means no zeros in $y \geq 0$ or in the parts of the third and fourth quadrant to the left of the line $x = -m^{1/2}$ and to the right of the line $x = m^{1/2}$, respectively. For $m < 0$ the segment of the imaginary axis between $0$ and $i|m|^{1/2}$ remains in doubt, while the rest of the upper half-plane is zero-free. The regions in the lower half-plane are now bounded by $x = \pm\tfrac{1}{4}(6|m|)^{1/2}$. Here we leave Bessel's equation.

### The Hermite-Weber Equation

This equation is written in the form

$$w'' + (2a + 1 - z^2)w = 0, \tag{8.4.23}$$

where $a$ is real and $> -\frac{1}{2}$. We recall the existence of four subdominant solutions, $E_a(z)$, $E_{-a-1}(-iz)$, $E_a(-z)$, $E_{-a-1}(iz)$, for the four sectors bounded by the lines $y = \pm x$.

We note first that a solution which is either even or odd cannot vanish anywhere in the two sectors that contain the imaginary axis. This follows from a construction of the zero-free star $S(0)$, which is left to the reader. Part of this result is true also for a solution whose initial parameter $\lambda(0)$ is real. If $\lambda(0) > 0$, the left halves of these sectors are zero-free. If, however, $\lambda(0) < 0$, we get the sectors $\frac{1}{2}\pi < \theta < \frac{2}{3}\pi$ and $-\frac{2}{3}\pi < \theta < -\frac{1}{2}\pi$. The first result is practically the best of its kind, but for $\lambda(0) < 0$ it is less than satisfactory, for the Liouville transformation shows that the indicatrices have the lines $y = \pm x$ as asymptotes, and this means that an oscillatory solution has infinitely many zeros forming four strings such that the distance of the $n$th zero from the asymptote goes to zero as $n$ becomes infinite. See also (8.2.27)–(8.2.30) and the adjacent discussion.

Let us now take up the zeros of $E_a(z)$. Here we see first that the right half-plane can contain no zeros of $E_a(z)E'_a(z)$ except real ones on the interval $[0, (2a + 1)^{1/2}]$. We use the Green transform and integrate along horizontal lines, applying the fact that $E_a(z)E'_a(z)$ goes exponentially to zero as $z \to \infty$ in the sector $|\arg z| \leq \frac{1}{4}\pi - \epsilon$. Since $E_a(0)E'_a(0) < 0$, we get in addition no zeros in $\frac{1}{2}\pi < \theta \leq \frac{2}{3}\pi$ and the symmetric sector in the lower half-plane. A more painstaking examination would enable us to extend the zero-free regions, but it is not worth the effort. As a matter of fact, the solution $E_{-a-1}(-iz)$ is easier to study from this point of view. Here there are no zeros in the upper half-plane. The zeros in the third quadrant lie in a region bounded by the hyperbola

$$x^2 - y^2 = a + \tfrac{1}{2}, \qquad 0 > x, \quad 0 > y,$$

by the cubic

$$x^3 - 3xy^2 - 3(2a + 1)x + 2(2a + 1)^{3/2} = 0, \qquad 0 > x, \quad 0 > y,$$

and a segment of the real axis from $-(a + \frac{1}{2})^{1/2}$ to $-(2a + 1)^{1/2}$.

Normally a solution of (8.4.22) has four strings of zeros, one in each quadrant. Let $S_j$ stand for the string in the $j$th quadrant. Then $E_a(z)$ lacks the strings $S_1$ and $S_4$, while $E_{-a-1}(-iz)$ lacks $S_1$ and $S_2$, and so on. If $a$ is a nonnegative integer, $E_a(z)$ is a constant multiple of $E_a(-z)$ and consequently lacks all four strings. Similar results hold for $E_{-a-1}(-iz)$ when $a$ is a nonnegative integer.

In the case of this equation is it possible to study the composition of the strings by the method of *variation of the parameter*, in this case variation of $a$. We recall that the string is the string of zeros of $w(z)w'(z)$, so that it contains both zeros and extrema of $w(z)$, and one problem is to decide whether or not they alternate. To fix the ideas, let us start with $a = -\frac{1}{2}$, i.e., the DE

$$w'' - z^2 w = 0, \tag{8.4.24}$$

and the solution, which is zero at $z = 0$. This solution is $z$ times a power series in $z^4$ with positive coefficients. Hence on the four rays $\arg z = (2k + 1)\frac{1}{4}\pi$ it is of real character (i.e., a complex constant times a real-valued function). If we set $z = \exp(\frac{1}{4}\pi i)t$, (8.4.3) becomes

$$\frac{d^2 v}{dt^2} + t^2 v = 0, \qquad v(t) = \exp(-\frac{1}{4}\pi i)w(t).$$

It follows that the solution is oscillatory on this ray, as well as on the three other rays. Since there is a zero at the origin, the nearest zeros of $w(z)w'(z)$ on the four rays are extrema of $w(z)$. Hence on all four rays the pattern is $E - Z - E - \cdots$, where E stands for an extremum and Z for a zero. Now the zeros and extrema are analytic functions of $a$—in particular, continuous functions, and they are also continuous functions of the initial parameter $\lambda(0)$. It follows that, if we vary either of these parameters, the configuration will change continuously, and since no two zeros or extrema can have the same (nonzero) abscissas or ordinates, the pattern is preserved unless $\lambda(0)$ tends to a limit for which two or four strings are missing, the corresponding points having moved off to infinity. New zeros are fed into the system, one at a time, from infinity. Extrema, on the other hand, appear on the real axis by coalescence of a pair of complex extrema at a point $\pm(2a + 1)^{1/2}$, where they separate and march off in opposite directions. It is quite a fascinating study, which should be pursued by any reader who is curious to know where zeros and extrema come from and go to.

### The Equation of Mathieu

In the normal form

$$w''(z) + (a + b \cos 2z)w(z) = 0, \tag{8.4.25}$$

we assume $0 < b < a$. The Liouville transformation will then map the half-strips $D_n^+$ and $D_n^-$ of (6.6.26) conformally on domains containing an upper or a lower half-plane, as in Figure 6.5. The equation (8.4.25) is real in nature on each of the lines $x = \frac{1}{2}k\pi$. If $k$ is odd, there are oscillatory solutions with infinitely many zeros and extrema on the line and the extrema alternate with the zeros. Variation of an initial parameter may push one of these points off the line. Then they will all go and to the same side of the line. According as the string is left of the line, on it, or to the right of it, we speak of $l$-, $r$-, or $d$-strings (*laevus, rectus*, or *dexter*). The Mathieu functions, $ce_n(z)$ and $se_n(z)$, have all their complex zeros on the lines $x = (k - \frac{1}{2})\pi$, so they are all $r$-strings. Zeros and extrema alternate, and the point nearest to but not on the real axis is an E or a Z according as $z = (k - \frac{1}{2})\pi$ is a Z or an E. In addition to the complex zeros, there are also infinitely many real zeros. If the string $S_n^+$ is a $d$-string and if $w(z)$ is real on the real axis, then $S_n^-$ is also a $d$-string, but to what extent these facts influence the nature of the other strings is unknown so far. The author (1924) has given a complete description for the case $a < b < 0$.

## EXERCISE 8.4

1. From (8.4.1) derive (8.4.2) and (8.4.3).

2. Verify (8.4.4) and (8.4.5). Why is it permissible to integrate to infinity?

3. In studying the variation of the argument of $P_a(z)$ the following path is considered: (i) real axis from 1 to $R$, (ii) semicircle $|z| = R$, $0 \le \arg z \le \pi$, (iii) real axis from $-R$ to $-1 - \epsilon$, (iv) semicircle $|z + 1| = \epsilon$, (v) real axis from $-1 + \epsilon$ to 1, avoiding the zeros by small semicircles in the upper half-plane. The total variation should be zero. Find the variation along (i)–(iv) with an error $< \pi$, and use this to determine the number of zeros of $P_a(z)$ in $(-1, 1)$.

4. Verify (8.4.8) and (8.4.9).

5. Show that it is permissible to use infinity as the upper limit in (8.4.11).

6. Verify (8.4.12) and the proof of Hurwitz's theorem.

7. Show that for $a > \frac{1}{2}$ the function $x \mapsto J_a(x)$ has no zeros in the half-open interval $(0, m^{1/2}]$, $m = a^2 - \frac{1}{2}$.

8. Suppose that $m > 0$ and that $z_n$ is a distant zero of $C_a(z)$ in the first quadrant. Set $m|z_n|^{-2} = P_n$, $\beta_n = \arcsin P_n$. Show that (i) $|\arg G(z_n)| \le \beta_n$, (ii) $\beta_{n+1} < \beta_n$, (iii) $\Sigma_1^\infty \beta_n < \infty$.

9. Discuss the shape of the indicatrices of (8.4.17) in a neighborhood of the origin.

10. Derive (8.4.21) from (8.4.20), and determine the various parts of the inflection locus.

11. Verify the statements about the zeros of the Hankel-Bessel function.

12. Discuss the zero-free regions of a solution of (8.4.23) for which the initial parameter $\lambda(0)$ is real.

13. Discuss the zero-free regions of $E_a(z)$.

14. Show that $E_{-a-1}(-iz)$ is of real nature on the imaginary axis, and show for this solution $\lambda(0) = bi$, $b > 0$.

15. Discuss the zero-free regions of $E_{-a-1}(-iz)$.

16. Verify the results stated for (8.4.24).

17. Suppose that for this equation the initial parameter $\lambda(0) > 0$. For $\lambda(0) = +\infty$ there is a zero at the origin. How does this zero move when $\lambda(0)$ decreases from $+\infty$? How are the four strings pushed?

18. When $\lambda(0) = 0$, there is a triple extremum at the origin. What happened to the zero? Indicate how the triple extremum was generated.

19. Discuss the singular boundary value problem

$$w''(x) + (2a + 1 - x^2)w(x) = 0, \qquad w(x) \in L_2(-\infty, \infty).$$

20. In a discussion of the singular boundary value problem

$$w''(x) + (b - x^4)w(x) = 0, \qquad w(x) \in L_2(-\infty, \infty),$$

E. C. Titchmarsh (1899–1963) conjectured that the solutions would have real zeros in the interval $(-b^{1/4}, b^{1/4})$, infinitely many purely imaginary zeros, and no other complex zeros. Verify that this is so by discussing the equation in the complex plane. Verify the existence of six sectors bounded by the rays $\arg z = \frac{1}{6}(2k - 1)\pi$, in each of which there is a subdominant solution that goes to zero in the interior of the sector as $r^{-1}\exp(-\frac{1}{3}r^3)$ as $r \to \infty$ and consequently lacks the two adjacent strings. The solutions of the boundary value problem correspond to those values of $b$ for which the subdominants in the sectors $|\arg z| < \frac{1}{6}\pi$ and $|\arg(-z)| < \frac{1}{6}\pi$ are linearly dependent. In this case four strings are missing, *and only the strings in the direction of the imaginary axis are left. Fill in the details.

21. Show that, if $w(z)$ is a solution of (8.4.25) which is oscillatory in $D_0^+$, is real on the real axis, and has a single zero or extremum in $(-\frac{1}{2}\pi, \frac{1}{2}\pi)$, then the string in $D_0^+$ is $l$-, $r$-, or $d$-, according as the real zero or extremum $< 0$, $= 0$, or $> 0$.

22. Discuss the nature of the string in $D_0^-$ when that of the string in $D_0^+$ is known.

23. Can a singular boundary value problem

$$\frac{d^2y}{dt^2} + (a - b\cosh 2t)v = 0, \qquad v(t) \in L_2(-\infty, \infty),$$

have a solution? Here $b$ is a given positive number.

## LITERATURE

See, in particular, Chapter 11 of the author's LODE, which has an extensive bibliography. See also Chapter 21 of:

Ince, E. L. *Ordinary Differential Equations*. Dover, New York, 1944.

For the papers mentioned in the text see:

Beesack, P. E. Non-oscillation and disconjugacy in the complex domain. *Trans. Amer. Math. Soc.*, **81** (1956), 211–242.

—— and B. Schwarz. On the zeros of solutions of second order linear differential equations. *Can. J. Math.*, **9** (1956), 504–515.

Boutroux, P. Recherches sur les transcendentes de M. Painlevé et l'étude asymptotique des équations différentielles du second ordre. *Ann. École Norm. Supér.*, (3), **30** (1913), 255–375; **31** (1914), 99–159.

Forsyth, A. R. *Theory of Differential Equations.* Part I, Vol. II. Cambridge University Press, London 1900, pp. 275–276.

Hurwitz, A. Über die Nullstellen der Bessel'schen Funktion. *Math. Ann.*, **33** (1889), 246–266.

———. Über die Nullstellen der hypergeometrischen Funktion. *Math. Ann.*, **64** (1907), 517–560.

Nehari, Z. The Schwarzian derivative and schlicht functions. *Bull. Amer. Math. Soc.*, **55** (1949), 545–551.

———. On the zeros of solutions of second-order linear differential equations. *Amer. J. Math.*, **76** (1954), 689–697.

———. Some criteria of univalence. *Proc. Amer. Math. Soc.*, **5** (1954), 700–704.

———. Univalent functions and linear differential equations. *Lectures on Functions of a Complex Variable.* University of Michigan Press, Ann Arbor, 1955, pp. 148–151.

Pokarnyi, V. V. On some sufficient conditions for univalence (Russian). *Doklady Akad. Nauk SSSR*, N.S., **79** (1951), 743–746.

Schafheitlin, P. Über die Nullstellen der hypergeometrischen Funktion. *Sitzungsber. Berliner Math. Gesell.* **7** (1908), 19–28.

Schwarz, B. Complex non-oscillation theorems and criteria of univalence. *Trans. Amer. Math. Soc.*, **80** (1955), 159–186.

Taam, C-T. Oscillation theorems. *Amer. J. Math.*, **74** (1952), 317–324.

———. On the complex zeros of functions of the Sturm-Liouville type. *Pacific J. Math.*, **3** (1953), 837–845.

———. Schlicht functions and linear differential equations of the second order. *J. Rat. Mech. Anal.*, **4** (1955), 467–480.

———. On the solutions of second-order linear differential equations. *Proc. Amer. Math. Soc.*, **4** (1953), 876–879.

Van Vleck, E. B. A determination of real and imaginary roots of the hypergeometric series. *Trans. Amer. Math. Soc.*, **3** (1902), 110–131.

Mention should also be made of six papers of the author, results of which are cited in the text:

Hille, E. On the zeros of Legendre functions. *Ark. Mat., Astr., Fys.*, **17**, No. 22 (1923), 16 pp.

———. On the zeros of Mathieu functions. *Proc. London Math. Soc.*, (2), **23** (1924), 185–237.

———. On the zeros of the functions of the parabolic cylinder. *Ark. Mat., Astr., Fys.*, **18**, No. 26 (1924), 56 pp.

———. A note on regular singular points. *Ark. Mat., Astr., Fys.*, 19A, No. 2 (1925), 21 pp.

———. Non-oscillation theorems. *Trans. Amer. Math. Soc.*, **64** (1948), 234–252.

———. Remarks on a paper by Zeev Nehari. *Bull. Amer. Math. Soc.*, **55** (1949), 445–457.

# 9

# LINEAR $n$th ORDER AND MATRIX DIFFERENTIAL EQUATIONS

This chapter is devoted to the theory of linear DE's of order $n$, normally greater than 2. For many purposes it is more convenient to consider an equivalent matrix equation. The passage from an $n$th order linear DE to a first order matrix DE or vice versa is elementary. Among the various problems to be considered are (i) existence and independence of solutions, (ii) existence and analyticity in a star, (iii) the group or monodromy, (iv) global representation and the Riemann problem, (v) rate of growth for approach to a singular point, (vi) regular-singular points, (vii) the Fuchsian class, (viii) irregular singularities.

## 9.1. EXISTENCE AND INDEPENDENCE OF SOLUTIONS

We start with an $n$th order linear DE,

$$P_0(z)w^{(n)}(z) + P_1(z)w^{(n-1)}(z) + \cdots + P_n(z)w(z) = 0, \qquad (9.1.1)$$

where the $P_j$'s are holomorphic in the domain $D$ under consideration and, furthermore, $P_0(z) \neq 0$. Some DE's are conveniently expressed in terms of the differential operator:

$$\vartheta f(z) = zf'(z), \qquad \vartheta^n = \vartheta(\vartheta^{n-1}), \qquad (9.1.2)$$

with a corresponding DE

$$\vartheta^n w + Q_1(z)\vartheta^{n-1}w + \cdots + Q_n(z)w = 0. \qquad (9.1.3)$$

We can also replace the $n$th order DE (9.1.1) by a matrix DE:

$$\mathcal{W}'(z) = \mathcal{A}(z)\mathcal{W}(z), \qquad \mathcal{A}(z) = [a_{jk}(z)], \qquad \mathcal{W}(z) = [w_{jk}(z)]. \qquad (9.1.4)$$

Here $\mathcal{A}$ and $\mathcal{W}$ are elements of the Banach algebra $\mathfrak{M}_n$ of $n$-by-$n$ matrices, and the entries $a_{jk}(z)$ are holomorphic in $D$, while the $w_{jk}$'s are locally holomorphic there.

To pass from (9.1.1) to (9.1.3) we should note that

$$z^k f^{(k)}(z) = \vartheta(\vartheta - I)(\vartheta - 2I)\ldots[\vartheta - (k-1)I]f(z), \qquad (9.1.5)$$

where $I$ denotes the identity operator, and $k = 1, 2, \ldots, n$. We now divide the equation by $P_0(z)$, and in the term involving the $k$th derivative we write

$$w^{(k)}(z) = z^{-k}z^k w^{(k)}(z).$$

Using (9.1.5), we get a result of the form

$$\sum_{k=1}^{n} B_{n-k}(z)\vartheta(\vartheta - I)(\vartheta - 2I)\ldots[\vartheta - (k-1)I]w + B_n(z)w = 0, \qquad (9.1.6)$$

where the $B$'s are holomorphic functions of $z$ in $D$, except possibly for the origin, if $z = 0$ belongs to $D$. Here we expand the operator polynomials and collect terms in order to obtain (9.1.3).

The passing from (9.1.1) to (9.1.4) may be achieved in various ways. The simplest one is to set

$$w = w_1, \qquad w' = w_2, \qquad \ldots, \qquad w^{(n-1)} = w_n. \qquad (9.1.7)$$

We form the column vector $\mathbf{w}$, whose components are $w_1, w_2, \ldots, w_n$, and the vector-matrix DE,

$$\mathbf{w}' = \mathcal{A}(z)\mathbf{w}, \qquad (9.1.8)$$

where $\mathcal{A}(z) = (a_{jk}(z))$ and

$$\left.\begin{aligned} a_{j,j+1}(z) &= 1 \\ a_{j,k}(z) &= 0, \qquad k \neq j+1 \end{aligned}\right\}, \qquad j = 1, 2, \ldots, n-1,$$
$$a_{n,k}(z) = -\frac{P_{n-k}(z)}{P_0(z)}, \qquad k = 1, 2, \ldots, n. \qquad (9.1.9)$$

Other transformations are possible and may be preferable for particular problems. We shall encounter such cases in the following.

From the vector-matrix equation we can proceed to a pure matrix DE in the following manner. Take a point $z_0$ in $D$, and determine $n$ solutions of (9.1.8) by the initial conditions

$$w_k^{(j)}(z_0) = \delta_{j+1,k}, \qquad j = 0, 1, \ldots, n-1; \quad k = 1, 2, \ldots, n. \qquad (9.1.10)$$

These conditions determine uniquely $n$ distinct solution vectors of (9.1.8) in some neighborhood of $z = z_0$. The solution vectors $\mathbf{w}_k(z)$ are linearly independent as elements of $\mathbf{C}^n$. We take these $n$ column vectors as column vectors of a matrix $\mathcal{W}(z)$, and this matrix satisfies (9.1.4).

Equation (9.1.1) has the property that, if $w_0(z)$ and $w_{00}(z)$ are two solutions and $C_1$, $C_2$ any two constants, then $C_1 w_0(z) + C_2 w_{00}(z)$ is also a solution, and similarly for the vector-matrix equation. For the matrix DE we have a more general situation: whenever $\mathcal{W}_0(z)$ is a solution and $\mathcal{M}$ is an $n$-by-$n$ matrix, then $\mathcal{W}_0(z)\mathcal{M}$ is also a solution. This follows from the fact that

$$[\mathcal{W}_0(z)\mathcal{M}]' = \mathcal{W}_0'(z)\mathcal{M} = \mathcal{A}(z)[\mathcal{W}_0(z)\mathcal{M}],$$

since matrix multiplication is associative. It is normally noncommutative, so that $\mathcal{M}\mathcal{A}(z) \neq \mathcal{A}(z)\mathcal{M}$ and $\mathcal{M}\mathcal{W}_0(z)$ is normally not a solution.

A solution matrix $\mathcal{W}_0(z)$ of (9.1.4) is called *fundamental* if every solution of the DE is of the form $\mathcal{W}_0(z)\mathcal{M}$.

## THEOREM 9.1.1

*A solution $\mathcal{W}_0(z)$ of (9.1.4) is* fundamental *iff $\mathcal{W}_0(z)$ is a regular element of the matrix algebra, i.e., has a unique inverse, and this is the case iff $\omega(z) = \det [\mathcal{W}_0(z)] \neq 0$.*

*Proof.* Suppose that for some $z = z_0 \in D$ the determinant $\omega(z_0) \neq 0$. Since $\omega(z)$ is a holomorphic function of $z$ in $D$, there is a neighborhood $N(z_0)$ of $z_0$ in which $\omega(z) \neq 0$. It follows that $[\mathcal{W}(z)]^{-1}$ exists and is holomorphic in $N(z_0)$. Now let $\mathcal{W}(z)$ be any nontrivial solution of (9.1.4), and set $\mathcal{W}(z_0) = \mathcal{M}_0$. Here it is assumed that $\mathcal{W}(z)$ is defined and holomorphic in $N(z_0)$. Then

$$\mathcal{W}_0(z)[\mathcal{W}_0(z_0)]^{-1}\mathcal{W}(z_0) \qquad (9.1.11)$$

is a solution of (9.1.4) holomorphic in $N(z_0)$, and at $z = z_0$ it takes the value $\mathcal{W}(z_0)$. Now there is one and only one solution of (9.1.8) with these properties, namely, $\mathcal{W}(z)$, so we must have $\mathcal{W}(z)$ represented by (9.1.11); i.e., the arbitrary solution $\mathcal{W}(z)$ admits of a representation of the form $\mathcal{W}_0(z)\mathcal{M}$. Thus the condition $\omega(z) \neq 0$ is sufficient to make $\mathcal{W}_0(z)$ fundamental.

It is also necessary, for if every solution of (9.1.4) is of the form $\mathcal{W}_0(z)\mathcal{M}$, this applies also to the solution whose initial value at $z = z_0$ is $\mathscr{E}$, the unit matrix of the algebra $\mathfrak{M}_n$. Then

$$\mathcal{W}_0(z_0)\mathcal{M}_0 = \mathscr{E}, \qquad \det [\mathcal{W}_0(z_0)] \det [\mathcal{M}_0] = 1. \qquad (9.1.12)$$

Since $\omega(z)$ is holomorphic at $z = z_0$, we must have $\omega(z_0) \neq 0$, so that the condition is also necessary. ∎

This theorem suggests that a solution matrix defined in the domain $D$ is regular either everywhere in $D$ or nowhere. That this is indeed the case follows from the explicit representation of $\omega(z)$, a formula which for the $n$th order linear DE goes back to N. H. Abel.

We shall consider a general matrix equation (9.1.8) not necessarily obtained from a linear $n$th order DE. The matrix $\mathcal{A}(z) = [a_{jk}(z)]$ shall be holomorphic in $D$; i.e., the $a_{jk}$ shall be holomorphic in the sense of Cauchy. We recall that the *trace* of a matrix is the sum of the diagonal elements:

$$\text{tr } \mathcal{A} = \sum_{j=1}^{n} a_{jj} \qquad \text{if } \mathcal{A} = [a_{jk}].\tag{9.1.13}$$

We have now

**THEOREM 9.1.2**

*Let* $\text{W}(z)$ *be a solution of* (9.1.8), *defined, holomorphic, and regular in some neighborhood* $\text{N}(z_0)$ *of* $z = z_0 \in D$. *Then*

$$det\ [\mathcal{W}(z)] \equiv \omega(z) = \omega(z_0)\ exp\left\{\int_{z_0}^{z} tr\ [\mathcal{A}(t)]\,\mathrm{d}t\right\}.\tag{9.1.13}$$

*The function* $\omega(z)$ *defined by this formula is locally holomorphic in all of* D, *and the formula is trivially valid also if* $\mathcal{W}(z)$ *is singular at* $z_0$.

*Proof.* Take $z$ in $N(z_0)$, and choose $\delta$ so small that all points $z + h$ with $|h| < \delta$ are also in $N(z_0)$. We have then

$$\mathcal{W}(z + h) = \mathcal{W}(z) + h\mathcal{W}'(z) + O(h^2) = \mathcal{W}(z) + h\mathcal{A}(z)\mathcal{W}(z) + O(h^2)$$
$$= [\mathcal{E} + h\mathcal{A}(z)]\mathcal{W}(z) + O(h^2)$$
$$= \{\mathcal{E} + h\mathcal{A}(z) + O(h^2)[\mathcal{W}(z)]^{-1}\}\mathcal{W}(z),$$

where we used the algebraic regularity of the solution. We now form determinants to obtain

$$\omega(z + h) = \omega[\mathcal{E} + h\mathcal{A}(z) + O(h^2)]\omega(z),$$

whence

$$\frac{1}{h}[\omega(z + h) - \omega(z)] = \frac{1}{h}\{\omega[\mathcal{E} + h\mathcal{A}(z)] + O(h^2) - 1\}\omega(z).$$

Here

$$\omega[\mathcal{E} + h\mathcal{A}(z)] = 1 + h\ \text{tr }[\mathcal{A}(z)] + O(h^2).$$

It follows that

$$\omega'(z) = \text{tr }[\mathcal{A}(z)]\omega(z),$$

and formula (9.1.13) results.

The function $\omega(z)$ defined by the formula evidently exists everywhere in $D$ and is locally holomorphic there. It is holomorphic in $D$ if $D$ is simply connected or, more generally, if $\omega(z)$ is single valued. If $\mathcal{W}(z_0)$ is a singular matrix, so that $\omega(z_0) = 0$, the formula still defines a function holomorphic in $D$, namely, the zero function.

It is not clear from the results obtained so far that $\omega(z)$, be it $\neq 0$ or $\equiv 0$, gives the determinant of $\mathcal{W}(z)$ in all of $D$ or in the domain of existence of $\mathcal{W}(z)$, whichever is larger. The main difficulty here is that our solution is defined only locally, namely, in $N(z_0)$. It is essentially a question of the analytic continuation of the matrix-valued function $z \mapsto \mathcal{W}(z)$, on the one hand, and the Cauchy holomorphic function, $z \mapsto \omega(z)$, on the other. In both cases we may appeal to the *law of permanency of functional equations*. The matrix function $\mathcal{W}(z)$ satisfies the DE (9.1.8) in $N(z_0)$. The matrix $\mathcal{A}(z)$ exists and is holomorphic in $D$. Now start from $z = z_0$ along a path $L$ in $D$. The analytic continuation of $\mathcal{A}(z)$ exists everywhere along $L$. Since $\mathcal{A}(z)$ is holomorphic everywhere on $L$, the existence theorems show that (9.1.8) has analytic solutions at all points of $L$. If $z = z_1$ is such a point, and if the analytic continuation of $\mathcal{W}(z)$ along $L$ gives the value $\mathcal{W}(z_1)$ at $z = z_1$, the analytic continuation of $\mathcal{W}(z)$ along $L$ agrees at $z = z_1$ with the local solution, which takes on the value $\mathcal{W}(z_1)$ at $z = z_1$. It follows that $\mathcal{W}(z)$ exists in all of $D$ and is a solution of (9.1.8). It is locally holomorphic and will be holomorphic in $D$ if it is single valued.

On the other hand, the function $z \mapsto \omega(z)$ defined by (9.1.13) exists and is locally holomorphic in $D$. If $\omega(z_0) \neq 0$, we have

$$\det [\mathcal{W}(z)] = \omega(z), \qquad z \in N(z_0). \qquad (9.1.14)$$

Now, if $\mathcal{A} \in \mathfrak{M}_n$, then $a = \det [\mathcal{A}]$ defines a mapping of the algebra $\mathfrak{M}_n$ onto the space $\mathbf{C}$ of complex numbers. The mapping function

$$a = f(\mathcal{A})$$

is called a *functional*—in the present case a *multiplicative functional*, since

$$f(\mathcal{A}\mathcal{B}) = f(\mathcal{A})f(\mathcal{B}).$$

Here also (9.1.14) is a functional equation to which the permanency principle applies. Both sides of the equation may be continued analytically everywhere in $D$. It follows that, if both sides are continued analytically along the same path $L$ in $D$, the determinant of the continuation of $\mathcal{W}(z)$ equals the continuation of the determinant, so that (9.1.14) holds everywhere in $D$.

The case $\omega(z_0) = 0$ remains. Here we are faced with two possibilities: either $\omega(z) \equiv 0$, or else there are points in $D$ where $\omega(z) \neq 0$, say, $\omega(z_1) \neq 0$. If this is the case, we can replace $z_0$ and $N(z_0)$ by $z_1$ and $N(z_1)$, respectively, in the preceding argument. It then follows that $\omega(z) \neq 0$ along all paths in $D$, (9.1.14) holds, and the assumption $\omega(z_0) = 0$, together with $\omega(z_1) \neq 0$, is untenable.

### COROLLARY

*The solution matrix* $\mathcal{W}(z)$ *is algebraically regular either everywhere in* D *or else nowhere.*

### EXERCISE 9.1

1. Write the equation $z^2(\vartheta - aI)[\vartheta + (a + 1)I]w = \vartheta(\vartheta - I)w$ in more conventional form.
2. Find a matrix DE equivalent to Legendre's equation.
3. Show that the solution $\mathcal{W}(z)\mathcal{M}$ is singular whenever $\mathcal{M}$ is.
4. Discuss any doubtful points in the proof of Theorem 9.1.2.
5. Show that any positive integral power of det $(\mathcal{A})$ is a multiplicative functional on $\mathfrak{M}_n$.

### 9.2. ANALYTICITY OF MATRIX SOLUTIONS IN A STAR

We consider the matrix DE

$$\mathcal{W}'(z) = \mathcal{A}(z)\mathcal{W}(z). \tag{9.2.1}$$

Here $\mathcal{A}(z)$ is assumed holomorphic in the finite plane save for a finite number $m$ of poles. The point at infinity may be an essential singulary. Let $z = z_0$ be a point where $\mathcal{A}(z)$ is holomorphic, and let $A[\mathcal{A}; z_0]$ be the Mittag-Leffler star of $\mathcal{A}(z)$ with respect to $z_0$. A point $z = z_1$ belongs to the star if $\mathcal{A}(z)$ is holomorphic at all points $z = \alpha z_0 + (1 - \alpha)z_1, 0 \leq \alpha \leq 1$. The boundary of the star is composed of at most $m$ rays joining the poles with the point at infinity so that the prolongations of the rays intersect at $z_0$. We have now

### THEOREM 9.2.1

*Equation* (9.2.1) *with initial condition* $\mathcal{W}(z_0) = \mathcal{E}$, *the unit matrix, has a unique solution holomorphic in* $A[\mathcal{A}; z_0]$.

*Proof.* Let $\mathbf{A}_0$ be a star with respect to $z = z_0$, which is a subset of $A[\mathcal{A}; z_0]$, bounded and bounded away from the boundary. Then there are two numbers such that

$$\|\mathcal{A}(z)\| \leq M, \qquad |z - z_0| \leq R \quad \text{for } z \in \mathbf{A}_0. \tag{9.2.2}$$

Here the matrix norm may be taken as

$$\|\mathcal{A}(z)\| = \max_j \sum_{k=1}^{n} |a_{jk}(z)|, \tag{9.2.3}$$

and similarly for other matrices in $\mathfrak{M}_n$. Then $\|\mathcal{E}\| = 1$. The matrix DE is equivalent to the matrix integral equation of the Volterra type,

$$\mathcal{W}(z) = \mathcal{E} + \int_{z_0}^{z} \mathcal{A}(t)\mathcal{W}(t)\,dt, \tag{9.2.4}$$

and this may be solved by the method of successive approximations. Thus we obtain a sequence $\{\mathcal{W}_m(z)\}$ of functions defined and holomorphic in $A[\mathcal{A}; z_0]$. For $z \in A_0$ the usual estimates give

$$\|\mathcal{W}_{m+1}(z) - \mathcal{W}_m(z)\| \leq \frac{(MR)^m}{m!}, \tag{9.2.5}$$

which shows that the sequence converges absolutely and uniformly in $A_0$ to a limit function $\mathcal{W}(z)$. By the arbitrariness of the choice of $A_0$, it follows that $\mathcal{W}(z)$ exists in all of $A[\mathcal{A}; z_0]$ and is holomorphic in any star domain like $A_0$. ∎

We denote this solution by $\mathcal{W}(z; z_0, \mathcal{E})$. Similarly, $\mathcal{W}(z; z_0, \mathcal{W}_0)$ stands for the solution of (9.2.1), which at $z = z_0$ takes the value $\mathcal{W}_0$. By (9.1.11) we have

$$\mathcal{W}(z; z_0, \mathcal{W}_0) = \mathcal{W}(z; z_0, \mathcal{E})\mathcal{W}_0. \tag{9.2.6}$$

This formula and its generalizations play an important role in the theory.

Suppose that $z = z_1$ is a point in $A[\mathcal{A}; z_0]$ and that $\mathcal{W}(z_1; z_1, \mathcal{E}) = \mathcal{W}_1$. The matrix function $z \mapsto \mathcal{A}(z)$ has a Mittag-Leffler star with respect to $z_1$, which is denoted by $A[\mathcal{A}; z_1]$.

**THEOREM 9.2.2**

*If* $z \in A[\mathcal{A}; z_0] \cap A[\mathcal{A}; z_1]$, *then*

$$\mathcal{W}(z; z_0, \mathcal{E}) = \mathcal{W}(z; z_1, \mathcal{E})\mathcal{W}(z_1, z_0, \mathcal{E}). \tag{9.2.7}$$

*Proof.* The second factor on the right is simply $\mathcal{W}_1$, and at $z = z_1$ both sides take on the value $\mathcal{W}_1$; hence the solutions are identical. ∎

Actually the right-hand side of this formula gives the analytic continuation of $\mathcal{W}(z; z_0, \mathcal{E})$ in $A[\mathcal{A}; z_1]$. Moreover, there is no need to stop at this point. Suppose that $z = z_2$ lies in $A[\mathcal{A}; z_1]$. We can then obtain the analytic continuation of $\mathcal{W}(z; z_0, \mathcal{E})$ in $A[\mathcal{A}; z_2]$ by forming

$$\mathcal{W}(z; z_0, \mathcal{E}) = \mathcal{W}(z; z_2, \mathcal{E})\mathcal{W}(z_2; z_1, \mathcal{E})\mathcal{W}(z_1; z_0, \mathcal{E}), \tag{9.2.8}$$

and this process can be repeated as often as we please. This is the basis for the method of analytic continuation, which will be discussed more fully in the next section.

Before taking up these matters, let us look at the matrix DE

$$\mathcal{V}'(z) = \mathcal{V}(z)\mathcal{B}(z), \qquad \mathcal{V}(z_0) = \mathcal{E}. \tag{9.2.9}$$

Here $\mathcal{B}(z)$ has the same analyticity properties as $\mathcal{A}(z)$. The method used above can be applied also to the present case and proves the existence and analyticity of $\mathcal{V}(z)$ in $\mathbf{A}[\mathcal{B}; z_0]$. Here we have

$$\mathcal{V}(z; z_0, \mathcal{V}_0) = \mathcal{V}_0\mathcal{V}(z; z_0, \mathcal{E}). \tag{9.2.10}$$

The analogue of (9.2.8) now proceeds with left-hand multipliers.

Consider the equations

$$\mathcal{W}'(z) = \mathcal{A}(z)\mathcal{W}(z), \qquad \mathcal{W}(z_0) = \mathcal{E},$$
$$\mathcal{V}'(z) = -\mathcal{V}(z)\mathcal{A}(z), \qquad \mathcal{V}(z_0) = \mathcal{E}. \tag{9.2.11}$$

We have

$$\mathcal{V}(z)\mathcal{W}(z) = \mathcal{E}, \tag{9.2.12}$$

as shown by W. A. Coppel (private communication). This shows that $\mathcal{V}(z)$ is the left inverse of $\mathcal{W}(z)$ in $\mathbf{A}[\mathcal{A}; z_0]$. On the other hand, from $\mathcal{W}(z_0) = \mathcal{E}$ it follows that $\mathcal{W}(z)$ has a two-sided inverse in some neighborhood of $z = z_0$. In this neighborhood

$$[\mathcal{W}(z)]^{-1} = \mathcal{V}(z), \qquad \mathcal{V}(z)\mathcal{W}(z) = \mathcal{W}(z)\mathcal{V}(z) = \mathcal{E}. \tag{9.2.13}$$

We now appeal to the law of permanency of functional equations, which shows that the relation must hold in the common domain of analyticity of $\mathcal{V}(z)$ and $\mathcal{W}(z)$, particularly in $\mathbf{A}[\mathcal{A}; z_0]$.

### EXERCISE 9.2

1.  Carry through the proof of Theorem 9.2.1 in detail.
2.  Fill in details in the discussion of (9.2.7) and (9.2.8).
3.  Obtain the analogue of (9.2.8) for (9.2.9).
4.  Prove (9.2.12) and (9.2.13).
5.  $\mathcal{Q}$ is a nilpotent element of $\mathfrak{M}_n$ if some power of $\mathcal{Q}$ is the zero matrix. Prove that the singular DE $\mathcal{Q}\mathcal{W}'(z) = \mathcal{W}(z)$ has the zero matrix as its only solution.

### 9.3.  ANALYTIC CONTINUATION AND THE GROUP OF MONODROMY

In Section 9.2 we presented a method of analytic continuation. The solution $\mathcal{W}(z; z_0, \mathcal{E})$ of the equation

$$\mathcal{W}'(z) = \mathcal{A}(z)\mathcal{W}(z) \tag{9.3.1}$$

can be extended consecutively in a sequence of Mittag-Leffler stars,

$$\mathbf{A}[\mathcal{A}; z_0], \qquad \mathbf{A}[\mathcal{A}; z_1], \qquad \mathbf{A}[\mathcal{A}; z_2], \qquad \ldots, \qquad \mathbf{A}[\mathcal{A}; z_n], \qquad (9.3.2)$$

where each $z_{j+1}$ is a point in the preceding star, $\mathbf{A}[\mathcal{A}; z_j]$. There is no need of considering analytic continuation along an arbitrary rectifiable oriented arc $C$ since such an arc may be approximated arbitrarily closely by a polygonal line.

Suppose now that after $n$ steps we find that $z_0$ lies in $\mathbf{A}[\mathcal{A}; z_n]$, and at $z = z_0$ the continuation takes on the value $\mathcal{W}_0$. Here $\mathcal{W}_0$ may be equal to $\mathcal{E}$, but also it may very well be different. If this is the case, the analytic continuation along the closed polygon $\Pi$: $z_0, z_1, z_2, \ldots, z_n, z_0$ multiplies the solution matrix on the right by $\mathcal{W}_0$:

$$\mathcal{W}(z; z_0, \mathcal{E}) \overset{\Pi}{\longrightarrow} \mathcal{W}(z; z_0, \mathcal{E})\mathcal{W}_0. \qquad (9.3.3)$$

We see that with each closed polygon $\Pi$ with $z_0$ as beginning and end, and not going through any of the singular points $a_1, a_2, \ldots, a_m$, there is a unique associated *transition matrix* $\mathcal{W}_\Pi$, and the continuation of $\mathcal{W}(z; z_0, \mathcal{E})$ along $\Pi$ carries it into the multiple $\mathcal{W}(z; z_0, \mathcal{E})\mathcal{W}_\Pi$. It is necessary to assign an orientation to the path $\Pi$. If $\Pi$ does not intersect itself, then by the Jordan curve theorem $\Pi$ separates the plane into two parts, an interior and an exterior. If $z = c$ belongs to the interior, $\Pi$ has a *winding number* of either $+1$ or $-1$ with respect to $z = c$. If the winding number is $+1$, then $\Pi$ will be said to have the *positive orientation*; if $-1$, the *negative orientation*. The winding number is by definition $(1/2\pi)$ times the change in the argument of $z - c$ as $z$ describes $\Pi$. A path $\Pi$ described in the opposite sense will be denoted by $-\Pi$, and (9.3.3) is replaced by

$$\mathcal{W}(z; z_0, \mathcal{E}) \overset{-\Pi}{\longrightarrow} \mathcal{W}(z; z_0, \mathcal{E})(\mathcal{W}_\Pi)^{-1}. \qquad (9.3.4)$$

It should be noted that a transition matrix is necessarily algebraically regular, since $\mathcal{W}(z; z_0, \mathcal{E})$ is regular and regularity is preserved under analytic continuation.

So much for the case in which $\Pi$ is a simple closed polygon; if this is not so, several possibilities arise. If $\Pi$ returns to $z_0$ several times before ending there, each subpolygon formed by consecutive passages of $z$ through $z_0$ has to be treated separately. We can thus restrict ourselves to polygons beginning and ending at $z = z_0$ with no intermediary passages through $z_0$. Such a polygon may still intersect itself either at a vertex or between two vertices. In any case there is a least subpolygon $\Pi_1$ with $z_0$ as beginning and end point. This will be a simple polygon for which an orientation is given by the previous rules; this orientation is assigned to all of $\Pi$. In this manner all polygons with $z_0$ as a vertex are oriented. They form the set $\Pi(z_0)$.

We now consider the set $\Sigma$ of all transition matrices $W_\Pi$. We say that two paths, $\Pi_0$ and $\Pi_1$, are *topologically equivalent* with respect to the DE (9.3.1), the initial point $z = z_0$, and the initial value $\mathscr{C}$ if all the members of the family of closed polygons,

$$\Pi_\alpha = \alpha \Pi_0 + (1 - \alpha)\Pi_1, \qquad 0 \le \alpha \le 1, \qquad (9.3.5)$$

are admissible paths in the sense that none of them passes through one of the singular points $a_j$ of the equation. If this is the case, the path $\Pi_0$ can be deformed continuously into $\Pi_1$ without striking a singular point. In the set $\Pi(z_0)$ we introduce the notion of an *equivalence class*: two paths belong to the same class iff they are topologically equivalent in the sense defined above. For each equivalence class we select in an arbitrary manner one of its elements as the representative of the class. Let $P(z_0)$ be the set of equivalence classes. To each equivalence class corresponds a unique transition matrix, since all the paths of an equivalence class produce the same transition matrix. The set of all transition matrices, one for each equivalence class of paths, is a group of matrices under the operation of matrix multiplication. This group $\mathscr{G}$ is *the group of monodromy* of (9.3.1). It is justifiable to say *the* group of the equation, for, as we shall see later, groups based on different initial points $z_0$ are *isomorphic*.

At the present juncture it is perhaps more important to show that we are really dealing with a group. Now the unit matrix is evidently an element of the set. For every matrix $\mathscr{W}_0$ in the set there is an inverse matrix $(\mathscr{W}_0)^{-1}$, for if $\mathscr{W}_0$ corresponds to the path $\Pi_0$ its inverse corresponds to $-\Pi_0$. Finally the product of two transition matrices, $\mathscr{W}_1$ and $\mathscr{W}_2$, is a transition matrix, for if $\mathscr{W}_1$ goes with the path $\Pi_1$ and $\mathscr{W}_2$ with the path $\Pi_2$, then $\mathscr{W}_1\mathscr{W}_2$ (in this order) goes with the path $\Pi_2$ followed by the path $\Pi_1$. This shows that $\mathscr{G}$ is indeed a group.

The group is *finitely generated*; i.e., there is a finite number of elements, the *generators*, such that every element of the group is the product of generators in some order. Factors are repeated in the product any (finite!) number of times, but since matrices normally do not commute, it is not permissible to combine equal matrices if they are separated by other matrices. Thus

$$\mathscr{G}_1\mathscr{G}_2\mathscr{G}_1 \ne \mathscr{G}_1^2\mathscr{G}_2 \ne \mathscr{G}_2\mathscr{G}_1^2.$$

We may assume without loss of generality that $\arg(a_j - z_0) \ne \arg(a_k - z_0)$, for $j \ne k$, and that the singularities are numbered so that $\arg(a_j - z_0)$ is an increasing function of $j$. We now take a set of simple loops $L_j$, $[z_0, a_j)$ in our previous notation, starting at $z = z_0$ and surrounding $z = a_j$ once in the positive sense, leaving all other singularities on the outside. We may take $L_j$ as composed of a line segment described twice in opposite senses

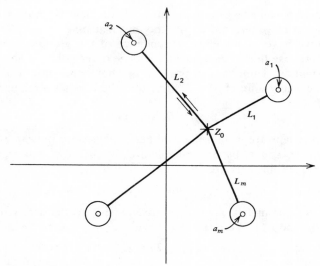

**Figure 9.1**

plus a small circle with center at $a_j$. See Figure 9.1. With each $L_j$ there is associated a unique transition matrix $\mathfrak{G}_j$. We assume $\mathfrak{G}_j \neq \mathfrak{G}_k$ if $j \neq k$. These matrices form the set of $m$ generators $\mathfrak{G}_1, \mathfrak{G}_2, \ldots, \mathfrak{G}_m$.

Suppose now that the path $\Pi$ belongs to $\Pi(z_0)$. There exists then an equivalent path made up of linear combinations of the loops $L_j$. We write such a combination as

$$k_1 L_{j_1} + \cdot k_2 L_{j_2} + \cdots + \cdot k_p L_{j_p}. \tag{9.3.6}$$

Here the order is essential, as is indicated by the dot following the plus sign. The $k$'s are integers, positive if the loop is to be described in the positive sense and negative if described in the negative sense. The formula is to be understood in the following manner. First the loop $L_{j_1}$ is to be described $k_1$ times, then the loop $L_{j_2}$ described $k_2$ times, and so on. The subscripts $j_\alpha$ take values from the set $1, 2, \ldots, m$, and consecutive loops are distinct but may be repeated later. To this representation of the path $\Pi$ corresponds a representation of the transition matrix,

$$\mathcal{W}_\Pi = (\mathfrak{G}_{j_p})^{k_p} (\mathfrak{G}_{j_{p-1}})^{k_{p-1}} \ldots (\mathfrak{G}_{j_1})^{k_1}. \tag{9.3.7}$$

This representation serves to establish the fact that the group is finitely generated, since every transition matrix admits of a representation in terms of the $\mathfrak{G}$'s. It is not claimed that (9.3.7) is minimal or unique, since there may be relations between the generators. If the generators are of finite order, a power of a generator equals the unit matrix; more generally, some product of generators equals the unit matrix.

Finally we have to settle the dependence of the group on the initial point $z_0$. Suppose that $z_0$ is replaced by some other point $z_1$, nonsingular for the equation. Then we can proceed as above and get a group $\mathfrak{G}(z_1)$ which is normally different from $\mathfrak{G}(z_0)$. But there exists a regular matrix $\mathcal{M}$ such that

$$\mathfrak{G}(z_1) = \mathcal{M}^{-1}\mathfrak{G}(z_0)\mathcal{M}, \qquad (9.3.8)$$

so the two groups are isomorphic and hence abstractly the same. To see this we join $z_0$ and $z_1$ by a rectifiable arc $C$, a straight line segment if possible. Describing this arc from $z_1$ to $z_0$ carries

$$\mathcal{W}(z; z_1, \mathscr{E}) \qquad \text{into} \qquad \mathcal{W}(z; z_0, \mathscr{E})\mathcal{M}$$

for some regular matrix $\mathcal{M}$, which may depend on the arc. Now let $z$ describe the loop $L_j$. We return to $z_0$ with the new determination

$$\mathcal{W}(z; z_0, \mathscr{E})\mathfrak{G}_j\mathcal{M},$$

and then return to $z_1$ along $C$. We return with the determination

$$\mathcal{W}(z; z_1, \mathscr{E})\mathcal{M}^{-1}\mathfrak{G}_j\mathcal{M}.$$

Now $C +\cdot L_j +\cdot (-C)$ is a loop $L_j^*$ with respect to $z = z_1$. This shows that the transition matrices

$$\mathcal{M}^{-1}\mathfrak{G}_j\mathcal{M}, \qquad j = 1, 2, \ldots, m,$$

form a set of generators for the group $\mathfrak{G}(z_1)$. Thus the group $\mathfrak{G}(z_1)$ is the *transform* of the group $\mathfrak{G}(z_0)$ under the transformation $\mathcal{M}$. The two groups are isomorphic, since for all $j$ and $k$

$$\mathcal{M}^{-1}\mathfrak{G}_j\mathfrak{G}_k\mathcal{M} = \mathcal{M}^{-1}\mathfrak{G}_j\mathcal{M}\mathcal{M}^{-1}\mathfrak{G}_k\mathcal{M}. \qquad (9.3.9)$$

Thus the underlying abstract group is unique.

**EXERCISE 9.3**

1. Consider the equation in $\mathfrak{M}_2$:

$$\mathcal{W}'(z) = \mathcal{A}(z - 1)^{-1}\mathcal{W}(z) \qquad \text{with } \mathcal{A} = \begin{bmatrix} 0 & a \\ a & 0 \end{bmatrix}.$$

   Show that

$$\tfrac{1}{2}\Big(\mathscr{E} + \frac{1}{a}\mathcal{A}\Big)(z - 1)^a + \tfrac{1}{2}\Big(\mathscr{E} - \frac{1}{a}\mathcal{A}\Big)(z - 1)^{-a}$$

   is a fundamental solution.

2. Show that the group of the equation in Problem 1 has a single generator $\mathfrak{G} = \exp(2\pi i \mathcal{A})$, where the exponential function is defined by the exponential series. Sum the series. If $a$ is an integer, the solution is single valued and the

group reduces to the identity $\mathscr{E}$. If $a$ is not an integer, the group is *cyclic* and is finite iff $a$ is rational.

3. In Problem 1 replace $\mathscr{A}$ by the nilpotent matrix

$$\mathscr{B} = \begin{bmatrix} 0 & 1 \\ 0 & 0 \end{bmatrix}.$$

Find a solution matrix and the group.

4. Suppose that (9.3.1) has the singular points $0, 1, \infty$, and no others. Take $z_0 = \frac{1}{2}$ and lay loops $L_0$ and $L_1$ around 0 and 1. Let the corresponding transition matrices be $\mathscr{G}_0$ and $\mathscr{G}_1$. Find a path corresponding to the transition matrix $\mathscr{G}_1\mathscr{G}_0^{-1}\mathscr{G}_1^{-1}\mathscr{G}_0$. Draw a figure.

### 9.4. APPROACH TO A SINGULARITY

For the study of this problem it is desirable to allow the matrix equation to have somewhat more general coefficients. In the DE

$$\mathscr{W}'(z) = \mathscr{A}(z)\mathscr{W}(z) \tag{9.4.1}$$

the entries $a_{jk}(z)$ shall be analytic but not necessarily single valued. The singularities shall be isolated, however, and we place one at the origin. Let a solution $\mathscr{W}(z)$ of (9.4.1) be determined by its initial value at $z = r_0$, and let the interval $(0, r_0)$ be free of singularities. We intend to approach the origin—to start with, along the positive real axis. It turns out that the behavior of $\|\mathscr{W}(r)\|$ for $r \downarrow 0$ is essentially determined by the rate of growth and the integrability properties of $\|\mathscr{A}(r)\|$ for $r \downarrow 0$. The basic result is

### THEOREM 9.4.1

*For* $r \downarrow 0$ *we have*

$$\|\mathscr{W}(r_0)\| \, exp \left\{ -\int_r^{r_0} \|\mathscr{A}(t)\| \, dt \right\} \leqslant \|\mathscr{W}(r)\| \leqslant \|\mathscr{W}(r_0)\| \, exp \left\{ \int_r^{r_0} \|\mathscr{A}(t)\| \, dt \right\}. \tag{9.4.2}$$

*Proof.* Let $0 < a < b < r_0$, and note that

$$\mathscr{W}(b) - \mathscr{W}(a) = \int_a^b \mathscr{A}(t)\mathscr{W}(t) \, dt, \tag{9.4.3}$$

whence

$$\|\mathscr{W}(b)\| \leqslant \|\mathscr{W}(a)\| + \int_a^b \|\mathscr{A}(t)\| \, \|\mathscr{W}(t)\| \, dt,$$

and by Gronwall's lemma (Theorem 1.6.5) this implies that

$$\|\mathscr{W}(b)\| \leqslant \|\mathscr{W}(a)\| \exp \left\{ \int_a^b \|\mathscr{A}(t)\| \, dt \right\}.$$

In particular, for $a = r$, $b = r_0$ we get

$$\|W(r_0)\| \exp\left\{-\int_r^{r_0} \|\mathcal{A}(t)\| \, dt\right\} \leq \|W(r)\|,$$

which is the lower bound in (9.4.2).

To get the upper bound we make a change of variables. We set

$$r = r_0 - x, \qquad W(r) = \mathcal{Y}(x), \qquad \mathcal{A}(r) = \mathcal{B}(x)$$

to obtain

$$\mathcal{Y}'(x) = -\mathcal{B}(x)\mathcal{Y}(x) \tag{9.4.4}$$

and

$$\mathcal{Y}(x) = \mathcal{Y}(0) - \int_0^x \mathcal{B}(s)\mathcal{Y}(s) \, ds. \tag{9.4.5}$$

This implies that

$$\|\mathcal{Y}(x)\| \leq \|\mathcal{Y}(0)\| + \int_0^x \|\mathcal{B}(s)\| \, \|\mathcal{Y}(s)\| \, ds,$$

and again by Gronwall's lemma

$$\|\mathcal{Y}(x)\| \leq \|\mathcal{Y}(0)\| \exp\left\{\int_0^x \|\mathcal{B}(s)\| \, ds\right\}.$$

Changing back to the old variables, we obtain the upper bound in (9.4.2). ■

## COROLLARY

*If as* $r \downarrow 0$

$$\|W(r_0)\| \exp\left\{\int_r^{r_0} \|\mathcal{A}(t)\| \, dt\right\} \to 0, \tag{9.4.6}$$

*then* $W(r) \equiv 0$.

Similar results hold if $z$ tends to zero in a sector

$$S: \alpha < \arg z < \beta, \qquad 0 < |z| \leq R,$$

in which $\mathcal{A}(z)$ is holomorphic. We can apply Theorem 9.4.1 to each ray, $\arg z = \theta$, in $S$. This gives

$$\|W(R e^{i\theta})\| \exp\left\{-\int_r^R \|\mathcal{A}(t e^{i\theta})\| \, dt\right\} \leq \|W(r e^{i\theta})\|$$

$$\leq \|W(R e^{i\theta})\| \exp\left\{\int_r^R \|\mathcal{A}(t e^{i\theta})\| \, dt\right\}. \tag{9.4.7}$$

Similar relations hold for the approach to infinity in a sector or in a strip. One can usually find a compact subset of $(\alpha, \beta)$ on which $\|W(R e^{i\theta})\|$

is bounded away from zero and infinity, and on such a subset it may be possible to find majorants and minorants for $\|\mathcal{A}(r\,e^{i\theta})\|$, independent of $\theta$, and thus obtain simpler estimates of $\|\mathcal{W}(r\,e^{i\theta})\|$.

Before discussing several such cases, let us consider briefly approach to the point at infinity in a strip. This becomes significant when $\mathcal{A}(z)$ is an entire periodic function of $z$ so that the solution $\mathcal{W}(z)$ is an entire function of $z$ of infinite order. Without loss of generality we may assume that the period equals one and study the bounds of $\|\mathcal{W}(x+iy)\|$ for $0 \leq x \leq 1$ as $y \to +\infty$. If $\mathcal{A}(z)$ is of finite order, a reasonable assumption on the norm of $\mathcal{A}(z)$ is

$$a\,e^{ky} \leq \|\mathcal{A}(x+iy)\| \leq b\,e^{ky} \tag{9.4.8}$$

for some $k \geq 0$ and some positive numbers $a$ and $b$ and for $0 \leq x \leq 1$, $0 \leq y$. We take an initial matrix $\mathcal{W}(x_0)$ for an $x_0$ in $[0, 1]$. The general result is given by

### THEOREM 9.4.2

*If $\mathcal{A}(z)$ is holomorphic in the closed half-strip $0 \leq x \leq 1$, $0 \leq y$, and $\mathcal{W}(z)$ is a solution of (9.4.1), then*

$$\|\mathcal{W}(x)\| \exp\left\{-\int_0^y \|\mathcal{A}(x+it)\|\,dt\right\} \leq \|\mathcal{W}(x+iy)\|$$

$$\leq \|\mathcal{W}(x)\| \exp\left\{\int_0^y \|\mathcal{A}(x+it)\|\,dt\right\}. \tag{9.4.9}$$

The proof is left to the reader. It follows, *mutatis mutandis*, the argument used above for Theorem 9.4.1.

Suppose, in particular, that the norm of $\mathcal{A}(z)$ is bounded by (9.4.8) and that

$$\min \|\mathcal{W}(x)\| = A, \qquad \max \|\mathcal{W}(x)\| = B \quad \text{for } 0 \leq x \leq 1.$$

Then for $k > 0$

$$A \exp\left[-\frac{b}{k}(e^{ky}-1)\right] \leq \|\mathcal{W}(x+iy)\| \leq B \exp\left[\frac{b}{k}(e^{ky}-1)\right], \tag{9.4.10}$$

while for $k = 0$ we have

$$A\,e^{-by} \leq \|\mathcal{W}(x+iy)\| \leq B\,e^{by}. \tag{9.4.11}$$

Let us return to the case in which the singular point is at the origin. The preceding theorems show that the case in which the norm of $\mathcal{A}(z)$ is integrable down to the origin plays an exceptional role. We proceed to an analysis of this case. The basic result is

**THEOREM 9.4.3**

*If* $\|\mathcal{A}(t)\| \in L(0, R)$, *then the equation*

$$\mathcal{W}'(t) = \mathcal{A}(t)\mathcal{W}(t), \qquad 0 < t \leq R, \tag{9.4.12}$$

*with the boundary condition* $\mathcal{W}(t) \to \mathcal{E}$ *as* $t \downarrow 0$, *has a unique solution.*

*Proof.* Apply the method of successive approximations to the integral equation

$$\mathcal{W}(t) = \mathcal{E} + \int_0^t \mathcal{A}(s)\mathcal{W}(s)\, ds \tag{9.4.13}$$

with

$$\mathcal{W}_0(t) = \mathcal{E}, \qquad \mathcal{W}_n(t) = \mathcal{E} + \int_0^t \mathcal{A}(s)\mathcal{W}_{n-1}(s)\, ds.$$

Here the integrability condition implies the existence of $\mathcal{W}_1(t)$. The existence of the higher approximations and the convergence proof are left to the reader. ∎

**COROLLARY**

*Under the assumption that* $\|\mathcal{A}(t)\| \in L(0, R)$, *all solutions of* (9.4.12) *tend to a finite limit as* $t \downarrow 0$, *and this limit is the zero matrix iff* $\mathcal{W}(t) \equiv 0$.

*Proof.* Consider the solution $\mathcal{W}(t; R, \mathcal{E})$. It is a fundamental solution, and so is the solution $\mathcal{W}(t; 0, \mathcal{E})$ constructed above. It follows that each one is a multiple of the other, giving

$$\mathcal{W}(t; R, \mathcal{E}) = \mathcal{W}(t; 0, \mathcal{E})[\mathcal{W}(R; 0, \mathcal{E})]^{-1},$$

since this is true for $t = R$. Hence

$$\lim \mathcal{W}(t; R, \mathcal{E}) = [\mathcal{W}(R; 0, \mathcal{E})]^{-1}. \tag{9.4.14}$$

Since $\mathcal{W}(t; R, \mathcal{W}_0) = \mathcal{W}(t; R, \mathcal{E})\mathcal{W}_0$, the existence of a limit holds for all solutions. ∎

The extension to the complex plane simplifies if one assumes the existence of a uniform dominant for the integral. This is what is done in

**THEOREM 9.4.4**

*Suppose that* $\mathcal{A}(z)$ *is holomorphic in the sector* S *defined above and there exists a function* $r \mapsto K(r)$ *belonging to* $C(0, R] \cap L(0, R)$ *such that*

$$\|\mathcal{A}(r\, e^{i\theta})\| \leq K(r), |\theta| < \alpha. \tag{9.4.15}$$

*Then* (9.4.1) *has a unique solution* $\mathcal{W}(z; 0, \mathcal{E})$ *such that*

$$\lim_{r \downarrow 0} \mathcal{W}(r \, e^{i\theta}; 0, \mathcal{E}) = \mathcal{E} \qquad (9.4.16)$$

*uniformly in* S. *Moreover, every solution of* (9.4.1) *tends to a finite limit as* r $\downarrow$ 0, *and the limit is independent of* $\theta$ *and exists uniformly with respect to* $\theta$.

The proof is left to the reader. It follows the pattern of previous arguments. The uniform convergence follows from the estimate

$$\|\mathcal{W}(z; 0, \mathcal{E}) - \mathcal{E}\| \leq \exp\left\{\int_0^{|z|} K(r) \, dr\right\} - 1. \qquad (9.4.17)$$

As the last item under the integrable case we list

**THEOREM 9.4.5**

*If* $\|\mathcal{A}(t)\| \in L(R, \infty)$, *then every solution of* (9.4.12) *tends to a finite limit as* t $\uparrow \infty$ *and the limit is the zero matrix iff* $\mathcal{W}(t) \equiv 0$.

The essential step in the proof is to show that the integral equation

$$\mathcal{W}(t) = \mathcal{E} - \int_t^\infty \mathcal{A}(s)\mathcal{W}(s) \, ds \qquad (9.4.18)$$

has a unique solution $\mathcal{W}(t; \infty, \mathcal{E})$, which tends to $\mathcal{E}$ as $t \uparrow \infty$. This solution is fundamental, and every solution of (9.4.12) is a multiple thereof, the multiplier figuring on the right and being an element of $\mathfrak{M}_n$.

Suppose now that $\|\mathcal{A}(z)\|$ is not integrable down to zero. The possibilities are legion, but outstanding cases correspond to $\mathcal{A}(z)$ having a pole at the origin. Here there is a great difference between a simple pole and a multiple one. For a simple pole we have.

**THEOREM 9.4.6**

*Let* $\mathcal{B}(z) = z\mathcal{A}(z)$ *be holomorphic in a disk* $|z| < R$, *and let* $\mathcal{B}(0) \neq 0$, *so that* $\mathcal{A}(z)$ *has a simple pole at* $z = 0$ *with residue* $\mathcal{B}(0)$. *For* $|\arg z| < \alpha$ *we can find positive numbers* b($\alpha$), c($\alpha$), *and* M *such that*

$$b(\alpha)|z|^M \leq \|\mathcal{W}(z)\| \leq c(\alpha)|z|^{-M}. \qquad (9.4.19)$$

*Proof.* This follows from (9.4.7). For R sufficiently small and $|z| = r \leq R$ we have

$$\|\mathcal{A}(z)\| \leq \tfrac{3}{2}\|\mathcal{B}(0)\|\frac{1}{r},$$

so we can take $M = \tfrac{3}{2}\|\mathcal{B}(0)\|$, $b(\alpha) = R^{-M} \inf \|\mathcal{W}(r \, e^{i\theta})\|$, and $c(\alpha) = R^M \sup \|\mathcal{W}(r \, e^{i\theta})\|$ for $\theta$ restricted to the interval $(-\alpha, \alpha)$. ∎

An estimate of this type was first given in 1913 by G. D. Birkhoff (1884–1944), who was inspired by ideas of Alexander Mikailovič Liapoonov (1857–1919). Similar estimates based on product integration were presented by L. Schlesinger at about the same time.

We have similar results at infinity.

## THEOREM 9.4.7

If $\mathscr{B}(z) = z\mathscr{A}(z)$ is holomorphic at $z = \infty$ and if $\mathscr{B}(\infty) \neq 0$, then $z = \infty$ is a regular-singular point of (9.4.1) and the solutions satisfy estimates of the form

$$b(\alpha)|z|^{-M} \leq \|\mathscr{W}(z)\| \leq c(\alpha)|z|^{M}, \qquad (9.4.20)$$

where $M \geq \|\mathscr{B}(\infty)\|$ and $|\arg z| < \alpha$.

We shall return to regular-singular points in the next two sections. But first we have to consider growth properties at an irregular-singular point, which it is convenient to place at infinity. It is also convenient to consider the situation in a sector and to allow a more general rate of growth of $\|\mathscr{A}(r\,e^{i\theta})\|$ so as to permit fractional powers of $r$. The basic result follows immediately from Theorems 9.4.1 and 9.4.2 by a suitable modification of the argument.

## THEOREM 9.4.8

If $\mathscr{A}(z)$ is holomorphic in a truncated sector $S|\arg z| < \alpha, R < |z| < \infty$, if

$$\|\mathscr{A}(r\,e^{i\theta})\| \leq Mr^{\mu-1}, \qquad 0 < \mu, \qquad (9.4.21)$$

and if $\mathscr{W}(z)$ is any solution of (9.4.1) in $S$, then there are positive numbers $b(\alpha)$ and $c(\alpha)$ such that

$$b(\alpha)\,exp\left(-\frac{M}{\mu}r^{\mu}\right) \leq \|\mathscr{W}(z)\| \leq c(\alpha)\,exp\left(\frac{M}{\mu}r^{\mu}\right). \qquad (9.4.22)$$

The proof is left to the reader.

We now consider the H. von Koch–O. Perron theorem, an extension of Theorem 5.4.1. We start with a lemma needed for the proof.

## LEMMA 9.4.1

Let $m_1, m_2, \ldots, m_n$ be $n$ given real numbers, and set

$$m = \max_{1 \leq j \leq n} \frac{m_j}{j}. \qquad (9.4.23)$$

Then for any choice of $n$ real numbers $a_1, a_2, \ldots, a_n$, the largest of the $2n - 1$ numbers,

$$a_2 - a_1, \qquad a_3 - a_2, \qquad \ldots, \qquad a_n - a_{n-1},$$

$$m_1, \qquad m_2 - (a_n - a_{n-1}), \qquad m_3 - (a_n - a_{n-2}), \qquad \ldots, \qquad (9.4.24)$$

$$m_n - (a_n - a_1),$$

*is not less than* m *and there is at least one choice of the* a's *for which the maximum equals* m.

*Proof.* Let $M$ be the infimum of the maximum of (9.4.24) for any choice of the $a$'s. It is clear that $M \geq m_1$. Now, if $j \geq 2$, the larger of the two numbers

$$m_j - (a_n - a_{n-j+1}) \qquad \text{and} \qquad a_{n-j+2} - a_{n-j+1}$$

is at least equal to

$$\tfrac{1}{2}[m_j - (a_n - a_{n-j+2})],$$

for if both were less than this number, adding the resulting inequalities would give $0 < 0$. Next, if $j \geq 3$, the larger of this number and $a_{n-j+3} - a_{n-j+2}$ at least equals

$$\tfrac{1}{3}[m_j - (a_n - a_{n-j+3})].$$

Again, if both were less than this bound, the first number plus half the second would give $0 < 0$, and so on. In this manner we show that $M \geq m_j/j$ for $j = 1, 2, \ldots, n$.

Suppose now that

$$m = \frac{m_k}{k}$$

and that, if there are several possible values for $k$, we take the least one. Next, we choose

$$a_j = (j - 1)m, \qquad j = 1, 2, \ldots, n. \qquad (9.4.25)$$

Thus

$$m_j - (a_n - a_{n-j+1}) = m_j - (j - 1)m \leq m,$$

i.e., a choice for which $M = m$. ∎

This is now to be applied to the $n$th order linear DE

$$w^{(n)}(z) + \sum_{j=1}^{n} P_j(z) w^{(n-j)}(z) = 0, \qquad (9.4.26)$$

where the $P_j$'s are holomorphic for $R < |z| < \infty$ and have at most a pole at infinity. The order of $P_j(z)$ at infinity is to be defined. If $P_j(z) \equiv 0$, we set the order $m_j = -\infty$; otherwise the order $m_j$ is defined so that

$$P_j(z) = z^{m_j} p_j(z), \qquad (9.4.27)$$

where $p_j(z)$ is holomorphic at infinity and different from zero there.

Setting

$$g = 1 + \max \frac{m_j}{j}, \tag{9.4.28}$$

we have now

**THEOREM 9.4.9**

*Every solution of (9.4.26) satisfies*

$$|\mathbf{w}(z)| \leq B \ exp \ (K|z|^g) \tag{9.4.29}$$

*for* $|z|$ *large and suitably chosen constants* B *and* K.

*Proof.*   We change the $n$th order DE (9.4.26) into a vector matrix DE by setting

$$w_j = z^{-(j-1)m} w^{(j-1)}, \qquad j = 1, 2, \ldots, n, \tag{9.4.30}$$

where $m$ is defined by (9.4.23). If $\mathbf{w}(z)$ is a column vector with the $w_j$'s as entries, then $\mathbf{w}(z)$ satisfies the equation

$$\mathbf{w}'(z) = \mathcal{A}(z)\mathbf{w}(z), \tag{9.4.31}$$

where the matrix $\mathcal{A}(z)$ has the elements

$$a_{jj}(z) = -(j-1)\frac{m}{z}, \qquad a_{j,j+1}(z) = z^m, \quad j = 1, 2, \ldots, n-1,$$

$$a_{nj}(z) = -z^{-(n-j)m} P_{n-j+1}(z), \qquad\qquad j = 1, 2, \ldots, n-1, \tag{9.4.32}$$

$$a_{n,n}(z) = -(n-1)\frac{m}{z} - P_1(z),$$

all other elements being zero. We may assume that $m > -1$.

The order at infinity of the elements of $\mathcal{A}(z)$ in the first $n-1$ rows is at most $m$, and of the elements above the main diagonal exactly $m$. The order of $a_{n,j}(z)$ is

$$m_{n-j+1} - (n-j)m \leq m$$

by lemma 9.4.1. This holds also for $j = n$. Hence we can find $M$ and $R$ so that $a_{jk}(z)$ is holomorphic for $R < |z| = r < \infty$, $|\arg z| < \pi$, and

$$|a_{jk}(z)| \leq Mr^m, \tag{9.4.33}$$

so that

$$\|\mathcal{A}(z)\| \leq nMr^m. \tag{9.4.34}$$

Now let $z_0$ be chosen with $R < |z_0|$, let the initial value $\mathbf{w}(z_0)$ be given, and integrate radially from $z_0$ to $z$ to obtain

$$\mathbf{w}(z) = \mathbf{w}(z_0) + \int_{z_0}^{z} \mathcal{A}(s)\mathbf{w}(s) \ ds$$

so that

$$\|\mathbf{w}(z)\| \leq \|\mathbf{w}(z_0)\| + \int_{z_0}^{z} \|\mathcal{A}(s)\| \|\mathbf{w}(s)\| \, |ds|$$

$$\leq \|\mathbf{w}(z_0)\| + nM \int_{z_0}^{z} |s|^{m} \|\mathbf{w}(s)\| \, |ds|.$$

In passing, let us observe that it is immaterial here what norm we use for the vectors: $l_1$, $l_2$, or $l_\infty$. By Gronwall's lemma the inequality gives

$$\|\mathbf{w}(z)\| \leq \|w(z_0)\| \exp\left\{ \frac{nM}{m+1} [|z|^{m+1} - |z_0|^{m+1}] \right\}, \qquad (9.4.35)$$

and this implies (9.4.29) if we take

$$K = \frac{nM}{m+1}, \qquad B = \max_{\theta} \|w(r_0 e^{i\theta})\| \exp(-Kr_0^{m+1}). \quad \blacksquare$$

Lemma 9.4.1 implies that the choice of multipliers in (9.4.30) is the best possible one in the sense of producing the lowest estimate for $\|\mathbf{w}(z)\|$. It should be noted that the estimate for the norm of the solution vector is also an upper bound for $|w_1(z)|$ and that $w_1(z) = w(z)$ is a solution of (9.4.26).

We call $g$ the *grade* of the irregular-singular point of (9.4.26) at infinity, as well as of the essential singularity of the entire functions which satisfy the equation. It is evidently a positive rational number $p/k$ with $1 \leq k \leq n$, and simple examples show that the minimal value $1/n$ can be taken on. There is no difficulty in constructing a DE of given order and polynomial coefficients and a preassigned grade $g$. There is, however, a Riemann-Birkhoff problem which does not seem perfectly trivial: given the order of the equation and the degrees of the polynomial coefficients so that the number $g$ is also given, find an equation for which the grade of the solutions is exactly $g$.

## EXERCISE 9.4

Fill in missing details or give proofs of the following:

1. Corollary of Theorem 9.4.1.
2. Theorem 9.4.2, including formulas (9.4.10) and (9.4.11).
3. Theorem 9.4.3.
4. Theorem 9.4.4, particularly (9.4.17).
5. Theorem 9.4.5, including (9.4.18).
6. Theorem 9.4.6.
7. Theorem 9.4.7.

8. Theorem 9.4.8.
9. Lemma 9.4.1.
10. Theorem 9.4.9.
11. Transform the DE $w'' + (1 - z^2)w = 0$ into vector-matrix form. Find the estimate of $\|\mathcal{W}(z)\|$, and determine $g$ and $K$. The DE has a particular solution which is an exponential function. Find it, and compare the estimate with the actual rate of growth. How good is your value of $K$?
12. Given the family of DE's

$$w''' + Azw'' + Bz^2w' + Cz^3w = 0,$$

Show that the solutions are of grade 2. If $a$ is given, $a \neq 0$, show that there are values of $A$, $B$, $C$ for which $e^{az^2}$ is a solution. Here $A$ is uniquely determined, while $B$ and $C$ have to satisfy a linear equation.

## 9.5.  REGULAR-SINGULAR POINTS

A point $z = a$ is a regular singularity of the matrix DE

$$\mathcal{W}'(z) = \mathcal{A}(z)\mathcal{W}(z) \tag{9.5.1}$$

if $(z - a)\mathcal{A}(z)$ is a holomorphic matrix function at $z = a$:

$$(z - a)\mathcal{A}(z) = \sum_{k=0}^{\infty} \mathcal{A}_k(z - a)^k, \tag{9.5.2}$$

where $\mathcal{A}_0$ is not the zero matrix. The point at infinity is a regular-singular point of (9.5.1) if $z\mathcal{A}(z)$ is holomorphic at infinity and

$$z\mathcal{A}(z) = \sum_{k=0}^{\infty} \mathcal{B}_k z^{-k}, \tag{9.5.3}$$

where $\mathcal{B}_0$ may be the zero matrix.

These conventions are of course of much more recent date than the original definitions for linear $n$th order DE's, which go back to Lazarus Fuchs (1866) and G. Frobenius (1873). Suppose that

$$w^{(n)}(z) + P_1(z)w^{(n-1)}(z) + \cdots + P_n(z)w(z) = 0, \tag{9.5.4}$$

where the coefficients are meromorphic in some neighborhood of $z = a$ and at least one has a pole. We can put (9.5.4) into matrix form by setting

$$w_1 = w, \qquad w_2 = (z - a)w', \qquad w_3 = (z - a)^2 w'', \qquad \ldots, $$
$$w_n = (z - a)^{n-1} w^{(n-1)}. \tag{9.5.5}$$

The resulting matrix $\mathcal{A}(z) = (a_{jk}(z))$ then has the entries:

$$a_{jj}(z) = (j - 1)(z - a)^{-1}, \qquad a_{j,j+1}(z) = (z - a)^{-1}, \quad j = 1, 2, \ldots, n - 1;$$
$$a_{n,k}(z) = -(z - a)^{n-k} P_{n-k+1}(z), \qquad k = 1, 2, \ldots, n - 1; \tag{9.5.6}$$
$$a_{n,n}(z) = (n - 1)(z - a)^{-1} - P_1(z);$$

all other entries being zero. This matrix will have a simple pole at $z = a$ iff

$$(z - a)^k P_k(z) \qquad (9.5.7)$$

is holomorphic at $z = a$, for $k = 1, 2, \ldots, n$. Hence this condition is necessary in order that $z = a$ be a regular-singular point. This does not imply that the point is a function-theoretical singularity of any of the solutions. They may all be holomorphic there, and this implies that the corresponding solution matrix is a holomorphic function of $z$. The point $z = a$ is, however, an algebraic singularity: the matrix $\mathcal{W}(a)$ turns out to be singular. For (9.5.4) it is customary to speak of an *inessential singularity*. We shall see that the Wronskian vanishes at $z = a$.

We have similar results at infinity. Suppose that the coefficients of (9.5.4) are holomorphic for large values of $|z|$ except possibly for poles at $z = \infty$. We can reduce to the matrix case via the substitution (9.5.5) provided that we take $a = 0$. The matrix $\mathcal{A}(z)$ is the same at infinity as at zero. The matrix $z\mathcal{A}(z)$ will be holomorphic at infinity iff each of the functions

$$z^k P_k(z) \qquad (9.5.8)$$

is holomorphic there. The matrix $\mathcal{B}_0$ is clearly not the zero matrix (why?). The singularity may be inessential in the sense that all solutions are holomorphic at $z = \infty$. We have proved ∎

### THEOREM 9.5.1

*The point* $z = a$ *is a regular-singular point, possibly inessential, if* $(z - a)^k P_k(z)$ *is holomorphic at* $z = a$ *for each* $k$. *The point at infinity is a regular-singular point, possibly inessential, if* $z^k P_k(z)$ *is holomorphic at* $\infty$ *for each* $k$.

We shall make a study of the solutions in a neighborhood of a regular-singular point, which, for convenience, we place at the origin. It is also convenient to rewrite the equation in the form

$$z^n w^{(n)}(z) + \sum_{j=1}^{n} Q_j(z) z^{n-j} w^{(n-j)}(z) = 0, \qquad (9.5.9)$$

where

$$Q_j(z) = z^j P_j(z). \qquad (9.5.10)$$

By assumption the coefficients $Q_j(z)$ are holomorphic at $z = 0$.

If every $Q_j(z)$ is a constant $c_j$, the DE becomes an Euler equation with solutions of the form $z^r$, where $r$ is a root of the algebraic equation

$$r(r - 1)(r - 2) \cdots (r - n + 1) + \sum_{j=1}^{n} c_j r(r - 1) \cdots (r - n + j + 1) = 0. \qquad (9.5.11)$$

If this equation has multiple roots, say $r = r_0$ is a $k$-tuple root, an aggregate of the form

$$z^{r_0}[C_0 + C_1 \log z + \cdots + C_{k-1}(\log z)^{k-1}]  \qquad (9.5.12)$$

appears in the solution. These facts were of course well known to Fuchs and could serve as a pattern. Equation (9.5.9) is a perturbed Euler equation, and perhaps the solution could be obtained by a similar perturbation of the solution of the Euler equation. Be that as it may, Fuchs substituted a series

$$w(z) = \sum_{k=0}^{\infty} a_k z^{r+k}  \qquad (9.5.13)$$

in the equation and proceeded to a determination of the coefficients. Let us denote the left member of (9.5.11) by $f(r)$. Fuchs found that for $r$ in (9.5.13) one must take a root of the equation

$$f(r) = 0,  \qquad (9.5.14)$$

which he referred to as *determinierende Fundamentalgleichung* and which in English is known as *indicial equation*. He found that, if the roots of this equation are distinct and do not differ by integers, it is possible to determine $n$ linearly independent convergent series solutions. The case of integral differences of the roots and, in particular, of equal roots was settled by Frobenius 7 years later. A number of arguments are available in the literature all of them more or less "corny" (slang or Milton). What I shall give here is not the corniest; it is adapted to the matrix case and is closely related to arguments given by J. A. Lappo-Danilevskij (1896–1931) for the matrix case and by the present writer for DE's in Banach algebras. It involves the notion of the *commutator operator* in the matrix algebra $\mathfrak{M}_n$, which is next on our agenda.

Let $\mathcal{A}$ be an arbitrary matrix in $\mathfrak{M}_n$, distinct from the zero and the unit matrices. We define three operators on $\mathfrak{M}_n$ into itself; *left-hand multiplication* by $\mathcal{A}$, *right-hand multiplication* by $\mathcal{A}$, and the *commutator*, which is the difference between the two. Thus

$$\mathbf{L}_{\mathcal{A}}\mathcal{X} = \mathcal{A}\mathcal{X}, \qquad \mathbf{R}_{\mathcal{A}}\mathcal{X} = \mathcal{X}\mathcal{A}, \qquad \mathbf{C}_{\mathcal{A}}\mathcal{X} = \mathbf{L}_{\mathcal{A}}\mathcal{X} - \mathbf{R}_{\mathcal{A}}\mathcal{X} = \mathcal{A}\mathcal{X} - \mathcal{X}\mathcal{A}.  \qquad (9.5.15)$$

These are obviously bounded operators, and

$$\|\mathbf{L}_{\mathcal{A}}\| = \|\mathbf{R}_{\mathcal{A}}\| = \|\mathcal{A}\|, \qquad \|\mathbf{C}_{\mathcal{A}}\| \leqslant 2\|\mathcal{A}\|.  \qquad (9.5.16)$$

Next, we have to consider the spectra of these operators and of $\mathcal{A}$. The spectrum of $\mathcal{A}$, written as $\sigma(\mathcal{A})$, is the set of numbers $\lambda_j$ which are roots of the equation

$$\det [\lambda \mathcal{E} - \mathcal{A}] = 0  \qquad (9.5.17)$$

or, equivalently, the values of $\lambda$ for which the matrix $\lambda\mathscr{E} - \mathscr{A}$ is singular. The spectrum of the operator $\mathbf{L}_\mathscr{A}$ is the set of values $\lambda$ for which the matrix equation

$$\lambda\mathscr{X} - \mathscr{A}\mathscr{X} = \mathscr{Y} \tag{9.5.18}$$

cannot be solved with respect to $\mathscr{X}$ for all $\mathscr{Y}$. A similar definition holds for $\sigma(\mathbf{R}_\mathscr{A})$:

**LEMMA 9.5.1**

$$\sigma(\mathbf{L}_\mathscr{A}) = \sigma(\mathbf{R}_\mathscr{A}) = \sigma(\mathscr{A}).$$

*Proof.* If $\lambda \in \sigma(\mathscr{A})$, then $\lambda\mathscr{E} - \mathscr{A}$ is a singular matrix, so (9.5.18) can be solvable for $\mathscr{X}$ only when $\mathscr{Y}$ belongs to a linear subspace of singular elements. This shows that $\lambda \in \sigma(\mathbf{L}_\mathscr{A})$ and, by the same token, $\lambda \in \sigma(\mathbf{R}_\mathscr{A})$. On the other hand, if $\lambda \in \sigma(\mathbf{L}_\mathscr{A})$, then $\lambda\mathscr{E} - \mathscr{A}$ cannot have an inverse, for if it did one would have $\mathscr{X} = (\lambda\mathscr{A} - \mathscr{E})^{-1}\mathscr{Y}$ for all $\mathscr{Y}$, which is absurd. It follows that $\lambda \in \sigma(\mathscr{A})$, and the lemma is proved. ∎

The spectrum of $\mathbf{C}_\mathscr{A}$ calls for more elaborate considerations. Here the *buffer* condition of N. Jacobson is basic. In his terminology $\mathfrak{M}_n$ is a *prime ring*. What we need is

**LEMMA 9.5.2**

*If $\mathscr{X}$ and $\mathscr{Y}$ are nonzero elements of $\mathfrak{M}_n$, then there exist elements $\mathscr{Z}$ such that $\mathscr{X}\mathscr{Z}\mathscr{Y}$ is not the zero element.*

*Proof.* This is trivial if $\mathscr{X}\mathscr{Y}$ is not the zero element, for then any multiple of $\mathscr{E}$ will do as a buffer. If $\mathscr{X}\mathscr{Y} = 0$, then $\mathscr{X}$ and $\mathscr{Y}$ as *divisors of zero* are both singular. But we can find an entry $x_{jm}$ in the matrix $\mathscr{X}$ and an entry $y_{pk}$ in $\mathscr{Y}$ which are not zero. Hence the matrix $\mathscr{Z}$, which has a one in the entry $(m, p)$, will do, for then $\mathscr{X}\mathscr{Z}\mathscr{Y}$ has the entry $x_{jm}y_{pk} \neq 0$ in the entry $(j, k)$. ∎

**LEMMA 9.5.3**

*Let $\alpha$ and $\beta$ belong to $\sigma(\mathscr{A})$, and let $\mathscr{X}$ and $\mathscr{Y}$ be nonzero matrices such that $\mathscr{A}\mathscr{X} = \alpha\mathscr{X}$, $\mathscr{Y}\mathscr{A} = \beta\mathscr{Y}$; then either $\mathscr{X}\mathscr{Y} = 0$ or $\alpha - \beta \in \sigma(\mathbf{C}_\mathscr{A})$ and $\mathbf{C}_\mathscr{A}\mathscr{X}\mathscr{Y} = (\alpha - \beta)\mathscr{X}\mathscr{Y}$.*

*Proof.* This follows immediately from

$$\mathbf{C}_\mathscr{A}(\mathscr{X}\mathscr{Y}) = \mathscr{A}\mathscr{X}\mathscr{Y} - \mathscr{X}\mathscr{Y}\mathscr{A} = (\alpha - \beta)\mathscr{X}\mathscr{Y}. \quad\blacksquare \tag{9.5.19}$$

Since

$$\mathbf{C}_\mathscr{A}(\mathscr{X}\mathscr{Z}\mathscr{Y}) = \mathscr{A}\mathscr{X}\mathscr{Z}\mathscr{Y} - \mathscr{X}\mathscr{Z}\mathscr{Y}\mathscr{A} = (\alpha - \beta)\mathscr{X}\mathscr{Z}\mathscr{Y},$$

it follows from the buffer condition that the spectrum of $\mathbf{C}_\mathscr{A}$ is the *difference set* of the spectrum of $\mathscr{A}$:

**LEMMA 9.5.4**

$\sigma(\mathbf{C}_{\mathcal{A}}) = \sigma(\mathcal{A}) - \sigma(\mathcal{A})$, *i.e., the set of* $n^2$ *differences of the spectral values of* $\mathcal{A}$.

Here each difference is repeated as often as it occurs. In particular, $\lambda = 0$ is at least an *n*-fold spectral value of the commutator, and if $\lambda$ is a spectral value so is $-\lambda$. Various properties of the characteristic values and elements are found in Exercise 9.5.

We shall need further analytic matrix theory for our considerations. They center around the *resolvent* of $\mathcal{A}$, i.e., the inverse of $\lambda \mathcal{E} - \mathcal{A}$, which is denoted by $\mathfrak{R}(\lambda; \mathcal{A})$. It is a rational matrix function with poles at the roots of (9.5.17). These numbers have several names: latent roots, *characteristic values*, and eigenvalues are the most common. The equation is known as the *characteristic equation*. The Cayley-Hamilton theorem asserts that $\mathcal{A}$ satisfies its own characteristic equation. Actually it may satisfy an equation of lower degree, known as the *minimal equation*. The distinct roots of the characteristic equation satisfy the minimal equation, but multiple roots may occur with a lower multiplicity in the minimal equation. Suppose that $\lambda = \alpha$ is a characteristic value which is a root of multiplicity $\mu$ of the minimal equation. Then at $\lambda = \alpha$ the principal part of the resolvent is

$$\frac{\mathcal{P}}{\lambda - \alpha} + \frac{\mathcal{Q}}{(\lambda - \alpha)^2} + \cdots + \frac{\mathcal{Q}^{\mu-1}}{(\lambda - \alpha)^{\mu}}. \tag{9.5.20}$$

Here $\mathcal{P}$ is an *idempotent*, also known as a *projection*, i.e., $\mathcal{P}^2 = \mathcal{P}$, $\mathcal{Q}$ is a *nilpotent* such that $\mathcal{Q}^{\mu-1} \neq 0$, and $\mathcal{Q}^{\mu} = 0$. The resolvent vanishes at $\lambda = \infty$ and equals the sum of its principal parts:

$$\mathfrak{R}(\lambda; \mathcal{A}) = \sum_{j=1}^{p} \left[ \frac{\mathcal{P}_j}{\lambda - \lambda_j} + \frac{\mathcal{Q}_j}{(\lambda - \lambda_j)^2} + \cdots + \frac{\mathcal{Q}_j^{\mu_j-1}}{(\lambda - \lambda_j)^{\mu_j}} \right]. \tag{9.5.21}$$

The number of projections equals the number of distinct roots. Here

$$\mathcal{P}_1 + \mathcal{P}_2 + \cdots + \mathcal{P}_p = \mathcal{E}, \qquad \mathcal{P}_j \mathcal{P}_k = \delta_{jk} \mathcal{P}_j,$$

known as a *resolution of the identity*. Furthermore,

$$\mathcal{A} = \lambda_1 \mathcal{P}_1 + \lambda_2 \mathcal{P}_2 + \cdots + \lambda_p \mathcal{P}_p + \mathcal{Q}_1 + \mathcal{Q}_2 + \cdots + \mathcal{Q}_p.$$

Only multiple roots $\lambda_j$ of the minimal equation contribute to the $\mathcal{Q}$-sum. It should be noted that

$$\mathcal{P}_j \mathcal{Q}_k = \mathcal{Q}_k \mathcal{P}_j = \delta_{jk} \mathcal{Q}_j, \qquad \mathcal{Q}_j \mathcal{Q}_k = 0 \quad \text{if } j \neq k. \tag{9.5.22}$$

With the aid of these formulas an exponential function of $z\mathcal{A}$ becomes

$$\exp{(z\mathscr{A})} = \sum_{j=1}^{p} \left[ \mathscr{P}_j + \mathscr{Q}_j z + \frac{\mathscr{Q}_j^2}{2!} z^2 + \cdots + \frac{\mathscr{Q}_j^{\mu_j - 1}}{(\mu_j - 1)!} z^{\mu_j - 1} \right] e^{\lambda_j z}. \quad (9.5.23)$$

Replacing $z$ by $\log u$, we obtain an expression for the power $u^{\mathscr{A}}$:

$$u^{\mathscr{A}} = \sum_{j=1}^{p} \left[ \mathscr{P}_j + \mathscr{Q}_j \log u + \frac{\mathscr{Q}_j^2}{2} (\log u)^2 + \cdots + \frac{\mathscr{Q}_j^{\mu_j - 1}}{(\mu_j - 1)!} (\log u)^{\mu_j - 1} \right] u^{\lambda_j}.$$

$$(9.5.24)$$

We have also to consider the resolvent of $\mathbf{C}_{\mathscr{A}}, \mathscr{R}(\lambda; \mathbf{C}_{\mathscr{A}})$. This is an operator-valued rational function of $\lambda$ with poles at the points of the difference set of $\sigma(\mathscr{A})$, i.e., the points $\alpha_j - \beta_k$ where $\alpha_j$ and $\beta_k \in \sigma(\mathscr{A})$. That these points are actually poles of the resolvent follows, for instance, from results due to Yu. L. Daletzky (1953) and S. R. Foguel (1957). Since $\mathbf{C}_{\mathscr{A}}\mathscr{X} = \mathscr{A}\mathscr{X} - \mathscr{X}\mathscr{A} \in \mathfrak{M}_n$, when $\mathscr{X}$ does, $\mathscr{R}(\lambda; \mathbf{C}_{\mathscr{A}})\mathscr{X} \in \mathfrak{M}_n$ whenever it exists, i.e., except for $\lambda$ in the difference set of $\sigma(\mathscr{A})$. This makes it rather plausible that $\mathscr{R}(\lambda; \mathbf{C}_{\mathscr{A}})$ has no other singularities than poles, and since it vanishes at infinity it is a rational operator-valued function of $\lambda$. Daletzky gave an explicit representation of $\mathscr{R}(\lambda; \mathbf{C}_{\mathscr{A}})$, from which this can be proved rigorously. It is also evident from the results of Foguel, from which one can read off that the rational character of $\mathscr{R}(\lambda; \mathscr{A})$ implies the same property of $\mathscr{R}(\lambda; \mathbf{C}_{\mathscr{A}})$. Moreover, Foguel proved that, if $\lambda_0$ is a pole of $\mathscr{R}(\lambda; \mathbf{C}_{\mathscr{A}})$, its order equals $\max{(m_j + n_k - 1)}$ if $\lambda_0 = \alpha_j - \beta_k$ and $\alpha_j$ is a pole of $\mathscr{R}(\lambda; \mathscr{A})$ of order $m_j$ while $\beta_k$ is a pole of order $n_k$; the "max" refers to the fact that $\lambda_0$ may be expressible in several ways as a difference of spectral values of $\mathscr{A}$. See Exercise 9.5 for some further details. For the following discussion we need only the fact that the spectral singularities of $\mathscr{R}(\lambda; \mathbf{C}_{\mathscr{A}})$ are poles.

We return now to (9.5.1) with a regular-singular point at the origin. By analogy with (9.5.9), where we attempted the series solution (9.5.13), we shall here attempt a matrix series solution of the form

$$\mathscr{W}(z) = \sum_{k=0}^{\infty} \mathscr{C}_k z^{k+\mathscr{A}}. \quad (9.5.25)$$

Here the coefficients $\mathscr{C}_k$ and $\mathscr{A}$ (all elements of $\mathfrak{M}_n$) are to be determined. The power $z^{\mathscr{A}}$ is defined by (9.5.24), and the exponent $k + \mathscr{A}$ is to be read as $k\mathscr{E} + \mathscr{A}$ whenever needed. We have $z\mathscr{A}(z) = \mathscr{B}(z)$ represented by (9.5.3), where $\mathscr{B}_0 \neq 0$. Substitution of (9.5.25) in the DE gives

$$\sum_{m=0}^{\infty} \mathscr{C}_m (m\mathscr{E} + \mathscr{A}) z^{m+\mathscr{A}} = \sum_{k=0}^{\infty} \mathscr{B}_k z^k \sum_{m=0}^{\infty} \mathscr{C}_m z^{m+\mathscr{A}}$$

$$= \sum_{m=0}^{\infty} \left( \sum_{k=0}^{m} \mathscr{B}_{m-k} \mathscr{C}_k \right) z^{m+\mathscr{A}},$$

assuming convergence in norm of the series involved. This must of course be justified *a posteriori*.

The term of lowest order on the left is $\mathscr{C}_0 \mathscr{A} z^{\mathscr{A}}$; on the right, $\mathscr{B}_0 \mathscr{C}_0 z^{\mathscr{A}}$. The convenient choice of $\mathscr{C}_0$ is $\mathscr{C}_0 = \mathscr{E}$, though later on we have to replace this by a scalar multiple of $\mathscr{E}$. This shows that

$$\mathscr{A} = \mathscr{B}_0.$$

The equations for the determination of the coefficients then become

$$\mathscr{C}_0 \mathscr{B}_0 = \mathscr{B}_0 \mathscr{C}_0,$$

$$m\mathscr{C}_m - (\mathscr{B}_0 \mathscr{C}_m - \mathscr{C}_m \mathscr{B}_0) = \sum_{k=0}^{m-1} \mathscr{B}_{m-k} \mathscr{C}_k, \qquad m = 1, 2, 3, \ldots. \qquad (9.5.26)$$

Here the first equation is satisfied by taking $\mathscr{C}_0 = \mathscr{E}$, as already mentioned. In trying to solve the second set of equations for $\mathscr{C}_m$, we are confronted by the spectral properties of the operator $\mathbf{C}_{\mathscr{B}_0}$. If the spectrum contains no positive integers, all is plain sailing and the coefficients $\mathscr{C}_m$ may be obtained successively by applying the operators $\mathscr{R}(m; \mathbf{C}_{\mathscr{B}_0})$. If, on the other hand, there are positive integers in $\sigma(\mathbf{C}_{\mathscr{B}_0})$, such a procedure breaks down. Now there will be positive integers in the spectrum of $\mathbf{C}_{\mathscr{B}_0}$ iff there are spectral values of $\mathscr{B}_0$ which differ by an integer. This is the matrix version of the difficulty which beset Fuchs in his discussion of the linear $n$th order DE. There it was the roots of the indicial equation which, when they differed by integers, caused trouble. It is advisable to separate the two cases and give first the analogue of Fuchs's result.

**THEOREM 9.5.1**

*Suppose that (9.5.1) has a regular-singular point at $z = 0$, where $\mathscr{A}(z)$ has a simple pole with residue $\mathscr{B}_0$. Suppose that no two spectral values of $\mathscr{B}_0$ differ by a positive integer. Then the equation has a unique solution of form (9.5.25) with $\mathscr{C}_0 = \mathscr{E}$, and the series converges in norm in the largest punctured disk, $0 < |z| < R$, in which $\mathscr{A}(z)$ is holomorphic. The solution is fundamental for $0 < |z| < R$.*

*Proof.* We have already seen that the coefficients $\mathscr{C}_m$ are uniquely determined, so all that is to be done is to prove convergence of the series. This can be done by the usual device of a majorant series. Set

$$B(z) = \sum_{k=0}^{\infty} b_k z^k, \qquad b_k = \|\mathscr{B}_k\|. \qquad (9.5.27)$$

By (9.5.16) we have

$$\|\mathbf{C}_{\mathscr{B}_0}\| \leq 2\|\mathscr{B}_0\| = 2b_0; \qquad (9.5.28)$$

and since

$$(m\mathscr{E} - \mathbf{C}_{\mathscr{B}_0})\mathscr{C}_m = \sum_{k=0}^{m-1} \mathscr{B}_{m-k}\mathscr{C}_k,$$

we get

$$(m - 2b_0)\|\mathscr{C}_m\| \leq \sum_{k=0}^{m-1} b_{m-k}\|\mathscr{C}_k\|,$$

which is nontrivial for $m > 2b_0$. Let $r$ be chosen so that $0 < r < R$. Then

$$(m - 2b_0)\|\mathscr{C}_m\|r^m \leq \sum_{k=0}^{m-1} b_{m-k}r^{m-k}\|\mathscr{C}_k\|r^k$$

$$\leq [B(r) - b_0] \max_{0 \leq k \leq m-1} (\|\mathscr{C}_k\|r^k).$$

We now choose $N > B(r) + b_0$, and see that for $m > N$

$$\|\mathscr{C}_m\|r^m \leq \max_{k \leq N} (\|\mathscr{C}_k\|r^k) \equiv M(r).$$

This shows that for all $k$

$$\|\mathscr{C}_k\|r^k \leq M(r), \tag{9.5.29}$$

or the terms of the formal solution series are bounded in norm for $|z| = r$. Since $r$ is arbitrary, $0 < r < R$, a well-known theorem due to Weierstrass shows that the series converges in norm for $0 \leq |z| < R$ and the formal solution is an actual solution. This proves the theorem. ∎

We have now to consider the case in which spectral values of $\mathscr{B}_0$ differ by an integer. We treat this by a suitable modification of the method of Frobenius. We shall prove

**THEOREM 9.5.2**

*Suppose that the spectrum of* $\mathbf{C}_{\mathscr{B}_0}$ *contains the positive integers* $n_1, n_2, \ldots, n_k$ *and no others, where*

$$1 \leq n_1 < n_2 < \cdots < n_k < 2b_0,$$

*to which correspond poles of* $\Re(\lambda; \mathbf{C}_{\mathscr{B}_0})$ *of orders* $m_1, m_2, \ldots, m_k$, *respectively, and set*

$$m_1 + m_2 + \cdots + m_k = p. \tag{9.5.30}$$

*Then there are* $p + 1$ *power series such that*

$$\sum_{j=0}^{p} (\log z)^j \sum_{m=0}^{\infty} \mathscr{C}_{mj} z^{m+\mathscr{B}_0} \tag{9.5.31}$$

*is a solution of (9.5.1). The series converge in norm for* $0 < |z| < R$, *and the solution is fundamental in the punctured disk.*

*Proof.* There exists a positive $\delta$ such that each disk $|\lambda - n| < \delta$ for $n > 0$ contains at most one spectral singularity of $\mathbf{C}_{\mathscr{B}_0}$. For $n = n_1, n_2, \ldots, n_k$ there is a pole of $R(\lambda; \mathbf{C}_{\mathscr{B}_0})$ at the center of the disk, and for all other values of $n$ there is no point of the spectrum in the disk.

Now take a complex number $\sigma$ with $0 < |\sigma| < \delta$, and consider the series

$$\mathscr{W}(z, \sigma) = \sum_{m=0}^{\infty} \mathscr{C}_m(\sigma) z^{\mathscr{B}_0 + m + \sigma}. \tag{9.5.32}$$

Since $\sigma \neq 0$, this cannot be a solution of the matrix DE, but we may choose the coefficients $\mathscr{C}_m(\sigma)$ so that $\mathscr{W}(z, \sigma)$ satisfies a nonhomogeneous linear equation such that the perturbation term is small in norm when $|\sigma|$ is small. To this end, let us form

$$[\vartheta - \mathscr{B}(z)]\mathscr{W}(z, \sigma), \quad \vartheta = z\frac{d}{dz},$$

and substitute the relevant series, multiply out, and rearrange according to ascending powers of $z$. This gives

$$\sum_{m=0}^{\infty} \left[ \mathscr{C}_m(\sigma)(\mathscr{B}_0 + m + \sigma) - \sum_{j=0}^{m} \mathscr{B}_{m-j}\mathscr{C}_j(\sigma) \right] z^{\mathscr{B}_0 + m + \sigma}.$$

For $m = 0$ the coefficient of the power is

$$\mathscr{C}_0(\sigma) + \mathscr{C}_0(\sigma)\mathscr{B}_0 - \mathscr{B}_0\mathscr{C}_0(\sigma).$$

We choose

$$\mathscr{C}_0(\sigma) = \sigma^p \mathscr{C}, \tag{9.5.33}$$

with $p$ given by (9.5.30). For $m > 0$ we choose $\mathscr{C}_m(\sigma)$ so that the coefficient of $z^{\mathscr{B}_0 + m + \sigma}$ is zero. This calls for

$$(m + \sigma)\mathscr{C}_m(\sigma) - \mathscr{B}_0\mathscr{C}_m(\sigma) + \mathscr{C}_m(\sigma)\mathscr{B}_0 = \sum_{j=0}^{m-1} \mathscr{B}_{m-j}\mathscr{C}_j(\sigma). \tag{9.5.34}$$

Since the resolvent exists for $0 < |\sigma| < \delta$, we get

$$\mathscr{C}_m(\sigma) = \mathscr{R}(m + \sigma; \mathbf{C}_{\mathscr{B}_0}) \left[ \sum_{j=0}^{m-1} \mathscr{B}_{m-j}\mathscr{C}_j(\sigma) \right], \quad m = 1, 2, 3, \ldots. \tag{9.5.35}$$

These coefficients can be determined successively, and if the resulting series converges it satisfies the equation

$$z\frac{\partial}{\partial z}\mathscr{W}(z, \sigma) = \mathscr{B}(z)\mathscr{W}(z, \sigma) + \sigma^{p+1}z^{\mathscr{B}_0 + \sigma}. \tag{9.5.36}$$

The time has come to examine the analytical nature of the coefficients $\mathscr{C}_m(\sigma)$ as functions of $\sigma$. For $1 < n_1$ we observe that $\mathscr{R}(1 + \sigma; \mathbf{C}_{\mathscr{B}_0})$ is a holomorphic function of $\sigma$ in $|\sigma| < \delta$ so that

$$\mathscr{C}_1(\sigma) = \mathscr{R}(1 + \sigma; \mathbf{C}_{\mathscr{B}_0})[\mathscr{B}_1 \sigma^p] = \sigma^p \mathscr{R}(1 + \sigma; \mathbf{C}_{\mathscr{B}_0})[\mathscr{B}_1]$$

is holomorphic and has a zero of order $p$ at $\sigma = 0$. The same result holds for $m = 2, 3, \ldots, n_1 - 1$. Something new happens for $m = n_1$, where we have

$$\mathscr{C}_{n_1}(\sigma) = \mathscr{R}(n_1 + \sigma; \mathbf{C}_{\mathscr{B}_0}) \left[ \sum_{j=0}^{n_1-1} \mathscr{B}_{n_1-j} \mathscr{C}_j(\sigma) \right].$$

Here the operand has a zero at $\sigma = 0$ at least of order $p$, but the resolvent operator has a pole of order $m_1$. This shows that $\mathscr{C}_{n_1}(\sigma)$ is holomorphic in the disk and has a zero of order at least $p - m_1$ at the origin. For $n_1 \leq m \leq n_2 - 1$, the resolvent is holomorphic and operates on a holomorphic function of $\sigma$ which has a zero of order $\geq p - m_1$ at the origin. Thus these coefficients are holomorphic for $\sigma$ in the disk and have a zero at the origin of order at least $p - m_1$. At $m = n_2$ the resolvent has a pole of order $m_2$ at the origin, and it operates on a matrix-valued function, holomorphic in $\sigma$ and with a zero of order $p - m_1$ at $\sigma = 0$. The result is a coefficient $\mathscr{C}_{n_2}(\sigma)$ holomorphic in the disk and with a zero of order $\geq p - m_1 - m_2$. In this manner we proceed, until finally for $m = n_k$ the orders of poles and zeros just balance and for $m \geq n_k$ the coefficients $\mathscr{C}_m(\sigma)$ normally do not vanish at the origin.

It is desired to show that the series

$$\sum_{m=0}^{\infty} \mathscr{C}_m(\sigma) z^m \tag{9.5.37}$$

converges for $|z| < R$, $|\sigma| < \delta$, uniformly on compact subsets. To this end it suffices to show that the sequence of terms is bounded for $|z| = r < R$, uniformly with respect to $\sigma$ for $|\sigma| \leq \epsilon < \delta$. This can be proved by the method used to prove (9.5.29). For $|\sigma| \leq \epsilon$ we have

$$(m - \epsilon - 2b_0)\|\mathscr{C}_m(\sigma)\|r^m \leq [B(r) - b_0] \max_{k<m} [\|\mathscr{C}_k(\sigma)\|r^k],$$

and for $m > N \geq B(r) + b_0 + \epsilon$ we obtain

$$\|\mathscr{C}_m(\sigma)\|r^m \leq \max_{|\sigma| \leq \epsilon} \max_{k \leq N} \|\mathscr{C}_k(\sigma)\|r^k \equiv M(r, \epsilon).$$

This shows that the terms of the series (9.5.37) are bounded, uniformly with respect to $z$ and $\sigma$ for $|z| \leq r < R$, $|\sigma| \leq \epsilon < \delta$. This in turn implies convergence of the series and shows that $\mathscr{W}(z, \sigma)$ satisfies (9.5.36). The uniform convergence with respect to $z$ and $\sigma$ implies that $\mathscr{W}(z, \sigma)$ is a holomorphic function of both variables and, moreover, that the series (9.5.37) may be differentiated termwise as often as we please with respect to $z$ as well as with respect to $\sigma$. Furthermore, we have mixed partials and relations of the form

$$\frac{\partial}{\partial z}\frac{\partial^p}{\partial \sigma^p}W(z, \sigma) = \frac{\partial^p}{\partial \sigma^p}\frac{\partial}{\partial z}W(z, \sigma). \tag{9.5.38}$$

The question of how we can utilize these various facts in the search for a solution of (9.5.1) now arises. We can let $\sigma \to 0$ in the series for $W(z, \sigma)$. The uniform convergence of the series, as well as of its partials, shows that there is a limit and that this limit satisfies the DE. The chances are, however, that the limit is identically zero, and, if not, it is likely to be a nonfundamental solution. Here we resort to the time-honored device of differentiation with respect to a parameter, $\sigma$ in the present case. In fact, this device was used by Frobenius in his memoir of 1873.

The perturbation term $\sigma^{p+1}z^{\mathfrak{R}_0+\sigma}$ vanishes for $\sigma = 0$, together with its partials with respect to $\sigma$ of order $\leqslant p$. Differentiating (9.5.36) $p$ times with respect to $\sigma$ and using (9.5.38), we get

$$z\frac{\partial^{p+1}}{\partial \sigma^p \partial z}W(z, \sigma) = \mathfrak{B}(z)\frac{\partial^p}{\partial \sigma^p}W(z, \sigma) + \frac{\partial^p}{\partial \sigma^p}(\sigma^{p+1}z^{\mathfrak{R}_0+\sigma}).$$

Here we let $\sigma$ tend to zero, the perturbation term vanishes, and it is seen that

$$\left[\frac{\partial^p}{\partial \sigma^p}W(z, \sigma)\right]_{\sigma=0} \equiv \mathcal{W}(z) \tag{9.5.39}$$

is a solution (and actually a fundamental solution) for $z \neq 0$.

Obviously (?) $\mathcal{W}(z)$ has the structure predicted in (9.5.31). In fact, each time we differentiate a power $z^{\mathfrak{R}_0+m+\sigma}$ with respect to $\sigma$ a factor $\log z$ appears. Since

$$\lim_{\sigma \to 0}\frac{\partial^p}{\partial \sigma^p}(\sigma^p z^{\mathfrak{R}_0+\sigma}) = p\,!z^{\mathfrak{R}_0},$$

it is seen that

$$\mathcal{C}_{00} = p\,!\mathcal{E} \neq 0 \tag{9.5.40}$$

and a regular matrix. The first logarithmic factor appears when the term with index $n_1$ is differentiated; for $m = n_2$ a further factor will appear and $(\log z)^p$ will correspond to $m = n_k$. No higher powers of $\log z$ will appear. Our solution $\mathcal{W}(z)$ is fundamental since the leading term $p\,!z^{\mathfrak{R}_0}$ is a regular matrix for $z \neq 0$. This completes the proof of Theorem 9.5.2. ■

### EXERCISE 9.5

1. Verify (9.5.6).
2. Verify the corresponding results at $\infty$.
3. Derive the indicial equation (9.5.11).
4. How is (9.5.16) obtained, and what does it mean?

5.  Show that, if $C_\mathcal{A}\mathcal{X} = \lambda\mathcal{X}$, $\mathcal{X} \neq 0$, then $\mathcal{X}$ is nilpotent.

6.  If $\lambda_1$ and $\lambda_2$ are nonzero points in the spectrum of $C_\mathcal{A}$, and if $C_\mathcal{A}\mathcal{X} = \lambda_1\mathcal{X}$ and $C_\mathcal{A}\mathcal{Y} = \lambda_2\mathcal{Y}$, then either $\mathcal{X}\mathcal{Y} = 0$ or $\lambda_1 + \lambda_2 \in \sigma(C_\mathcal{A})$ and $C_\mathcal{A}\mathcal{X}\mathcal{Y} = (\lambda_1 + \lambda_2)\mathcal{X}\mathcal{Y}$.

    Given the matrices

    $$\mathcal{A} = \begin{bmatrix} 2 & 0 & 0 \\ 0 & 1 & 0 \\ 0 & 1 & 1 \end{bmatrix}, \qquad \mathcal{B} = \begin{bmatrix} 2 & 0 & 0 \\ 0 & 0 & 0 \\ 0 & 0 & 2 \end{bmatrix},$$

    find (i) the characteristic and (ii) the minimal equations of these matrices.

7.  Obtain the corresponding resolutions of the identity.

8.  What form does exponential formula (9.5.23) take for these special matrices?

9.  Derive (9.5.23).

10. Verify Daletzky's formula,

    $$\mathcal{R}(\lambda ; C_\mathcal{A})\mathcal{X} = \frac{1}{(2\pi i)^2} \int_{\Gamma_1} \int_{\Gamma_2} \frac{\mathcal{R}(\alpha ; \mathcal{A})\mathcal{X}\mathcal{R}(\beta ; \mathcal{A})}{\lambda - \alpha + \beta} \, d\alpha \, d\beta,$$

    where the contours of integration surround the spectrum of $\mathcal{A}$ once in the positive sense. [*Hint:* Operate with $\lambda\mathcal{E} - C_\mathcal{A}$ to verify that $(\lambda\mathcal{E} - C_\mathcal{A})\mathcal{R}(\lambda ; C_\mathcal{A})\mathcal{X} = \mathcal{X}$ for all $\mathcal{X}$. Then replace $\mathcal{X}$ in the formula by $\lambda\mathcal{X} - \mathcal{A}\mathcal{X} + \mathcal{X}\mathcal{A}$ to get the inverse formula. Remember further that $(\alpha\mathcal{E} - \mathcal{A})\mathcal{R}(\alpha ; \mathcal{A})\mathcal{X} = \mathcal{X}$.]

11. For the matrix $\mathcal{B}$ defined in Problem 6 the operator $C_\mathcal{B}$ has a positive integer in its spectrum. What does Foguel's theorem say about the multiplicity of this pole? Try to verify this result with the aid of Daletzky's formula.

12. Verify (9.5.26).

13. Why does (9.5.29) imply convergence?

14. Verify (9.5.36).

15. Supply missing details in convergence proofs.

16. Discuss the differentiation process to get (9.5.31).

17. Are there any coefficients $\mathscr{C}_{jk}$ which have to be zero?

18. Solve $z\mathcal{W}'(z) = (\lambda\mathcal{E} + \mathcal{A}z)\mathcal{W}(z)$, where $\lambda$ is a complex number and $\mathcal{A} \in \mathfrak{M}_n$.

19. Find the resolvent of $\mathcal{A}$ and $z^\mathcal{A}$ if

    $$\mathcal{A} = \begin{bmatrix} 0 & 1 \\ 1 & 0 \end{bmatrix}.$$

20. With this $\mathcal{A}$ study the commutator $C_\mathcal{A}$. Show that $(C_\mathcal{A})^3 = 4C_\mathcal{A}$, and use this to find the resolvent $\mathcal{R}(\lambda ; C_\mathcal{A})$.

## 9.6. THE FUCHSIAN CLASS; THE RIEMANN PROBLEM

A linear DE belongs to the Fuchsian class if all its singularities are regular singular points. This implies that there are only a finite number of such points and that the point at infinity is either regular or regular singular. For the case of a matrix DE this tells us right away the equation is of the form

$$\mathcal{W}'(z) = \left\{ \sum_{j=1}^{m} \frac{\mathcal{B}_j}{z - a_j} \right\} \mathcal{W}(z), \tag{9.6.1}$$

where the $a_j$'s are distinct complex numbers, and the $\mathcal{B}_j$'s are elements of $\mathfrak{M}_n$, not necessarily distinct.

In the $n$th order linear case the situation is less obvious. Theorem 9.5.1 gives us, however, the structure at each of the singularities; and, piecing these items of information together, we get the following picture. Set

$$P(z) = (z - a_1)(z - a_2) \cdots (z - a_m), \tag{9.6.2}$$

and let $Q_p(z)$ denote a polynomial of degree not exceeding $p$; then an equation of the class must have the form

$$w^{(n)}(z) + \frac{Q_{m-1}(z)}{P(z)} w^{(m-1)}(z) + \frac{Q_{2(m-1)}(z)}{[P(z)]^2} w^{(m-2)}(z)$$

$$+ \cdots + \frac{Q_{n(m-1)}(z)}{[P(z)]^n} w(z) = 0. \tag{9.6.3}$$

This form is necessary as well as sufficient for all the singularities to be regular singular.

In the matrix case the point at infinity is not a singular point if

$$\mathcal{B}_1 + \mathcal{B}_2 + \cdots + \mathcal{B}_m = 0, \tag{9.6.4}$$

the zero matrix. For the $n$th order linear case the condition is much more involved. The standard method of studying this case is to use an inversion $z = 1/s$. If the transformed DE in $s$ has the origin as a regular point, then by definition the original DE admits the point at infinity as a regular point. The calculations become more complicated the larger $n$ is, but for $n = 3$ the work is manageable. Set

$$p_1(z) = \frac{Q_{m-1}(z)}{P(z)}, \qquad p_2(z) = \frac{Q_{2m-2}(z)}{[P(z)]^2}, \qquad p_3(z) = \frac{Q_{3m-3}(z)}{[P(z)]^3}. \tag{9.6.5}$$

Then the transformed equation is

$$\frac{d^3 w}{ds^3} + \left[ \frac{6}{s} - \frac{1}{s^2} p_1\left(\frac{1}{s}\right) \right] \frac{d^2 w}{ds^2}$$

$$+ \left[ \frac{6}{s^2} - \frac{2}{s^3} p_1\left(\frac{1}{s}\right) + \frac{1}{s^4} p_2\left(\frac{1}{s}\right) \right] \frac{dw}{ds} + \frac{1}{s^6} p_3\left(\frac{1}{s}\right) w = 0. \tag{9.6.6}$$

Here the three coefficients have to be holomorphic at $s = 0$. This condition places rather severe restrictions on the coefficients. Thus

$$p_1(z) = \frac{6}{z} + \frac{a}{z^2} + \cdots,$$

$$p_2(z) = \frac{6}{z^2} - \frac{2a}{z^3} + \cdots, \tag{9.6.7}$$

$$p_3(z) = O(z^{-6}).$$

For the $Q$-polynomials this implies that

$$Q_{m-1}(z) = 6z^{m-1} + bz^{m-2} + \cdots,$$
$$Q_{2m-2}(z) = 6z^{2m-2} - 2bz^{2m-3} + \cdots, \tag{9.6.8}$$

while $Q_{3m-3}(z)$ actually has to reduce to $Q_{3m-6}(z)$, which presupposes $m > 1$.

For (9.6.1) it is possible to obtain a representation of the solutions valid in the entire domain of analyticity. Indications of such a possibility occur in papers of Fuchs and of Poincaré, and the representation was worked out in detail by J. A. Lappo-Danilevskij. The work of this brilliant author was collected and published by the Akademia Nauk after his death.

Mark the points $a_1, a_2, \ldots, a_m$ in the complex plane and denote by $S$ the universal covering surface of the extended plane, punctured at $z = a_1, a_2, \ldots, a_m$ and infinity. On this surface each of the functions $\log(z - a_j)$ is holomorphic, and it is the least surface with this property. Let $b$ be a point on the surface, and define the family of *hyperlogarithms* $L_b(a_{j_1}, a_{j_2}, \ldots, a_{j_p}; z)$, where

$$L_b(a_j; z) = \int_b^z \frac{ds}{s - a_j} = \log \frac{z - a_j}{b - a_j},$$

$$L_b(a_{j_1}, \ldots, a_{j_p}, a_{j_{p+1}}; z) = \int_b^z L_b(a_{j_1}, \ldots, a_{j_p}; s) \frac{ds}{s - a_{j_{p+1}}}. \tag{9.6.9}$$

The subscripts $j_1, \ldots, j_p$ take on the values $1, 2, \ldots, m$. These functions are zero at $z = b$ and are single valued and holomorphic on $S$. If $C$ is any rectifiable path on $S$ from $b$ of length $L$ and if its distance from the branch points of $S$ is $\delta$, then

$$|L_b(a_{j_1}, \ldots, a_{j_p}; z)| \leq \frac{1}{p!} \left( \frac{L}{\delta} \right)^p. \tag{9.6.10}$$

This is clearly true for $p = 1$ and is easily verified by complete induction. The *representation theorem of Lappo-Danilevskij* (1928) reads as follows:

**THEOREM 9.6.1**

*For any* $z$ *on* S

$$\mathcal{W}(z; b, \mathscr{C}) = \mathscr{C} + \sum_{p=1}^{\infty} \left\{ \sum \mathscr{B}_{j_1} \mathscr{B}_{j_2} \cdots \mathscr{B}_{j_p} L_b(a_{j_p}, a_{j_{p-1}}, \ldots, a_{j_1}; z) \right\}, \quad (9.6.11)$$

*where the second summation is extended over all ordered products of* p *factors formed from the matrices* $\mathscr{B}_1, \mathscr{B}_2, \ldots, \mathscr{B}_m$.

*Proof.* For the convergence proof we start by estimating the sum for a fixed value of $p$. We have

$$\left\| \sum \mathscr{B}_{j_1} \mathscr{B}_{j_2} \cdots \mathscr{B}_{j_p} L_b(a_{j_p}, a_{j_{p-1}}, \ldots, a_{j_1}; z) \right\|$$

$$\leq \frac{1}{p!} \left( \frac{L}{\delta} \right)^p \sum \|\mathscr{B}_{j_1} \mathscr{B}_{j_2} \cdots \mathscr{B}_{j_p}\| \leq \frac{1}{p!} \left( \frac{L}{\delta} \right)^p \left[ \sum_1^m \|\mathscr{B}_j\| \right]^p.$$

This shows that the series is convergent and that

$$\|\mathcal{W}(z; b, \mathscr{C}) - \mathscr{C}\| \leq \exp \left[ \frac{L}{\delta} \sum_1^m \|\mathscr{B}_j\| \right] - 1. \quad (9.6.12)$$

That this function $\mathcal{W}(z; b, \mathscr{C})$ is a solution of (9.6.1) is shown by the following argument. By (9.6.9)

$$\frac{d}{dz} L_b(a_{j_p}, \ldots, a_{j_2}, a_{j_1}; z) = \frac{1}{z - a_{j_1}} L_b(a_{j_p}, \ldots, a_{j_2}; z).$$

The series can obviously be differentiated term by term, giving

$$\mathcal{W}'(z; b, \mathscr{C}) = \sum \frac{1}{z - a_{j_1}} \mathscr{B}_{j_1} \mathscr{B}_{j_2} \cdots \mathscr{B}_{j_p} L_b(a_{j_p}, \ldots, a_{j_1}; z).$$

In this multiple series we collect the terms for which $j_1$ has the fixed value $k$. It is seen that the coefficient of $(z - a_k)^{-1}$ equals $\mathscr{B}_k \mathcal{W}(z; b, \mathscr{C})$, so that the series (9.6.11) satisfies (9.6.1) and has the initial value $\mathscr{C}$ at $z = b$. ∎

At this juncture it is appropriate to make some remarks on the Riemann problem for DE's of the Fuchsian class. The original Riemann problem of 1857 concerned his $P$-function. It is required to determine a function

$$z \mapsto P \begin{Bmatrix} a & b & c \\ \alpha & \beta & \gamma & z \\ \alpha' & \beta' & \gamma' \end{Bmatrix} \quad (9.6.13)$$

such that (i) it is analytic throughout the whole plane except for the singular points $a$, $b$, $c$; (ii) any three determinations of this function are linearly dependent; and (iii) at each of the singular points $a$, $b$, $c$ there are two distinct determinations,

$$(z - z_0)^{\rho_1} f_1(z; z_0), \qquad (z - z_0)^{\rho_2} f_2(z; z_0),$$

where $f_1$ and $f_2$ are holomorphic at $z = z_0$ and $\neq 0$ there. Here for

$$
\begin{aligned}
z_0 &= a, & \rho_1 &= \alpha, & \rho_2 &= \alpha', \\
z_0 &= b, & \rho_1 &= \beta, & \rho_2 &= \beta', \\
z_0 &= c, & \rho_1 &= \gamma, & \rho_2 &= \gamma'.
\end{aligned}
\tag{9.6.14}
$$

These conditions determine the $P$-function uniquely, provided that

$$\alpha + \alpha' + \beta + \beta' + \gamma + \gamma' = 1.$$

This satisfies a linear second order DE reducible to the hypergeometric equation. The six special determinations will be of the prescribed type if none of the differences $\alpha - \alpha'$, $\beta - \beta'$, $\gamma - \gamma'$ is zero.

For the early history of the Riemann problem, see Section 10.5. It did not appear in print until the publication of Riemann's *Gesammelte Werke: Nachlass* in 1876, and for almost 30 years afterward the Riemann problem slumbered in peace. In 1905 it caught the attention of David Hilbert (1862–1943), who may have been inspired by a lecture given by Wirtinger at the 1904 Heidelberg Congress on part of the Riemann *Nachlass*. Be that as it may, Hilbert found the problem amenable to the new theory of integral equations. He first solved as a preparation the problem of constructing inner and outer matrix functions with respect to a given analytic matrix $\mathscr{A}(z)$ and a given simple closed rectifiable curve $C$ such that $\mathscr{I}(z) = \mathscr{A}(z)\mathscr{U}(z)$ on $C$. Here $\mathscr{I}(z)$ is holomorphic inside $C$, while $\mathscr{U}(z)$ is holomorphic outside. $\mathscr{A}(z)$ is holomorphic in a ribbon containing $C$, and $C$ has to satisfy fairly severe restrictions. These, however, made no difference to the original Riemann problem, which Hilbert solved for the case of 2-by-2 matrices. The problem was then to determine a DE of the Fuchsian class, knowing the location of its singular points and the generators of its group of monodromy.

Soon afterward (1906, 1907) Josip Plemelj (1873–1967) took up both problems in $\mathfrak{M}_n$ with simpler methods and fewer restrictions on $C$. George David Birkhoff (1884–1944) tackled the problems in 1913; moreover, he generalized them so as to deal also with DE's having irregular-singular points. He also considered Riemann problems for linear difference and $q$-difference equations.

A new mode of attack was found by Lappo-Danilevskij in 1928. For a DE of the Fuchsian class his general representation theorem (Theorem 9.6.1) gives the value of the solution at every nonsingular point of the Riemann surface $S$. We are dealing here with the matrix DE (9.6.1) and the solution $\mathscr{W}(z; b, \mathscr{C})$. The representation is in terms of a power series in $m$ noncommuting matrix variables with complex coefficients. The generators of the monodromic group are the values of $\mathcal{W}$ at $z = b_1, b_2, \ldots, b_m$, where $b_j$ is congruent to $b$ modulo a closed circuit $L_j$, beginning and ending at $z = b$ (in the plane) so that $L_j$ surrounds the singular point $z = a_j$ and no other singularity. In this setting the problem becomes one of inverting $m$ simultaneous power series in $m$ unknowns—obviously not a trivial task.

If the generators of the group $\mathfrak{G}$ are denoted by $\mathcal{G}_1, \mathcal{G}_2, \ldots, \mathcal{G}_m$, we have the system of simultaneous equations

$$\mathcal{G}_j = \mathcal{E} + 2\pi i \mathcal{B}_j + \sum_{p=2}^{\infty} \left\{ \sum \mathcal{B}_{j_1} \mathcal{B}_{j_2} \cdots \mathcal{B}_{j_p} L_b(a_{j_p}, \ldots, a_{j_1}; b_j) \right\},$$
$$j = 1, 2, \ldots, m. \qquad (9.6.15)$$

Note that for each $j$ there is a single first degree term. Its multiplier is the increment of $L_b(a_j; z)$ when $z$ describes the loop $L_j$; all the other $L_b(a_k; z)$ are assigned the increment zero when $L_j$ is described if $k \neq j$. To make the system more amenable to a function-theoretical argument, we set

$$\mathcal{G}_j - \mathcal{E} = \mathcal{M}_j, \qquad j = 1, 2, \ldots, m. \qquad (9.6.16)$$

Equation (9.6.15) then becomes

$$\mathcal{B}_j = \frac{1}{2\pi i} \left( \mathcal{M}_j - \sum_{p=2}^{\infty} \left\{ \sum \mathcal{B}_{j_1} \mathcal{B}_{j_2} \ldots \mathcal{B}_{j_p} L_b(a_{j_p}, \ldots, a_{j_1}; b_j) \right\} \right), \qquad (9.6.17)$$

where $j$ takes on the values from 1 to $m$.

It pays to rewrite this equation in vector form. Set

$$\mathbf{V} = (\mathcal{V}_1, \mathcal{V}_2, \ldots, \mathcal{V}_m) \qquad \text{with } \|\mathbf{V}\| = \max_{j=1}^{m} \|\mathcal{V}_j\|, \qquad (9.6.18)$$

so that the vectors $\mathbf{M}$ and $\mathbf{B}$ are well defined. We also define a vector-valued function of vectors $\mathbf{F}(\mathbf{B})$, whose $k$th component is the matrix-valued function

$$\sum_{p=2}^{\infty} \left\{ \sum \mathcal{B}_{j_1} \mathcal{B}_{j_2} \ldots \mathcal{B}_{j_p} L_b(a_{j_p}, \ldots, a_{j_1}; b_k) \right\}. \qquad (9.6.19)$$

In this manner we obtain the functional equation

$$\mathbf{B} = \frac{1}{2\pi i} [\mathbf{M} - \mathbf{F}(\mathbf{B})]. \qquad (9.6.20)$$

Here $\mathbf{F}(\mathbf{B})$ satisfies a Lipschitz condition,

$$\|\mathbf{F}(\mathbf{B}) - \mathbf{F}(\mathbf{C})\| \leq (1/N)[\exp(2mLN/\delta) - 1 - 2mLN/\delta] \|\mathbf{B} - \mathbf{C}\|, \qquad (9.6.21)$$

where $N$ is an upper bound for the matrix vectors to be considered. We denote the multiplier by $K(N)$.

We now define a complete metric space $\mathfrak{X}$ and a contraction operator $\mathbf{T}$, whose unique fixed point is the desired solution. We consider all vectors $\mathbf{V} = (\mathcal{V}_1, \ldots, \mathcal{V}_m)$ subject to the following conditions. Let

$$\mathbf{T}[\mathbf{V}] = \frac{1}{2\pi i} [\mathbf{M} - \mathbf{F}(\mathbf{V})]. \tag{9.6.22}$$

Let $N_0$ be the positive root of the equation

$$K(N) = 2\pi, \tag{9.6.23}$$

and let $N_1$ be a fixed positive number $< N_0$.

We have then

$$\|\mathbf{T}(\mathbf{U}) - \mathbf{T}(\mathbf{V})\| \le \frac{1}{2\pi} K(N_1)\|\mathbf{U} - \mathbf{V}\| \equiv k\|\mathbf{U} - \mathbf{V}\|, \tag{9.6.24}$$

where $k < 1$. Thus the mapping $\mathbf{V} \mapsto \mathbf{T}(\mathbf{V})$ is a pure contraction. Let $\mathfrak{X}$ be the space of all vectors $\mathbf{V}$ of norm $< N_1$. If $\mathbf{V} \in \mathfrak{X}$, then by (9.6.22)

$$\|\mathbf{T}[\mathbf{V}]\| \le \frac{1}{2\pi} [\|\mathbf{M}\| + \|\mathbf{F}(\mathbf{V})\|] \le \frac{1}{2\pi} [\|\mathbf{M}\| + k\|\mathbf{V}\|].$$

Here $\|\mathbf{V}\| < N_1$, so $\|\mathbf{T}[\mathbf{V}]\| < N_1$ provided that

$$\|\mathbf{M}\| < (2\pi - k)N_1. \tag{9.6.25}$$

If $\mathbf{M}$ satisfies (9.6.25), then $\mathbf{V} \in \mathfrak{X}$ implies that $\mathbf{T}[\mathbf{V}] \in \mathfrak{X}$. Since the space $\mathfrak{X}$ is complete (why?), there is a unique fixed point and (9.6.20) has a unique solution.

It remains to prove the Lipschitz condition (9.6.21). Consider the terms of weight $p$ in (9.6.19), i.e., disregarding scalar factors, terms of the form

$$\mathcal{B}_{j_1}\mathcal{B}_{j_2} \ldots \mathcal{B}_{j_p}. \tag{9.6.26}$$

Here the first factor can be selected in $m$ different ways, and the choice does not affect the choice of the following factors. Thus we have $m^p$ terms of weight $p$. Furthermore, the difference

$$\mathcal{B}_{j_1}\mathcal{B}_{j_2} \ldots \mathcal{B}_{j_p} - \mathcal{C}_{j_1}\mathcal{C}_{j_2} \ldots \mathcal{C}_{j_p}$$

can be estimated by writing $\mathcal{B}_{j_k} = \mathcal{C}_{j_k} + \mathcal{D}_{j_k}$ so that $\|\mathcal{D}_{j_k}\| \le \|\mathbf{B} - \mathbf{C}\| \equiv D$. Since $\|\mathcal{B}_{j_k}\|$ and $\|\mathcal{C}_{j_k}\| < N_1$ we have for $D < N_1$

$$\|\mathcal{B}_{j_1} \cdots \mathcal{B}_{j_p} - \mathcal{C}_{j_1} \cdots \mathcal{C}_{j_p}\| = \|(\mathcal{C}_{j_1} + \mathcal{D}_{j_1}) \cdots (\mathcal{C}_{j_p} + \mathcal{D}_{j_p}) - \mathcal{C}_{j_1} \cdots \mathcal{C}_{j_p}\|$$

$$\le (N_1 + D)^p - N_1^p = D[(N_1 + D)^{p-1} + (N_1 + D)^{p-2}N_1 + \cdots + N_1^{p-1}]$$

$$< N_1^{p-1}[2^{p-1} + 2^{p-2} + \cdots + 1]D = N_1^{p-1}2^p D.$$

Thus in the norm of $F(B) - F(C)$ the terms of weight $p$ contribute at most

$$(2m)^p N^{p-1} \frac{1}{p!} \left(\frac{L}{\delta}\right)^p \|B - C\|.$$

Summing for $p$ from 2 to $\infty$, we get (9.6.21).

We have thus proved ■

**THEOREM 9.6.2**

*For $\|M\| < (2\pi - \mathrm{k})N_1$ the functional equation (9.6.20) has a unique solution $B = (\mathcal{B}_1, \mathcal{B}_2, \ldots, \mathcal{B}_m)$. To this solution corresponds a DE (9.6.1) with coefficients $\mathcal{B}_j$ and having $\mathcal{G}_j = \mathcal{E} + \mathcal{M}_j$, $j = 1, 2, \ldots, m$, as generators of its group of monodromy.*

This result seems to be the best obtainable by this method. Every transit matrix is regular, but not every regular matrix can figure as a transit matrix for a particular DE of the Fuchsian class. Here the evidence is somewhat conflicting. On the one hand, L. Bieberbach (1953, p. 251) claims that the matrices

$$\begin{bmatrix} 1 & 0 \\ 0 & a \end{bmatrix}, \quad a \neq 1, \quad \text{together with} \quad \begin{bmatrix} 1 & 0 \\ 0 & c \end{bmatrix}, \quad c \neq 1, ac \neq 1, \quad (9.6.27)$$

cannot be transit matrices of a hypergeometric equation. On the other hand, for $m = 2$, $n = 2$, Lappo-Danilevskij seems to have proved that the functional equation (9.6.20) is solvable for $B$ in terms of $M$ no matter how $M$ is chosen. The solution is no longer necessarily unique and need not be single valued (Lappo-Danilevskij, *Mémoires*, 1936, Vol. III, pp. 77–97).

**EXERCISE 9.6**

1. Verify (9.6.3).
2. Verify condition (9.6.4).
3. Obtain the transformed equation (9.6.6), and verify condition (9.6.8).
4. Verify (9.6.10).
5. Expand $L_b(a_{j_1}, a_{j_2}, \ldots, a_{j_p}; z)$ in powers of $z - b$.
6. Prove that

$$L_b(a_{j_1}, a_{j_2}, \ldots, a_{j_p}; z) = \sum_{k=1} L_b(a_{j_1}, \ldots, a_{j_k}; c) L_c(a_{j_{k+1}}, \ldots, a_{j_p}; z).$$

## 9.7. IRREGULAR-SINGULAR POINTS

Let us now consider a matrix linear DE with an irregular singularity which we place at infinity. Thus the equation is

$$z\,\mathcal{Y}'(z) = z^p \sum_{j=0}^{\infty} \mathcal{M}_j z^{-j} \mathcal{Y}(z), \tag{9.7.1}$$

where the $\mathcal{M}_j$'s are constant matrices in $\mathfrak{M}_n$ and $\mathcal{M}_0 \neq 0$. The series is supposed to be convergent in norm for large values of $|z|$. Also, $p$ is a positive integer, the *rank* of the singularity. It is desired to find a solution in $\mathfrak{M}_n$ holomorphic in some sector abutting at infinity. We shall take the equation in the *Birkhoff canonical form*,

$$z\,\mathcal{W}'(z) = z^p \sum_{j=0}^{p} \mathcal{A}_j z^{-j} \mathcal{W}(z). \tag{9.7.2}$$

In two papers (1909, 1913) G. D. Birkhoff attempted to show that it is always possible to reduce (9.7.1) to the form (9.7.2) by a change of dependent variable. In both cases he put

$$\mathcal{W} = \mathcal{B}(z)\mathcal{Y}, \tag{9.7.3}$$

where the matrix $\mathcal{B}(z)$ is holomorphic in some neighborhood of infinity. In the first case he assumed $\det \mathcal{B}(\infty) \neq 0$. In the second case

$$\mathcal{B}(z) = z^g \sum_{j=0}^{\infty} \mathcal{B}_j z^{-j} \tag{9.7.4}$$

convergent for $R \leq |z| < \infty$. Here $\det \mathcal{B}_0$ may $= 0$ but $\det \mathcal{B}(z)$ should $\neq 0$ for $R \leq |z|$. Furthermore, $g$ is an integer.

Birkhoff's argument was questioned; in 1953 F. R. Gantmacher and in 1959 P. Masani gave counterexamples for $p = 0$ (the regular-singular case). In 1963 H. L. Turrittin came to the rescue. He showed that it is always possible to find an equivalent matrix,

$$\mathcal{A}(z) = z^p \sum_{j=0}^{q} \mathcal{A}_j z^{-j}, \tag{9.7.5}$$

with a finite $q$ which, however, may have to exceed $p$. If $p = 0$, Birkhoff's first method would have given $q = 1$ while the second one could have given $q = p = 0$. More generally, when $p > 0$ and the second type of transformation is used, Turrittin could show that $q = p$ is obtainable provided that the characteristic roots of $\mathcal{A}_0$ are distinct. When there are multiple roots the situation is still obscure. Turrittin could prove, however, the following theorem, which we state without proof.

**THEOREM 9.7.1**

*If* $n = 2$, *(9.7.1) may be reduced to the form (9.7.2) with* $q = p$ *by a transformation of the second type, provided that the asymptotic solution of (9.7.1) does not contain fractional powers of* $z^{-1}$.

This is true for $p = 0$, and Turrittin's proof is essentially induction on the rank.

Turrittin has also proved that it is always possible to reduce the rank from $p$ to 1 at the price of increasing the order of the matrix algebra from $n$ to $np$.

Fractional powers may very well appear. Thus the equation

$$z\mathcal{W}'(z) = \begin{bmatrix} 0 & z \\ \frac{1}{2} & \frac{1}{4} \end{bmatrix} \mathcal{W}(z) \tag{9.7.6}$$

is of rank 1, and a solution matrix is given by

$$\begin{bmatrix} \exp(z^{1/2}) & \exp(-z^{1/2}) \\ \frac{1}{2}z^{-1/2}\exp(z^{1/2}) & -\frac{1}{2}z^{-1/2}\exp(-z^{1/2}) \end{bmatrix}. \tag{9.7.7}$$

Formal, (possibly asymptotic) solutions have figured in the study of irregular-singular points since 1872, when Ludwig Wilhelm Thomé (1841–1910) introduced what he called *normal series*. For a second order linear DE such a solution would be of the form

$$e^{P(z)} z^{\rho} (c_0 + c_1 z^{-1} + \cdots). \tag{9.7.8}$$

Here $P(z)$ is a polynomial in $z$ of degree $p$, the rank of the singularity, $\rho$ is a constant, and the coefficients $c_j$ have to be determined. This can be done step by step if the coefficients of the DE are rational functions of $z$. The expansion is normally divergent, but the series $\sum_0^\infty c_j z^{-j}$ represents

$$e^{-P(z)} z^{-\rho} w(z)$$

asymptotically in the sense of Poincaré.

Now we recall that, when $n = 2$, the *grade* of the singularity in the sense of Section 5.4 is an integral multiple of $\frac{1}{2}$. If in the present case the grade is an odd multiple of $\frac{1}{2}$, the asymptotic behavior is sharply changed. Now $P(z)$ will be a polynomial in $z^{1/2}$, and the series will proceed by powers of $z^{-1/2}$. Such variants of the normal series, known as *subnormal series*, date back to the 1885 Paris thesis of Eugène Fabry (1856–1944), whose name is also attached to a theorem on noncontinuable gap series.

For the matrix case (9.7.2) we have the following result, which will be stated without proof:

**THEOREM 9.7.2**

*If the characteristic roots of $\mathcal{A}_0$ are all distinct, say* $s_1, s_2, \ldots, s_n$, *there exist* n *scalar polynomials* $p_k(z)$ *and* n *exponents* $\sigma_1, \sigma_2, \ldots, \sigma_n$ *such that*

$$\mathcal{W}(z) = \mathcal{N}(z)\mathcal{D}\{\ldots, e^{p_k(z) + \sigma_k \log z}, \ldots\} \tag{9.7.9}$$

*is a fundamental solution of (9.7.2). Here*

$$P_k(z) = \frac{1}{p} s_k z^p + \cdots + s_{k,1} z^{p-1} + \cdots + s_{k,k-1} z. \tag{9.7.10}$$

$\mathcal{D}\{\ldots\}$ *is a diagonal matrix with the stated element in the kth place. The curvilinear sectors are bounded by infinite branches of the curves*

$$C_{jk}: Re[P_j(z)] = Re[P_k(z)]. \tag{9.7.11}$$

*For each sector there is a matrix series $\mathcal{N}(z)$ in powers of $1/z$ which is asymptotic to the function.*

If there are multiple characteristic roots of $\mathcal{A}_0$, the representation changes in nature. There is a polynomial $P_k$ for each of the distinct roots, but this may be a polynomial in a fractional power of $z$. The $\mathcal{N}(z)$ series may proceed after negative fractional powers. Finally, instead of the diagonal matrix $\mathcal{D}$ we obtain a diagonal block matrix. We omit further details.

The divergent series $\mathcal{N}(z)$ may be replaceable by a convergent *factorial series*,

$$F(z) = a_0 + \sum_{m=1}^{\infty} \frac{m! a_m}{z(z+c)\ldots[z+(m-1)c]}, \qquad c > 0. \tag{9.7.12}$$

Such a series has a half-plane of absolute convergence, $\text{Re}(z) > \sigma_a$, and in addition possibly a half-plane of ordinary convergence, $\text{Re}(z) > \sigma_0$, where $0 \le \sigma_a - \sigma_0 \le c$. Such a representation of $F(z)$ presupposes that $F(z)$ can be represented by a Laplace integral,

$$F(z) = \int_0^{\infty} e^{-zt} G(t)\, dt + a_0, \tag{9.7.13}$$

with

$$G(t) = \sum_{m=1}^{\infty} a_m (1 - e^{-ct})^m, \tag{9.7.14}$$

where the series converges in a domain $D$ extending to infinity, the conformal image under the mapping $v = e^{-ct}$ of the disk $|v - 1| < 1$ and having the lines $\text{Im}(t) = \pm\frac{1}{2}\pi c^{-1}$ as asymptotes. If $F(z)$ is not bounded in a half-plane but has this property in a sector, we may get a valid series representation by replacing $z$ by $z^k$ in (9.7.12) and (9.7.13) for some positive integer $k$.

The use of factorial series for the purpose of obtaining convergent representation in a partial neighborhood of an irregular singularity dates back to Jakob Horn (1867–1946), who considered such representations in 1915. In 1924 he also examined *binomial series* for the same purpose. Such representations were again studied in great detail by the Russo-

American mathematician W. J. Trjitzinsky in 1935. He showed that not all formal series arising in formal integration of a linear DE at an irregular-singular point could be replaced by factorial series in some variable $z^s$.

In this part of the discussion we have dealt with series having scalar coefficients. We get similar results if the coefficients are $n$-by-$n$ matrices. Turrittin (1964) considered linear DE's of the form

$$\mathcal{W}'(z) = z^p \left[ \sum_{m=0}^{\infty} \frac{m! \mathcal{A}_m}{z(z+1) \ldots (z+m)} \right] \mathcal{W}(z), \qquad (9.7.15)$$

where $p = 0$ or 1. The $\mathcal{A}$'s are constant matrices, and the lead matrix $\mathcal{A}_0$ has, for instance, the form

$$\mathcal{A}_0 = (\delta_{ij}(\rho_i \mathcal{I}_i + \mathcal{J}_i)). \qquad (9.7.16)$$

Then (9.7.15) has a fundamental solution of the form

$$\mathcal{W}(z) = \left[ \mathcal{W}_0 + \sum_{k=0}^{\infty} \frac{k! \mathcal{W}_{k+1}}{z(z+1) \ldots (z+k)} \right] [\delta_{ij} z^{\rho_i} \exp(\mathcal{J}_i \log z)], \qquad (9.7.17)$$

where the series converges in norm in some half-plane if the series in the right member of (9.7.15) has such a half-plane of convergence. Here $p = 0$, the $\mathcal{I}_i$'s are unit matrices, one for each distinct root $\rho_i$ of the characteristic equation of $\mathcal{A}_0$, and the sum of the orders of the $\mathcal{I}_k$'s is equal to $n$. Finally the $\mathcal{J}_i$'s are square matrices of the same order as $\mathcal{I}_i$ with zeros and ones running down the first subdiagonal, all other elements being zero.

Mention was made in Theorem 9.7.1 of sectors in the plane abutting at infinity, in which the solution has different asymptotic behavior. This is *Stokes's phenomenon*, named after the Cambridge don George Gabriel Stokes (1819–1903), who also made important contributions to optics and hydrodynamics. We have already encountered this phenomenon for second order linear DE's—more precisely, in our study of Bessel's, the Hermite-Weber, and Laplace's equations.

Stokes's phenomenon leads to further difficult questions—in particular, the problem of finding the relation between different asymptotic solutions. The general theory of equations of type (9.7.1) tells us that, if $\mathcal{W}_1(z)$ and $\mathcal{W}_2(z)$ are distinct fundamental solutions, $\mathcal{W}_2(z) = \mathcal{W}_1(z)\mathcal{C}$, where $C$ is a nonsingular constant matrix. This applies also to the Stokes problem.

Let us, with Turrittin (1964), consider (9.7.2), the Birkhoff canonical form. Suppose that the neighborhood of infinity is covered by $m$ sectors $S_1, S_2, \ldots, S_m$, where the numbering is counterclockwise and $S_1$ contains the end of the positive real axis. Two adjacent sectors, $S_j$ and $S_{j+1}$ (with $S_{m+1} = S_1$), have a narrow strip in common. For each sector $S_j$ a fundamental solution $\mathcal{W}_j$ will be given, and the problem is to determine the Stokes multiplier $\mathcal{C}_j$:

$$\mathcal{W}_{j+1}(z) = \mathcal{W}_j(z)\mathcal{C}_j. \tag{9.7.18}$$

In principle, though but seldom in practice, this is possible since $S_j$ and $S_{j+1}$ intersect. If $z = z_0$ is a point common to both sectors, we find that

$$\mathcal{C}_j = [\mathcal{W}_j(z_0)]^{-1}\mathcal{W}_{j+1}(z_0). \tag{9.7.19}$$

Equation (9.7.2) has $z = 0$ as its only finite singularity, and this is a regular-singular point (why?). It follows that the equation has a fundamental solution $\mathcal{W}_0(z)$ which involves powers of $z$ and possibly also powers of $\log z$, each multiplied by an entire function. This means that $\mathcal{W}_0(z)$ is defined for all finite values of $z \neq 0$. The domain of definition of $\mathcal{W}_0$ thus contains each of the sectors $S_j$. We have now Stokes's matrices such that $\mathcal{W}_j(z) = \mathcal{W}_0(z)\mathcal{C}_{j0}$, and we have $\mathcal{C}_j = (\mathcal{C}_{j0})^{-1}\mathcal{C}_{j+1,0}$. Again this is more easily said than done. In some cases, however, we have seen that the divergent asymptotic series may be replaced by convergent factorial series, so at least the determination of the Stokes matrices will involve convergent expansions.

We shall now discuss the application of the Laplace transformation to the case $p = 1$ and the Birkhoff canonical form

$$z\mathcal{W}'(z) = [\mathcal{A}_0 z + \mathcal{A}_1]\mathcal{W}(z), \tag{9.7.20}$$

where $\mathcal{A}_0$ and $\mathcal{A}_1$ are $n$-by-$n$ constant matrices and a solution in $\mathfrak{M}_n$ is desired. We set

$$\mathcal{W}(z) = \int_C \mathcal{U}(t)\, e^{zt}\, dt, \tag{9.7.21}$$

where the weight function $\mathcal{U}(t)$ and the path of integration are to be determined. Substitution and integration by parts gives

$$(t\mathcal{I} - \mathcal{A}_0)\, e^{zt}\big|_C = \int_C [t\mathcal{I} - \mathcal{A}_0)\mathcal{U}'(t) + (\mathcal{A}_1 + \mathcal{I})\mathcal{U}(t)]\, e^{zt}\, dt$$

To satisfy this we can find a solution of

$$(t\mathcal{I} - \mathcal{A}_0)\mathcal{U}'(t) + (\mathcal{A}_1 + \mathcal{I})\mathcal{U}(t) = 0 \tag{9.7.22}$$

and a path $C$ such that the integrated part

$$(t\mathcal{I} - \mathcal{A}_0)\, e^{zt}\big|_C = 0. \tag{9.7.23}$$

Now, if $t$ does not belong to the spectrum of $\mathcal{A}_0$, the matrix $(t\mathcal{I} - \mathcal{A}_0)$ has a unique inverse, the resolvent $\mathcal{R}(t; \mathcal{A}_0)$, and the auxiliary DE becomes

$$\mathcal{U}'(t) = -\mathcal{R}(t; \mathcal{A}_0)(\mathcal{A}_1 + \mathcal{I})\mathcal{U}(t). \tag{9.7.24}$$

To satisfy (9.7.23) the path $C$ would normally surround a part of the spectrum of $\mathcal{A}_0$ and extend to infinity in a direction such that

$$\lim e^{zt} = 0. \tag{9.7.25}$$

This condition can be satisfied for any given value of $z \neq 0$.

The singular points of (9.7.24) are the point at infinity and part or all of the spectrum of $\mathcal{A}_0$. Normally the whole spectrum would be singular, but the factor $\mathcal{A}_1 + \mathcal{I}$ may possibly annihilate the projection operator $\mathcal{P}$ and hence also the nilpotent $\mathcal{Q}$ that belong to a particular spectral value. See formula (9.5.21) and the following ones. We omit the trivial case $\mathcal{A}_1 = -\mathcal{I}$, where the solution is elementary and may be read off by inspection. If $R$ is so large that the spectrum of $\mathcal{A}_0$ lies in the disk $|t| < R$, we can expand the coefficient of $\mathcal{U}(t)$ in (9.7.24) in powers of $t^{-1}$ to obtain

$$t\mathcal{U}'(t) = -\left[\mathcal{A}_1 + \mathcal{I} + \sum_{m=1}^{\infty} \mathcal{A}_0^m(\mathcal{A}_1 + \mathcal{I})t^{-m}\right]\mathcal{U}(t). \tag{9.7.26}$$

Referring to Section 9.5, we see that, if the difference set

$$[\gamma;\, \gamma = \alpha - \beta,\, \alpha \in \sigma(\mathcal{A}_1),\, \beta \in \sigma(\mathcal{A}_1)]$$

does not contain any positive integers, (9.7.26) has a solution

$$\mathcal{U}(t) = \left(\mathcal{I} + \sum_{k=1}^{\infty} \mathcal{C}_k t^{-k}\right) t^{-\mathcal{A}_1 - \mathcal{I}}, \tag{9.7.27}$$

where the series converges for $|t| > r(\mathcal{A}_0)$, the spectral radius of $\mathcal{A}_0$, i.e., the radius of the least circle with center at $t = 0$ which contains the spectrum of $\mathcal{A}_0$. Furthermore, $\sigma(\mathcal{A}_1)$ is the spectrum of $\mathcal{A}_1$ and the power $t^{-\mathcal{A}_1 - \mathcal{I}}$ is defined as in (9.5.27).

We can use this solution to form the integral (9.7.21). Suppose that $\mathrm{Re}(z) > 0$, and take for $C$ a loop surrounding the spectrum of $\mathcal{A}_0$ and the negative real axis. More precisely, let $C$ be the semicircle $|t| = r(\mathcal{A}_0) + 1$, $\mathrm{Re}(t) > 0$, and the two lines $\mathrm{Im}(t) = \pm[r(\mathcal{A}_0) + 1]$, $\mathrm{Re}(t) \leq 0$, described in the positive sense so that arg $t$ goes from $-\pi$ to $+\pi$. The integral will then exist, for on $C$

$$\|\mathcal{U}(t)\| \leq M|t|^b, \qquad b = \|\mathcal{A}_1\| - 1 \tag{9.7.28}$$

and

$$\|t\mathcal{I} - \mathcal{A}_0\|\, \|\mathcal{U}(t)\| \exp[\mathrm{Re}(zt)] \to 0.$$

To derive the series expansions for this solution, we need to define the *reciprocal gamma function* of a matrix $\mathcal{A} \in \mathfrak{M}_n$. In the scalar case $[\Gamma(\lambda)]^{-1}$ is representable by a power series convergent for all $\lambda$, an infinite product, and Hankel's integral (6.4.23). In each of these expressions it is permissible to replace the scalar $\lambda$ by a matrix $\mathcal{A}$ in $\mathfrak{M}_n$ and thus obtain a definition of $(1/\Gamma)(\mathcal{A})$. The infinite product shows that $(1/\Gamma)(\mathcal{A})$ is a singular element of

$\mathfrak{M}_n$ iff one of the elements $\mathcal{A}, \mathcal{A} + \mathcal{I}, \ldots, \mathcal{A} + n\mathcal{I}, \ldots$ is singular. It also gives relations like

$$\frac{1}{\Gamma}(\mathcal{A}) \frac{1}{\Gamma}(\mathcal{I} - \mathcal{A}) = \frac{1}{\pi} \sin(\pi\mathcal{A}), \qquad (9.7.29)$$

where the sine function is defined by the power series or the infinite product. It is also seen that

$$\mathcal{A} \frac{1}{\Gamma}(\mathcal{A} + \mathcal{I}) = \frac{1}{\Gamma}(\mathcal{A}). \qquad (9.7.30)$$

Hankel's integral is what is particularly needed here. It gives

$$\frac{1}{\Gamma}(\mathcal{A}) = \frac{1}{2\pi i} \int_C t^{-\mathcal{A}} e^t \, dt, \qquad (9.7.31)$$

where $C$ is a loop surrounding the spectrum of $\mathcal{A}$ and the negative real axis; $t^{-\mathcal{A}}$ is defined as in (9.5.27). It should be noted that the three definitions of $1/\Gamma$ given above are consistent. Moreover, the extension of $1/\Gamma$ from $C$ to $\mathfrak{M}_n$ is unique in the sense that this is the only definition that leads to matrix functions with components holomorphic in the sense of Cauchy.

Let us now substitute the series (9.7.27) in the integral (9.7.21) and integrate termwise, as is permitted by the uniform convergence. The result is

$$W(z) = 2\pi i \sum_{k=0}^{\infty} \mathcal{C}_k \frac{1}{\Gamma} [\mathcal{A}_1 + (k+1)\mathcal{I}] z^{\mathcal{A}_1 + k}. \qquad (9.7.32)$$

This is an entire function of $z$ of order 1 multiplied by $z^{\mathcal{A}_1}$. It is a fundamental solution unless $\mathcal{A}_1 + \mathcal{I}$ is singular. The order is obtained by a direct estimate of the terms of the series, using (9.7.31). Although it is a solution, the representation gives no idea of its asymptotic behavior.

To explore this we have to examine the spectrum of $\mathcal{A}_0$. The method to be used presupposes that the spectrum has certain properties to be stated, and we shall see later what simpler assumptions will ensure that these properties hold. We assume the following:

$S_1$. The spectrum of $\mathcal{A}_0$ contains a value $s_1$ which is a simple pole of $\mathfrak{R}(t; \mathcal{A}_0)$.

$S_2$. If $\mathcal{P}_1$ is the corresponding idempotent, the difference set of the spectrum of $\mathcal{P}_1(\mathcal{A}_1 + \mathcal{I})$ contains no integer $\neq 0$.

We have then

$$\mathfrak{R}(t; \mathcal{A}_0)(\mathcal{A}_1 + \mathcal{I}) = \frac{\mathcal{P}_1(\mathcal{A}_1 + \mathcal{I})}{t - s_1} + \sum_{m=0}^{\infty} \mathcal{B}^{m+1}(\mathcal{A}_1 + \mathcal{I})(s_1 - t)^m. \qquad (9.7.33)$$

Here $\mathcal{P}_1\mathcal{B} = \mathcal{B}\mathcal{P}_1 = 0$. This follows from (9.5.21), for we have now

$$\mathcal{R}(t\,;\mathcal{A}_0) = \frac{\mathcal{P}_1}{t - s_1} + \sum_{j=2}^{n} \mathcal{P}_j\mathcal{S}_j(t).$$

Since $\mathcal{P}_1\mathcal{P}_j = 0$ for $j \neq 1$, we see that the series in (9.7.33) is annihilated by $\mathcal{P}_1$, and this will hold also if the factor $(\mathcal{A}_1 + \mathcal{S})$ is omitted. Since this holds for all $t$ for which the series converges, it also holds for $t = s_1$, which means that $\mathcal{P}_1\mathcal{B} = 0$. Since $\mathcal{P}_1$ commutes with $\mathcal{B}$, we have also $\mathcal{B}\mathcal{P}_1 = 0$.

Now the auxiliary DE (9.7.24) may be written as

$$\mathcal{U}'(t) = \left[\frac{\mathcal{P}_1(\mathcal{A}_1 + \mathcal{S})}{t - s_1} + \sum_{m=0}^{\infty} \mathcal{B}^{m+1}(\mathcal{A}_1 + \mathcal{S})(s_1 - t)^m\right]\mathcal{U}(t). \qquad (9.7.34)$$

This shows that $t = s_1$ is a regular-singular point of the DE, and condition $S_2$, by Theorem 9.5.1, ensures that the equation has a solution of the form

$$\mathcal{U}(t) = \sum_{m=0}^{\infty} \mathcal{B}_m(t - s_1)^{-\mathcal{P}_1(\mathcal{A}_1 + \mathcal{S}) + m}, \qquad \mathcal{B}_0 = \mathcal{S}. \qquad (9.7.35)$$

Here the series converges in the same disk as the resolvent series (9.7.33). Now (9.7.24) has the point at infinity as a regular-singular point. The finite singularities are poles at most $n$ in number. It follows that the solution (9.7.35) may be continued analytically to infinity along any path that avoids the finite singularities. Here the part of the path outside a large circle is at our disposal and may be chosen so that $\operatorname{Re}(zt) \to -\infty$.

We can take $C$ as a loop in the resolvent set of $\mathcal{A}_0$, i.e., the set where $t\mathcal{S} - \mathcal{A}_0$ is regular. If $\operatorname{Re}(z) > 0$, we can take the infinite part of the loop to be two horizontal lines going to the left. This will ensure that $\operatorname{Re}(zt) \to -\infty$, and condition (9.7.23) is ensured, so that the corresponding integral (9.7.21) is a solution.

To get the asymptotic representation of this solution we proceed as in the discussion of Laplace's equation in Section 6.4, formulas (6.4.24)–(6.4.34). In other words, we substitute the series (9.7.35) in the integral (9.7.21) and integrate termwise. Since the series that enters in the integral diverges for all large values of $t$, the series obtained by integration will also diverge, but it gives an asymptotic representation of the solution in the sense of Poincaré.

We start by assuming $s_1$ to be real positive. This is not as restrictive as it may seem, for (9.7.21) admits of transformations which preserve the form but shift the "parameters" $\mathcal{A}_0$ and $\mathcal{A}_1$. The substitution

$$\mathcal{W}(z) = \mathcal{Y}(z)\, e^{\beta_0 z} z^{\beta_1} \qquad (9.7.36)$$

has this property and replaces $\mathcal{A}_0$ and $\mathcal{A}_1$ by

$$\mathcal{A}_0 - \beta_0\mathcal{S} \qquad \text{and} \qquad \mathcal{A}_1 - \beta_1\mathcal{S}, \qquad (9.7.37)$$

respectively. Here the $\beta$'s are arbitrary complex numbers. The spectral values of $\mathscr{A}_0 - \mathscr{B}_0 \mathscr{I}$ are $s_j - \beta_0$. Here $\beta_0$ is at our disposal and may be chosen so that the singularity under consideration is real positive. The transformation obviously carries simple poles into simple poles. Consider now one of the integrals obtained by the process,

$$\int_C (t - s_1)^{-\mathscr{P}_1(\mathscr{A}_1 + \mathscr{I}) + k} e^{zt}\, dt.$$

We take Re $(z) > 0$ and deform $C$ into a loop surrounding the line segment $(-\infty, s_1)$ of the real axis. We set $t = s_1 + s/z$ to obtain

$$z^{\mathscr{P}_1(\mathscr{A}_1 + \mathscr{I}) - k - 1} e^{s_1 z} \int_C s^{-\mathscr{P}_1(\mathscr{A}_1 + \mathscr{I}) + k} e^s\, ds$$

$$= 2\pi i \frac{1}{\Gamma} [\mathscr{P}_1(\mathscr{A}_1 + \mathscr{I}) - k\mathscr{I}] z^{\mathscr{P}_1(\mathscr{A}_1 + \mathscr{I}) - k - 1} e^{s_1 z},$$

where we use (9.7.31) and, if necessary, a suitable rotation of the path of integration. Multiplying by $\mathscr{B}_k$ and summing for $k$, we obtain the formal series

$$\mathscr{W}(z) \sim 2\pi i \sum_{k=0}^{\infty} \mathscr{B}_k \frac{1}{\Gamma} [\mathscr{P}_1(\mathscr{A}_1 + \mathscr{I}) - k\mathscr{I}] z^{\mathscr{P}_1(\mathscr{A}_1 + \mathscr{I}) - k - 1} e^{s_1 z}. \quad (9.7.38)$$

The series is likely to be divergent, for the factor involving the reciprocal gamma function grows as a factorial. The series is asymptotic in the sense of Poincaré and represents the solution in the right half-plane. In the first approximation it behaves as

$$z^{\mathscr{P}_1(\mathscr{A}_1 + \mathscr{I}) - 1} e^{s_1 z}. \quad (9.7.39)$$

The proof that this is indeed an asymptotic representation follows the same general lines as the argument in Section 6.4 for the special scalar case. It is long and involved and will be omitted here.

The same type of argument applies to the left half-plane; the same series is obtained, but normally it represents a different solution. The solutions may be expected to be singular elements of $\mathfrak{M}_n$, since $\mathscr{P}_1(\mathscr{A}_1 + \mathscr{I})$ is a singular matrix and hence also the reciprocal gamma function. If there are several spectral values of $\mathscr{A}_0$ satisfying conditions $S_1$ and $S_2$, each will give an asymptotic representation of a particular integral, making the appropriate change of the coefficients $\mathscr{B}_m$, the idempotent $\mathscr{P}_1$, and the spectral value $s_1$ in (9.7.35). If, in particular, the $n$ spectral values of $\mathscr{A}_0$ are all distinct and hence simple poles of $\mathscr{R}(t; \mathscr{A}_0)$, then we have $n$ linearly independent solutions which may be combined to form a fundamental system. In this case the solution (9.7.32) will be the sum of solutions of known asymptotic behavior.

Suppose now that the matrix $\mathscr{A}_0$ has distinct characteristic roots

$s_1, s_2, \ldots, s_n$. We can then transform $\mathcal{A}_0$ into a diagonal form. There exists a regular matrix $\mathcal{M}$ such that $\mathcal{M}\mathcal{A}_0\mathcal{M}^{-1} = \mathcal{D}_0 = \mathcal{D}(\ldots s_j \ldots)$, and $\mathcal{Y}(z) = \mathcal{M}\mathcal{W}(z)\mathcal{M}^{-1}$ satisfies

$$z\mathcal{Y}'(z) = (\mathcal{D}_0 z + \mathcal{C}_1)\mathcal{Y}(z), \qquad \mathcal{C}_1 = \mathcal{M}\mathcal{A}_1\mathcal{M}^{-1}. \tag{9.7.40}$$

We have now

$$-\mathcal{P}_j(\mathcal{A}_1 + \mathcal{I}) = -\mathcal{E}_j(\mathcal{C}_1 + \mathcal{I}), \tag{9.7.41}$$

where $\mathcal{E}_j$ is the matrix which has a one in the place $(j, j)$ and zeros elsewhere. If $\mathcal{C}_1 = (c_{jk})$, the right member of (9.7.41) is the matrix whose $j$th row is

$$-\delta_{jk} - c_{jk},$$

while all other elements are zero. The characteristic values of this matrix are $0$, repeated $n - 1$ times, and $-1 - c_{jj}$. It is permitted to assume that the last number is not an integer or zero, for if this should turn out not to be the case we could use a transformation of type (9.7.36), now affecting only $\mathcal{A}_1$. The preceding discussion applies and gives $n$ linearly independent solution vectors for the right half-plane and $n$ for the left. The elements of the $j$th column vector behave asymptotically as constant multiples of

$$z^{\beta_j} e^{s_j z}, \qquad \beta_j = c_{jj}.$$

Equations of type (9.7.20) are of considerable importance for the set of applicable DE's where the associated matrix algebra is normally $\mathfrak{M}_2$. This includes the confluent hypergeometric case. For $n > 2$, K. Okubo (1963) has done valuable work, but nontrivial problems remain unsolved in this field.

For the case $p > 1$ we refer to Section 6.9 of LODE and the memoirs of Turrittin, in particular.

## EXERCISE 9.7

In 1913 G. D. Birkhoff examined as a by-product the case $p = 1$, $n = 2$ and, in particular, the canonical form of the second order DE

$$zw'' + (z - a)w' + bzw = 0.$$

Let $\alpha$ and $\beta$ be two roots of the equation

$$\lambda^2 + (a - 1)\lambda + b = 0,$$

where $\alpha - \beta$ is not zero or an integer.
1.  Show that the equation has the solutions

$$z^\alpha {}_1F_1(\alpha, \alpha - \beta; z) \qquad \text{and} \qquad z^\beta {}_1F_1(\beta, \beta - \alpha; z).$$

2.  If $\alpha$ and $\alpha - \beta$ are real positive, estimate the maximum modulus of the first solution.

3. Prove the existence of asymptotic solutions of the form

$$ {}_2F_0\left(\alpha, \beta ; -\frac{1}{z}\right) \quad \text{and} \quad e^z z^{-a} {}_2F_0\left(1 - \alpha, 1 - \beta ; \frac{1}{z}\right). $$

4. Show that, if the series (9.7.12) converges for $z = x_0 > 0$, it converges for every $z$ with $\operatorname{Re}(z) > x_0$.

5. Why is the width of the strip of nonabsolute convergence at most $c$?

6. Use the properties of the gamma function to show that, if the series converges for $z = x_0 > 0$, then

$$ a_m = o\left\{ c^m \exp\left[\left(\frac{x_0}{c} - 1\right)\log m\right]\right\}. $$

7. Verify the form of $G(t)$ in (9.7.14).

8. With $G(t)$ as in (9.7.14) find the Laplace transform of $G'(t)$.

9. Why is the origin a regular-singular point of (9.7.20)?

10. Show that the point at infinity is a regular-singular point of (9.7.24).

11. Verify (9.7.32).

12. Try to verify the asymptotic character of (9.7.38).

13. Take $\mathfrak{M}_2$ and the matrices

$$ \mathscr{A}_0 = \begin{bmatrix} 0 & 1 \\ \beta^2 & 0 \end{bmatrix}, \qquad \mathscr{A}_1 = \begin{bmatrix} 0 & 0 \\ 0 & 1 \end{bmatrix}, $$

where $\beta$ is real positive. Determine the spectrum of $\mathscr{A}_0$, and find the residue idempotents, the spectrum of $\mathscr{P}(\mathscr{A}_1 + \mathscr{I})$, and the leading terms in the asymptotic expansions.

Solve Problem 13 if

14. $\mathscr{A}_0 = \begin{bmatrix} 0 & 0 \\ 0 & a \end{bmatrix}, \qquad \mathscr{A}_1 = \begin{bmatrix} 0 & 1 \\ a & -b \end{bmatrix}.$

15. $\mathscr{A}_0 = \begin{bmatrix} 0 & 1 \\ -1 & 0 \end{bmatrix}, \qquad \mathscr{A}_1 = \begin{bmatrix} 0 & 0 \\ 0 & -2a - 1 \end{bmatrix}.$

## LITERATURE

The relevant parts of the author's LODE are Appendix B and Chapter 6. See also Chapters 3–5 of:

Coddington, E. A. and Norman Levinson. *Theory of Ordinary Differential Equations.* McGraw-Hill, New York, 1955.

Also see Chapters 15–20 of:

Ince, E. L. *Ordinary Differential Equations.* Dover, New York, 1944.

This treatise has a brief discussion of the monodromic group on p. 389.
In connection with Theorem 9.4.6 see:

Birkhoff, G. D. A simplified treatment of the regular singular point. *Trans. Amer. Math. Soc.*, **11** (1910), 199–202.

For Lemma 9.4.1 and Theorem 9.4.9 see:

Koch, H. von. Un théorème sur les intégrales irrégulières des équations differentielles linéaires et son application au problème de l'intégration. *Ark. Mat., Astr., Fys.*, **13**, No. 15, (1918), 18 pp.

Perron, O. Über einen Satz des Herrn Helge von Koch über die Integrale linearer Differentialgleichungen. *Math. Zeit.*, **3** (1919), 161–174.

The discussion in Section 9.5 is an adaptation to the matrix case of Section 6.6 of LODE, and this goes back to three papers of the author. See:

Hille, E. Linear differential equations in Banach algebras. *Proceedings of the International Symposium on Linear Spaces*. Jerusalem, 1960, pp. 263–273.

――. Remarks on differential equations in Banach algebras. *Studies in Mathematical Analysis and Related Topics: Essays in Honor of George Pólya*. Stanford University Press, 1962, pp. 140–145.

――. Some aspects of differential equations in B-algebras. *Functional Analysis: Proceedings of the Conference at University of California, Irvine, 1966*, edited by B. R. Gelbaum. Thompson Book Co., Washington, D.C., 1967, pp. 185–194.

For a fuller discussion of the commutator operator, including the integral of Yu. L. Daletskij and the theorem of S. R. Foguel, see Section 4.6 of LODE.

There is an extensive literature on the Fuchsian class and the Riemann-Hilbert and Riemann-Birkhoff problems. The matrix theory is developed in great detail in:

Lappo-Danilevskij, I. A. *Mémoires sur la théorie des systèmes des équations différentielles linéaires*. 3 vols. Trav. Inst. Phys.-Math. Stekloff, 1934, 1935, 1936; Chelsea, New York, 1953.

Though not perfect, a convenient, readable presentation of the Riemann-Hilbert and Riemann-Birkhoff problems is found in:

Birkhoff, G. D. The generalized Riemann problem for linear differential equations and the allied problems for linear difference and *q*-difference equations. *Proc. Amer. Acad. Arts Sci.*, **49**, No. 9 (1913), 521–568.

This paper has a good account of the earlier work on these problems. Of later papers we mention:

Garnier, R. Sur le problème de Riemann-Hilbert. *Compositio Math.*, **8** (1951), 185–204.

Helson, H. Vectorial function theory. *Proc. London Math. Soc.*, (3), **17** (1967), 499–504.

Helson extends Birkhoff's equivalence problem for analytic matrices to analytic operator functions operating in a Hilbert space.

References for Section 9.7 are:

Birkhoff, G. D. Singular points of ordinary differential equations. *Trans. Amer. Math. Soc.*, **10** (1909), 436–470.

———. Equivalent singular points of ordinary linear differential equations. *Math. Ann.*, **74** (1913), 134–139.

———. A simple type of irregular singular point. *Trans. Amer. Math. Soc.*, **14** (1913), 462–476.

Fabry, E. Sur les intégrales des équations différentielles à coefficients rationnels. Thesis, Paris, 1885.

Gantmacher, F. R. *The Theory of Matrices.* Vol. 2, Chelsea, New York, 1959; Russian original, Moscow, 1953.

Horn, J. Integration linearer Differentialgleichungen durch Laplacesche Integrale und Fakultätenreihen. *Jahresb. Deut. Math. Ver.*, **24** (1915), 309–329.

———. Laplacesche Integrale, Binomialkoeffizientenreihen und Gammaquotientenreihen. *Math. Zeit.*, **21** (1924), 85–95.

Hukuhara, M. Sur les points singuliers des équations différentielles linéaires. II. *J. Fac. Sci. Hokkaido Imp. Univ.*, **5** (1937), 123–166. III. *Mem. Fac. Sci. Univ. Kyusyu*, **A2** (1942), 127–137.

Masani, P. On a result of G. D. Birkhoff on linear differential systems. *Proc. Amer. Math. Soc.*, **10** (1959), 696–698.

Okubo, K. A global representation of a fundamental set of solutions and a Stokes phenomenon for a system of linear ordinary differential equations. *J. Math. Soc. Japan*, **15** (1963), 268–288.

Thomé, L. W. Zur Theorie der linearen Differentialgleichungen. *J. Math.*, **74** (1872); **75, 76** (1873); and many later volumes. Summaries in **96** (1884) and **122** (1900).

Trjitzinsky, W. J. Analytic theory of linear differential equations. *Acta Math.*, **62** (1934), 167–227.

———. Laplace integrals and factorial series in the theory of linear differential and difference equations. *Trans. Amer. Math. Soc.*, **37** (1935), 80–146.

Turrittin, H. L. Stokes multipliers for asymptotic solutions of certain differential equations. *Trans. Amer. Math. Soc.*, **68** (1950), 304–329.

———. Convergent solutions of linear homogeneous differential equations in the neighborhood of an irregular singular point. *Acta Math.*, **93** (1955), 27–66.

———. Reduction of ordinary differential equations to the Birkhoff canonical form. *Trans. Amer. Math. Soc.*, **107** (1963), 485–507.

———. Reducing the rank of ordinary differential equations. *Duke Math. J.*, **30** (1963), 271–274.

———. Solvable related equations pertaining to turning point problems. *Symposium on Asymptotic Solutions of Differential Equations and Their Applications.* Publication No. 13, Mathematical Research Center, U.S. Army, University of Wisconsin, John Wiley, New York, 1964, pp. 27–52.

# 10
# THE SCHWARZIAN

The Schwarzian derivative enters into several branches of complex analysis. It plays an important role in the theory of linear second order DE's, in conformal mapping, and in the study of univalence and of uniformization. We shall have something to report on all these aspects. First a historical note is in order: the first "Schwarzian" seems to have occurred in a paper by Kummer listed in a Liegnitz Gymnasialprogramm of 1834. The paper, "De generali aequatione differentiali tertii ordinis," was reprinted in Crelle, Vol. 100, 1887. As we shall see in Section 10.5, the operator was also known to Riemann as early as 1857.

## 10.1. THE SCHWARZIAN DERIVATIVE

Hermann Amandus Schwarz (1843–1921) was Weierstrass's most brilliant pupil and ultimately became his successor at the University of Berlin. In the late 1860's he was engrossed in the problem of mapping circular polygons conformally. This led him to his famous *principle of symmetry* and also, in 1869, to the differential operator which later became known as the *Schwarzian derivative* or the *differential parameter* or, simply, the *Schwarzian*. What he wanted was a differential operator invariant under the *group of all fractional linear transformations*, also known as the *projective group*.

We denote the Schwarzian of $w$ with respect to $z$ by $\{w, z\}$. It is defined by

$$\{w, z\} = \left(\frac{w''}{w'}\right)' - \frac{1}{2}\left(\frac{w''}{w'}\right)^2. \tag{10.1.1}$$

This is obviously

$$\frac{d^2}{dz^2}\left(\log \frac{dw}{dz}\right) - \frac{1}{2}\left[\frac{d}{dz}\left(\log \frac{dw}{dz}\right)\right]^2 \tag{10.1.2}$$

and may also be written as

$$\frac{w'''}{w'} - \frac{3}{2}\left(\frac{w''}{w'}\right)^2.$$

The primes denote differentiation with respect to $z$. There are of course other differential operators invariant under the projective group, but in 1969 the late Israeli mathematician Meira Lavie showed that such an operator of order $n$ operating on a function $f$, under mild restrictions on $f$, is expressible rationally in terms of $\{f, z\}$ and its derivatives of order $\leq n - 3$. There are various considerations which lead to this operator, but in a treatise on ordinary differential equations perhaps the following mode of approach is the most natural one.

**THEOREM 10.1.1**

*Let $y_1$ and $y_2$ be two linearly independent solutions of the equation*

$$y'' + Q(z)y = 0, \tag{10.1.3}$$

*which are defined and holomorphic in some simply connected domain* $D$ *in the complex plane. Then*

$$w(z) = \frac{y_1(z)}{y_2(z)} \tag{10.1.4}$$

*satisfies the equation*

$$\{w, z\} = 2Q(z) \tag{10.1.5}$$

*at all points of* $D$ *where $y_2(z) \neq 0$. Conversely, if $w(z)$ is a solution of* (10.1.5), *holomorphic in some neighborhood of a point $z_0 \in D$, then one can find two linearly independent solutions, $u(z)$ and $v(z)$, of* (10.1.3) *defined in* $D$ *so that*

$$w(z) = \frac{u(z)}{v(z)}, \tag{10.1.6}$$

*and if $v(z_0) = 1$ the solutions $u$ and $v$ are uniquely defined.*

*Proof.* We may assume that the Wronskian of $y_1$ and $y_2$ is identically one. Then

$$w'(z) = [y_2(z)]^{-2},$$

so that

$$\frac{w''(z)}{w'(z)} = -2\frac{y_2'(z)}{y_2(z)}$$

and

$$\left[\frac{w''(z)}{w'(z)}\right]' = -2\frac{y_2''(z)}{y_2(z)} + 2\left[\frac{y_2'(z)}{y_2(z)}\right]^2 = 2Q(z) + \frac{1}{2}\left[\frac{w''(z)}{w'(z)}\right]^2,$$

from which the first assertion follows.

Next suppose that a solution of (10.1.5) is given by its initial values $w(z_0)$, $w'(z_0)$, $w''(z_0)$ at a point $z_0$ in $D$. Here we may assume that $w'(z_0) \neq 0$ since otherwise $Q(z)$ could not be holomorphic at $z = z_0$. We can now choose two linearly independent solutions, $u(z)$ and $v(z)$, of (10.1.3) so that at $z = z_0$ the quotient $w_1(z) = u(z)/v(z)$ has the initial values $w(z_0)$, $w'(z_0)$, $w''(z_0)$, the same as $w(z)$. If $v(z_0) = 1$, the solutions $u(z)$ and $v(z)$ are uniquely determined, and we must have $w(z) \equiv w_1(z)$. The details are left to the reader. ■

Since (10.1.5) is of the third order, its general solution should involve three arbitrary constants. Such a solution can be found via (10.1.3). If $y_1(z)$ and $y_2(z)$ are linearly independent solutions, if $a, b, c, d$ are constants such that $ad - bc = 1$, then $ay_1(z) + by_2(z)$ and $cy_1(z) + dy_2(z)$ are also linearly independent, and their quotient

$$\frac{ay_1(z) + by_2(z)}{cy_1(z) + dy_2(z)} = w(z) \tag{10.1.7}$$

is a solution of (10.1.5) involving three arbitrary constants. This situation suggests the truth of

**THEOREM 10.1.2**

*The Schwarzian is invariant under fractional linear transformations acting on the first argument*

$$\left\{\frac{aw + b}{cw + d}, z\right\} = \{w, z\}. \tag{10.1.8}$$

The proof is a matter of a computation and is left to the reader. Thus the Schwarzian is a differential invariant of the projective group. A change of independent variable obeys the law

$$\{w, z\} = \{w, t\}\left(\frac{dt}{dz}\right)^2 + \{t, z\}. \tag{10.1.9}$$

We may apply a Möbius transformation to the second argument instead of the first. The result is

$$\left\{w, \frac{az + b}{cz + d}\right\}\frac{(ad - bc)^2}{(cz + d)^4} = \{w, z\}. \tag{10.1.10}$$

The second factor on the left is the square of the derivative of the second argument with respect to $z$.

We have expressed the general solution of (10.1.5) in terms of solutions of (10.1.3). The converse is also possible, for if $y_1$, $y_2$ are solutions of (10.1.3) of Wronskian equal to one, and if $w = y_1/y_2$, then, as we have seen,

$$y_2(z) = [w'(z)]^{-1/2},$$

and $y_1(z) = w(z)y_2(z) = [w'(z)]^{-1/2}w(z)$, so that the general solution of (10.1.3) is

$$Ay_1(z) + By_2(z) = [w'(z)]^{-1/2}[Aw(z) + B]. \tag{10.1.11}$$

In the domain $D$ where $Q(z)$ is holomorphic, the solutions of (10.1.5) are evidently meromorphic with simple poles (why?) and these are movable singularities.

### EXERCISE 10.1

1. If $\{w, z\}$ is identically 0, show that $w$ is a linear fractional function of $z$.
2. If $f(z)$ is holomorphic at infinity, show that $z^4\{f, z\}$ is holomorphic there.
3. In the second half of the proof of Theorem 10.1.1, determine the initial values of $u(z)$ and $v(z)$ subject to the stated conditions.
4. Prove Theorem 10.1.2.
5. Prove (10.1.9).
6. If $f(z) = 4z(1 + z)^{-1}$, show that $\{f, z\} = -6(1 - z^2)^{-2}$ and in the unit disk $z_1 \ne z_2$ implies $f(z_1) \ne f(z_2)$, so that $f$ is univalent there.
7. If $f(z) = \log[(z - a)/(z - b)]$, $a \ne b$, find $\{f, z\} \equiv 2Q(z)$.
8. What is the general solution of $y'' + Q(z)y = 0$ with $Q$ as in Problem 7?
9. Prove that $\{w, z\} = -\{z, w\}\left(\dfrac{dw}{dz}\right)^2$.
10. Prove (10.1.9).

## 10.2. APPLICATIONS TO CONFORMAL MAPPING

The conformal mapping problem was formulated by Bernhard Riemann in the (later) famous dissertation of 1851. He based his existence proof on Dirichlet's principle in the calculus of variations. Weierstrass showed that the existence of a minimizing function could not be concluded in this way, and the mapping problem remained an open question until 1913, when rigorous proofs were given by Carathéodory, Koebe, and Osgood. In the meantime numerous attempts were made either to provide an existence proof under less general assumptions or to find explicit solutions for important special cases. In both attempts Schwarz was a pioneer.

Particularly important for the later development was his solution of the problem of mapping the upper half-plane conformally on the interior of a circular polygon (1869). It revealed a close connection between mapping problems and linear DE's of the Fuchsian class, which in its turn led to Henri Poincaré's discovery of *automorphic functions*, i.e., analytic functions $F(z)$ such that

$$F(Sz) = F(z), \tag{10.2.1}$$

where $S$ belongs to a given group $\mathfrak{G}$ of fractional linear transformations on $z$. Special examples already known before Poincaré's work in the 1880's are doubly periodic functions and elliptic modular functions. Under suitable assumptions regarding $\mathfrak{G}$, there exists a circular polygon $\Pi$, the interior of which acts as a *fundamental region* of the group, in the sense that the images of this region under the elements of the group cover without overlapping the domain of existence of $z \mapsto F(z)$. Moreover, the inverse of the mapping function of the fundamental region is an automorphic function under the group.

With this background information, let us proceed to the specific mapping problem. There is given a simple, closed polygon $\Pi$ in the finite $w$-plane with vertices $A_1, A_2, \ldots, A_n$ numbered in such a way that the interior of $\Pi$ lies to the left of an observer proceeding from $w = A_j$ to $A_{j+1}$, where $A_{n+1} = A_1$ and $j = 1, 2, \ldots, n$. The sides of the polygon are circular arcs or straight line segments. At $w = A_j$ the two adjacent sides make an angle of $\alpha_j \pi$ with each other in the interior of $\Pi$, where $0 < \alpha_j < 2$ and $\alpha_j \neq 1$.

It is possible to map the upper half of the $z$-plane conformally on the interior of $\Pi$, and the mapping function is unique up to a fractional linear transformation which leaves the upper half-plane invariant. Here the boundaries are in a one-to-one correspondence. To the vertex $A_j$ corresponds a point $a_j$ on the real axis, and, without loss of generality, we may suppose that

$$a_1 < a_2 < \cdots < a_n.$$

Actually it is possible to fix three of these points, say the first three, at preassigned points by a suitable choice of the mapping function. The remaining points are then uniquely determined. Let this choice of the mapping function be $z \mapsto f(z)$.

Whereas $f(z)$ is normally rather complicated, its Schwarzian $\{f, z\}$ is a simple rational function, with double poles at the points $a_j$, which vanishes at least to the fourth order at infinity. This implies that $f(z)$ is the quotient of two linearly independent solutions of a second order linear DE of the Fuchsian class. More precisely, we have

**THEOREM 10.2.1**

*If the* a's *and* $\alpha$'s *satisfy the above restrictions, then there exist real numbers* $\beta_j$ *such that*

$$\{f, z\} = \sum_{j=1}^{n} \left[ \frac{1}{2} \frac{1 - \alpha_j^2}{(z - a_j)^2} + \frac{\beta_j}{z - a_j} \right], \tag{10.2.2}$$

*where*

$$\sum_{1}^{n} \beta_j = 0, \sum_{1}^{n} (2\beta_j a_j + 1 - \alpha_j^2) = 0,$$

$$\sum_{1}^{n} [\beta_j a_j^2 + (1 - \alpha_j^2) a_j] = 0. \tag{10.2.3}$$

*Proof (sketch).* The correspondence between the upper $z$-half-plane and the interior of $\Pi$ is one-to-one. This implies that $f'(z) \neq 0$ for Im $(z) > 0$ and, thus, that $\{f, z\}$ exists and is holomorphic in the upper half-plane. On the open intervals $(a_j, a_{j+1})$ the mapping function $f(z)$ is continuous and takes values on the arc $(A_j, A_{j+1})$ of $\Pi$. The original formulation of Schwarz's principle of symmetry states that, if a function $z \mapsto g(z)$ is holomorphic in a domain $D$ in the upper half-plane bounded below by a segment $(a, b)$ of the real axis on which the boundary values are real and continuous, then $g$ admits of an analytic extension across $(a, b)$ into the symmetric domain $\bar{D}$ and $g(\bar{z}) = \overline{g(z)}$. The extended symmetry principle is obtained by applying a Möbius transformation to the $w$-plane. The principle then states that $f(z)$ can be continued into the lower half-plane across any of the line segments $(a_j, a_{j+1})$, and its value at $\bar{z}$ is obtained from its value at $z$ by *reciprocation* in the arc $(A_j, A_{j+1})$. Thus, if the center of the circle is at $w = b$ and its radius is $R$, then

$$f(\bar{z}) = b + \frac{R^2}{\overline{f(z) - b}}. \tag{10.2.4}$$

The values of $f(z)$ in the lower half-plane obtained by crossing $(a_j, a_{j+1})$ fill out the interior of a polygon $\Pi_j$ similar to $\Pi$ and obtained from it by a reciprocation in the arc $(A_j, A_{j+1})$. We can obtain the analytic continuation of $f(z)$ across each of the arcs $(A_j, A_{j+1})$, including $(A_n, A_1)$, and the result is a set of $n$ polygonal regions attached to the $n$ sides of $\Pi$. Each of these regions is the conformal image of a copy of the lower $z$-half-plane. The derivative $f'(z) \neq 0$ in each of these half-planes and, since the correspondence of the boundaries is one to one, $f'(z)$ also $\neq 0$ in each of the open intervals $(a_j, a_{j+1})$. This means that $\{f, z\}$ is holomorphic for each continuation.

This reciprocation process can now be repeated for each of the $n$ lower half-planes. From the $j$th lower half-plane we can cross $n - 1$ intervals $(a_k, a_{k+1})$ with $k \neq j$. This gives rise to $n - 1$ polygons attached to the previously free sides of the polygon $\Pi_j$. Over the $z$-plane a Riemann surface is generated, normally with infinitely many sheets, each sheet consisting of an upper and a lower half-plane joined along one of the line segments $(a_j, a_{j+1})$. The values of $f(z)$ at conjugate points are related by a reciprocation in one of the circular arcs. If $z$ describes a closed path in the plane surrounding one or more of the points $a_j$, the mapping function will normally not return to its original value but $f(z)$ is locally holomorphic in the finite plane except at the points $a_j$. It is also holomorphic at infinity, which on the sphere is an interior point of the arc of the great circle which joins $a_n$ with $a_1$. The Schwarzian $\{f, z\}$ is holomorphic everywhere in the finite plane, the points $a_j$ excepted. It is also holomorphic at infinity, since $f(z)$ has this property. Moreover, by Problem 10.1:2, $z^4\{f, z\}$ is holomorphic at infinity, and this requires that $\{f, z\}$ have a zero at infinity at least of order 4.

We have to show that $\{f, z\}$ is holomorphic in the whole plane, again excepting the points $a_j$—in fact, that $\{f, z\}$ is a rational function of $z$ with poles at $z = a_j, j = 1, 2, \ldots, n$. First let us note that $\{f, z\}$ is real on each of the intervals $(a_j, a_{j+1})$ regardless of the determination of the mapping function. This follows from Theorem 10.1.2, for the values of $f$ in the interval $(a_j, a_{j+1})$ lie on some circular arc, and we can find constants $a, b, c, d$ with $ad - bc = 1$ such that

$$\frac{af(z) + b}{cf(z) + d} \equiv F(z) \tag{10.2.5}$$

is real on the interval under consideration. We recall that a Möbius transformation maps circles and straight lines into circles and straight lines. In particular, we can find a transformation which carries the circular arc that is the carrier of the values of $f(z)$ into the real axis. But if $F$ is real on $(a_j, a_{j+1})$, so are all its derivatives and hence also $\{F, z\} = \{f, z\}$.

We come finally to a discussion of the singular points $a_j$, which correspond to the vertices $A_j$ of the original polygon. Here we know that the adjacent arcs meet at $w = A_j$, forming an angle $\alpha_j \pi$ with each other. For the discussion of the local properties of $\{f, z\}$ at $z = a_j$ it would be simpler if the sides were straight line segments rather than circular arcs. Again this can be achieved with the aid of a transformation of type (10.2.5), which does not change the Schwarzian. We take

$$F(z) = e^{i\omega} \frac{f(z) - w_0}{f(z) - w_1}. \tag{10.2.6}$$

Here $w_0$ is the vertex under consideration, the original $A_j$ or one of its homologues in the repeated reciprocation process. The two circles intersect at $w_0$ and at $w_1$, and $\omega$ is so chosen that one of the arcs $(w_0, w_1)$ is mapped on the positive real axis, while the other is mapped on a ray which makes the angle $\alpha_j \pi$ with the positive real axis. The corresponding mapping function $F(z)$ is then of the form

$$F(z) = (z - a_j)^{\alpha_j} H_j(z), \qquad (10.2.7)$$

where $H_j(z)$ is holomorphic at $z = a_j$ and different from zero there, and the power series expansion of $H_j(z)$ in powers of $z - a_j$ has real coefficients. This gives

$$\frac{F''(z)}{F'(z)} = \frac{\alpha_j - 1}{z - a_j} + b_j + O(z - a_j),$$

where normally $b_j \neq 0$ and is real. This shows that in a neighborhood of $z = a_j$ we have

$$\{f, z\} = \{F, z\} = \frac{1}{2} \frac{1 - \alpha_j^2}{(z - a_j)^2} + \frac{\beta_j}{z - a_j} + h_j(z), \qquad (10.2.8)$$

where $h_j(z)$ is holomorphic at $z = a_j$, and its only singularities are the points $a_k$ with $k \neq j$. From this we conclude that $\{f, z\}$ is holomorphic in the extended plane except for double poles at the points $a_j$.

It remains to prove that $\{f, z\}$, which is a rational function of $z$, actually equals the sum of its principal parts. To this end we form the difference

$$D(z) = \{f, z\} - \frac{1}{2} \sum_{j=1}^{n} \frac{1 - \alpha_j^2}{(z - a_j)^2} - \sum_{j=1}^{n} \frac{\beta_j}{z - a_j}. \qquad (10.2.9)$$

This difference is holomorphic at each of the points $a_j$, regardless of the determination of the mapping function considered. It is also holomorphic at infinity. This means that $D(z)$ is a constant, and since $D(z)$ is zero at infinity, it must be identically zero.

Now $\{f, z\}$ vanishes at least to the fourth order at infinity, and if the right-hand side of (10.2.2) is expanded in powers of $1/z$, conditions (10.2.3) express the fact that the terms in $1/z^m$ are missing for $m = 1, 2, 3$.

If $\{f, z\}$ is given by (10.2.2), the equation

$$y''(z) + \tfrac{1}{2}\{f, z\} y(z) = 0 \qquad (10.2.10)$$

is clearly of the Fuchsian class. Its only singularities are the points $a_1, a_2, \ldots, a_n$, and they are regular-singular points. ∎

### EXERCISE 10.2

1. Find the general form of a fractional linear transformation which leaves the upper half-plane invariant. What is the image of the real axis?

2. Consider

$$f(z) = \int_0^z (1 - t^2)^{-1/2} (4 - t^2)^{-1/2} \, dt,$$

where the radical equals $+\frac{1}{2}$ at $t = 0$. Show that this function maps the upper half-plane conformally on the interior of a rectangle in the upper half-plane with two vertices on the real axis.

3. Find the Schwarzian of the mapping function in Problem 2.

4. Show that the inverse of the mapping function in Problem 2 is a doubly periodic elliptic function, and find integrals representing the primitive periods, one real, the other purely imaginary.

5. Make a similar study of the mapping function

$$f(z) = \int_0^z (1 - t^3)^{-2/3} \, dt,$$

which is associated with an equilateral triangle.

6. Generalize so as to obtain a mapping function for a general circular polygon.

7. Consider the function $f(z) = 2z(1 + z^2)^{-1}$. Show that it maps the upper half-plane conformally on the whole $w$-plane cut along the line segment $(-1, +1)$. This is a mapping of the upper half-plane on the exterior of a degenerate polygon, namely, the double line from $-1$ to $+1$. Here there are two angles, each of opening $2\pi$ in the "interior" of the degenerate polygon.

8. Show that

$$z \mapsto f(z) = \frac{1}{\log z + bi}, \qquad b > 0,$$

with $\log z = \log r + i\theta$, $0 \le \theta \le \pi$, maps the upper half-plane conformally on the interior of a circular polygon bounded by two tangent circles, one inside the other. This is a case with $\alpha_1 = \alpha_2 = 0$ (horn angles).

9. Find the Schwarzian of this mapping function.

10. Find the general solution of the equation

$$y''(z) + \tfrac{1}{2}\{f, z\} y = 0.$$

11. Find the general solution of

$$\{w, z\} = \{f, z\},$$

where again $f$ is defined as in Problem 8.

12. Construct a mapping function for a domain outside of two externally tangent circles.

13. Let $y_1(z) = F(\frac{1}{2}, \frac{1}{2}, 1; z)$, $y_2(z) = iF(\frac{1}{2}, \frac{1}{2}, 1; 1 - z)$. Verify that the quotient $w(z) = y_2(z)/y_1(z)$ satisfies

$$\{w, z\} = \frac{1}{2} \left[ \frac{1}{z^2} + \frac{1}{(1 - z)^2} + \frac{1}{z(1 - z)} \right]$$

(the Schwarzian equation for the modular function).

14. Show that $w(z)$ maps the upper half-plane conformally on a "triangle" bounded by two parallel vertical lines and a half-circle joining the end points of the lines on the real axis, the triangle lying in the upper half-plane. There are three horn angles. The inverse of the mapping function is an automorphic function invariant under the group generated by the transformations

$$z \to z + 1, \qquad z \to \frac{1}{z}.$$

## 10.3. ALGEBRAIC SOLUTIONS OF HYPERGEOMETRIC EQUATIONS

The problem of finding when a hypergeometric DE has algebraic solutions was the point of departure of Schwarz's investigations, which led to so many profound results. This is of course only a special case of the more general problem of determining when an equation of the Fuchsian class of any order has algebraic solutions. The results are more striking and easier to grasp, however, in the hypergeometric case, to which we restrict ourselves.

There are really two problems:

1. When does a hypergeometric DE admit of an algebraic solution?
2. When are all solutions algebraic?

We start with the first problem.

1. Let the equation

$$z(1 - z)w'' + [\gamma - (\alpha + \beta + 1)z]w' - \alpha\beta w = 0 \qquad (10.3.1)$$

have an algebraic integral $w(z)$. This would normally not be single valued, but we may assume that the various functional elements that belong to the same center $z = z_0$ have a constant ratio. This implies that the logarithmic derivative $w'(z)/w(z)$ is single valued, and being an algebraic function of $z$ it is in fact rational:

$$\frac{w'(z)}{w(z)} = R(z). \qquad (10.3.2)$$

There can obviously be no logarithmic terms in the expansions of $w(z)$ at the singular points 0, 1, $\infty$. If $w(z)$ should be holomorphic at $z = 0$ and 1, then $w(z)$ must be a polynomial, and the indicial equation at infinity shows that one of the parameters $\alpha$ or $\beta$ is then required to be a negative integer or zero. Suppose that $\alpha = -n$. Then the solution is $w(z) = CF(-n, \beta, \gamma; z)$. Various special cases occur in Exercise 10.3.

Suppose that $w(z)$ is not a polynomial. We can then find rational numbers $a$ and $b$ such that

$$w(z) = z^a(z - 1)^b p(z), \qquad (10.3.3)$$

where $p(z)$ is a polynomial, for the rational function $R(z)$ can only have simple poles and it vanishes at $z = \infty$. Hence

$$R(z) = \frac{a}{z} + \frac{b}{z - 1} + \sum_{j=1}^{k} \frac{\alpha_j}{z - a_j}. \tag{10.3.4}$$

Indeed a multiple pole would imply an irregular singularity, which is absurd. Integration gives

$$w(z) = z^a (z - 1)^b \prod_{j=1}^{k} (z - a_j)^{\alpha_j},$$

and here the exponents must be positive integers, since otherwise there would be additional singularities. It is clear that $a$ and $b$ must be rational if $w(z)$ is to be algebraic. Again the indicial equations impose some conditions on $a$, $b$ and the degree $N$ of $p(z)$. We leave these considerations to the reader.

2. Suppose now that all solutions are algebraic and that $w_1(z)$ and $w_2(z)$ are linearly independent solutions. Their quotient

$$W(z) = \frac{w_1(z)}{w_2(z)} \tag{10.3.5}$$

is then also algebraic and satisfies the Schwarzian equation

$$\{W, z\} = \frac{1 - \lambda^2}{2z^2} + \frac{1 - \mu^2}{2(1 - z)^2} + \frac{1 - \lambda^2 - \mu^2 + \nu^2}{2z(1 - z)}, \tag{10.3.6}$$

where the relations between the two sets of parameters are

$$\lambda = 1 - \gamma, \qquad \mu = \gamma - \alpha - \beta, \qquad \nu = \alpha - \beta, \tag{10.3.7}$$

and

$$\begin{aligned}
\alpha &= \tfrac{1}{2}(1 - \lambda - \mu + \nu), \\
\beta &= \tfrac{1}{2}(1 - \lambda - \mu - \nu), \\
\gamma &= 1 - \lambda.
\end{aligned} \tag{10.3.8}$$

The general solution of (10.3.6) is then also algebraic, and it has the form

$$\frac{aW(z) + b}{cW(z) + d}, \qquad ad - bc = 1. \tag{10.3.9}$$

If we start at some point $z = z_0$ where $W(z)$ is holomorphic and describe a closed path returning to $z_0$, all determinations of the solution are of the form (10.3.9). Now, if the solution is to be algebraic, only a finite number of determinations can exist at $z = z_0$. This implies and is implied by the group of monodromy of (10.3.1) being finite. To each element of the group of monodromy corresponds a linear fractional transformation

$$w \to \frac{aw + b}{cs + d}.$$

(10.3.10)

Now (and this is the crucial observation), a group of linear fractional transformations can be interpreted as a group of *rotations* of the sphere via the stereographic projection of the plane on the sphere. Here the word "rotation" can be taken in the sense of either Euclidean or non-Euclidean geometry. Finite groups of rotations of the sphere are of the following types: (i) the identity, (ii) cyclic groups, (iii) dihedral groups, (iv) the tetrahedral group, (v) the octahedral group, and (vi) the icosahedral group.

It would take us too far afield to attempt a rigorous proof of these statements, but we shall try to indicate the various considerations involved and will enumerate the sets of parameters $\alpha$, $\beta$, $\gamma$ for which the hypergeometric DE has all-algebraic solutions.

The point of departure is Schwarz's *triangular functions* $s(\lambda, \mu, \nu; z)$, i.e., the solutions of (10.3.6). From the discussion in Section 10.2 we know that any determination of such a function maps the upper half-plane into a circular triangle with angles $\lambda \pi$, $\mu \pi$, $\nu \pi$. The mapping is one to one and conformal if the (real) parameters satisfy the inequalities $|\lambda| \leq 2$, $|\mu| \leq 2$, $|\nu| \leq 2$. If all solutions of (10.3.1) are to be algebraic, there can be no logarithmic terms in the expansions at the singular points. This implies, in particular, that $1 - \gamma$, $\gamma - \alpha - \beta$, and $\alpha - \beta$ cannot be zero; i.e., the parameters $\lambda$, $\mu$, $\nu$ are not zero, so no horn angles can be present. The monodromy group of (10.3.1) now appears as the group of transformations which are composed of an even number of reflections in the three circles that carry the circular arcs bounding the triangle in question. We may assume that these circles are distinct. If all three should coincide, the group is the identity. The case with two coincident circles and a third distinct from them is also trivial. Therefore let all three circles be distinct.

We now map the plane on the surface of a sphere by stereographic projection. Then the three circles in the plane are mapped on three circles on the sphere, and the group of monodromy becomes a group of *collineations* leaving the sphere invariant. Our three circles on the sphere are obtainable as intersections of three planes with the surface of the sphere. Here there are three separate cases according as the planes intersect in a point outside, on, or inside the sphere. In the first case the three circles in the plane have a common orthogonal circle which is left invariant by the monodromy group, and the latter becomes a *group of motions in the non-Euclidean plane* (the Lobatševskij geometry as realized by Poincaré). In the second case the common point of the three circles can be sent to infinity by a suitable linear transformation. The monodromy group then becomes a *group of motions in the Euclidean plane*. In the third case the point of intersection of the three planes can be sent to the center of the sphere by a suitable collineation. Now the monodromy group appears as a *group of rotations of the sphere*. The exceptional cases in which two or all three circles coincide can be brought under this heading as a group of rotations of the sphere. Here the finite groups of rotations of a sphere are well known. They may be cyclic or leave a dihedron on one of the regular polyhedra invariant. Note that the cube and the octahedron have the same group, and so also is the case with the dodecahedron and the icosahedron.

The finite groups of motion of the non-Euclidian geometry are cyclic. This is seen as follows. Linear fractional transformations are classified according to the nature of their fixed points and the motion around these points. There are *loxodromic, hyperbolic, parabolic,* and *elliptic* transformations. Only the last type can occur if the group is to be finite. Moreover, the group is generated by a single element $\mathfrak{S}$. If there is an element of the group, say $\mathfrak{T}$, which does not commute with $\mathfrak{S}$, it may be shown that $\mathfrak{S}\mathfrak{T}\mathfrak{S}^{-1}\mathfrak{T}^{-1}$ has a fixed point on the orthogonal circle, and this element of the group is either hyperbolic or parabolic, which is nonadmissible. A similar argument applies to a group of motions in the plane: if a group is not cyclic, it must contain a translation and this violates the assumption that the group is finite.

Once the finite rotation groups of the sphere are known, one can proceed to further study of the problem on hand. The mapping of the $z$-plane on the $s$-plane is obtained by analytic continuation of the function which maps the upper half-plane conformally on the basic circular triangle with angles $\lambda\pi$, $\mu\pi$, $\nu\pi$, the continuation being achieved by repeated reflections in the sides of the original and the image triangles. The resulting Riemann surface over the $s$-plane must cover it a finite number of times if the solutions of (10.3.6) are algebraic. If the $s$-plane has nowhere more than one covering, the mapping is one-to-one and $z$ is a single-valued function of $s$. The values of $\lambda$, $\mu$, $\nu$ for which such a single-valued inverse exists are listed in the table.

| $\lambda$ | $\mu$ | $\nu$ | |
|------|------|------|----------------------------------|
| 1 | 1 | 1 | Identical group |
| 1 | $1/n$ | $1/n$ | Cyclic group ($n$ positive integer) |
| 1/2 | 1/2 | $1/n$ | Dihedron ($n$ positive integer) |
| 1/2 | 1/3 | 1/3 | Tetrahedron |
| 1/2 | 1/3 | 1/4 | Cube (octahedron) |
| 1/2 | 1/3 | 1/5 | Icosahedron (dodecahedron) |

It is now possible to handle the general situation in which the solutions are algebraic but $s(z)$ does not have a single-valued inverse. Suppose that the general integral of (10.3.1) is algebraic, and consider the corresponding Schwarzian equation

$$\{s, Z\} = R(Z), \qquad (10.3.11)$$

where $R(Z)$ is rational. If the general solution of (10.3.1) is algebraic, so is the general solution of (10.3.11). Let $s = A(Z)$ be an algebraic solution of (10.3.11). Together with these equations, consider an equation of type (10.3.6), say,

$$\{s, z\} = r(z). \qquad (10.3.12)$$

It is assumed that this equation has the same group as (10.3.11) and that the parameters $\lambda$, $\mu$, $\nu$ coincide with one of the sets listed in the table. It has then an algebraic integral $s = s(z)$ with single-valued inverse. Thus $s(z)$ is an automorphic function of the monodromy group of (10.3.11). We can then form

$$z = z[A(Z)] \equiv Q(Z), \qquad (10.3.13)$$

which is algebraic and single valued and hence is a rational function. This implies that (10.3.1) is obtainable from (10.3.12) by making the substitution $z = Q(Z)$. From these considerations we obtain

**THEOREM 10.3.1**

*All algebraically integrable DE's of the Fuchsian class and second order can be obtained by taking an equation of type (10.3.6) with $\lambda$, $\mu$, $\nu$ taken from the table and setting $z = Q(Z)$ in the equation, where now $Q(Z)$ is an arbitrary rational function.*

We have to restrict ourselves to these brief indications of this beautiful theory, discovered and developed by H. A. Schwarz around 1870.

**EXERCISE 10.3**

1. Why is a single-valued algebraic function necessarily rational?
2. What light does (10.3.3) throw on the roots of the indicial equations and on permissible values of $a$ and $b$?

Among the hypergeometric equations having polynomial solutions, the following should be noted:

3. $\alpha = -n$, $\beta = n + 1$, $\gamma = 1$: Legendre polynomial in $t = \frac{1}{2}(1 - z)$.
4. $\alpha = -n$, $\beta = p + n$, $\gamma = a$, $t = \frac{1}{2}(1 - z)$: Jacobi polynomial.
5. For $p = 0$, $q = \frac{1}{2}$ we obtain the $n$th Čebyšev polynomial of the first kind:

$$T_n(z) = 2^{1-n} F[-n, n, \tfrac{1}{2}; \tfrac{1}{2}(1 - z)].$$

6. What orthogonality properties do these polynomials have on the interval $(-1, 1)$?

7. Show that for all polynomials $P_n(x) = x^n + \cdots$ of degree $n$ and for $x$ restricted to $(-1, 1)$ the maximum of the absolute value of $P_n(x)$ is a minimum for $P_n = T_n$.

8. A "dihedron" is a double pyramid with the base a regular polygon of $n$ sides. The top and the bottom vertices are at a vertical distance 1 from the center of the polygon. The dihedral group are the rotations about the diameter joining the top and bottom vertices. What is the order of the group?

9. What are the orders of the octahedral and the icosahedral groups?

10. Why are the groups of the cube and of the octahedron the same?

11. Answer the question of Problem 10 for the dodecahedron and the icosahedron.

12. In principle it would seem to be possible for the planes which have circular intersections with the sphere to have a line in common instead of a single point. Why can this situation not arise?

13. Find the values of $\alpha$, $\beta$, $\gamma$ which correspond to the parameters $\lambda$, $\mu$, $\nu$ of the table.

## 10.4. UNIVALENCE AND THE SCHWARZIAN

Suppose that $f(z)$ is holomorphic and *univalent* in a domain $D$ in the sense that $z_1 \neq z_2$, both in $D$, implies that $f(z_1) \neq f(z_2)$. A necessary condition for univalence is that $f'(z) \neq 0$ in $D$. Now, if $f$ is holomorphic and univalent in $D$, then $\{f, z\}$ exists and is holomorphic in $D$. Under special assumptions on $D$, it is possible to find other necessary or sufficient conditions on $\{f, z\}$ which imply univalence.

There are two domains which are usually considered in problems involving univalence, namely, the interior $K$ and the exterior $Y$ of the unit circle. Let $\mathfrak{S}$ denote the class of functions

$$z \mapsto f(z) = z + a_2 z^2 + a_3 z^3 + \cdots \qquad (10.4.1)$$

which are univalent in $K$, and let $\mathfrak{U}$ denote the functions

$$z \mapsto F(z) = z + A_0 + A_1 z^{-1} + A_2 z^{-2} + \cdots \qquad (10.4.2)$$

which are holomorphic and univalent in $Y$. There is a one-to-one correspondence between the elements of the two classes, for if

$$f(z) \in \mathfrak{S}, \qquad \text{then } \left[ f\left(\frac{1}{z}\right) \right]^{-1} \in \mathfrak{U} \qquad (10.4.3)$$

and vice versa. If $F(z) = [f(1/z)]^{-1}$, then

$$A_1 = a_2^2 - a_3. \qquad (10.4.4)$$

This will be useful below. We shall also need

**LEMMA 10.4.1**

*If* $F(z) \in \mathfrak{U}$, *then* $|A_1| \leq 1$.

This follows from Gronwall's *area theorem*,

$$\sum_{n=1}^{\infty} n |A_n|^2 \leq 1. \tag{10.4.5}$$

Actually this inequality also shows that the inequality $|A_1| \leq 1$ is the best possible, since

$$z + A_0 + e^{i\theta} z^{-1} \in \mathfrak{U}$$

for any choice of $\theta$. For a proof of the lemma consult Exercise 10.4.
The basic result here is a theorem proved by Z. Nehari (1949).

**THEOREM 10.4.1**

*If* $f(z)$ *is univalent in* K, *then*

$$|\{f, z\}| \leq 6(1 - |z|^2)^{-2} \tag{10.4.6}$$

*and* 6 *is the best possible constant.*

*Proof.* Take an arbitrary point $z = a$ in $K$, and consider the Möbius transformation,

$$Z = \frac{z + a}{1 + \bar{a}z} = a + (1 - |a|^2)(z - \bar{a}z^2 - \bar{a}^2 z^3 + \cdots) \equiv a + w, \tag{10.4.7}$$

which maps $K$ in a one-to-one manner on itself. It follows that $f(Z)$ is univalent in $K$. It does not belong to $\mathfrak{S}$, but

$$g(z) = \frac{f(Z) - f(a)}{f'(a)(1 - |a|^2)} = z + a_2 z^2 + a_3 z^3 + \cdots \tag{10.4.8}$$

does. The assertion of the theorem follows from a comparison of $a_2$ and $a_3$. By Taylor's theorem

$$f(a + w) = f(a) + f'(a)w + \tfrac{1}{2}f''(a)w^2 + \tfrac{1}{6}f'''(a)w^3 + \cdots.$$

Here we substitute the power series for $w$, expand the powers of $w$, and collect terms. By the Weierstrass double series theorem this is allowed for small values of $z$. We obtain

$$a_2 = -\bar{a} + \frac{1}{2}\frac{f''(a)}{f'(a)}(1 - |a|^2),$$

$$a_3 = \bar{a}^2 - \bar{a}\frac{f''(a)}{f'(a)}(1 - |a|^2) + \frac{1}{6}\frac{f'''(a)}{f'(a)}(1 - |a|^2)^2. \tag{10.4.9}$$

Thus

$$a_2^2 - a_3 = -\tfrac{1}{6}(1 - |a|^2)^2 \left\{\frac{f'''(a)}{f'(a)} - \frac{3}{2}\left[\frac{f''(a)}{f'(a)}\right]^2\right\}$$

$$= -\tfrac{1}{6}(1 - |a|^2)^2\{f, a\}.$$

By (10.4.4) and Lemma 10.4.1 this expression is at most one in absolute value. Since $a$ is arbitrary in $K$, inequality (10.4.6) follows. ∎

That the constant 6 is the best possible follows from Problem 10.1:4, according to which $z \mapsto f(z) = z(1 + z)^{-2}$ is an element of $\mathfrak{S}$ and $\{f, z\} = -6(1 - z^2)^{-2}$.

## COROLLARY

If $f(z)$ is holomorphic and univalent in the upper half-plane $y > 0$, then

$$|\{f, z\}| \le \tfrac{3}{2}y^{-2}, \qquad z = x + iy, \tag{10.4.10}$$

and the constant $\tfrac{3}{2}$ is the best possible.

*Proof.* This follows from (10.1.9). The upper $z$-half-plane is mapped on the interior of the unit circle in the $s$-plane by

$$\frac{z - i}{z + i} = s.$$

Univalence of $f(z)$ in the upper $z$-half-plane implies and is implied by univalence of the transformed function in the disk $K$, and a simple calculation gives (10.4.10). The function $z \mapsto z^2$ is univalent in the upper half-plane, and $\{z^2, z\} = \tfrac{3}{2}z^{-2}$.

A simpler necessary condition for univalence in $K$ is

$$\left|\frac{f''(z)}{f'(z)}\right| \le \frac{6}{1 - |z|^2}. \tag{10.4.11}$$

Here also the constant is the best possible, for

$$f(z) = \frac{z}{(1 + z)^2} \qquad \text{gives} \qquad \frac{f''(z)}{f'(z)} = \frac{2z - 4}{1 - z^2}$$

and the numerator goes to $-6$ when $z$ goes to $-1$.

These various conditions are necessary for univalence in the domain under consideration. Sufficient conditions are normally obtained by lowering the constant figuring in the inequality. Thus Theorems 8.3.3 and 8.3.4 and the comments thereon show that

$$|\{f, z\}| \leqslant 2(1 - |z|^2)^{-2}, \qquad |z| < 1, \tag{10.4.12}$$

and

$$|\{f, z\}| \leqslant \tfrac{1}{2} y^{-2}, \qquad z = x + iy, \quad 0 < y, \tag{10.4.13}$$

are sufficient for univalence of $f$ in the indicated domain. Condition (10.4.12) was found to be of importance in the theory of quasi-conformal mapping by L. Ahlfors and G. Weill in 1962.

Again by (8.3.19) and comments

$$|\{f, z\}| \leqslant \tfrac{1}{2}\pi^2, \qquad |z| < 1, \tag{10.4.14}$$

is a sufficient condition, as shown by Nehari. In all these cases the sufficient condition is the best of its kind in the sense that for any larger numerical constant there are nonunivalent functions satisfying the condition in question. We also note the Pokarnyi-Beesack condition:

$$|\{f, z\}| \leqslant 2(3 - \alpha)(1 - |z|^2)^{-\alpha}, \qquad 1 \leqslant \alpha \leqslant 2, \quad |z| < 1 \tag{10.4.15}$$

for univalence in the unit disk. See also (8.3.22).

S. Friedland and Z. Nehari (1970) have generalized this range of ideas by proving that, if $R(r)$ is a nonincreasing nonnegative function for which

$$\int_0^1 R(r)\, dr \leqslant 1, \tag{10.4.16}$$

then the condition

$$|\{f, z\}| \leqslant 2R(r)(1 - r^2)^{-2}, \qquad |z| < r, \tag{10.4.17}$$

is sufficient for univalence of $f$ in the unit disk. This condition is also sufficient if $R(r)(1 - r^2)^{-1}$ is nonincreasing in $[0, 1)$.

Other relations between the Schwarzian and univalence were explored by Rolf Nevanlinna in 1932. The results may be regarded as a partial converse of his second fundamental theorem as expressed by (4.5.19). It is a construction of meromorphic functions with preassigned defects. He posed and solved the following problem:

*Given* q $(\geqslant 2)$ *points* $a_1, a_2, \ldots, a_q$ *and* q *positive integers* $m_1, m_2, \ldots, m_q$ *with* $\Sigma_1^q m_j = p$, *find all analytic function* z(w) *with the following properties:*

*The function* w$\rightarrow$z(w) *is meromorphic in the* w-*plane punctured at* $a_1$, $a_2, \ldots, a_q$.

2.   z(w) *is univalent.*

3. *Over each of the points* $a_j$ *there are* $m_j$ *distinct logarithmic elements and possibly also regular elements.*

4. *The range of* $z(w)$ *is a simply connected domain* D.

If $q = 2$, we must have $m_1 = m_2 = 1$, $p = 2$, and

$$z(w) = A \log \frac{w - a_1}{w - a_2} + B \tag{10.4.18}$$

with arbitrary constants $A$ and $B$. For $q > 2$ the problem is solvable iff $2m_j \le p$ for each $j$. In this case the range of $z(w)$ must be the finite plane, so a unique single-valued inverse function $w(z)$ is defined in the finite plane. Furthermore, there exists a polynomial $P(z)$ of degree $p - 2$ such that

$$\{w, z\} = 2P(z). \tag{10.4.19}$$

Conversely, any choice of such a polynomial leads to functions $z(w)$ with such a set of properties.

Thus we have

$$w(z) = \frac{y_1(z)}{y_2(z)}, \tag{10.4.20}$$

where $y_1$ and $y_2$ are two linearly independent solutions of the DE

$$y''(z) + P(z)y(z) = 0. \tag{10.4.21}$$

The discussion in Section 5.4 shows that $y_1$ and $y_2$ are entire functions of $z$ of order $[\frac{1}{2}p]$ and of the normal type of this order. Thus $w(z)$ is a meromorphic function of the same order. Any choice of $y_1$ and $y_2$ subject to the stated conditions will generate a function $w(z)$ with the desired properties. The corresponding numbers $a_j$ are simply the values of $C$ for which the solution

$$y_1(z) - Cy_2(z) \tag{10.4.22}$$

is subdominant in one of the $p$ sectors at infinity (see Section 5.6). If for $C = a_j$ the solution is subdominant in $m_j$ sectors, the inverse function $z(w)$ has $m_j$ distinct logarithmic elements over $w = a_j$. If $p > 2$, there are in addition always infinitely many regular elements at $a_j$. The condition $m_j \le \frac{1}{2}p$ expresses the fact that the solution can be subdominant in at most half of the sectors. For the equation

$$y''(z) = n[nz^{2n-2} + (n - 2)z^{n-2}]y(z) \tag{10.4.23}$$

we have $p = 2n$, and the solution $e^{z^n}$ is subdominant in $n$ sectors. Thus the bound $m_j \le \frac{1}{2}p$ is sharp.

The value of $q \leqslant p$ and the integers $m_i$ are uniquely determined by the coefficients of $P(z)$. On the other hand, the points $a_j$ depend on the choice of $y_1$ and $y_2$. In this case passing from one fundamental system to another involves a fractional linear transformation of the $w$-plane and the points $a_j$. If $p = q = 3$, the $a_j$'s can be assigned prescribed values and $P(z)$ may be chosen as $-z$. For $p = 4$ we have $q = 3$ or 4, and (10.4.21) may be reduced to the Hermite-Weber equation involving a single arbitrary parameter. The exceptional case $q = 3$ arises iff the equation is

$$y''(z) + (2n + 1 - z^2)y(z) = 0,$$

where $n$ is an integer. For nonintegral values the cross ratio of the four critical points $a_i$ is a simple function of the parameter.

For further details we refer the reader to Nevanlinna's paper.

## EXERCISE 10.4

1. Verify (10.4.3) and (10.4.4).

2. The area theorem may be proved as follows. Let $F(z) \in \mathfrak{U}$ and $r > 1$; then the circle $|z| = r$ is mapped by $F(z)$ on a simple closed analytic curve $C_r$. The area enclosed by $C_r$ can be obtained by the methods of the calculus, say by using polar coordinates. A somewhat more sophisticated expression is

$$A(r) = \frac{1}{2} \int_0^{2\pi} \left\{ U \frac{\partial V}{\partial \theta} - V \frac{\partial U}{\partial \theta} \right\} d\theta, \qquad F(z) = U(re^{i\theta}) + iV(re^{i\theta}).$$

Show that this integral is the imaginary part of

$$\frac{1}{2} \int_{|z|=r} \overline{F(z)} F'(z) \, dz.$$

3. Show that

$$A(r) = \pi \left[ r^2 - \sum_{n=1}^{\infty} n |A_n|^2 r^{-2n} \right] > 0$$

for all $r > 1$, and use this to prove (10.4.5).

4. Verify (10.4.7)–(10.4.9).

5. Verify the corollary.

6. How is (10.4.11) obtained?

7. Show that $f(z)$ is univalent in the upper half-plane if

$$[x^2 + (y + 1)^2]^2 |\{f, z\}| \leqslant 2, \qquad z = x + iy, \quad 0 < y.$$

8. Show that another sufficient condition is

$$y [x^2 + (y + 1)^2] |\{f, z\}| \leqslant 4.$$

9. For $q = 2$, $z(w)$ is given by (10.4.18). Verify that this function has the desired properties. Find the corresponding equation (10.4.21). Show that $P(z) = -A^{-2}$, independently of $a_1$, $a_2$, and $B$. Determine these quantities if

$y_1 = \sinh{(z/2A)}$, $y_2 = \cosh{(z/2A)}$. What happens if $y_1, y_2 = \exp{(\pm z/2A)}$ instead?

10. For what values of $A$ ($>\frac{1}{2}\pi^2$) can there exist a function which is holomorphic and univalent in the unit disk where the supremum of $|\{f, z\}| = A$?

## 10.5. UNIFORMIZATION BY MODULAR FUNCTIONS

In 1904 the Austrian mathematician Wilhelm Wirtinger (1865–1945) addressed the Third International Mathematical Congress at Heidelberg regarding Riemann's lectures on the hypergeometric series and their importance. Riemann lectured twice on this topic at Göttingen, in the Wintersemesters of 1856–1857 and 1858–1859, and lecture notes exist of both courses. As could be expected, the two courses abounded in new ideas, most of which were to bear fruit much later.

In the first course the $P$-function was introduced, represented by Euler transform integrals, which were to appear in print in a posthumous paper by Jacobi in 1859. In the second course Riemann solved the special Riemann problem by constructing a system of functions which transform in a given manner when $z$ makes a circuit around given singular points, two or three in number. This was preceded by a brief exposition of the composition and reduction of linear substitutions, *terra incognita* to most mathematicians of the time. He also used the Schwarzian; he probably did not know of Kummer's paper of 1834, and in any case this was 10 years before Schwarz used it. Riemann treated conformal mapping of curvilinear triangles on the sphere (he seems to have been the first to use stereographic projection) or in the plane, the mapping function being the quotient of two linearly independent solutions of the DE. He noticed that the independent variable $z$ can be a single-valued function of the quotient, and that what later became known as the monodromy group implies that $z$ is invariant under a group of linear fractional transformations of the quotient. Riemann used the symmetry principle long before Schwarz. In his lectures he carried out the mapping of the fundamental region for the elliptic modulus $k^2$ effected by the quotient of the periods of the elliptic integral of the first kind. In addition, he determined some cases in which the mapping properties indicate that the solutions are algebraic functions of $z$.

Wirtinger's address must have been a splendid lecture. The printed paper, however, puts rather heavy demands on the reader and suffers from a lack of formulas and bibliographical references.

We shall take up here an investigation dating from 1888 by Erwin Papperitz (1857–1938), who carried out part of the Riemann program of which he could have had no direct knowledge. He considered the hypergeometric DE for the $P$-function which involves six parameters, the sum of which is one. The quotient of two linearly independent solutions is a solution of the equation

$$\{s, z\} = \frac{(1 - \lambda^2) + (\lambda^2 + \mu^2 - \nu^2 - 1)z + (1 - \mu^2)z^2}{2z^2(1 - z)^2}, \qquad (10.5.1)$$

where the left member is the Schwarzian with respect to $z$ of the integral quotient $s(\lambda, \mu, \nu; z)$. Here the three parameters are the differences $\lambda = \alpha - \alpha'$, $\mu = \beta - \beta'$, $\nu = \gamma - \gamma'$.

Felix Klein proved in 1878 that, if a new variable $k^2(\tau) = z$ is introduced in the equation, $s(\lambda, \mu, \nu: z)$ becomes a single-valued function of $\tau = iK'/K$, where

$$K = \tfrac{1}{2}\pi F(\tfrac{1}{2}, \tfrac{1}{2}, 1; k^2), \qquad K' = \tfrac{1}{2}\pi F(\tfrac{1}{2}, \tfrac{1}{2}, 1; 1 - k^2)$$

by (7.3.22) and (7.3.23). Actually, $\tau(k^2)$ is an $s$-function corresponding to the parameters $0, 0, 0$. The inverse function $k^2(\tau)$ is single valued. This theorem also occurs in Riemann's second course, unknown to Klein and to Papperitz. The latter gave a simple direct proof for Klein's theorem based on the properties of the group $\mathfrak{G}$ operating on $\tau$ which leaves $z = k^2(\tau)$ unchanged. This group is a subgroup of the *modular group* $\mathfrak{M}$, which contains all mappings of the upper $\tau$-half-plane into itself of the form

$$\tau \to \frac{a\tau + b}{c\tau + d}, \qquad ad - bc = 1, \qquad (10.5.2)$$

where $a, b, c, d$ are integers. Here $\mathfrak{M}$ is generated by

$$S(\tau) = \tau + 1, \qquad U(\tau) = -\frac{1}{\tau},$$

and the subgroup $\mathfrak{G}$ is generated by

$$S^2(\tau) = \tau + 2, \qquad US^2U(\tau) = \frac{\tau}{1 - 2\tau} \qquad (10.5.3)$$

and $(US)^3(\tau) = \tau$. Here $a$ and $d$ are odd integers, while $b$ and $c$ are even and of course subject to the condition $ad - bc = 1$. The fundamental region $B$ of $\mathfrak{G}$ is depicted in Figure 10.1, which also shows the image of $B$ in the $z$-plane under the mapping

$$k^2(\tau) = z. \qquad (10.5.4)$$

We have

$$\text{Int}\,(B) = \{\tau; |\tau + \tfrac{1}{2}| > \tfrac{1}{2}, |\tau - \tfrac{1}{2}| > \tfrac{1}{2}, -1 < \text{Re}\,(\tau) < 1, 0 < \text{Im}\,(\tau)\}. \qquad (10.5.5)$$

To complete $B$ we add that part of the boundary where $\text{Re}\,(\tau) \leq 0$.

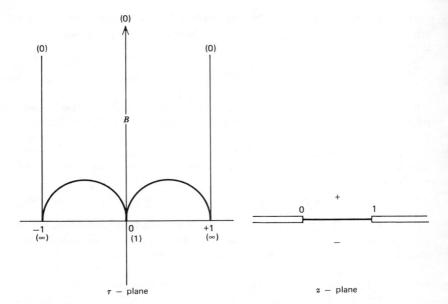

**Figure 10.1**

Incidentally this important figure occurs in Riemann's lecture notes for the second course.

That $B$ is indeed the fundamental region for $\mathfrak{G}$ follows from the fact that for any choice of $z$ in its plane there is one and only one root of (10.5.4).

Let us add some comments on the Klein-Riemann problem. For a given value of $\tau$ we have a unique value of $z$ but normally infinitely many determinations of $s(\lambda, \mu, \nu; z)$. The latter are related via elements of $\mathfrak{M}$. Any determination of $s$ can be obtained from a fixed other determination in one of two ways: either analytically by letting $z$ describe a suitable closed path in the $z$-plane surrounding one or more singular points of (10.5.1), or by applying a Möbius transformation belonging to $\mathfrak{G}$ operating on $\tau$. Now to each completed circuit in the $z$-plane corresponds an element $V$ of $\mathfrak{G}$ such that

$$k^2[V(\tau)] = k^2(\tau). \tag{10.5.6}$$

Thus, if there should be a value of $\tau$ such that to it would correspond two distinct values, $s_1$ and $s_2$, of $s[\lambda, \mu, \nu; k^2(\tau)]$, then we can carry one into the other, say $s_2$ into $s_1$, by one of the two operations indicated above. Here we must have $V(\tau) = \tau$. Papperitz concludes that this is impossible, since $\mathfrak{G}$ does not contain any periodic transformations. Both $U$ and $US$ are periodic, since $U^2 = I$ and $(US)^3 = I$, but neither $U$ nor $US$ belongs to $\mathfrak{G}$, though both belong to $\mathfrak{M}$.

Actually we have here a solution of the uniformization problem, since $k^2(\tau) = z$ and $s[\lambda, \mu, \nu; k^2(\tau)]$ are single-valued functions of $\tau$.

Papperitz goes much further, however. We denote the right member of (10.5.1) by $R(\lambda, \mu, \nu; z)$, so that the equation becomes

$$\{s, z\} = R(\lambda, \mu, \nu; z). \tag{10.5.7}$$

But we have also

$$\{\tau, z\} = R(0, 0, 0; z), \qquad z = k^2(\tau), \tag{10.5.8}$$

as observed above. Now we introduce $\tau$ as a new variable in (10.5.7) with the aid of (10.1.9), obtaining

$$\{s, z\} = \{s, \tau\}\left(\frac{d\tau}{dz}\right)^2 + \{\tau, z\}.$$

It follows that $s(\lambda, \mu, \nu; z)$ as a function of $\tau$ satisfies the DE

$$\{s, \tau\} = \{R[\lambda, \mu, \nu; k^2(\tau)] - R[0, 0, 0; k^2(\tau)]\}\left[\frac{dk^2(\tau)}{d\tau}\right]^2. \tag{10.5.9}$$

We denote the right member of this equation by $F(\lambda, \mu, \nu; \tau)$.

The analogue of (10.1.3) is now

$$\frac{d^2y}{dz^2} + \tfrac{1}{2}R(\lambda, \mu, \nu; z)y = 0. \tag{10.5.10}$$

If $y_1$ and $y_2$ are two linearly independent solutions of this equation, their quotient is a solution of (10.5.7). We can introduce $\tau$ as a new independent variable in this equation. We do not state the form of the resulting equation; instead we note the DE

$$\frac{d^2Y}{d\tau^2} + \tfrac{1}{2}F(\lambda, \mu, \nu; \tau)Y = 0. \tag{10.5.11}$$

If $Y$ is any solution of this equation, then

$$y = \left[\frac{dk^2(\tau)}{d\tau}\right]^{1/2} Y \tag{10.5.12}$$

is a solution of (10.5.11).

For the $F$-function we get by substitution

$$F(\lambda, \mu, \nu; \tau) = -\frac{\lambda^2(1-k^2) - \mu^2 k^2(1-k^2) + \nu^2 k^2}{2k^4(1-k^2)}\left(\frac{dk^2}{d\tau}\right)^2.$$

Here we need the theta null functions, for which see Exercise 7.3. We have

$$k^2 = \frac{\vartheta_2^4(0, \tau)}{\vartheta_3^4(0, \tau)}, \qquad 1 - k^2 = \frac{\vartheta_4^4(0, \tau)}{\vartheta_3^4(0, \tau)}. \qquad (10.5.13)$$

Furthermore, it may be shown that

$$\frac{d}{d\tau} \log k^2 = \pi i \vartheta_4^4(0, \tau). \qquad (10.5.14)$$

Combining, we find that

$$F(\lambda, \mu, \nu; \tau) = \tfrac{1}{2}\pi^2(\lambda^2 \vartheta_3^4 \vartheta_4^4 - \mu^2 \vartheta_4^4 \vartheta_2^4 + \nu^2 \vartheta_2^4 \vartheta_3^4), \qquad (10.5.15)$$

where we have omitted the $(0, \tau)$ for the sake of brevity.

The further substitution

$$q = \exp(\pi i \tau) \qquad (10.5.16)$$

leads to the equation

$$\frac{d^2 Y}{dq^2} + \frac{1}{q} \frac{dY}{dq} - \frac{1}{2\pi^2 q^2} F(\lambda, \mu, \nu; \tau) Y = 0. \qquad (10.5.17)$$

Here $F(\ldots)$ may be expanded in positive powers of $q$ convergent for $|q| < 1$; this means for $\tau$ in the upper half-plane. Indeed, the formulas in Exercise 7.3 show that the theta null functions are power series in $q$ with a factor $q^{1/4}$ added in the case of $\vartheta_1$ and $\vartheta_2$. Since we are concerned with fourth powers, we see that $F$ is a power series in $q$. Actually

$$\frac{1}{2\pi^2} F(\lambda, \mu, \nu; \tau) = \tfrac{1}{4}\lambda^2 + 4 \sum_{n=1}^{\infty} [\lambda^2(-q)^n + (-1)^n \mu^2 + \nu^2] n^3 \frac{q^n}{1 - q^{2n}}. \qquad (10.5.18)$$

It follows that the indicial equation of (10.5.17) is

$$\rho(\rho - 1) + \rho - \tfrac{1}{4}\lambda^2 = 0 \qquad \text{with roots } \tfrac{1}{2}\lambda \text{ and } -\tfrac{1}{2}\lambda.$$

Hence, if $\lambda$ is not an integer, equation (10.5.17) has two linearly independent solutions,

$$\begin{aligned} Y_1 &= q^{\lambda/2} G(\lambda, \mu, \nu, q), \\ Y_2 &= q^{-\lambda/2} G(-\lambda, \mu, \nu, q). \end{aligned} \qquad (10.5.19)$$

Here $G$ is a power series in $q$, convergent for $|q| < 1$, i.e., $\text{Im}(\tau) > 0$, and $G(\lambda, \mu, \nu, 0) = 1$. The coefficients are rational functions of $\lambda, \mu, \nu$ and

$$G(\lambda, \mu, \nu, q) = G(\lambda, \nu, \mu, -q)$$

identically.

Since

$$\frac{d}{d\tau} k^2 = (\pi i)^{1/2} \left\{ \frac{\vartheta_2 \vartheta_4}{\vartheta_3} \right\}^2, \tag{10.5.20}$$

by (10.5.13) and (10.5.14) we get two solutions of the hypergeometric equation (10.5.10):

$$y_1(\tau) = (\pi i)^{1/2} \left\{ \frac{\vartheta_2 \vartheta_4}{\vartheta_3} \right\}^2 q^{\lambda/2} G(\lambda, \mu, \nu, q),$$
$$y_2(\tau) = (\pi i)^{1/2} \left\{ \frac{\vartheta_2 \vartheta_4}{\vartheta_3} \right\}^2 q^{-\lambda/2} G(-\lambda, \mu, \nu, q). \tag{10.5.21}$$

Here the theta quotient is, aside from a factor $q^{1/2}$, a holomorphic function of $q$ in $|q| < 1$ for which we have the infinite product

$$4q^{1/2} \prod_1^\infty (1 - q^{2n})^2 \left[ \prod_1^\infty (1 + q^{2n})(1 - q^{2n-1}) \right]^4 \left[ \prod_1^\infty (1 + q^{2n-1}) \right]^{-4}. \tag{10.5.22}$$

Here all four products are absolutely convergent for $|q| < 1$, and, since the last factor does not become zero, the quotient is holomorphic for $|q| < 1$. As $\tau \to +i\infty$, $q \to 0$, and we see that

$$y_1(\tau) = q^{(1+\lambda)/2}[1 + O(q)],$$
$$y_2(\tau) = q^{(1-\lambda)/2}[1 + O(q)]. \tag{10.5.23}$$

Now we have

$$z = k^2(\tau) = 16q[1 + O(q)].$$

Combining these formulas, we see that, if we set

$$\alpha = \tfrac{1}{2}(1 + \lambda), \qquad \alpha' = \tfrac{1}{2}(1 - \lambda),$$

then $y_1$ and $y_2$ must be constant multiples of the two branches of the $P$-function which are prescribed at the origin. These two determinations of $P(\ldots)$ are then represented by formulas (10.5.21) in the whole $z$-plane, cut as indicated in the right half of Figure 10.1.

We can get similar representations for the two branches of $P$ at $z = 1$ and at $z = \infty$. For this purpose it is necessary to invoke more theory of theta functions, namely, how these functions transform if $\tau$ is replaced by $T(\tau)$, where $T$ is an element of the modular group. To study the solutions of (10.5.1) in a neighborhood of $z = 1$, we take $T = U$, i.e., $\tau \to -1/\tau$. We write $\vartheta_j(0, -1/\tau) = \theta_j$ and have the basic transformation formula

$$\vartheta_j = (-i\tau)^{-1/2} \theta_{6-j}. \tag{10.5.24}$$

This is now used to express $F(\lambda, \mu, \nu; \tau)$ as a function of the $\theta$'s. If we set

$$Q = \exp\left(-\frac{\pi i}{\tau}\right),$$

we can expand $F(\lambda, \mu, \nu; \tau)$ in powers of $Q$. In the resulting equation (10.5.17) $q$ is replaced by $Q$, and the corresponding indicial equation has the roots $\frac{1}{2}\mu$ and $-\frac{1}{2}\mu$. There will be solutions analogous to (10.5.19) with $q$ replaced by $Q$ and $\lambda$ by $\mu$. In the analogue of (10.5.21) the new theta quotient will also admit of a product expansion of type (10.5.22). Here we have

$$k^2(\tau) = 1 - 16Q[1 + O(Q)], \qquad (10.5.25)$$

which shows that we are dealing with the branches prescribed at $z = 1$. We refrain from further elaboration of this theme.

The function $s[\lambda, \mu, \nu; k^2(\tau)]$ has the real $\tau$-axis as a natural boundary. Papperitz examines in great detail the behavior of the function as $\tau$ approaches a rational real value $t$ along a path not tangent to the real axis. The $s$-function becomes infinite for such an approach, but the rate of growth depends on the arithmetical nature of $t$. Again we have to abandon the pursuit.

### EXERCISE 10.5

1. Show that the modular group is generated by $S$ and $U$.
2. Show that the subgroup $\mathfrak{G}$ is generated by $S^2$ and $US^2U$.
3. Try to show that the equation $k^2(\tau) = z$ has a solution in $B$ which is unique.
4. Verify (10.5.8).
5. Verify that (10.5.12) is a solution of (10.5.10).
6. Assuming the validity of (10.5.13) and (10.5.14), verify (10.5.15).
7. Verify (10.5.17).
8. Prove the absolute convergence of the infinite product in (10.5.22).
9. Verify (10.5.23) and the expression for $k^2(\tau)$.
10. Discuss $Q$ and (10.5.25).

### LITERATURE

We refer to Appendix D of the author's LODE.

The subject matter of Section 10.3 is treated at great length on pp. 40–147, Vol. II: 2, of:

Schlesinger, L. *Handbuch der Theorie der linearen Differentialgleichungen.* Teubner, Leipzig, 1898.

See also pp. 232–245 of:

Bieberbach, L. *Theorie der gewöhnlichen Differentialgleichungen.* Springer-Verlag, Berlin, 1953.

The references in the *Handbuch* are in the list of contents but lack titles and year; those of Bieberbach are very scanty, not to say, in some cases, lacking.

For Section 10.2 see also Section 17.6 of Vol. II of the author's AFT. Other references are:

Beesack, P. E. Non-oscillation and disconjugacy in the complex domain. *Trans. Amer. Math. Soc.*, **81** (1956) 211–242.

Friedland, S. and Z. Nehari. Univalence conditions and Sturm-Liouville eigenvalues. *Proc. Amer. Math. Soc.*, **24** (1970), 595–603.

Hille, E. Remarks on a paper by Zeev Nehari. *Bull. Amer. Math. Soc.*, **55** (1949), 552–553.

Lavie, M. The Schwarzian derivative and disconjugacy of *n*th order linear differential equations. *Can. J. Math.*, **21** (1969), 235–249.

Nehari, Z. The Schwarzian derivative and schlicht functions. *Bull. Amer. Math. Soc.*, **55** (1949), 544–551.

———. Some criteria of univalence. *Proc. Amer. Math. Soc.*, **5** (1954), 700–704.

———. Univalent functions and linear differential equations. *Lectures on Functions of a Complex Variable*. University of Michigan Press, Ann. Arbor, 1955, pp. 148–151.

Nevanlinna, R. Über Riemannsche Flächen mit endlich vielen Windungspunkten. *Acta Math.*, **58** (1932), 295–373.

Pokarnyi, V. V. On some sufficient conditions for univalence (Russian). *Doklady Akad. Nauk SSSR*, N.S., **79** (1951), 743–746.

Robertson, M. S. Schlicht solutions of $W'' + pW = 0$. *Trans. Amer. Math. Soc.*, **76** (1954), 254–274.

Schwarz, H. A. Über einige Abbildungsaufgaben. *J. reine angew. Math.*, **70** (1869), 105–120; *Ges. Math. Abhandl.*, **2** (1890), 65–83.

References for Section 10.5 are:

Klein, F. Ueber die Transformation der elliptischen Functionen und die Auflösung der Gleichungen fünften Grades. *Math. Ann.*, **14** (1878).

Papperitz, E. Ueber die Darstellung der hypergeometrischen Transcendenten durch eindeutige Functionen. *Math. Ann.*, **34** (1889), 247–296.

Wirtinger, W. Riemanns Vorlesungen über die hypergeometrische Reihe und ihre Bedeutung. *Sitzungsberichte der 3ten Internat. Math. Kongress, Heidelberg, 1904.* 19 pp.

The notation for theta functions varies; Papperitz used $\vartheta_0$ where we have $\vartheta_4$. Also, he wrote $SU$ where we have $US$. Papperitz has another paper fairly closely related to the paper here discussed.

Papperitz, E. Untersuchungen über die algebraische Transformation der hypergeometrischen Functionen. *Math. Ann.*, **27** (1886), 315–356.

# 11

# FIRST ORDER NONLINEAR DIFFERENTIAL EQUATIONS

We shall consider some problems relating to first order nonlinear DE's in this chapter. Second order nonlinear DE's will be discussed in Chapter 12. In both cases the theory is too extensive and has too many at least seemingly unrelated features to be covered in the space available; we have to be satisfied with a glance here and a nibble there. Before proceeding any further in this chapter, the reader is advised to review the first three sections of Chapter 3, where he or she will encounter a discussion of fixed and mobile singularities, the problem of continuing a solution analytically, and various results obtained by Paul Painlevé in respect to such questions. In the present chapter investigations by Charles Briot and Jean Claude Bouquet and by Pierre Boutroux will occupy the foreground.

## 11.1. SOME BRIOT-BOUQUET EQUATIONS

In Section 3.1 mention was made of equations of the form

$$\sum C_{jk} w^j (w')^k = 0 \tag{11.1.1}$$

with constant coefficients encountered by the brilliant French research team of Briot and Bouquet in their work on elliptic functions. We shall consider such equations again in this chapter, but at this juncture we shall examine another class of equations also considered by Briot and Bouquet

(1856). We take

$$zw' = pz + qw + F(z, w), \qquad (11.1.2)$$

where

$$F(z, w) = \sum_{j+k \geqslant 2} \sum c_{jk} z^j w^k, \qquad (11.1.3)$$

a function holomorphic in some dicylinder $|z| < R_1$, $|w| < R_2$. Here $z = 0$, $w = 0$ is a fixed singularity of the equation and is class 4 in the Painlevé classification (see Section 3.2). Here we have

$$P(z, w) = pz + qw + F(z, w), \qquad Q(z, w) = z. \qquad (11.1.4)$$

Both $P$ and $Q$ are holomorphic at $(0, 0)$, where both are zero.

Now the first question which the DE (11.1.2) poses is whether or not there are solutions holomorphic at $z = 0$. We prove

**THEOREM 11.1.1**

*If q is not a positive integer, then* (11.1.2) *has a unique solution* $z \mapsto w(z)$ *which is holomorphic at* $z = 0$ *and* $w(0) = 0$.

*Proof.* Let us first note that, if there is a solution holomorphic at $z = 0$, then $zw'(z)$ is also holomorphic and obviously zero there. If we replace $z$ by zero in the right member of (11.1.2), the result is a power series in $w$ starting with the term $qw$. It follows that a holomorphic solution at $z = 0$ must have the initial value $w(0) = 0$. Suppose that such a solution exists, and set

$$w(z) = \sum_{j=1}^{\infty} a_j z^j. \qquad (11.1.5)$$

Substitution gives

$$\sum_{j=1}^{\infty} j a_j z^j = pz + q \sum_{j=1}^{\infty} a_j z^j + \sum_{j+k \geqslant 2} \sum c_{jk} z^j \left( \sum_{n=1}^{\infty} a_n z^n \right)^k. \qquad (11.1.6)$$

Expanding, rearranging, and collecting terms, we get

$$\begin{aligned}
(1-q)a_1 &= p, \\
(2-q)a_2 &= c_{20} + c_{11}a_1 + c_{02}(a_1)^2, \\
&\cdots\cdots\cdots\cdots\cdots\cdots\cdots\cdots\cdots \\
(n-q)a_n &= M_n(c_{jk}; a_1, \ldots, a_{n-1}),
\end{aligned} \qquad (11.1.7)$$

where $M_n$ is multinomial in the indicated quantities, and $j + k \leqslant n$. Since $q$ is not a positive integer, the $a$'s may be found successively and uniquely.

Convergence of the formal series is proved by a Cauchy majorant argument involving an implicit function rather than a DE. Consider an equation of the form

$$BY = G(z, Y), \qquad (11.1.8)$$

where $B$ is a suitably chosen number, and

$$pz + qw + F(z, w) \ll G(z, w). \tag{11.1.9}$$

We can take

$$B \min_n |n - q| = 1, \tag{11.1.10}$$

$$G(z, w) = M\left(1 - \frac{z}{a}\right)^{-1}\left(1 - \frac{w}{b}\right)^{-1} - M. \tag{11.1.11}$$

Here $0 < a < R_1$, $0 < b < R_2$, and $M > \max(Bb, M_0)$, where

$$M_0 = \max_{j,k} |c_{jk}| a^i b^k, \qquad c_{10} = p, \quad c_{01} = q. \tag{11.1.12}$$

Equation (11.1.8) is now a quadratic in $Y$, and the root which is zero for $z = 0$ is

$$Y(z) = \frac{M - bB}{2B}\left\{-1 + (a - z)^{-1/2}\left[a - \left(\frac{M + bB}{M - bB}\right)^2 z\right]^{1/2}\right\}, \tag{11.1.13}$$

where the roots are real positive for $z = 0$. This function is holomorphic for

$$|z| < \bar{a} = a\left(\frac{M - bB}{M + bB}\right)^2 < a. \tag{11.1.14}$$

Let us now set

$$Y(z) = \sum_{n=1}^{\infty} A_n z^n \tag{11.1.15}$$

and determine the $A$'s so that (11.1.8) is satisfied by the series (11.1.15) identically. We have

$$BA_n = M_n(C_{jk}; A_1, \ldots, A_{n-1}). \tag{11.1.16}$$

Here $M_n$ is the same multinomial in the $C_{jk}$, and the $A_m$'s and all numerical coefficients are positive. We have $|c_{jk}| \le C_{jk}$ for all $j$ and $k$. Since

$$BA_1 = C_{10}, \qquad BA_2 = C_{20} + C_{11}A_1 + C_{02}(A_1)^2,$$

we have clearly $|a_1| \le A_1$, $|a_2| \le A_2$, and by complete induction

$$|a_n| \le A_n, \qquad \forall n. \tag{11.1.17}$$

It follows that the series (11.1.5) converges absolutely for $|z| < \bar{a}$ and is a solution of (11.1.2), holomorphic at the origin. Since the coefficients $a_i$ are uniquely determined, this is the only solution holomorphic at the origin. ∎

It is not, however, the only solution that is zero for $z = 0$. This is shown by the simple example

$$zw' = pz + qw, \qquad 0 < q. \tag{11.1.18}$$

Here the general solution is

$$-\frac{p}{q-1}z + Cz^a \qquad \text{if } q \neq 1,$$

$$Cz + pz \log z \qquad \text{if } q = 1.$$

If $q = 1$, $p \neq 0$, there is no solution holomorphic at $z = 0$, whereas if $q$ is a positive integer $> 1$, there are infinitely many solutions holomorphic at the origin and vanishing there.

We shall now examine the excluded case of $q$ a positive integer and $F(z, w) \not\equiv 0$. We start with $q = 1$.

**THEOREM 11.1.2**

*The equation*

$$zw' = pz + w + F(z, w) \qquad (11.1.19)$$

*has no solution holomorphic at* $z = 0$ *if* $p \neq 0$.

*Proof.* If $w(z) = \sum_0^\infty a_j z^j$ is a holomorphic solution, it must vanish for $z = 0$, so that

$$zw'(z) - w(z) = \sum_{j=2}^\infty (j - 1)a_j z^j = O(z^2).$$

Similarly $F(z, w) = O(z^2)$, so we must have $p = 0$ for the existence of a holomorphic solution. As we have seen above, there may exist nonholomorphic solutions which vanish at $z = 0$. ∎

We still have the problem of finding the general analytic, nonholomorphic solution at the origin. The special cases considered above make it plausible that the general solution admits of a representation in terms of *psi series* (see Section 7.1), a double series in $z$ and $z^a$ if $q$ is not a positive integer and a logarithmic psi series if $q$ is a positive integer. The existence of such expansions was proved in 1921 by Johannes Malmquist, whose work on the Riccati equation occupied us in Section 4.6. We state the result without proof.

**THEOREM 11.1.3**

*In a neighborhood of* $z = 0$ *the general solution of* (11.1.2) *is given by a convergent psi series of the form*

$$w(z) = \sum \sum a_{mn} c^n z^{m+nq} \qquad (11.1.20)$$

*if* $q$ *is not a positive integer and by*

$$w(z) = \sum \sum a_{mn} z^{m+nq} (c + h \log z)^n \qquad (11.1.21)$$

*in the integral case. Here* c *is an arbitrary constant and* h *is a fixed constant.*

Although we shall not prove this theorem (see Problem 14 below, however), we shall have more to say about the existence of holomorphic solutions in exceptional cases. We start with

**THEOREM 11.1.4**

*The equation*

$$zw' = w + F(z, w) \qquad (11.1.22)$$

*has infinitely many solutions holomorphic at the origin. Here* $w'(0)$ *is arbitrary.*

*Proof.* We assume the existence of a solution given by the power series (11.1.5). The coefficients can be determined from (11.1.7), but since $q = 1$ the leading coefficient $a_1$ is arbitrary. Once it has been chosen, all other coefficients are uniquely determined. The convergence proof carries over, and (11.1.10) shows that $B = 1$. The majorant requires $M$ to be chosen so large that $a|a_1| \leq M$. The rest of the proof carries over without change. ∎

If $q$ is a positive integer $> 1$, we can make a change of dependent variable so as to get an equation of the same type but with a lower value of $q$. We set in (11.1.2)

$$w = \frac{pz}{1-q} + zv. \qquad (11.1.23)$$

The result is

$$z^2 v' = (q-1)zv + F\left[z, \left(\frac{p}{1-q} + v\right)z\right].$$

Here the $F$-term is divisible by $z$ (actually by $z^2$), so we obtain

$$zv' = p_1 z + (q-1)v + F_1(z, v), \qquad (11.1.24)$$

an equation of the same form as (11.1.2) but with $q$ replaced by $q-1$. If $q > 2$, we repeat this type of substitution and after $q-1$ steps we have an equation with $q = 1$. Here there are two possibilities.

1. In the final equation the coefficient $p_{q-1} = 0$. In this case Theorem 11.1.4 applies and yields infinitely many solutions holomorphic at the origin. This gives, of course, infinitely many holomorphic solutions for the original equation (11.1.2). This case will arise iff the final coefficient $p_{q-1} = 0$—something that is not very likely to happen.

2. In the final equation $p_{q-1} \neq 0$. There are now no solutions which are holomorphic at the origin.

To round off the discussion, let us consider a very simple equation of the Briot-Bouquet type with an abominable solution. We take

$$z^2 w' = pz + qw, \qquad q \neq 0. \qquad (11.1.25)$$

The solution is elementary:

$$w(z) = \exp\left(-\frac{q}{z}\right)\left[p \int \exp\left(\frac{q}{t}\right)\frac{dt}{t} + C\right].$$

If $p = 0$, the solutions have an essential singularity at the origin. If $p \neq 0$, a logarithm is present and the solution takes the form

$$w(z) = \exp\left(-\frac{q}{z}\right)\left[p\left(\log z - \frac{q}{1!z} - \frac{q^2}{2!2z^2} - \cdots\right) + C\right].$$

Thus there are no solutions which are holomorphic in a domain containing the origin.

### EXERCISE 11.1

1. Is there any value of $p$ for which the equation

$$zw' = pz + 3w + w^2$$

has solutions holomorphic at the origin?

2. In utilizing the lower bound (11.1.14) for the radius of convergence of the series (11.1.5), the choice of $a$ and $b$ is usually quite arbitrary. In the case where the right member of (11.1.2) is a polynomial in $z$ and $w$, this is particularly the case and a problem of optimization presents itself. Try to find a reasonable bound for the equation $zw' = \frac{1}{2}(z + w) + (z + w)^2$.

3. The equation $zw' = aw + z^n$, where $n$ is a positive integer and $a \neq n$, has a unique solution holomorphic at the origin. Find this solution explicitly. What happens when $a \to n$?

4. Show that, if $w(z)$ is an entire function satisfying

$$w' = P(w),$$

where $P(w)$ is a polynomial in $w$ of degree $> 2$ having constant coefficients, $w(z)$ is a constant. Specify the constants which can give solutions.

In the following we consider some nonlinear matrix DE's, mostly generalized Riccati equations:

$$\mathcal{Y}'(z) = \mathcal{A}(z) + \mathcal{B}(z)\mathcal{Y}(z) + \mathcal{Y}(z)\mathcal{C}(z)\mathcal{Y}(z),$$

and the associate:

$$\mathcal{V}'(z) = \mathcal{C}(z) + \mathcal{V}(z)\mathcal{B}(z) + \mathcal{V}(z)\mathcal{A}(z)\mathcal{V}(z).$$

Here all matrices shall be elements of the algebra $\mathfrak{M}_n$ of $n$-by-$n$ matrices, and $\mathcal{A}$, $\mathcal{B}$, $\mathcal{C}$ shall be holomorphic except for poles in the finite plane which form a finite set $S$ of fixed singularities of the equations.

5. Show that, if $\mathcal{Y}(z)$ has an inverse in the domain of consideration, $\mathcal{V}(z) = -[\mathcal{Y}(z)]^{-1}$ is a solution of the associate equation.

6. The substitution $\mathcal{Y}(z) = -[\mathcal{C}(z)]^{-1}\mathcal{W}'(z)[\mathcal{W}(z)]^{-1}$ carries the first Riccati equation into a linear second order matrix equation for $\mathcal{W}(z)$. Find this DE.

7. The movable singularities of $\mathcal{Y}(z)$ are poles located at the points where $\mathcal{W}(z)$ is singular, and similarly for $\mathcal{V}(z)$, the poles of which occur where $\mathcal{W}'(z)$ is singular. $\mathcal{Y}(z)$ and $\mathcal{V}(z)$ may have poles in common. There is no limitation on the order of the poles. Try to verify these assertions.

8. Show that the equation $\mathcal{Y}'(z) = -[\mathcal{Y}(z)]^2$ is satisfied by all resolvents $\mathcal{R}(z, \mathcal{A}) = [z\mathcal{I} - \mathcal{A}]^{-1}$ wherever the inverse exists. Here $\mathcal{A}$ is any constant matrix in $\mathfrak{M}_n$, regular or singular.

9. If $\mathcal{B}$ is any regular matrix in $\mathfrak{M}_n$, find the solution of the equation in Problem 8 which takes the value $\mathcal{B}$ at $z = z_0$.

10. Show that the movable singularities of the equation in Problem 8 are poles and the sum of their orders is $n$.

11. If $\mathcal{P}$ is an idempotent matrix, $\mathcal{Q}$ a nilpotent one, $\mathcal{P}^2 = \mathcal{P}, \mathcal{P}\mathcal{Q} = \mathcal{Q}\mathcal{P} = \mathcal{Q}, \mathcal{Q}^n = 0$, show that a solution is given by

$$\mathcal{P}z^{-1} + \mathcal{Q}z^{-2} + \cdots + \mathcal{Q}^{n-1}z^{-n}.$$

12. The equation $(k - 1)\mathcal{U}'(z) = -[\mathcal{U}(z)]^k, k > 1$, may be reduced to the form of the equation in Problem 8 by setting $\mathcal{U}^{k-1} = \mathcal{Y}$. What is now the nature of the movable singularities?

13. With the notation of Problem 11, but with $n$ replaced by $m \leq n$ and $z$ by $z - a$, the resulting matrix $\mathcal{R}(z)$ is a solution of Problem 8. Find the expression for $[\mathcal{R}(z)]^{1/2}$. What DE does it satisfy?

14. If $q$ in (11.1.2) is neither a positive integer nor a real negative number and if $y(z)$ is the holomorphic solution at $z = 0$, then the solution (11.1.20) is obtainable by setting $w(z) = y(z) + z^q v(z)$. If

$$zv' = v \sum \sum a_{mn} z^{m+nq} v^n \tag{*}$$

and

$$\sum \sum |a_{mn}| a^{m+nQ} b^n \equiv M < \infty, \quad |q| < Q, \quad \text{integer,}$$

then (*) has a solution $v(z) = \sum \sum c_{mn} z^{m+nq}$ where the $c_{mn}$ are uniquely determined once $c_{00}$ has been chosen and the series for $v(z)$ has as a majorant the solution $V(z)$ with $V(O) = C_{00}$ of the algebraic equation

$$H(V - C_{00}) = MV\{(1 - z/a)^{-1}[1 - z^Q V/(a^Q b)]^{-1} - 1\}.$$

Here $H$ is suitably chosen and $C_{00} > |c_{00}|$. Verify!

## 11.2. GROWTH PROPERTIES

We shall now consider the equation

$$w' = R(z, w) = \frac{P(z, w)}{Q(z, w)} \tag{11.2.1}$$

with

$$P(z, w) = \sum_{j=0}^{p} P_j(z)w^j, \qquad Q(z, w) = \sum_{k=0}^{q} Q_k(z)w^k, \qquad (11.2.2)$$

where $P_j$ and $Q_k$ are polynomials in $z$. This is the equation which figured in Chapters 3 and 4. The fixed singularities of the equation are the zeros of $P_p(z)$ and $Q_q(z)$. These are finite in number, and $R$ is so chosen that all the finite fixed singularities will lie in the disk $|z| < R$. The nature of the movable singularities will depend on the relation between the orders $p$ and $q$. We start with the case

$$p = q + 1. \qquad (11.2.3)$$

The discussion in Section 4.6 [see in particular (4.6.7) and (4.6.8)] shows that there can be no movable infinitudes. On the other hand, there may be movable branch points where the solution approaches a finite limit.

Dividing $P(z, w)$ by $Q(z, w)$, we obtain an expression of the form

$$w' = \frac{P_{q+1}(z)}{Q_q(z)} w + S_0(z) + \frac{S_1(z, w)}{Q(z, w)}, \qquad (11.2.4)$$

where $S_0(z)$ is a rational function of $z$, and $S_1(z, w)$ is a polynomial in $z$ and $w$ of degree at most $q - 1$ in $w$. In the following it will be assumed that the degree of $P_{q+1}$ is at least equal to that of $Q_q$. We let $\mu \geq 0$ be the difference in the degrees and set

$$E_0(z) = \frac{P_{q+1}(z)}{Q_q(z)} = \frac{a}{b} z^{\mu} + \cdots, \qquad (11.2.5)$$

$$E(z) = \int^{z} E_0(s) \, ds = \frac{a}{b} \frac{z^{\mu+1}}{\mu + 1} + \cdots, \qquad (11.2.6)$$

where

$$P_{q+1}(z) = az^m + \cdots, \qquad Q_q(z) = bz^n + \cdots. \qquad (11.2.7)$$

Here $E_0(z)$ is clearly a rational function, and all its poles are inside the circle $|z| = R$. We may assume $R$ so large that the arcs

$$\text{Re}\,[E(z)] = 0 \qquad (11.2.8)$$

consist of $2\mu + 2$ separate arcs which proceed from the circle $|z| = R$ to infinity with no finite intersections. The arcs divide the $z$-plane beyond the circle into $2\mu + 2$ sectors $S_j$, where $\text{Re}\,[E(z)]$ keeps a constant sign. We number these cyclically $S_1, S_2, \ldots, S_{2\mu+2}$, where $\text{Re}\,[E(z)]$ is positive or negative in $S_j$, according as $j$ is even or odd.

We have now

$$\frac{w'(z)}{w(z)} - E_0(z) = \frac{S_0(z)}{w(z)} + \frac{S_1(z, w)}{w(z)Q(z, w)}. \qquad (11.2.9)$$

Let $z = z_0$ be a point in $S_{2j}$ for some $j$. Further restrictions will be imposed on $z_0$ when they become meaningful. Let $g$ be a positive number which will also be restricted later. Let $S_{2j}^*$ be the set

$$S_{2j}^* = \{z\,;\,s \in S_{2j},\, |w(z)| > |z|^g\}. \tag{11.2.10}$$

We shall see later that $S_{2j}^*$ is not void; in fact, it occupies the greater part of $S_{2j}$.

We now proceed to an estimate of the right member of (11.2.9), assuming that $z \in S_{2j}^*$. For $|z| > R_1 > R$,

$$S_0(z) = c_0 z^{k_1}[1 + o(1)], \tag{11.2.11}$$

so that

$$\frac{|S_0(z)|}{|w(z)|} < |z|^{\alpha - g}, \tag{11.2.12}$$

where $\alpha > \max(0, k_1 + 1)$.

The second fraction on the right in (11.2.9) is more intricate. The degree of $S_1(z, w)$ as a polynomial in $w$ is at most $q - 1$; to fix the ideas, suppose that it is exactly $q - 1$ and that

$$S_1(z, w) = F_0(z) + F_1(z)w + \cdots + F_{q-1}(z)w^{q-1}$$

and also

$$\max \deg [F_j] = k_2.$$

Then for large values of $z$, say for $|z| > R_1$,

$$\frac{|S_1(z, w)|}{|w(z)|\,|Q(z, w)|} < |w(z)|^{-2}|z|^\beta < z^{\beta - 2g}, \tag{11.2.13}$$

where $\beta > k_2$.

We now choose $g$ so that

$$g > \alpha + 2, \qquad 2g > \beta + 4, \tag{11.2.14}$$

and $z_0 \in S_{2j}^*$. The right-hand side of (11.2.9) is now less than $2|z|^{-2}$ for $z \in S_{2j}^*$. There is then a neighborhood of $z_0$ where (11.2.10) is satisfied. We now integrate (11.2.9) along a straight line from $z_0$ to $z$ in $S_{2j}^*$. Here we may assume that $E(z_0) = 0$. Thus

$$\log \frac{w(z)}{w(z_0)} - E(z) = h(z, z_0),$$

where the right-hand side is bounded and less than $4|z_0|^{-1}$ in absolute value if $|z| > |z_0|$. It follows that

$$w(z) = w(z_0) \exp[E(z)]H(z). \tag{11.2.15}$$

If $|z_0| > 4$, the factor $H(z)$ is bounded and lies between $e$ and $e^{-1}$ in absolute value. To start with, this holds in a neighborhood of $z_0$ where

$|w(z)| > |z|^g$. On the boundary of the subset of $S_{2j}^*$ where this inequality holds we must have equality. But we have also equality in (11.2.15). This requires that

$$|w(z_0)| \, |\exp [E(z)]| \, |H(z)| = |z|^g.$$

Here $w(z_0)$ is a fixed number, and $|H(z)|$ lies between $e$ and $e^{-1}$. It follows that the boundary of $S_{2j}^*$ must be asymptotic to the boundary of $S_{2j}$. Since $w(z)$ becomes infinite as an exponential function when $z$ recedes to infinity in the interior of the even-numbered sectors, Boutroux refers to this case as one of *exponential growth*. The analysis gives no information about what happens in the odd-numbered sectors.

Here we have supposed that $\mu = m - n \geqslant 0$. If this is not the case, the solutions may very well be bounded at infinity. An example is given by (11.1.25) with $p = 0$, i.e.,

$$z^2 w' = qw \qquad \text{with } w(z) = C \exp\left(-\frac{q}{z}\right).$$

Even if $p \neq 0$ in (11.1.25), the logarithmic rate of growth is not impressive. We state the proved results as

**THEOREM 11.2.1**

*If* $\mathrm{p} = \mathrm{q} + 1$ *and if* $\mu = \deg \mathrm{P}_{q+1} - \deg \mathrm{P}_q \geqslant 0$, *then all solutions of* (11.2.1) *are of exponential growth in the even-numbered sectors* $S_{2j}$:

$$w(z) = \exp [E(z)]H(z, z_0), \tag{11.2.16}$$

*where* $H(z, z_0)$ *is bounded away from zero and infinity, provided that*

$$|\exp [E(z)]| > C|z|^g. \tag{11.2.17}$$

*If* $\mu < 0$, *the solutions may be bounded in the neighborhood of infinity.*

We now turn to the case in which $p \neq q + 1$. There are actually two cases according as $p < q + 1$ or $p > q + 1$. In the first instance there are no mobile infinitudes, while in the second there are such. See Section 4.6, where the case $p = q + 2$ is treated.

We start with

Case 1: $q = p + m$, $m > 1$.

Here

$$w^m w' = \frac{P_p(z)}{Q_{p+m}(z)} \frac{1+S}{1+T}, \tag{11.2.18}$$

where

$$S = \sum_{j=0}^{P-1} \left(\frac{P_j}{P_p}\right) w^{j-p}, \qquad T = \sum_{k=0}^{p+m-1} \left(\frac{Q_k}{Q_{p+m}}\right) w^{k-m-p}. \tag{11.2.19}$$

We set

$$R_0(z) = \frac{P_p(z)}{Q_{p+m}(z)}, \qquad R(z) = \int^z R_0(s)\, ds. \qquad (11.2.19)$$

Here $R_0(z)$ is a rational function of $z$, all the poles of which are inside the circle $|z| = R$. If necessary, we make $R(z)$ single valued by radial cuts from the poles to infinity. We suppose that

$$\deg P_p - \deg Q_{p+m} = \mu \geq 0. \qquad (11.2.20)$$

To simplify matters still further, we restrict ourselves to the case in which

$$\deg P_j \leq \deg P_p, \qquad \deg Q_k \leq \deg Q_{p+m} \qquad (11.2.21)$$

for all $j$, $k$ under consideration.

Equation (11.2.18) may now be written as

$$w^m w' - R_0(z) = R_0(z)\frac{S - T}{1 + T}. \qquad (11.2.22)$$

Let $g$ be a positive integer to be disposed of later, and consider the set

$$\mathfrak{S} = \{z\,; R < |z|, |w(z)| > |z|^g\}. \qquad (11.2.23)$$

Let

$$M = \sup \left\{\frac{|P_j(z)|}{|P_p(z)|}, \frac{|Q_k(z)|}{|Q_{p+m}(z)|}\right\}, \qquad \begin{array}{l} j = 0, 1, \ldots, p - 1, \\ k = 0, 1, \ldots, p + m - 1 \end{array} \qquad (11.2.24)$$

and $z$ be restricted to $|z| > 2R$. Since the ratios tend to constant limits when $z \to \infty$, it is seen that $M$ is finite. Hence for $z \in \mathfrak{S}$

$$|S - T| < 2M[|z|^{-g} + |z|^{-2g} + \cdots + |z|^{-(p+m)g}]$$
$$< 2M|z|^{-g}[1 - |z|^{-g}]^{-1} < 2M(|z|^g - 1)^{-1} < 4M|z|^{-g}$$

if $|z|^g > 2$. Furthermore,

$$|1 + T| \geq 1 - |T| > 1 - 2M[|z|^g - 1]^{-1} > \tfrac{1}{2}$$

if $|z|^g > 4M + 1$. It follows that the right member of (11.2.22) for $z$ in $\mathfrak{S}$ does not exceed

$$8M|R_0(z)|\,|z|^{-g}. \qquad (11.2.25)$$

Suppose that $z_0 \in \mathfrak{S}$ and $z$ is a point of $\mathfrak{S}$ visible from $z_0$. We integrate (11.2.22) along the line segment from $z_0$ to $z$ and obtain

$$\frac{1}{m + 1}\{[w(z)]^{m+1} - [w(z_0)]^{m+1}\} = O[|z|^{\mu+1-g}] \qquad (11.2.26)$$

so that

$$w(z) = O[|z|^{(\mu+1-g)/(m+1)}]. \qquad (11.2.27)$$

This estimate will contradict our basic hypothesis (11.2.23) unless

$$\frac{\mu + 1 - g}{m + 1} \ge g.$$

Here we can take equality so that $g = (\mu + 1)/(m + 2)$ and

$$w(z) = O[|z|^{(\mu+1)/(m+1)}]. \tag{11.2.28}$$

This estimate holds in $\mathfrak{S}$. Outside $\mathfrak{S}$ the inequality (11.2.23) is reversed. Thus we have proved

**THEOREM 11.2.2**

*If* $q = p + m$, $1 < m$, *and if the degrees of the polynomials* $P_j$ *and* $Q_k$ *satisfy the conditions* (11.2.20) *and* (11.2.21), *then as z becomes infinite the solutions satisfy*

$$|w(z)| < C|z|^{(\mu+1)/(m+1)}. \tag{11.2.29}$$

Boutroux refers to this case as one of *rational growth*. We shall say that the equation is of the *power type*.

We turn now to

Case 2: $p = q + m$, $1 < m$.
Here we have movable branch points where the solutions become infinite. By (4.6.9) at such a point

$$w(z) = a(z - z_0)^{-1/(m-1)}[1 + o(1)], \tag{11.2.30}$$

where $a$ depends on $m$, $P_{q+m}(z_0)$, and $Q_q(z_0)$. We divide $P(z, w)$ by $Q(z, w)$ and obtain

$$w' = A_0(z)w^m + \cdots + A_m(z) + \frac{P_1(z, w)}{Q(z, w)}. \tag{11.2.31}$$

Here the $A$'s are rational functions of $z$, and $P_1(z, w)$ is a polynomial in $z$ and $w$ of degree at most $q - 1$ in $w$. Compare Section 4.6, where $m = 2$. Since $w(z)$ is not single valued, we have to apply a system of radial cuts from the fixed singularities and from the branch points of the solution. The cuts extend to infinity.

We have now

$$\frac{w'}{(w)^m} = A_0\left[1 + \frac{A_1}{A_0 w} + \cdots + \frac{A_m}{A_0(w)^m} + \frac{P_1(z, w)}{A_0(w)^m Q(z, w)}\right] \tag{11.2.32}$$

and proceed as in Section 4.6, the object being to make the right-hand side of (11.2.32) into an expression of the form $A_0[1 + h(z)]$, where $|h(z)| < \frac{1}{2}$ in a subset of the distant part of the cut plane. Let $g$ be a constant to be determined, and set

$$S = [z ; R < |z|, |w(z)| > |z|^g].$$                                     (11.2.33)

This set $S$ obviously contains the infinitudes in the domain under consideration. To find the desirable limitations on $g$ we proceed as in earlier discussions of this nature. We can find numbers $a_1, a_2, \ldots, a_{m+1}$ such that for $z \in S$

$$\frac{|A_j|}{|w^j| |A_0|} < |z|^{a_j - jg}, \qquad j = 1, 2, \ldots, m,$$     (11.2.34)

$$\frac{|P_1(z, w)|}{|w|^m |A_0| |Q(z, w)|} < |z|^{a_{m+1} - (m+1)g}.$$

It is now assumed that $g$ is so large that (i) all the exponents in the right members of these inequalities are negative, and (ii)

$$R^{a_j - jg} < \frac{1}{2(m + 1)}$$

Then for $z \in S$

$$\frac{w'(z)}{[w(z)]^m} = A_0(z)[1 + h(z)],$$                                (11.2.35)

where $|h(z)| < \frac{1}{2}$. Here we assume that

$$A_0(z) = \frac{P_{q+m}(z)}{Q_q(z)} = Az^\mu [1 + o(1)], \qquad \mu > -1,$$   (11.2.36)

so that

$$\int_{z_0}^z A_0(a) \, ds = \frac{A}{\mu + 1} (z^{\mu+1} - z_0^{\mu+1})[1 + o(1)].$$

Thus for $z_0$ and $z$ in $S$, $z$ visible from $z_0$ in $S$, we get

$$\tfrac{1}{2}(m - 1)|A(z) - A(z_0)| < |[w(z)]^{1-m} - [w(z_0)]^{1-m}|$$
$$< \tfrac{3}{2}(m - 1)|A(z) - A(z_0)|.$$                                   (11.2.37)

In particular, if $z_0$ is an infinitude of $w(z)$, the term $[w(z_0)]^{1-m}$ drops out, and we have an inequality for $[w(z)]^{1-m}$.

We can now introduce *infinitudinal neighborhoods* of the form

$$U_n = [z; |z - z_n| < K |z_n|^{-(m-1)g - \mu}].$$                          (11.2.38)

It is possible to choose constants $K$ and $R_0$ so that these neighborhoods do not overlap. The argument is analogous to that given in Section 4.6 for $m = 2$ but considerably more complicated. The only case of interest is that in which there are infinitudes clustering at infinity. We can then find a point $z_0$ where

$$|w(z_0)| > |z_0|^{g+1}.$$

If all the points on the circle $|z - z_0| = |z_0|^{-g}$ satisfy the inequality $|w(z)| > |z|^g$, we may assume that the inequality holds also inside the circle. This means that the disk $|z - z_0| < |z_0|^{-g}$ belongs to $S$. If this is not the case, we can find a concentric circle with a smaller radius passing through a point $z = t$ such that $|w(z)| > |z|^g$ in the smaller disk and $|w(t)| = |t|^g$. Then the smaller disk belongs to $S$, and we can use (11.2.37) with $z = t$. Here

$$|A(t) - A(z_0)| < 2|A|[|z_0| + |t - z_0|]^\mu |t - z_0|$$
$$< 2|A|[|z_0| + |z_0|^{-g}]^\mu |t - z_0|$$
$$= 2|A| |z_0|^\mu [1 + |z_0|^{-g-1}]^\mu |t - z_0|.$$

From (11.2.37) we then get

$$|t - z_0| > (3|A|)^{-1}|z_0|^{-\mu}[1 + |z_0|^{-g-1}]^{-\mu}$$
$$\times (m - 1)^{-1}|z_0|^{-(m-1)g}\left\{\left|\frac{t}{z_0}\right|^{-(m-1)g} - |z_0|^{-(m-1)g}\right\}.$$

Here the factor $[1 + |z_0|^{-g-1}]^{-\mu}$ exceeds $\frac{1}{2}$ when $|z_0|$ exceeds some number depending on $g$ and $\mu$. The same is true for the last factor, since the maximum of $|t/z_0| = 1 + |z_0|^{-g}$ and $|z_0|^{-(m-1)g}$ is small. This means that we can find an $R_0$ depending on $g$, $m$, $\mu$ such that for $|z_0| > R_0$ (11.2.38) will hold with $K > [12|A|(m - 1)]^{-1}$. If $K$ and $R_0$ have been chosen in this manner, then $U_n$, the infinitudinal neighborhood of the infinitude at $z_n$, belongs to $S$ and by (11.2.37) these neighborhoods cannot overlap. We have thus proved ∎

**THEOREM 11.2.3**

*If* $p = q + m$, $1 < m$, *and if* deg $P_{q+m} -$ deg $Q_q = \mu > -1$, *then the solutions of* (11.2.1) *have (movable) infinitudes where* w(z) *becomes infinite as indicated by* (11.2.30). *With each infinitude* $z = z_n$ *there is an infinitudinal neighborhood given by* (11.2.38). *These neighborhoods do not overlap in the same sheet of the Riemann surface on which the solution is single valued. The diameter of* $U_n$ *goes to zero as* $n \to \infty$. *Outside these neighborhoods* w(z) *satisfies an inequality of the form* $|w(z)| < |z|^g$. *The infinitudes are branch points where* $m - 1$ *branches commute.*

This is again a case of equations of the power type, in the modified terminology of Boutroux.

**EXERCISE 11.2**

1. For the equation $w' = z^3 - w^3$ we have $p = 3$, $q = 0$. Show that the finite singularities are algebraic branch points where $(z - z_0)^{1/2}w(z) \to \pm 2^{-1/2}$ as $z \to z_0$.

2. Show that the equation in Problem 1 has a solution which tends to $+\infty$ when $z = x \downarrow 0$. Show that this solution is positive for $x > 0$, has a single positive minimum, and becomes infinite with $x$.

3. More precisely, show that $w(x) = x - \frac{1}{3}x^{-2} + O(x^{-5})$.

4. Find a suitable $g$ for this equation, and estimate $K$ and $R_0$ in (11.2.38).

5. Try to estimate the rate of growth of $|w(z)|$ as $z \to \infty$, avoiding the branch points. Problem 3 suggests $O(|z|)$ for the growth. Is this right or wrong?

6. Another equation of Boutroux's power type is $w' = z^2 w - w^3$. This is a Bernoulli equation and hence reducible to a linear equation and solvable by elementary methods. The solution with $w(0) = 1$ is

$$w(z) = \exp\left(\tfrac{1}{3}z^3\right)\left[1 + 2\int_0^z \exp\left(\tfrac{2}{3}s^3\right) ds\right]^{-1/2}.$$

Here the branch points are given by the zeros of the second factor. Define six sectors, $S_k = [z ; \tfrac{1}{6}(2k - 3)\pi < \arg z < \tfrac{1}{6}(2k - 1)\pi]$, $k = 1, 2, \ldots, 6$. Show that the branch points lie in $S_1, S_3, S_5$ and that the sequences of the infinitudes approach the boundary lines of the sectors. Show that $w(z)$ goes exponentially to zero in $S_2, S_4, S_6$ and $w(z)/z \to 1$ in the interior of the remaining sectors, omitting small neighborhoods of the branch points.

Next we present two DE's with elementary integrals which are of the exponential type.

7. Take

$$w' = \frac{zw^2 + 1}{2w}$$

with

$$w(z) = \exp\left(\tfrac{1}{4}z^2\right)\left[\int_0^z \exp\left(-\tfrac{1}{2}s^2\right) ds + C\right]^{1/2}.$$

Verify. The lines $y = \pm x$ divide the plane into four sectors. Determine the behavior of the solutions in each sector. Where are the branch points? Determine their nature, and find roughly how many there are in the disk $|z| < r$ as $r \to \infty$.

8. Take

$$w' = \frac{1}{z}\frac{w^2 + zw + z^2}{w + z} \quad \text{with } w(z) = z[-1 \pm (2 \log Cz + 1)^{1/2}].$$

Verify. Determine the behavior of the solutions for large values of $z$. Where are the branch points, and what is their nature?

## 11.3. BINOMIAL BRIOT-BOUQUET EQUATIONS OF ELLIPTIC FUNCTION THEORY

We consider equations of the form

$$(w')^n = P(w), \tag{11.3.1}$$

where $P(w)$ is a polynomial in $w$ of degree $m$ and with constant coefficients. It is assumed that $P(w)$ is not the $k$th power of a polynomial where $k$ is a divisor of $m$. We want to determine the equations of this type which have a single-valued solution and, in particular, the equations which are satisfied by elliptic functions (= doubly periodic meromorphic functions). There are a number of trivial cases to put aside.

For $n = 1$ we know that the equation must be a Riccati equation. There are essentially three subcases:

1:1.  $P(w) = A(w - a)$ with the solution

$$w(z) = a + (w_0 - a) \exp [A(z - z_0)],  \tag{11.3.2}$$

which is single valued and periodic and is an entire function.

1:2.  $P(w) = A(w - a)(w - b)$, $a \neq b$, with a solution which is periodic and meromorphic but not elliptic.

1:3.  $P(w) = A(w - a)^2$, with a solution which is rational with a single pole.

For $n = 2$ we have a greater variety and even encounter elliptic functions. By Yosida's extension of Malmquist's theorem, $m$ is at most 4 if the solution is to be single valued and meromorphic. For $m = 1$, the solution is a quadratic polynomial. For $m = 2$, $P(w) = A(w - a)(w - b)$, $a \neq b$, the solution is simply periodic and meromorphic. For $m = 3$, we can discard the DE's

$$(w')^2 = w^3 \quad \text{and} \quad (w')^2 = (w - a)(w - b)^2.  \tag{11.3.3}$$

The first one is satisfied by a rational function $4(z - C)^{-2}$, while the second has the solution

$$w(z) = b + \{\exp [(b - a)^{1/2}(z - z_0)] - 1 + (w_0 - b)^{-1/2}\}^{-2}.$$

If the cubic polynomial $P(w)$ has three distinct roots, a linear transformation on $z$ and on $w$ will bring the equation into the form

$$(w')^2 = 4(w - e_1)(w - e_2)(w - e_3), \quad e_1 + e_2 + e_3 = 0,  \tag{11.3.4}$$

the general solution of which is $\wp(z - z_0; g_2, g_3)$, the Weierstrass $\wp$-function where $g_2$ and $g_3$ are symmetric functions of the roots $e_j$. In all three cases there are double poles.

We can also have $m = 4$; the poles are now simple. Here the case $P(w) = (w - a)(w - b)^3$ is trivial, a rational function of $z$ with a single pole. The case with two double roots reduces to subcase 1:2. If $P(w)$ has four distinct roots, we can use a linear transformation to reduce the equation to the Jacobi normal form:

$$(w')^2 = (1 - w^2)(1 - k^2 w^2),  \tag{11.3.5}$$

where the modulus $k \neq \pm 1$. The solution is the Jacobi *sine amplitude*, sn $(z - z_0; k)$ in the Gudermann notation. Christopher Gudermann (1798–1851), among other distinctions, was the teacher of Weierstrass.

Actually Yosida's extension of Malmquist's theorem is not needed to prove that $m \le 4$ when $n = 2$. This follows from the test-power principle of Section 3.3. See (4.6.47), which here becomes

$$(p - 2)\alpha = 2, \tag{11.3.6}$$

$\alpha$ a positive integer. If $\alpha = 1$, then $p = 4$, and if $\alpha = 2$, then $p = 3$; these are the only possibilities, in agreement with the statements made above.

We can continue in this manner to test the various possibilities for the nonexistence of mobile branch points, but it is helpful to derive some general results. We start with a binomial equation

$$(w')^n = R(w), \qquad R(w) = \frac{P(w)}{Q(w)}, \tag{11.3.7}$$

where $P$ and $Q$ are polynomials in $w$ without a common factor, $P$ of degree $p$ and $Q$ of degree $q$. Here we find

**THEOREM 11.3.1**

A necessary condition in order for (11.3.7) to have no mobile branch points is that $R(w) = P(w)$, a polynomial of degree $\le 2n$.

*Proof.* If this is not true, $Q(w)$ will admit at least one factor, say $Q(w) = (w - c)^k Q_1(w)$, where $Q_1(c) \ne 0$ and $k$ is a positive integer. For small values of $|w - c|$ we have then

$$w' = (w - c)^{-k/n} \sum_{j=0}^{\infty} a_j (w - c)^j$$

or

$$\sum_{j=0}^{\infty} b_j (w - c)^{j+k/n} w' = 1$$

and

$$\sum_{j=0}^{\infty} \frac{nb_j}{k + (j + 1)n} (w - c)^{j+1+k/n} = z - z_0,$$

whence by inversion

$$w - c = \sum_{j=1}^{\infty} c_j (z - z_0)^{nj/(k+n)}.$$

Thus, if $Q(z)$ is not identically one, there are mobile branch points.

If $Q(w) \equiv 1$, we make the inversion $w = 1/v$ and obtain

$$(-1)^n (v')^n = v^{2n} P\left(\frac{1}{v}\right).$$

Here the right member must be a polynomial in $v$, to avoid branch points, and this is the case iff $\deg(P) \leqslant 2n$. ∎

Furthermore, the factors of $P(w)$ are subject to severe restrictions. Suppose that $(w - a)^k$ is a factor of $P(w)$ so that

$$P(w) = (w - a)^k P_1(w), \qquad P_1(a) \neq 0.$$

Then for small values of $|w - a|$

$$(w - a)' = (w - a)^{k/n} \sum_{j=0}^{\infty} c_j (w - a)^j, \qquad c_0 \neq 0.$$

Here we set $w - a = u^n$ to obtain

$$nu^{n-1}u' = \sum_{j=0}^{\infty} c_j u^{k+jn}. \tag{11.3.8}$$

There are three different possibilities. The first case is $k = n - 1$, and then

$$nu'(z) = c_0 + \sum_{j=1}^{\infty} c_j u^{jn}, \qquad c_0 \neq 0,$$

so that

$$u(z) = \sum_{j=1}^{\infty} b_j (z - z_0)^j, \qquad w(z) = a + [u(z)]^n.$$

This function is holomorphic in a neighborhood of $z = z_0$ where $w(z_0) = a$.

The second case is $k \geqslant n$, and then

$$nu'(z) = \sum_{j=0}^{\infty} c_j u^{k+1+(j-1)n}$$

with the unique holomorphic solution $u(z) \equiv 0$, so that $w(z) \equiv a$.

Finally, suppose that $k < n - 1$. Then

$$b_0 u^{n-k-1} + b_1 u^{2n-k-1} + \cdots = z'(u),$$

so that

$$z - z_0 = u^{n-k} \left( \frac{b_0}{n-k} + \frac{b_1}{2n-k} u^n + \cdots \right)$$

or

$$(z - z_0)^{1/(n-k)} = u(d_0 + d_1 u^n + \cdots).$$

This gives

$$u(z) = d_0^{-1}(z - z_0)^{1/(n-k)} \left[ 1 + \sum_{j=1}^{\infty} e_j (z - z_0)^{jn/(n-k)} \right],$$

$$w(z) = a + d_0^{-n}(z - z_0)^{n/(n-k)} \left[ 1 + \sum_{j=1}^{\infty} f_j (z - z_0)^{jn/(n-k)} \right]. \tag{11.3.9}$$

This solution has a branch point at $z = z_0$ unless

$$\frac{n}{n-k} = p,$$

an integer $\geq 2$. This gives

$$k = \frac{p-1}{p}\, n \geq \tfrac{1}{2}n, \tag{11.3.10}$$

where $p$ is a divisor of $n$. Thus we have proved ∎

**THEOREM 11.3.2**

*A factor* $(w - a)^k$ *leads to solutions with movable branch points unless either* $k \geq m - 1$ *or* $k = [(p-1)/p]n$, *where* $p \geq 2$ *is a divisor of* n.

Suppose now that the degree of $P(w)$ is exactly $2m$, and let $(w - a)^k$ be a factor of $P(w)$ with $k \geq n$. The equation $P(w) = 0$ can have only one root distinct from $a$, for if there were two other roots, then

$$P(w) = (w - a)^k (w - b)^l (w - c)^m.$$

Here $k \geq n$, and by (11.3.10) both $l$ and $m$ must be at least $\tfrac{1}{2}n$, so the degree of $P(w)$ would exceed $2n$. This would lead to movable branch points and hence is not admissible. Therefore $l = 2n - k$ and $m = 0$. There are two possibilities: (i) $l = n - 1$, and (ii) $l = [(p-1)/p]n$, where $p \geq 2$ is a divisor of $n$. The second alternative leads to a contradiction, for we would have

$$\frac{n}{n - (n - k)} = \frac{n}{k},$$

an integer $> 1$, and this is impossible since $k \geq n$.

The resulting equation is thus

$$(w')^n = (w - a)^{n+1}(w - b)^{n-1}. \tag{11.3.11}$$

The general solution of this equation is a rational function of $z$ of degree $n$ and with simple poles. This follows from setting

$$w - a = \frac{1}{W}, \tag{11.3.12}$$

leading to the equation

$$(W')^n = (-1)^n [1 + (a - b)W]^{n-1}, \tag{11.3.13}$$

which has polynomial solutions. If in (13.3.12) we replace $a$ by $b$, we obtain an equation of the form

$$(W')^n = A(W - c)^{n+1}, \tag{11.3.14}$$

which also has polynomial solutions.

We can take $k = n$, but this offers very few admissible equations. In this case with $(w - a)^n$ as a factor, the other factors of $P(w)$ are either

$$(w - b)^n \qquad \text{or} \qquad (w - b)^{n/2}(w - c)^{n/2}.$$

The first choice works iff $n = 1$ and the corresponding equation is subcase 1:2. Larger values of $n$ are ruled out, for the corresponding equations are reducible. The second choice is admissible iff $n = 2$; this leads to the equation

$$(w')^2 = (w - a)(w - b), \tag{11.3.15}$$

which was mentioned above. The solutions are simply periodic entire functions.

We have left the case in which the exponents are of the form (11.3.10). Here the problem is to find positive fractions of the form $j/n$ which add up to 2. There are only four possibilities:

$$\tfrac{1}{2}, \tfrac{1}{2}, \tfrac{1}{2}, \tfrac{1}{2}, \quad n = 2; \qquad \tfrac{2}{3}, \tfrac{2}{3}, \tfrac{2}{3}, \quad n = 3;$$
$$\tfrac{3}{4}, \tfrac{3}{4}, \tfrac{1}{2}, \quad n = 4; \qquad \tfrac{5}{6}, \tfrac{2}{3}, \tfrac{1}{2}, \quad n = 5.$$

The corresponding equations are as follows:

$$(w')^2 = C(w - a)(w - b)(w - c)(w - d), \tag{11.3.16}$$

$$(w')^3 = C(w - a)^2(w - b)^2(w - c)^2 \tag{11.3.17}$$

$$(w')^4 = C(w - a)^3(w - b)^3(w - c)^2, \tag{11.3.18}$$

$$(w')^6 = C(w - a)^5(w - b)^4(w - c)^3. \tag{11.3.19}$$

This exhausts the list of binomial equations of the form $(w')^n = P(w)$, where $P$ is a polynomial of degree $2n$ and the solution has no branch points and is hence single valued. We shall discuss some of these equations below in more detail.

First let us observe, however, that there are also admissible binomial equations where the degree $m$ of $P(w)$ is less than $2n$. We have seven admissible equations at the outset, namely, subcase 1:2, i.e.,

$$w' = C(w - a)(w - b), \tag{11.3.20}$$

(11.3.11), and (11.3.15)–(11.3.19). From each of these seven equations we can get more admissible equations by suppressing one of the factors in the right member. The suppression is achieved with the aid of the transformation (11.3.12), the power of which we have already seen in carrying (11.3.11) into (11.3.13). We take a more general case. Suppose that we are given

$$(w')^n = C_0(w - a)^j(w - b)^k(w - c)^l, \qquad j + k + l = 2n, \tag{11.3.21}$$

and make the change of dependent variable defined by (11.3.12). The result is

$$(W')^n = K(W - B)^k (W - C)^l,$$    (11.3.22)

where

$$K = (-1)^n c_0 (a - b)^k (a - c)^l, \qquad B = (b - a)^{-1}, \qquad C = (c - a)^{-1}.$$

This proves the assertion.

Of the seven original equations only four, (11.3.16)–(11.3.19), lead to elliptic functions. So do the reduced equations. Quite generally the equations are formally invariant under fractional linear transformations on $w$, and this can be used to bring the equations into a desirable normal form. For (11.3.16) we have already shown the Jacobi normal form (11.3.5). In this setting the Weierstrass normal form (11.3.4) can be regarded as a reduced form of (11.3.16) followed by a translation $W \mapsto W + D$ to ensure that the sum of the roots of the cubic polynomial is zero. The factor 4 can be introduced by a dilation of the independent variable $z \mapsto kz$.

All the equations (11.3.16)–(11.3.19) have solutions which are rational functions of a suitably chosen $\wp$-function and its derivative. We shall study (11.3.17) and (11.3.19) from this point of view. We start with

$$(w')^3 = (w - a)^2 (w - b)^2 (w - c)^2,$$    (11.3.23)

where $a, b, c$ are distinct complex numbers. We remove the third factor on the right by setting $w - c = 1/W$. The result is

$$(W')^3 = K(W - A)^2 (W - B)^2$$    (11.3.24)

with

$$K = -(c - a)^2 (c - b)^2, \qquad A = (a - c)^{-1}, \qquad B = (b - c)^{-1}.$$

The solutions of this equation have triple poles, and so has the derivative of the $\wp$-function. Our next step calls for the differential equation satisfied by $\wp'$. Now differentiating

$$(\wp')^2 = 4\wp^3 - g_2 \wp - g_3,$$

we obtain, after cancellation,

$$\wp'' = 6\wp^2 - \tfrac{1}{2}g_2.$$    (11.3.25)

The case in which $g_2 = 0$ is known as *equiharmonic*. Here the roots $e_1, e_2, e_3$ form the vertices of an equilateral triangle, and so do the points $0, 2\omega_1, 2\omega_3$, where $2\omega_1, 2\omega_3$ form a pair of primitive periods of $\wp$. In this case we have

$$\wp'' = 6\wp^2, \qquad (\wp')^2 = 4\wp^3 - g_3.$$    (11.3.26)

We eliminate $\wp$ between these two equations and obtain

$$(\mathfrak{p}'')^3 = \tfrac{27}{2}[(\mathfrak{p}')^2 + g_3]^2. \tag{11.3.27}$$

This shows that $\mathfrak{p}'(z; 0, g_3)$ satisfies the equation

$$(v')^3 = \tfrac{27}{2}(v^2 + g_3)^2. \tag{11.3.28}$$

This can be brought to the form (11.3.24) by a linear transformation of both variables. The result is that

$$m\,\mathfrak{p}'(z + k: 0, g_3) + \tfrac{1}{2}(A + B) \quad \text{with } m = \frac{27}{2K}, \quad m^2 g_3 = -\tfrac{1}{4}(A - B)^2$$

is a solution of (11.3.24), and to get the solution of (11.3.23) we have simply to note that $w = c + 1/W$, which is clearly also an elliptic function but with simple poles.

We now consider (11.3.19), which we first reduce to the form

$$(W')^6 = K(W - B)^4(W - C)^3. \tag{11.3.29}$$

Here we set

$$W = B + U^3$$

and get

$$3^6 U^{12}(U')^6 = KU^{12}(U^3 + B - C)^3.$$

We cancel the factor $U^{12}$ and extract a cube root to obtain

$$(U')^2 = \tfrac{1}{9}K^{1/3}(U^3 + B - C), \tag{11.3.30}$$

which is satisfied by an elliptic function of the type $\alpha\,\mathfrak{p}(\beta z + \gamma, 0, g_3)$. This shows that all nonconstant solutions of (11.3.19) are fractional linear transforms of $\mathfrak{p}^3(\beta z + g; 0, g_3)$.

Equation (11.3.18) is integrable in terms of the Jacobi sine amplitude. Details are left to the reader.

We end this discussion by showing that every elliptic function satisfies a Briot-Bouquet equation. This is a by-product of a result, also proved by Briot and Bouquet, to the effect that any two elliptic functions with the same periods are algebraic functions of each other. In other words, if $f(z)$ and $g(z)$ are elliptic functions of periods $2\omega_1$ and $2\omega_3$, there exists a polynomial $F(Z_1, Z_2)$ in two variables with constant coefficients such that

$$F[f(z), g(z)] \equiv 0. \tag{11.3.31}$$

Now, if $w(z)$ is an elliptic function with periods $2\omega_1, 2\omega_3$, so is $w'(z)$. Hence we have a DE

$$F[w(z), w'(z)] = 0. \tag{11.3.32}$$

This equation has a somewhat special structure.

**THEOREM 11.3.3**

*The differential equation satisfied by an elliptic function of order* m *is of the form*

$$(w')^m + \sum_{j=1}^{m} P_j(w)(w')^{m-j} = 0, \qquad (11.3.33)$$

*where* $P_j(w)$ *is a polynomial in* w *with constant coefficients and of degree* $\leq 2j$. *Furthermore,* $P_{m-1}(w) \equiv 0$.

*Proof.* For properties of elliptic functions used in the following see the Appendix of this section, and for further details Chapter 13 of Vol. 2 of AFT.

The order of an elliptic function $f(z)$ is the number of roots of the equation $f(z) = a$ in the period parallelogram. It is independent of $a$. Suppose now that the DE satisfied by $w(z)$ is

$$\sum_{j=0}^{n} P_j(w)(w')^{n-j} = 0. \qquad (11.3.34)$$

That $n$ has to be equal to $m$ will become apparent from the proof.

Our first task is to show that $P_0(w)$ must be a constant. If not, the equation $P_0(w) = 0$ has at least one root, say $w = a$. This implies by a well-known property of algebraic functions that the equation $F(Z_1, Z_2) = 0$ has a solution $Z_2 = G(Z_1)$ which becomes infinite as $Z_1 \to a$. But this is impossible, for in our case $Z_1 = w(z)$, $Z_2 = w'(z)$ and they are *cofinite*, i.e., if one is finite so is the other. This shows that $P_0(w)$ must be a constant which we may take to be equal to one.

Next we have the degrees of the coefficient polynomials. These can be estimated in one way or another by the infinitary behavior of the solution. Here a difficulty arises from the fact that an elliptic function may have poles of various degrees so that the order of the poles is not unique, a phenomenon which we have not encountered before. That it can happen is shown by the elliptic function

$$\mathfrak{p}'(z) + [\mathfrak{p}(z) - e_1]^{-1}, \qquad \mathfrak{p}(\omega_1) = e_1, \qquad (11.3.35)$$

which is of order 5 and has a triple pole at the origin, a double pole at $z = \omega_1$, and congruent points modulo period. We can use the test-power test of Section 3.3 to study this matter, but for a complete discussion we have to invoke the method of Puisseux or a similar argument.

Suppose that $z = z_0$ is a pole of $w(z)$ of order $\alpha$ so that

$$w(z) \sim a(z - z_0)^{-\alpha}, \qquad w'(z) \sim -\alpha a(z - z_0)^{-\alpha-1}.$$

We denote the degree of $P_j(w)$ by $p_j$ and substitute in (11.3.34). It is seen that we have to balance expressions of the form

$$b_0(z - z_0)^{-n(1+\alpha)}, \qquad \ldots, \qquad b_j(z - z_0)^{-p_j-(n-j)(1+\alpha)}, \qquad \ldots.$$

Suppose initially that the first term is the heaviest. There will then be an integer $j$ such that

$$n(1 + \alpha) = p_j\alpha + (n - j)(1 + \alpha), \qquad (11.3.36)$$

whence

$$j < p_j = \frac{1+\alpha}{\alpha} j \leqslant 2j. \qquad (11.3.37)$$

If the terms with subscripts 0 and $j$ are the only ones which are of maximum order, we find that $p_k < 2k$ for all other subscripts.

The situation is less satisfactory when the first term is not of maximum order. Then there are two integers, $j$ and $k$, such that

$$n(1 + \alpha) < p_j\alpha + (n - j)(1 + \alpha) = p_k\alpha + (n - k)(1 + \alpha),$$

and the inequalities give

$$p_j > \frac{\alpha + 1}{\alpha} j, \qquad p_k > \frac{\alpha + 1}{\alpha} k.$$

Thus, if the theorem is correct, this contingency could arise only for multiple poles. The equality and the other inequalities do not provide enough information to give useful upper bounds for the $p$'s. Here we get more help from the method of Puiseux, which was also sketched at the end of Section 3.3.

In an $(x, y)$-plane we mark the points $(j, k)$ corresponding to the terms $w^j(w')^k$ actually present in the DE. Next we plot the least convex polygon $\Pi$ which encloses these points. Here $\Pi$ has points on the positive axes; the point $X = (p_n, 0)$ is furthest away on the $x$-axis, and the point $Y = (0, n)$ furthest on the $y$-axis. We are interested in that part of $\Pi$ which faces infinity. It is a broken line segment joining $X$ and $Y$ and lying above or on the straight line segment $(X, Y)$. See Figure 11.1.

Now the equation $F(Z_1, Z_2) = 0$ has all its infinitary branches of the form

$$Z_2 = CZ_1^{(\alpha+1)/\alpha}[1 + o(1)], \qquad (11.3.38)$$

since at a pole $w'(z) = C[w(z)]^{(\alpha+1)/\alpha}[1 + o(1)]$ and $w(z)$ has only polar infinitudes. Here, as observed in connection with (3.3.18), the negative of the reciprocal of the exponent is the slope of one of the boundary lines of $\Pi$. The polygon lies above or on the line $XY$ of the equation

$$y - n = -\frac{n}{P_n} x. \qquad (11.3.39)$$

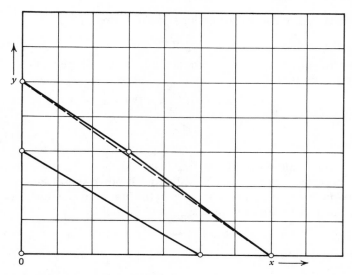

**Figure 11.1.** Puiseux diagram for $(w')^5 - (1 + w^3)(w')^3 - w^7 - w^5 = 0$, $\alpha_1 = 2$, $\alpha_2 = 3$, $m = 5$.

Now, if the point $(P_j, n - j)$ lies above the line, it is seen that

$$\frac{P_j}{P_n} < \frac{j}{n}.$$

But $\Pi$ also lies below or on the line through $Y$ parallel to the side of $\Pi$ of largest slope. This is the line

$$y - n = -\frac{\alpha_0}{\alpha_0 + 1} x, \tag{11.3.40}$$

where $\alpha_0 = \inf \alpha$, the infimum of the order of the poles of $w(z)$. This shows that the point $(p_j, n - j)$ lies below or on the line (11.3.40), so that

$$n - \frac{\alpha_0}{\alpha_0 + 1} p_j \geqslant n - j \qquad \text{or} \qquad p_j < \frac{\alpha_0 + 1}{\alpha_0} j \leqslant 2j,$$

and this now holds for all $j$. Note that no side of $\Pi$ can have slope $-1$ (why?).

Next, the number of poles is $m = \Sigma \alpha_j$, where the $\alpha$'s are the numbers read off from the slopes $-\alpha/(\alpha + 1)$ of the sides of $\Pi$, starting from $Y$. On the other hand, the geometry of the polygon shows that $\Sigma \alpha_j = n$, so that $n = m$.

It remains to show that $P_{m-1}(w) \equiv 0$. We have

$$\pm \frac{P_{m-1}(w)}{P_m(w)}$$

equal to the sum of the reciprocals of the roots of (11.3.34) regarded as an algebraic equation for $w'$. Thus

$$\frac{P_{m-1}(w)}{P_m(w)} = \sum_{j=1}^{m} \frac{1}{w'(z_j)},$$

where $z_1, z_2, \ldots, z_m$ are the points in the period parallelogram where $w(z)$ assumes the value $w$, an arbitrary complex number. Now

$$\frac{1}{w'(z_j)} = z'_j(w),$$

and by a basic property of elliptic functions

$$\sum z_j(w)$$

is independent of $w$ and hence a constant. Its derivative with respect to $w$ then is zero or $P_{m-1}(w) \equiv 0$, as asserted. This completes the proof.


# Appendix

### ELLIPTIC FUNCTIONS

An elliptic function is a single-valued doubly periodic function whose only singularities in the finite plane are poles. That $f(z)$ is doubly periodic means the existence of two complex numbers, $2\omega_1$ and $2\omega_3$, such that

$$f(z + 2\omega_1) = f(z), \qquad f(z + 2\omega_3) = f(z). \tag{A.1}$$

The ratio $\omega_3/\omega_1 = \tau$ is necessarily nonreal, and it is customary to assume its imaginary part to be positive. If the ratio is real, the function is simply periodic if $\tau$ is a rational number, and a constant if $\tau$ is irrational. There are no triply periodic functions except constants. The numbers

$$2m\omega_1 + 2n\omega_3, \qquad m, n = 0, \pm 1, \pm 2, \ldots$$

are *periods* of $f(z)$. The parallelogram with vertices at

$$0, \qquad 2\omega_1, \qquad 2\omega_2 = 2\omega_1 + 2\omega_3, \qquad 2\omega_3$$

is known as the *period parallelogram* of $f(z)$. The series

$$\sum \sideset{}{'}\sum |2m\omega_1 + 2n\omega_3|^{-s}, \tag{A.2}$$

where the prime indicates that $(m, n) \neq (0, 0)$, is convergent for $s > 2$ and diverges for $s \leq 2$.

It follows that the series

$$z^{-3} + \sum \sum{}' (z - 2m\omega_1 - 2n\omega_3)^{-3} \tag{A.3}$$

is absolutely convergent for all values of $z$ different from one of the points $2m\omega_1 + 2n\omega_3$. It clearly admits these numbers as periods, and it has no other finite singularities than triple poles at the periods. Thus it is an elliptic function and, in fact, minus one half of the derivative of the Weierstrass $\wp$-function. The latter has the representation

$$z^{-2} + \sum \sum{}' [(z - \omega_{m,n})^{-2} - (\omega_{m,n})^{-2}], \tag{A.4}$$

where $\omega_{m,n} = 2m\omega_1 + 2n\omega_3$.

Thus we know that there are elliptic functions. In studying their properties Briot and Bouquet found Cauchy's integral theorem very useful. In applying the latter, it is convenient to assume that there are no poles on the boundary of the period parallelogram. We have then

$$\frac{1}{2\pi i} \int \frac{f'(z)}{f(z)} \, dz = 0, \tag{A.5}$$

where the integral is taken along the boundary. The integral is zero by the double periodicity. On the other hand, the value of the integral is the sum of the residues of the poles inside the contour. This says that the sum of the residues is zero. One consequence thereof is that the order of $f(z)$ must exceed one, the order $m$ being the sum of the orders of the poles. Consideration of the integral

$$\frac{1}{2\pi i} \int \frac{f'(z)}{f(z) - a} \, dz$$

shows that the function assumes every value $a$ in the period parallelogram $m$ times if multiple roots are counted with their proper multiplicities. Similarly the integral

$$\frac{1}{2\pi i} \int z \frac{f'(z)}{f(z) - a} \, dz = \sum c_i - \sum p_i \tag{A.6}$$

equals a period of $f(z)$. Here the $c$'s are the $m$ roots of the equation $f(z) = a$ and the $p$'s are the $m$ poles of $f(z)$. This implies that $\Sigma c_i$ is independent of $a$, a fact used above.

We have more to say about the $\wp$-function. We expand (A.4) in powers of $z$. Since

$$\wp(-z) = \wp(z), \tag{A.7}$$

the expansion will involve only even powers of $z$:

$$\wp(z) = z^{-2} + A_0 + A_1 z^2 + A_2 z^4 + \cdots.$$

Now

$$\frac{1}{(z-\omega)^2} - \frac{1}{\omega^2} = \frac{2z}{\omega^3} + \frac{3z^2}{\omega^4} + \frac{4z^3}{\omega^5} + \frac{5z^4}{\omega^6} + \cdots,$$

so that

$$A_0 = 0, \qquad A_1 = 3 \sum \sum {}' (\omega_{m,n})^{-4}, \qquad A_2 = 5 \sum \sum {}' (\omega_{m,n})^{-6}, \ldots.$$

Next, we define

$$g_2 = 60 \sum \sum {}' (\omega_{m,n})^{-4}, \qquad g_3 = 140 \sum \sum {}' (\omega_{m,n})^{-6}, \qquad (A.8)$$

so that

$$\wp(z) = \frac{1}{z^2} + \tfrac{1}{20} g_2 z^2 + \tfrac{1}{28} g_3 z^4 + \cdots,$$

$$\wp'(z) = -\frac{2}{z^3} + \tfrac{1}{10} g_2 z + \tfrac{1}{7} g_3 z^3 + \cdots.$$

Our object is to derive the DE satisfied by $\wp(z)$. To this end we look for a combination of $\wp$ and $\wp'$ which is nonsingular at $z = 0$. Here $[\wp'(z)]^2 - 4[\wp(z)]^3$ looks promising. We have

$$[\wp'(z)]^2 = \frac{4}{z^6} - \tfrac{2}{5} g_2 \frac{1}{z^2} - \tfrac{4}{7} g_3 + O(z^2),$$

$$4[\wp(z)]^3 = \frac{4}{z^6} + \tfrac{3}{5} g_2 \frac{1}{z^2} + \tfrac{3}{7} g_3 + O(z^2),$$

the difference of which is

$$-g_2 \frac{1}{z^2} - g_3 + O(z^2) = -g_2 \wp(z) - g_3 + O(z^2).$$

It follows that

$$F(z) = [\wp'(z)]^2 - \{4[\wp(z)]^3 - g_2 \wp(z) - g_3\}$$

is not singular at $z = 0$. Since $F(z)$ is obviously doubly periodic, it follows that $F(z)$ has no singularities and is hence a constant. Since $F(0) = 0$, it is seen that $F(z) \equiv 0$, and we conclude that $\wp(z)$ is a solution of the DE

$$(w')^2 = 4w^3 - g_2 w - g_3, \qquad (A.9)$$

as expected.

If we set $z = \omega_j$, $j = 1, 2, 3$, in (A.3), it is seen that all terms cancel, so that

$$\wp'(\omega_1) = \wp'(\omega_2) = \wp'(\omega_3) = 0.$$

The corresponding values of $\wp$ are denoted by $e_1$, $e_2$, $e_3$. Their sum is zero. It is possible to recover the periods from the values of $g_2$ and $g_3$ via the theta functions. For us, however, it is more natural to use the DE. For the sake of simplicity suppose that the $e$'s are real and that $e_3 < e_2 < e_1$. We have then

$$\omega_1 = \int_{e_1}^{\infty} [4(s - e_1)(s - e_2)(s - e_3)]^{-1/2} \, ds,$$
$$\omega_3 = \int_{-\infty}^{e_3} [4(s - e_1)(s - e_2)(s - e_3)]^{-1/2} \, ds,$$

(A.10)

where the argument of the integrand is zero in the first integral and $\frac{1}{2}\pi$ in the second. These formulas are valid for all values of $e_1$ and $e_3$ with $e_2 = -e_1 - e_3$ and define analytic functions of $e_1$, $e_3$. We mention (but have no need of the fact) that $\wp(z)$ has an algebraic addition theorem in the sense that $f(u + v)$ is a rational function of $f(u), f'(u), f(v)$, and $f'(v)$.

The Jacobi elliptic functions can be defined as the solutions of the system of nonlinear first order equations:

$$\begin{aligned} w_1'(z) &= w_2(z)w_3(z), & w_1(0) &= 0, \\ w_2'(z) &= -w_3(z)w_1(z), & w_2(0) &= 1, \\ w_3'(z) &= -k^2 w_1(z)w_2(z), & w_3(0) &= 1. \end{aligned}$$

(A.11)

Here $k$ is an external parameter known as the *modulus*. The solutions are analytic functions of $z$ as well as of $k$. The *complementary modulus* is defined as

$$k' = +(1 - k^2)^{1/2}.$$

(A.12)

By elementary manipulations we obtain

$$[w_1(z)]^2 + [w_2(z)]^2 = 1, \qquad k^2[w_1(z)]^2 + [w_3(z)]^2 = 1.$$

(A.13)

Combining with the DE's, it is seen that $w_1(z)$ satisfies the DE

$$(w')^2 = (1 - w^2)(1 - k^2 w^2).$$

(A.14)

We have $w_1(z) = \operatorname{sn}(z, k), w_2(z) = \operatorname{cn}(z, k), w_3(z) = \operatorname{dn}(z, k)$ in the Gudermann notation. The periods are $4K$ and $2iK'$, where

$$K = \int_0^1 [(1 - s^2)(1 - k^2 s^2)]^{-1/2} \, ds,$$
$$K' = \int_0^{\infty} [(1 + s^2)(1 + k^2 s^2)]^{-1/2} \, ds.$$

(A.15)

As observed earlier, they are hypergeometric functions of $k^2$. The points $iK'$ and $2K + iK'$ are poles of $\operatorname{sn} z$ with residues $1/k$ and $-1/k$, respectively. The poles are simple, and the sum of the residues in the period parallelogram is zero, as it should be.

This brief survey should give the reader some idea of the properties of elliptic functions.

## EXERCISE 11.3

1. Solve $w' = A(w - a)(w - b)$ and $(w')^2 = A^2(w - a)(w - b)$.
2. Solve $(w')^2 = (w - a)(w - b)^2$.
3. How is the equation $(w')^2 = (w - a)(w - b)(w - c)(w - d)$ reduced to the Jacobi normal form?
4. Solve (11.3.13) and (11.3.14).
5. Solve (11.3.18).
6. Transform Jacobi's and Weierstrass's normal forms into each other.
7. Use the result of Problem 6 to express the $\wp$-function in terms of a sine amplitude. Keep good track of the parameters.
8. Derive (11.3.28), and reduce it to the form (11.3.24).
9. If the order of an elliptic function is $m$, what are the bounds for the order of the derivative? Give examples.
10. Suppose that weights are assigned to the terms of equations of type (11.3.24) or higher order in such a manner that $w$ is of weight 1, $w'$ of weight 2, ..., $w^{(k)}$ of weight $k + 1$. The weight of a product is to be the sum of the weights of the factors; the weight of a sum, equal to the maximal weight of any term (if there is a single dominating term, otherwise $\leq$ the maximum). Show that Theorem 11.3.1 expresses the fact that no term can have a heavier weight than that of the leading term $(w')^m$. This is a special case of a theorem of Jean Chazy (1911) for DE's with fixed critical points.
11. $\wp(2z)$ is a rational function of $\wp(z)$. The function $[\wp(z)]^4[\wp'(z)]^{-2}$ has the same poles as $\wp(2z)$, but the principal parts do not coincide. How should the expression be modified in order to represent $\wp(2z)$?
12. Given $\wp(z + a) + \wp(a)$, where $a$ is not congruent to zero modulo period, construct an expression involving $\wp(z)$, $\wp'(z)$, $\wp(a)$, and $\wp'(a)$ with the same poles and principal parts, so that the difference is a constant and the addition theorem results.
13. Prove the formulas (A.13).
14. Find the DE's for $w_2(z)$ and $w_3(z)$.
15. Prove that $w_1(z)$ becomes infinite as $z$ tends to $iK'$ along the imaginary axis, if $0 < k < 1$.
16. Show that $\operatorname{sn}(K) = 1$ and $\operatorname{sn}(2K) = 0$.

## LITERATURE

The author's LODE, Section 12.3, and AFT, Vol. II, Chapter 13, have a bearing on Section 11.3.

There is an extensive literature relating to Section 11.1. We cite:

Briot, Ch. and J. Cl. Bouquet. Recherches sur les propriétés des fonctions définies par des équations différentielles. *J. École Imp. Polytech.*, **21** (1856), 133–197.

Dulac, H. Solutions d'un système d'équations différentielles dans le voisinage de valeurs singulières. *Bull. Soc. Math. France*, **40** (1912), 324–383.

———. *Points singuliers des équations différentielles.* Mémorial des Sci. Math., No. 61. Gauthier-Villars, Paris, 1934.

Hukuhara, M. Sur la généralisation des théorèmes de M. T. (*sic*!) Malmquist. *J. Fac. Sci. Univ. Tokyo,* **6** (1949), 77–84.

———, T. Kimura, and T. Matuda. Équations différentielles du premier ordre dans le champ complexe. *Publ. Math. Soc. Japan,* **7** (1961).

Malmquist, J. Sur les points singuliers des équations différentielles. *Ark Mat., Astr., Fys.,* **15**, Nos. 3, 27 (1921).

Sansone, G. and R. Conti. *Non-linear Differential Equations* (translated by A. H. Diamond). Pergamon Press, Macmillan, New York, 1964.

Trjitzinsky, W. J. *Analytic Theory of Non-linear Differential Equations.* Mémorial des Sci. Math., No. 90, Gauthier-Villars, Paris, 1938.

Valiron, G. *Équations fonctionnelles: Applications.* 2nd ed. Masson, Paris, 1950, pp. 268–274.

For Section 11.2 the main source is Chapters 1 and 2 of:

Boutroux, P. *Leçons sur les fonctions définies par les équations différentielles du premier ordre.* Gauthier-Villars, Paris, 1908.

The use of infinitudinal neighborhoods, implicit in Boutroux's work, was strongly influenced by the work of Hans Wittich; see Section 4.6. Section 11.3 goes back to:

Briot, Ch. and J. Cl. Bouquet. Intégration des équations différentielles au moyen des fonctions élliptiques. *J. École Imp. Polytech.,* **21**: 36 (1856).

# 12

# SECOND ORDER NONLINEAR DIFFERENTIAL EQUATIONS AND SOME AUTONOMOUS SYSTEMS

This chapter also considers an assortment of loosely related topics which center on nonlinearity, usually of higher order than the first. Some attention will be paid to Briot-Bouquet equations of second and higher order. At the end of the nineteenth century Painlevé determined all nonlinear second order DE's with single-valued solutions. He found some 50 different types, all but 6 of which could be reduced to equations with already known solutions. Six types led to new transcendental meromorphic functions; we shall discuss briefly the simplest of these types. Some 60 years ago Boutroux showed that the solutions of this equation are asymptotic in a certain sense to the Weierstrass ℘-function. Here again our discussion will just skim the surface. We shall take up the Emden and the Thomas-Fermi equations, which lead to rather peculiar solutions of the psi-series type. Finally we shall consider quadratic and higher order polynomial autonomous systems. They may be regarded as nonlinear first order equations, but some are closely related to the second order Emden DE. Here work by the author and by the British mathematician Russell Smith provides most of the material. These investigations have appeared in the *Proceedings of the Royal Society of Edinburgh*; the Society has graciously permitted the use of this material here.

## 12.1. GENERALITIES; BRIOT-BOUQUET EQUATIONS

For first order equations Painlevé proved that mobile singularities are necessarily algebraic (see Section 3.3), provided that the coefficients satisfy certain restrictions. This property no longer holds, however, for second or higher order equations, again as demonstrated by Painlevé. Here we can have movable essential singularities, as shown by the equation

$$\left[\frac{d}{dz}\left(\frac{w'}{w}\right)\right]^2 + 4\left(\frac{w'}{w}\right)^3 = 0 \qquad \text{with } w = C_1 \exp (z - C_2)^{-1}. \quad (12.1.1)$$

Here there is a movable essential singularity at $z = C_2$. Similarly the DE

$$(a - 1)(w')^2 - aww'' = 0 \qquad \text{with } w(z) = C_1(z - C_2)^a \quad (12.1.2)$$

has a movable singularity at $z = C_2$ which is a transcendental critical point unless $a$ is a rational number. We can also find second order equations with movable logarithmic branch points or movable cluster points of poles.

The situation becomes considerably more complicated for third order equations. Here we may have movable natural boundaries. To show this, we may use the properties of the modular function. Consider the $w$-plane, and mark the points $0, 1, \infty$. Let $S$ be the universal covering surface of the $w$-plane punctured at these points. This surface can be mapped conformally on any simply connected domain of the $z$-plane. In particular, we may map $S$ on the upper $z$-half-plane. Let the mapping function be $z(w)$. Its inverse $w(z)$ is the modular function. Then $z(w)$ satisfies

$$\{z, w\} = R(w), \qquad (12.1.3)$$

where $\{z, w\}$ is the Schwarzian and $R(w)$ is a rational function of $w$ with poles at $w = 0$ and $w = 1$. The Schwarzian is invariant under any fractional linear transformation applied to the first argument. Hence for any choice of constants $a, b, c, d$ with $ad - bc = 1$ the function

$$z \mapsto Z(w) = \frac{az(w) + b}{cz(w) + d} \qquad (12.1.4)$$

is a solution of (12.1.3). This function maps the real axis in the $z$-plane on a circle $C$, and the upper $z$-half-plane goes into the interior or the exterior of $C$, depending on the choice of the constants. Since they are at our disposal, $C$ is an arbitrary circle in the $z$-plane. By Problem 10.1:9

$$\{w, z\} = - R(w)(w')^2, \qquad (12.1.5)$$

and this is a third order nonlinear DE satisfied by the functions

$$w(Z) = w\left(\frac{az+b}{cz+d}\right) \tag{12.1.6}$$

for any choice of the constants $a$, $b$, $c$, $d$ satisfying $ad - bc = 1$. The domain of existence of $w(Z)$ has the circle $C$ as a natural boundary. This means that the domain of existence of the solutions varies with the three essential constants of integration.

Returning to the second order case, we start with some remarks in connection with Section 11.3. There it was observed that, if two elliptic functions $f$ and $g$ have the same periods, then one is an algebraic function of the other. This was used to show that each elliptic function $w(z)$ satisfies a first order nonlinear DE because $w(z)$ and $w'(z)$ are coelliptic and have the same periods. But the same evidently holds for $w(z)$ and $w''(z)$ or, for that matter, for $w(z)$ and $w^{(k)}(z)$ for any $k$. We sketch a proof of an analogue of Theorem 11.3.1 for the second order case.

**THEOREM 12.1.1**

*The second order nonlinear DE satisfied by an elliptic function* w(z) *of order* m *is of the form*

$$(w'')^n + \sum_{j=1}^{n} P_j(w)(w'')^{n-j} = 0. \tag{12.1.7}$$

*Here* $p_j = deg\ (P_j) \leqslant 3j$ *and* $n \leqslant m$.

*Proof.* That the coefficient of the highest power of $w''$ is a constant which may be taken to be one is proved in the same manner as in the first order case. The statements about $p_j$ can be proved in various ways. If the theorem of Chazy is accepted (see Problem 11.3:10), then

$$3n \geqslant p_j + 3(n-j) \qquad \text{or} \qquad p_j \leqslant 3j, \forall j.$$

We can also use the test-power test if

$$n(\alpha+2) \geqslant p_j\alpha + (n-j)(\alpha+2), \qquad \forall j$$

with the same result.

For the inequality between $m$ and $n$ we need the method of Puisseux. Let $p = \max p_j$, and let $X = (p, 0)$, $Y = (0, n)$. There is a least convex polygon $\Pi$ that encloses all the points $(j, k)$ for which there are terms $w^j(w'')^k$ in the equation. As usual, we are interested only in that part of the polygon which joins $X$ and $Y$ and faces infinity. The various sides of this part have slopes, going from the left to the right,

$$-\frac{\alpha_1}{\alpha_1+2}, \qquad -\frac{\alpha_2}{\alpha_2+2}, \qquad \cdots, \qquad -\frac{\alpha_\nu}{\alpha_\nu+2}$$

and

$$\alpha_1 < \alpha_2 < \cdots < \alpha_\nu, \qquad (12.1.8)$$

by the convexity of the polygon. The $\alpha$'s are the different orders of the poles in the period parallelogram of $w(z)$. Their sum is $m$. The polygon $\Pi$ lies on or below the line

$$y - n = -\frac{\alpha_1}{\alpha_1 + 2} x \qquad (12.1.9)$$

and above or on the line

$$y - n = -\frac{n}{p} x, \qquad (12.1.10)$$

which joins $X$ and $Y$. All the terms involved in $P_j(w)(w'')^{n-j}$ give rise to lattice points interior to $\Pi$ with the possible exception of $(p_j, n - j)$, which may be a vertex of $\Pi$ or lie on one of the sides. From the fact that the point $(p_j, n - j)$ lies on or below the line (12.1.9) it follows again that $p_j \leq 3j$, and now for all $j$.

Suppose that all poles are of the same order $k \geq 1$. The part of the polygon that faces infinity now reduces to a single straight line from $Y$ to $X$ having the equation

$$y - n = -\frac{k}{k+2} x. \qquad (12.1.11)$$

It intersects the $x$-axis at $x = p = (k+2)n/k$. This must be an integer. Hence there are three possibilities: (i) $k = 1$, (ii) $k = 2$, (iii) $k$ divides $n$. The line passes through the lattice points

$$[(k+2)j, n - kj], \qquad j = 0, 1, \ldots, \frac{n}{k}. \qquad (12.1.12)$$

Here the last point must be $(p, 0)$, so that $k$ divides $n$. This makes sense as long as $n \geq k$. But if $k > n$, we must have $k = 1$ or 2. Here the case $k = 1$, $n < k$ is clearly excluded, so we are left with $k = 2$, $n = 1$. This means an equation of the form

$$w'' = a_0 + a_1 w + a_2 w^2 \qquad (12.1.13)$$

with constant $a$'s, which is satisfied by $c_0 + c_1 \wp(c_2 z)$ for suitable choice of the constants. We cannot allow a third order term, since this would lead to simple poles whereas we want double poles. Any higher power would lead to branch points.

Let us return to the case of $k \leq n$ and $k$ a divisor of $n$. Now $m$, the order of $w(z)$, equals the number $p$ of distinct poles in the period parallelogram multiplied by their common order $k$, so that $m = pk$. On the other hand, the lattice points (12.1.12) with $j > 0$ are $n/k$ in number and are connected with the algebraic meaning of the Puisseux diagram. The equation $F(Z_1, Z_2) = 0$ has $n/k$ solutions of the form

$$Z_2 = Z_1^{(k+2)/k}\left(a_\lambda + \sum_{j=1}^{\infty} a_{\lambda j} Z_1^{-j/k}\right), \qquad (12.1.14)$$

$\lambda = 1, 2, \ldots, n/k$, at infinity. Now there are $p$ distinct poles in the period parallelogram, and if

$$w(z) \sim A(z - z_0)^{-k}, \qquad w''(z) \sim Ak(k+1)(z - z_0)^{-k-2},$$

we should have

$$a_\lambda A^{2/k} = k(k+1), \qquad A = \left[\frac{k(k+1)}{a_\lambda}\right]^{k/2}.$$

This implies that $n/k \leqslant p$ and hence $n \leqslant m$. It is seen that both $m$ and $n$ are divisible by $k$ and $n \geqslant k$.

Here it is assumed that all poles have the same order $k$. Now, if the polygon has $\nu$ sides facing infinity and the enumeration is such that the orders $\alpha_j$ satisfy (12.1.8), we can essentially repeat the argument for each side. If there are $\pi_j$ poles of order $\alpha_j$ and $k_j$ lattice points on the $j$th side, we have $k_j \leqslant \pi_j$ for each $j$ and

$$m = \sum_{j=1}^{\nu} \pi_j \alpha_j \geqslant \sum_{j=1}^{\nu} k_j \alpha_j = n,$$

so the theorem is proved. ■

As illustrations we take up some special cases. We have already considered the Weierstrass $\wp$-function, which leads to a DE with $k = 3$, $m = 2$, $n = 1$. For the elliptic functions defined by

$$(w')^4 = (w - a)^3(w - b)^3(w - c)^2 \qquad (12.1.15)$$

we find $(w'')^2 = P_2(w)$, where $P_2$ is of degree 6. Here $k = 1$, $m = 4$, $n = 2$. Similarly for the functions satisfying

$$(w')^6 = (w - a)^3(w - b)^4(w - c)^5, \qquad (12.1.16)$$

we find that $(w'')^3 = P_3(w)$, where $P_3$ is a polynomial of degree 9, not a perfect cube; in fact, $P_3(w) = (w - b)(w - c)^2[Q(w)]^3$, where $Q$ is a quadratic polynomial. Here $k = 1$, $m = 6$, $n = 3$. Finally for

$$(w')^3 = (w - a)^2(w - b)^2 \qquad (12.1.17)$$

we obtain

$$(w'')^3 = \tfrac{64}{27}(w - a)(w - b)[w - \tfrac{1}{2}(a + b)]^3$$

with $k = m = n = 3$.

We can extend the discussion to equations of higher order than the second and prove analogues of Theorem 12.1.1 for equations

$$[w^{(k)}]^n + \sum_{j=1}^{n} P_j(w)[w^{(k)}]^{n-j} = 0. \tag{12.1.18}$$

We observe only that deg $(P_j) = p_j \leq (k+1)j$. For the binomial equations

$$[w^{(k)}]^n = P(w) \tag{12.1.19}$$

the discussion is still not too complicated, but we refrain from further excursions in this direction.

Equations of the form

$$w'' = R(z, w) \tag{12.1.20}$$

have been studied from various points of view. If the problem is to find necessary conditions for the existence of solutions holomorphic in the distant part of the plane save for movable poles, we can start with the test-power test. If

$$R(z, w) = \frac{P(z, w)}{Q(z, w)},$$

where $P$ and $Q$ are polynomials in $z$ and $w$, and if $P$ is of degree $p$, $Q$ of degree $q$ in $w$, the substitution $w(z) \sim A(z - z_0)^{-\alpha}$ leads to the condition

$$(p - q - 1)\alpha = 2. \tag{12.1.21}$$

If $\alpha$ is to be an integer, there are two possibilities: (i) $\alpha = 2$, $p = q + 2$, and (ii) $\alpha = 1$, $p = q + 3$.

Though it is not obvious that we must have $q = 0$ in order to exclude branch points, we shall assume this to be the case. Thus we have the equations

$$w'' = \sum_{j=0}^{p} A_j(z)w^j, \tag{12.1.22}$$

where $p = 2$ or 3. Paul Painlevé, who made a general study of the problem of finding nonlinear second order DE's without movable branch points, did find two equations of type (12.1.22) which define new transcendental meromorphic functions, one corresponding to $p = 2$ and the other to $p = 3$. See Sections 12.2 and 12.3.

It should be observed, however, that our conditions are merely necessary for the existence of poles, not sufficient. This is shown in a glaring manner by the special Emden equation

$$zw'' = w^2, \tag{12.1.23}$$

where the solutions have movable logarithmic pseudopoles. This case and related ones will be discussed in Section 12.4 and following.

**EXERCISE 12.1**

1. Verify (12.1.1) and (12.1.2).

2. How should the constants $a$, $b$, $c$, $d$ be chosen in order that the circle $C$ will be $|z - 2i| = 1$?

3. Why does (12.1.8) hold?

4. Why do the points $(a, n - j)$, $a$ integer $< p_j$, lie inside the polygon $\Pi$ if $(p_j, n - j)$ lies inside or on $\Pi$?

5. Find the second order DE satisfied by Jacobi's sine amplitude.

6. Find the first order DE satisfied by $\wp'(z; 0, 4)$. Here, to simplify, we take $g_2 = 0, g_3 = 4$. Show that the second order equation satisfied by this function is

$$(w'')^3 = \tfrac{27}{4}w^3(w^2 + 4),$$

so that $k = m = n = 3$.

7. Verify the conclusion for (12.1.15).

8. Solve Problem 7 for (12.1.16).

9. Solve Problem 7 for (12.1.17).

10. Verify that in (12.1.18) we have $p_j \le (k + 1)j$. The task is trivial on the basis of Chazy's theorem, so do it the hard way.

11. Verify (12.1.21).

## 12.2. THE PAINLEVÉ TRANSCENDENTS

Consider DE's of the form

$$w'' = F(z, w, w'), \tag{12.2.1}$$

where $F$ is rational in $w$ and $w'$, and analytic in $z$. In 1887 Émile Picard raised the question of when such an equation has fixed critical points. Here "critical points" includes branch points and essential singular points for the sake of convenience. The problem was solved by P. Painlevé in a series of papers starting in 1893. R. Gambier made important contributions from 1906 to 1910, and R. Fuchs (the son of Lazarus Fuchs) in 1905. Extensions to third order equations were made by J. Chazy (1911) and René Garnier (1912–16). The general method, due to Painlevé, consists in finding two necessary conditions (this is fairly easy) and then verifying that these conditions are also sufficient. This part of the argument involves considering a large number of separate cases and showing that the corresponding equations either are reducible to equations known to have the desired property or define new transcendents. This part is very laborious.

Painlevé's basic device is his $\alpha$-method. He introduced a parameter $\alpha$ in the equation such that for $\alpha = 0$ the resulting equation can be seen by inspection to be with or without movable critical points. If now the equation has single-valued solutions for small $|\alpha|$ with the possible exception of $\alpha = 0$, then $\alpha = 0$ can be no exception. Using this artifice, Painlevé showed that $F$ must actually be a polynomial in $w'$ of degree $\leq 2$ so that the equation is of the form

$$w'' = L(z, w)(w')^2 + M(z, w)w' + N(z, w), \qquad (12.2.2)$$

where $L$, $M$, $N$, are rational in $w$. If now $z = c$ is not a fixed singular point, he showed that the equation

$$W'' = L(c, W)(W')^2$$

must have fixed critical points. This will be the case if $L \equiv 0$, but there are also five other possibilities, obtained by applying logarithmic differentiation to the Briot-Bouquet equations of Section 11.3 and replacing the constants by functions of $z$. If $L \equiv 0$, then $M$ must be linear in $w$, while $N$ is a polynomial in $w$ of degree $\leq 3$. If $L \neq 0$, $M$ and $N$ take the forms

$$M(z, w) = \frac{m(z, w)}{D(z, w)}, \qquad N(z, w) = \frac{n(z, w)}{D(z, w)}. \qquad (12.2.3)$$

Here $D(z, w)$ is the least common denominator of the partial fractions in $L(z, w)$ and is a polynomial of $w$ of degree $d$, $2 \leq d \leq 4$, while $m$ and $n$ are polynomials in $w$ of degrees $\leq d + 1$ and $\leq d + 3$, respectively. This exhausts the necessary conditions and leads to 50 possible types. Only 6 of these lead to new classes of functions, the Painlevé transcendents. These equations are as follows:

(1)  $w'' = 6w^2 + z,$

(2)  $w'' = 2w^3 + zw + a,$

(3)  $w'' = \dfrac{1}{w}(w')^2 - \dfrac{1}{z}w' + \dfrac{1}{z}(aw^2 + b) + cw^3 + \dfrac{d}{w},$

(4)  $w'' = \dfrac{1}{2w}(w')^2 + \dfrac{3}{2}w^3 + 4zw^2 + 2(z^2 - a)w + \dfrac{b}{w},$

(5)  $w'' = \left(\dfrac{1}{2w} + \dfrac{1}{w-1}\right)(w')^2 - \dfrac{1}{z}w' + z^{-2}(w-1)^2\left(aw + \dfrac{b}{w}\right)$

$\qquad + c\dfrac{w}{z} + d\dfrac{w(w+1)}{w-1},$

(6)  $w'' = \dfrac{1}{2}\left(\dfrac{1}{w} + \dfrac{1}{w-1} + \dfrac{1}{w-z}\right)(w')^2 - \left(\dfrac{1}{z} + \dfrac{1}{z-1} + \dfrac{1}{w-z}\right)w'$

$\qquad + \dfrac{w(w-1)(w-z)}{z^2(z-1)^2}\left[a + \dfrac{bz}{w^2} + c\dfrac{(z-1)}{(w-1)^2} + d\dfrac{z(z-1)}{(w-z)^2}\right].$

Equation (6) is the master type and was added to the list by R. Fuchs. The other equations may be obtained from it by passage to the limit and coalescence. Equations (4) and (5) were added by Gambier. There are no critical points at all in the first three cases. If the substitution $z = e^t$ is made in (4) and (5), the solutions become single-valued functions of $t$. For type (6) $0, 1, \infty$ are fixed critical points.

We shall discuss equation (1), which is the simplest and has very interesting properties. Here we start by showing that the only algebraic singularities are movable double poles. There is a unique solution $w(z; z_0, w_0, w_0')$, which satisfies the equation and is holomorphic in some neighborhood of $z = z_0$, where it takes on the value $w_0$ while its derivative equals $w_0'$. It follows that, if $z = z_0$ is to be an algebraic singularity, at least one of the quantities $w(z)$ and $w'(z)$ must become infinite when $z$ approaches $z_0$. Now the equation shows that the three quantities $w(z)$, $w'(z)$, $w''(z)$ are cofinite, so for an algebraic branch point we must have both $\lim w(z)$ and $\lim w'(z)$ infinite.

This being the case, we may postulate a formal series solution

$$w(z) = \sum_{n=0}^{\infty} a_n (z - z_0)^{\alpha_n}, \qquad (12.2.4)$$

where the $\alpha$'s form a strictly increasing sequence tending to $\infty$ and $\alpha_0 < 0$. Such an expansion will exist at an algebraic branch point but also if a nonlogarithmic psi series is hidden. The argument to be given will show, however, that the $\alpha$'s have to be integers so that at least some pathological phenomena cannot occur. Since we are not going to give Painlevé's proof of the meromorphic character of the solutions, and psi series will occur in Section 12.4 and later, some reassurance is due the reader.

Substitution in the equation gives

$$\sum_{n=0}^{\infty} \alpha_n (\alpha_n - 1) a_n (z - z_0)^{\alpha_n - 2} = 6 \left[ \sum_{n=0}^{\infty} a_n (z - z_0)^{\alpha_n} \right]^2 + z_0 + (z - z_0).$$

Here the terms of lowest order must cancel. Thus

$$\alpha_0(\alpha_0 - 1) a_0 (z - z_0)^{\alpha_0 - 2} = 6 a_0^2 (z - z_0)^{2\alpha_0}$$

so that

$$\alpha_0 = -2, \qquad a_0 = 1.$$

This is in agreement with our expectations. After these terms have been canceled, the lowest term on the left is

$$\alpha_1(\alpha_1 - 1) a_1 (z - z_0)^{\alpha_1 - 2}, \qquad \alpha_1 > -2,$$

while on the right we have two terms,

$$12 a_1 (z - z_0)^{\alpha_1 - 2} + z_0.$$

If $z_0 \neq 0$, the two terms on the right are of the same order iff $\alpha_1 = 2$, and for this value the term on the left is also constant. This gives $\alpha_1 = 2$, $a_1 = -\frac{1}{10}z_0$, which is true also if $z_0 = 0$.

At the next stage we should have

$$\alpha_2(\alpha_2 - 1)a_2(z - z_0)^{\alpha_2-2} = 12a_0(z - z_0)^{\alpha_0-2} + (z - z_0),$$

which gives

$$\alpha_2 = 3, \qquad a_2 = -\tfrac{1}{6}.$$

For $n = 3$ we should have

$$\alpha_3(\alpha_3 - 1)a_3(z - z_0)^{\alpha_3-2} = 12a_3(z - z_0)^{\alpha_3-2},$$

and this shows that $\alpha_3 = 4$ and leaves $a_3 = h$ arbitrary. We are thus going to get a one-parameter family of solutions. In this manner we prove successively that all exponents $\alpha_n$ are integers. Moreover we have $\alpha_n \geq n + 1$ for $n > 0$. The expansion of the solution starts out

$$w(z) = (z - z_0)^{-2} - \tfrac{1}{10}z_0(z - z_0)^2 - \tfrac{1}{6}(z - z_0)^3 + h(z - z_0)^4$$
$$+ \tfrac{1}{300}z_0^2(z - z_0)^6 + \tfrac{1}{150}z_0(z - z_0)^7 + \cdots. \qquad (12.2.5)$$

This formal series should now be proved convergent in some neighborhood of $z = z_0$. We start by choosing a number $M > 1$ such that

$$|z_0| < 10M, \qquad |h| < M^3. \qquad (12.2.6)$$

A look at the series (12.2.5) shows that we have then

$$|a_k| < M^k, \qquad 0 \leq k \leq 6. \qquad (12.2.7)$$

Now the coefficients $a_k$ are determined from the recurrence relation

$$[\alpha_n(\alpha_n - 1) - 12]a_n = 6\sum a_j a_k, \qquad (12.2.8)$$

where $j$ and $k$ are positive integers $< n$ whose sum $j + k \leq n$. The sum on the right in (12.2.8) contains at most $n - 1$ terms because for a given $j$, $1 \leq j \leq n - 1$, there is at most one $k$ in the same range such that $a_j a_k$ is actually present in the sum. The bracket in the left member equals $(n - 3)(n + 4)$ at least, since $\alpha_n \geq n + 1$. If (12.2.7) holds for the $j$'s and $k$'s under consideration, we have

$$|a_n| \leq \frac{6(n - 1)}{(n - 3)(n + 4)} M^n \leq M^n \qquad \text{for } 6 \leq n.$$

Since the inequality holds for $n < 6$, it holds for all $n$. This shows that the formal series (12.2.5) converges for

$$O < |z - z_0| < M^{-1}$$

and represents a solution of the Painlevé equation (1) in this punctured disk.

From this point on we shall accept Painlevé's result that the solutions of the equation

$$w'' = 6w^2 + z \tag{12.2.9}$$

are transcendental meromorphic functions with double poles. Hans Wittich (1953) considered these functions from the point of view of the Nevanlinna theory. The following remarks have a similar trend but veer off in a slightly different direction. To fix the ideas, let $z_0 \neq 0$ and consider the solution $w(z; z_0, 0, 0)$, which is holomorphic in some neighborhood of $z = z_0$. Multiply the equation by $2w'$, and integrate from $z_0$ to $z$ along a path which avoids the poles. In view of the initial conditions the result is

$$[w'(z)]^2 = 4[w(z)]^3 + zw(z) - \int_{z_0}^{z} w(t)\, dt. \tag{12.2.10}$$

Here we divide by $4[w(z)]^3$ to obtain

$$\frac{[w'(z)]^2}{4[w(z)]^3} = 1 + \frac{1}{4}\frac{z}{[w(z)]^2} - \frac{1}{4}\left\{ \int_{z_0}^{z} w(t)\, dt \Big/ [w(z)]^3 \right\}. \tag{12.2.11}$$

Consider a set $S$, defined as follows:

$$S = \{z\,;\, |w(z)| > |z|^{1/2}, |J(z)| < 1\}, \tag{12.2.12}$$

where $J(z) = [w(z)]^{-3} \int_{z_0}^{z} w(t)\, dt$. All poles of $w(z)$ belong to $S$. It is clear that the first condition is satisfied at a pole. Since $J(z) \to 0$ as $z$ tends to a pole, the second condition is also satisfied. If $z \in S$, both fractions in the right member of (12.2.11) are less than $\frac{1}{4}$ in absolute value, so the right-hand side is of the form $1 + h_0(z)$, where $|h_0(z)| < \frac{1}{2}$. Next we extract the square root and obtain

$$\frac{w'(z)}{2[w(z)]^{3/2}} = 1 + h(z), \qquad |h(z)| < \frac{1}{4} \quad \text{for } z \in S. \tag{12.2.13}$$

If now $z_1$ and $z_2$ are two points in $S$ which may be joined by a line segment in $S$, integration gives

$$[w(z_1)]^{-1/2} - [w(z_2)]^{-1/2} = \int_{z_1}^{z_2} [1 + h(t)]\, dt. \tag{12.2.14}$$

If now $z_2 = z_n$, a pole of $w(z)$, we obtain the inequality

$$\tfrac{16}{25}|z - z_n|^{-2} < |w(z)| < \tfrac{16}{9}|z - z_n|^{-2} \tag{12.2.15}$$

for any point $z$ which is "visible" in $S$ from $z_n$.

We can now introduce polar neighborhoods,

$$U_n = \{z\,;\, |z - z_n| < \tfrac{4}{3}|z_n|^{-1/2}\}. \tag{12.2.16}$$

That the constant $\frac{4}{3}$ is admissible is proved in the same manner as was used in Section 4.6, and the proof may be left to the reader. No two polar neighborhoods can intersect, as is shown using (12.2.14).

We can then use the polar statistics of Section 4.6 with $g = \frac{1}{2}$. This gives $n(r, \infty; w) = O(r^3)$, $m(r, \infty; w) = O(\log r)$, and

$$T(r; w) = O(r^3). \tag{12.2.17}$$

This estimate is surprisingly good for, as will be shown in the next section, $w(z)$ is of order $\frac{5}{2}$. We have found the estimate $\rho(w) = 3$ for the solution $w(z; z_0, 0, 0)$. Although the assumption $w(z_0) = 0$, $w'(z_0) = 0$ makes the formulas neater, the same result holds if this assumption is dropped.

Wittich has proved that the Nevanlinna defect $\delta(a) = 0$ for all values of $a$. The value $a = \infty$ is completely ramified for all solutions, but no other value can have this property.

**EXERCISE 12.2**

1. Show that a solution of (12.2.9) can have a triple zero at $z = z_0$ iff $z_0 = 0$.

2. Let $z = x > 0$, and take a solution $w(z; 0, w_0, w_1)$ with $w_0 > 0$, $(w_1)^2 > 4(w_0)^3$. Show that $w(x)$ becomes infinite for some $x = x_1$ with $0 < x_1 < (w_0)^{-1/2}$.

3. Let $w(z)$ be a solution of (12.2.9), $z_0$ not a pole of $w(z)$, and form $E(z) = \exp[-\int_{z_0}^{z} \int_{z_0}^{s} w(t) \, dt \, ds]$. Show that $E(z)$ is an entire function of $z$, and express $w(z)$ in terms of $E(z)$ and its derivatives. Show that a pole of $w(z)$ is a zero of $E(z)$. This function $E(z)$ plays a role in the Painlevé theory analogous to that of the sigma function in the theory of the Weierstrass $\wp$-function.

4. Consider a Painlevé equation of type (2). Show that its movable singularities are simple poles. Find the Laurent expansion at one of the poles, and give a convergence proof for the expansion. Take $a = 1$.

5. Show that the neighborhoods defined by (12.2.16) are actually contained in $S$ and hence do not overlap.

## 12.3. THE ASYMPTOTICS OF BOUTROUX

In 1913–1914 Boutroux published a memoir (180 pp.!) in which he examined the properties of the Painlevé transcendents from the point of view of complex function theory and also developed a method of asymptotic integration for a class of second order nonlinear DE's including those of Painlevé. We shall discuss briefly his results for equations of type (1), to which he gave the form

$$w'' = 6w^2 - 6z. \tag{12.3.1}$$

This may be obtained from (12.2.9) by dilation of the variables.

The equation is unchanged under the rotation

$$z \to \omega z, \qquad w \to \omega^3 w, \qquad \omega = \exp\frac{2\pi i}{5}. \tag{12.3.2}$$

This implies that, if $w(z)$ is a solution, so is

$$\omega^3 w(\omega z).$$

It also shows that (12.3.1) admits of the fifth roots of unity as distinguished directions, so that a study of the equation in one of these directions gives global information.

This observation suggests the following change of variables:

$$Z = \tfrac{4}{5} z^{5/4}, \qquad W = z^{1/2} w, \tag{12.3.3}$$

and the resulting equation

$$W'' - 6W^2 + 6 = -\frac{W'}{Z} + \frac{4}{25} \frac{W}{Z^2}. \tag{12.3.4}$$

This equation is asymptotic to the elliptic equation

$$W_0'' - 6W_0^2 + 6 = 0, \tag{12.3.5}$$

which is satisfied by $\wp(Z - Z_0; 12, g_3)$ for any choice of $Z_0$ and $g_3$. This is analogous to the situation encountered in Section 5.6, where the sine equation

$$Y_0'' + Y_0 = 0 \tag{12.3.6}$$

was found to be asymptotic to the generalized Bessel equation

$$Y'' + [1 - F(Z)]Y = 0 \tag{12.3.7}$$

under suitable assumptions on $F(Z)$.

The first integral of (12.3.5) is

$$(W_0')^2 = 4W_0^3 - 12W_0 - g_3 \tag{12.3.8}$$

with integrals as stated above. Normally the solution is doubly periodic with an array of poles at the points

$$Z_0 + 2m\omega_1 + 2n\omega_3,$$

where the periods $\omega_1$, $\omega_3$ may be computed from the value of $g_3$. The case where the cubic in the right member of (12.3.8) has equal roots, $g_3 = \pm 8$, is exceptional. Here solutions degenerate into simply periodic solutions:

$$1 + 3[\sinh 3^{1/2}(Z - Z_0)]^{-2} \quad \text{or} \quad -1 + 3[\sin 3^{1/2}(Z - Z_0)]^{-2} \tag{12.3.9}$$

with a single string of poles instead of a double array. This behavior of $W_0(Z)$ is reflected as asymptotic properties of $W(Z)$.

Now take a solution $W(Z)$ of (12.3.4) given by its initial values, $\eta_0$ and $\eta_1$, at a distant point $Z_0$, set

$$D = (\eta_1)^2 - 4(\eta_0)^3 + 12\eta_0 \neq \pm 8,$$

and let $W_0(Z)$ be the solution of

$$(W_0')^2 = 4(W_0)^3 - 12W_0 + D, \qquad (12.3.10)$$

which has the same initial values as $W(Z)$ at $Z = Z_0$. This is an elliptic function of $Z$. Let $a, b, c$ be the distinct roots of the cubic

$$4s^3 - 12s + D = 0.$$

At this stage it is convenient to work with the inverse functions $Z(W)$ and $Z_0(W)$. Now let $W$ describe a loop $L_1$, starting and ending at $\eta_0$ and surrounding $a$ and $b$ but not $c$. At $W = \eta_0$ both inverses start with the value $Z_0$, and at the end point of $L_1$ we have

$$Z_0(W) = Z_0 + 2\omega_1, \qquad Z(W) = Z_{01},$$

where $2\omega_1$ is the period

$$2\omega_1 = \int_{L_1} (4s^3 - 12s + D)^{-1/2} \, ds$$

of $W_0(Z; 12, D)$. Moreover,

$$|Z(W) - Z_0(W)| \qquad (12.3.11)$$

is small for all $W$ on $L_1$. In the same manner we can let $W$ describe a loop $L_3$ surrounding $b$ and $c$ but not $a$. Here the terminal values are, respectively,

$$Z_0 + 2\omega_3 \quad \text{and} \quad Z_{10}, \quad 2\omega_3 = \int_{L_3} (4s^3 - 12s + D)^{-1/2} \, ds.$$

These operations may now be repeated at $Z_{01}$ and $Z_{10}$, where in each case the initial values, the parameter $D$, the roots of the cubic, and the loops are given their local values. It follows from (12.3.11) that these values change by small amounts only when we go from $Z_0$ to $Z_{10}$ or $Z_{01}$. Suppose that at $Z_{01}$ we use the $L_3$-operation and at $Z_{10}$ the $L_1$-operation. For the class of DE's asymptotic to (12.3.5) distinct though nearby points would normally be obtained. But in the case under consideration we get the same point, $Z_{11}$. Now the four points

$$Z_0, \qquad Z_{01}, \qquad Z_{11}, \qquad \text{and} \qquad Z_{10}$$

form the vertices of a curvilinear quadrilateral $Q_{11}$, the sides of which are traced by different determinations of $Z(W)$ when the loops are traced.

This construction can now be repeated at the new vertices and leads to an array of lattice points $\{Z_{jk}\}$, obtained by $j$ applications of the $L_1$-operation and $k$ applications of the $L_3$-operations. The four points

$$Z_{j,k}, \qquad Z_{j+1,k}, \qquad Z_{j+1,k+1}, \qquad \text{and} \qquad Z_{j,k+1}$$

form the vertices of the curvilinear quadrilateral $Q_{j,k}$. These quadrilaterals form *approximate period parallelograms*. Each $Q_{j,k}$ contains a single double pole of $W(Z)$ and three zeros of $W'(Z)$. Furthermore,

$$|W(Z) - W_0(Z; 12, D_{j,k})| \qquad (12.3.12)$$

is small everywhere in $Q_{j,k}$ except for small neighborhoods of the poles. We have

$$D_{j,k} = [W'(Z_{j,k})]^2 - 4[W(Z_{j,k})]^3 + 12W(Z_{j,k}).$$

Here $j$ and $k$ are positive; but if the imaginary part of $Z_0$ is large, $Z_{j,k}$ can be defined for negative values of $j$, and if the real part is large, for negative values of $k$.

It should be noticed that the approximation is local and shifts from one quadrilateral to the next. It is true, however, that

$$\lim_{j \to \infty} D_{j,k} \qquad \text{and} \qquad \lim_{k \to \infty} D_{j,k}$$

exist. Furthermore,

$$\lim_{j \to \infty} (Z_{j+1,k} - Z_{j,k}) \qquad \text{and} \qquad \lim_{k \to \infty} (Z_{j,k+1} - Z_{j,k})$$

exist, and the first is real, the second purely imaginary.

If $D_{j,k}$ keeps away from the critical values $\pm 8$, the poles of $W(Z)$ form a net $\{Z_{j,k}\}$ which at a distance from the origin is approximately rectangular. The number of poles inside a big circle $|Z| = R$ satisfies $AR^2 < n(r) < BR^2$. Now, if $D$ has one of the critical values $\pm 8$, Boutroux shows that there is an ultimate string of poles approximately horizontal or vertical, beyond which there are no poles and $W(Z)$ tends to $+1$ or $-1$ as $Z \to \infty$ in the pole-free half-plane.

Now the mapping $Z = \frac{4}{3} z^{5/4}$ takes $Z$-half-planes into sectors. There are five such sectors,

$$k\tfrac{2}{5}\pi < \arg z < (k+1)\tfrac{2}{5}\pi, \qquad k = 0, 1, 2, 3, 4. \qquad (12.3.13)$$

Since the order of $W(Z)$ is 2, that of $w(z)$ will be $\frac{5}{2}$. Normally there will be a curvilinear net of poles of $w(z)$ in each of the sectors (12.3.13). There are, however, infinitely many *truncated* solutions where an ultimate string of poles exists in one sector or such a string extends through two adjacent sectors. Such a truncated solution ordinarily lacks one fifth of the normal allotment of poles. But it is not merely shorn of poles: every value $a$ is taken on with the same lowered frequency. There are, in addition, five *triply truncated solutions* related by the transformation (12.3.2). Such a solution has a net of poles in only one of the five sectors; each of the remaining sectors has at most a finite number of poles, and there

$$\lim_{z \to \infty} z^{-1/2} w(z) = +1. \qquad (12.3.14)$$

**EXERCISE 12.3**

1. Show that (12.3.14) is consistent with (12.2.12).
2. Verify (12.3.4).
3. The equation $w'' = 2w^3 - 2zw + a$ is reducible to Painlevé's type (2). How? Show that the transformation

$$Z = \tfrac{2}{3}z^{3/2}, \qquad W = z^{-1/2}w$$

transforms the equation into

$$W'' = 2W^3 - 2W - \frac{W'}{Z} + \frac{1}{9}\frac{W}{Z^2} + \frac{a}{Z}.$$

4. The $(Z, W)$-equation in Problem 3 has, as asymptotic elliptic equation,

$$W_0'' = 2W_0^3 - 2W_0 \qquad \text{or} \qquad (W_0')^2 = W_0^4 - 2W_0^2 + D.$$

Express the solution of the second equation in terms of a Jacobi elliptic function. What are the critical values of $D$ for which the solutions become degenerate, and what are the degenerate solutions?

5. Show that the order of the Jacobi functions is two and hence that the order of $w(z)$ is three.

6. Discuss Nevanlinna defects and complete ramifications for $w(z)$.


## 12.4. THE EMDEN AND THE THOMAS-FERMI EQUATIONS

The equations encountered so far in this chapter have fairly simple solutions. We shall now consider a class of equations which presents a more complicated picture: the Emden-Fowler and Thomas-Fermi equations. First we have something to say about the background of these equations.

In 1907 R. Emden published a book entitled *Gaskugeln*, in which he developed a theory of gas spheres with applications to cosmology and stellar dynamics. In this theory the equation

$$y''(z) = -x^{1-m}[y(x)]^m, \qquad 1 < m < 3, \tag{12.4.1}$$

plays a fundamental role. Further developments are due to R. H. Fowler in 1914 and 1931.

In the following we shall attach Emden's name to the equation

$$y''(x) = x^{1-m}[y(x)]^m, \tag{12.4.2}$$

an equation with much more interesting and spectacular properties. Set

$$m = \frac{p+2}{p}, \qquad 1 < p < \infty, \tag{12.4.3}$$

so that the equation becomes

$$w''(z) = z^{-2/p} [w(z)]^{(p+2)/p}. \tag{12.4.4}$$

In the physical theory the variables are normally real positive, but the various series which satisfy the equation (and which have been developed largely by physicists) call for complex variables.

The special case $p = 4$ is known as the Thomas-Fermi equation. This DE was encountered in 1927 by the American physicist L. H. Thomas, then working at Cambridge, England, on the potential field around an atom, and, independently, by the Italian Enrico Fermi, later of atom bomb fame, but in 1927 still working in Rome. Fermi encouraged two of his Italian mathematical colleagues, A. Mambriani and G. Scorza Dragoni, to provide rigorous proofs of the existence of solutions of a singular boundary value problem which he had encountered in the work. The Thomas-Fermi equation

$$w'' = z^{-1/2} w^{3/2} \tag{12.4.5}$$

has quite an extensive literature. Additional references are given below and at the end of the chapter.

Equation (12.4.4) is algebraic iff $p > 1$ is an integer, when it may be written as

$$(w'')^p = z^{-2} w^{p+2}.$$

Since it does not belong to any of the Painlevé classes, its movable singularities cannot be poles, though there are algebraic branch points. The only rational case occurs for $p = 2$:

$$w'' = z^{-1} w^2, \tag{12.4.6}$$

which repeatedly has served as a bugbear in earlier chapters.

If (12.4.4) has a solution $w(z)$ and if $k$ is any constant, then

$$k^{1-p} w(kz) \tag{12.4.7}$$

is also a solution. This can be used to place movable singularities where it is convenient, to normalize boundary conditions, etc.

The equation has an elementary solution,

$$w_e(z) = [p(p-1)]^{p/2} z^{1-p}, \tag{12.4.8}$$

which normally is singular at both the fixed singularities, zero and infinity.

This solution may be embedded in two systems of solutions, one valid in a neighborhood of infinity, and the other in a neighborhood of zero. The first system has the form

$$w_e(z) \left\{ 1 + \sum_{n=1}^{\infty} A_n c^n z^{n\sigma} \right\}. \tag{12.4.9}$$

Here $c$ is an arbitrary parameter, and $\sigma$ is the negative root of the equation

$$s^2 - (2p - 1)s - 2(p - 1) = 0, \qquad (12.4.10)$$

i.e.,

$$\sigma = p - \tfrac{1}{2} - [(p + \tfrac{1}{2})^2 - 2]^{1/2}. \qquad (12.4.11)$$

The series converges for large values of $|z|$. At $z = 0$ we have the second embedding system,

$$w_e(z)\Big\{1 + \sum_{n=1}^{\infty} B_n c^n z^{n\tau}\Big\}, \qquad (12.4.12)$$

where $\tau$ is the positive root of (12.4.10). The series converges for small values of $|z|$. For the Thomas-Fermi case these series are associated with the names of C. A. Coulson and N. H. March (1950).

We shall sketch how the series (12.4.9) is obtained. Set

$$w(x) = w_e(x)v(x). \qquad (12.4.13)$$

Here $v(x)$ is monotone and converges to $+1$ as $x \to \infty$. Furthermore, $v(x)$ satisfies the DE

$$x^2 v'' - 2(p - 1)xv' + p(p - 1)v = p(p - 1)v^{(p+2)/p}, \qquad (12.4.14)$$

which has the solution $v \equiv 1$. To get other solutions we set $v = 1 + u$. The resulting equation can be brought to the form

$$x^2 u'' - 2(p - 1)xu' - p(p - 1)u = F(u), \qquad (12.4.15)$$

where $F(u)$ is holomorphic for $|u| < 1$, and its Taylor series at the origin starts with a term in $u^2$. Now the homogeneous DE

$$x^2 U'' - 2(p - 1)xU' - p(p - 1)U = 0$$

is an Euler equation with the two linearly independent solutions

$$x^\sigma \quad \text{and} \quad x^\tau.$$

To get the system (12.4.9) we expand the solution of (12.4.15) in powers of $x^\sigma$. The system (12.12) is obtained by expanding $u$ in powers of $x^\tau$. Convergence is shown by a suitable application of Lindelöf's majorant method.

There is no real-valued integral which tends to $+\infty$ with $x$. It is not known whether there are other unbounded integrals at infinity. At zero we have a large number of additional integrals besides (12.4.12). For $p > 2$ the boundary value problem

$$\lim_{x \downarrow 0} y(x) = 1, \qquad \lim_{x \downarrow 0} y'(x) = a \qquad (12.4.16)$$

has a unique solution given by a series of the form

$$y(x) = 1 + ax + \sum_{1}^{\infty} c_n x^{\mu_n}, \tag{12.4.17}$$

where the $\mu$'s form an increasing sequence taken from the semimodule $k_1 + 2k_2[1 - (1/p)]$, where $k_1$ and $k_2$ are nonnegative integers. For $p = 4$, the Thomas-Fermi case, this series is known as the Baker series (E. B. Baker, 1930). The coefficients $c_n$ are rational functions of $p$, and the set of coefficients has infinitely many poles. In particular, the values

$$p = \frac{2k}{2k - 1}$$

are poles, and these are precisely the values for which the initial value problem $\lim_{x \downarrow 1} y(x) = 1$ does not have a unique solution and a logarithmic psi series appears. This series is of the form

$$y(x) = \sum P_n (\log x) x^{n/k}, \qquad p = \frac{2k}{2k - 1}. \tag{12.4.18}$$

Here $P_n(t)$ is a polynomial in $t$ of degree $[(n + 1)/(k + 1)]$, $P_0(t) = y_0 > 0$, and $P_{k+1}(0) = b$ is arbitrary. The series is convergent for small values of $|x|$. Actually there is a solution which is holomorphic at the origin and has the form

$$w(z) = \sum_{0}^{\infty} b_k z^{2k+1}. \tag{12.4.19}$$

This solution is unique.

So much for the fixed singularities. The point where a solution vanishes is obviously a singularity, and since the initial-value problem

$$w(z_0) = 0, \qquad w'(z_0) = b \tag{12.4.20}$$

has a unique solution, we have movable branch points. The corresponding expansion is

$$w(z) = \sum_{1}^{\infty} a_n (z - z_0)^{\lambda_n}; \tag{12.4.21}$$

$a_1 = b$, $\lambda_1 = 1$, $\lambda_2 = (3p + 2)/p$, and generally $\{\lambda_n\}$ is an increasing sequence taken from the same semimodule as the $\mu$'s.

But we have also movable infinitudes, and if $2p$ is an integer they involve logarithmic psi series. Of the real-valued solutions only those whose graphs lie below or on that of the elementary solution (12.4.8) do not become infinite somewhere. The following case is typical.

**THEOREM 12.4.1**

Let $y(x)$ be a solution of (12.4.4) with $y(a) = b$, $y'(a) = c$, where $a, b, c$ are positive numbers such that

$$(p + 1)c^2 \geqslant p a^{-2/p} b^{2(p+1)/p}. \tag{12.4.22}$$

*Then there exists an* X *with* $a < X$ *such that* $\lim_{x \uparrow X} y(x) = +\infty$ *and*

$$X^{1-1/p} < a^{1-1/p} + (p-1)\left(\frac{p+1}{p}\right)^{1/2} b^{-1/p}, \qquad (12.4.23)$$

*and for* $a < x < X$

$$y(x) < [p(p+1)]^{p/2} X(X-x)^{-p}. \qquad (12.4.24)$$

*Proof (sketch).*  Since

$$2y''(t)y'(t) = 2t^{-2/p}[y(t)]^{(p+2)/p}y'(t),$$

integration from $t = a$ to $t = x$ gives

$$[y'(x)]^2 - c^2 = 2\int_a^x t^{-2/p}[y(t)]^{(p+2)/p}y'(t)\,dt.$$

Here we integrate by parts to obtain

$$[y'(x)]^2 = c^2 - \frac{p}{p+1}a^{-2/p}b^{2(p+1)/p} + \frac{p}{p+1}x^{-2/p}[y(x)]^{2(p+1)/p}$$
$$+ \frac{2}{p+1}\int_a^x t^{-(p+1)/p}[y(t)]^{2(p+1)/p}\,dt.$$

By virtue of (12.4.22) we have

$$[y'(s)]^2 > \frac{p}{p+1}s^{-2/p}[y(s)]^{2(p+1)/p}, \qquad a \leq s, \qquad (12.4.25)$$

and since $y'(s) > 0$ we get

$$[y(s)]^{-1-1/p}y'(s) > \left(\frac{2}{p+1}\right)^{1/2}s^{-1/p}. \qquad (12.4.26)$$

Integration gives

$$\{b^{-1/p} - [y(x)]^{-1/p}\} > \left(\frac{2}{p+1}\right)^{1/2}\frac{1}{p-1}(x^{1-1/p} - a^{1-1/p}).$$

Here the left-hand side is bounded for all $x > a$, whereas the right side is not. It follows that there is a finite $X > a$ to the left of which the inequalities hold, but not to the right. This shows that $\lim_{x \uparrow X} y(x) = +\infty$. If in (12.4.25) we integrate from $x$ to $X$ instead, we get an inequality; and if we apply the mean-value theorem of the differential calculus to the right member, (12.4.24) results. The solution ceases to be real for $x > X$. The constant in the right member of (12.4.24) is the best one possible. ∎

We come now to the expansions at a movable infinitude. In view of (12.4.7) we can place this infinitude anywhere we choose. It simplifies the work somewhat to take $X = 1$. We now set

$$x = 1 - e^t, \qquad t < 0, \qquad (12.4.27)$$

in the DE and obtain

$$Y'' - Y' = e^{2t}(1 - e^t)^{-2/p}Y^{1+2/p}, \qquad (12.4.28)$$

where the primes denotes differentiation with respect to $t$. We postulate the existence of a solution of the form

$$Y(t) = \sum_{n=0}^{\infty} P_n(t) e^{(n-p)t}, \qquad (12.4.29)$$

where the $P_n$'s are polynomials in $t$ which may possibly reduce to constants. The degree and coefficients of $P_n$ are to be determined. If the series (12.4.29) has a domain of absolute convergence extending to $-\infty$, we can differentiate the series, substitute the result in (12.4.28), multiply by $e^{pt}$, expand $(1 - e^t)^{-2/p}$ in powers of $e^t$, collect terms, and equate the coefficient of $e^{nt}$ to zero for all $n$. For $n = 0$ we get

$$P_0''(t) - (2p + 1)P_0'(t) + p(p + 1)P_0(t) = [P_0(t)]^{1+2/p},$$

which has a unique polynomial solution, namely, the constant

$$P_0(t) = [p(p + 1)]^{p/2},$$

which we recognize as the factor occurring in the right member of (12.4.24).

We are finally led to an infinite system of linear nonhomogeneous second order DE's for the successive determination of the $P_n$'s. These equations are of the form

$$P_n''(t) + (2n - 2p - 1)P_n'(t) + (n + 1)(n - 2p - 2)P_n(t) = Q_n(t), \qquad (12.4.30)$$

where $Q_n$ is a multinomial in $P_1, P_2, \ldots, P_{n-1}$, $0 < n$. Here it makes a difference whether or not $2p$ is an integer. Suppose that $2p$ is not an integer, and that for $n < k$ the coefficients $P_n$ have been found to be specific constants. This means that the $Q_n$'s with $n \leq k$ are constants, and (12.4.30) then gives for $n = k$ the unique polynomial solution

$$P_k(t) = \frac{Q_k}{(k + 1)(k - 2p - 2)},$$

again a well-determined constant since the denominator cannot vanish for any value of $k$. The solution is then of the form

$$Y(t) = \sum_{k=0}^{\infty} P_k e^{(k-p)t}, \qquad y(x) = \sum_{k=0}^{\infty} P_k(1 - x)^{k-p}.$$

This solution has a branch point at $x = 1$ which is algebraic iff $p$ is rational.

If $2p$ is an integer, the situation changes radically. Then the coefficients are constants only for $n < 2p + 2$. For $n = 2p + 2$ (12.4.30) takes the form

$$P''_{2p+2}(t) + (2p + 1)P'_{2p+2}(t) = Q_{2p+2}(t).$$

Here the right member is a constant, and the equation has the polynomial solution

$$P_{2p+2}(t) = \frac{Q_{2p+2}}{2p+1} t + B,$$

where $B$ is an arbitrary constant. For $2p + 2 < n < 4p + 4$ the $Q_n$'s are first degree polynomials, and so are the $P_n$'s. For $n = 4p + 4$ there is a change, since $Q_{4p+4}(t)$ involves the square of $P_{2p+2}$ and hence $P_{4p+4}(t)$ is a quadratic polynomial and so on. Each time that $n$ passes a multiple of $2p + 2$ the degree of $P_n(t)$ increases by 1 unit.

The resulting series (12.4.29) may be shown to have a domain of absolute convergence extending to $-\infty$ in the $t$-plane or a punctured neighborhood of $z = 1$ on the Riemann surface of $\log(1 - z)$. The various proofs are nasty, and considerations of space require their omission. We formulate the results as

**THEOREM 12.4.2**

*The equation* (12.4.4) *has movable singularities. If* 2p *is not an integer* $> 2$, *then there is a solution at* $z = 1$ *of the form*

$$w(z) = \sum_{k=0}^{\infty} P_k(1 - z)^{k-p}. \tag{12.4.31}$$

*If* 2p *is an integer* $> 2$, *then the solution is given by the logarithmic psi series*

$$w(z) = \sum_{n=0}^{\infty} P_n[log\ (1 - z)](1 - z)^{n-p}, \tag{12.4.32}$$

*where* $P_n(t)$ *is a polynomial in* t *of degree* $[n/(2p + 2)]$. *Here* $P_{2p+2}(0) = B$ *is an arbitrary constant, and* $P_n(t)$ *is a polynomial in* B *as well as in* t.

**EXERCISE 12.4**

1. Verify (12.4.7).
2. Verify (12.4.8).
3. Prove that the function $x \mapsto v(x)$ defined by (12.4.13) has the stated property of monotony and existence of limit.
4. Verify (12.4.14) and (12.4.15), and find the leading term of $F(u)$.
5. Show that the graphs of two solutions of (14.4.14) cannot intersect and hence that a solution whose graph contains a point $(a, b)$ above the graph of $y = w_e(x)$ stays above the latter and is monotone decreasing to its limit.

6. Find the first two coefficients and exponents in Baker's series.
7. Show that the coefficients $c_n$ are rational functions of $p$ and that the point $2k/(2k-1)$ is a pole for some coefficients.
8. Discuss the series (12.4.12). Show that this solution exists also for $p < 1$, $p \neq 0$.
9. Verify doubtful points in the proof of Theorem 12.4.1.
10. Solve Problem 9 for Theorem 12.4.2.

## 12.5. QUADRATIC SYSTEMS

The term *quadratic system* was introduced by the Australian mathematician W. A. Coppel in 1966 for a pair of autonomous DE's of the form

$$\dot{x} = Q_1(x, y), \qquad \dot{y} = Q_2(x, y), \tag{12.5.1}$$

where $Q_1$ and $Q_2$ are quadratic polynomials in $x$ and $y$ with constant coefficients, and the dot indicates differentiation with respect to the variable $t$. We shall take the equations in the normal form

$$\dot{x} = xL_1(x, y), \qquad \dot{y} = yL_2(x, y), \tag{12.5.2}$$

where $L_1$ and $L_2$ are first degree polynomials,

$$L_1 = a_0 + a_1 x + a_2 y, \qquad L_2 = b_0 + b_1 x + b_2 y. \tag{12.5.3}$$

The coefficients and the variables are assumed to be real.

Coppel's paper was essentially devoted to the topological nature of the integral curves, singularities, closed cycles, etc. Although such questions will be considered in the following, our main task will be the existence and nature of the movable singularities. In particular, we shall consider the question, Are there solutions which are singular when $t \to t_0$, an arbitrary finite value?

Quadratic systems are not artificial constructs; as a matter of fact, Coppel lists no fewer than 13 problems in applied mathematics that lead to such systems, either directly or as the result of applying a suitable transformation to a nonlinear DE, usually of the second order. Thus quadratic systems arise in the theory of the struggle for life of two competing species (Lotka, 1925; Volterra, 1931). The second order nonlinear DE

$$w'' = z^{j-1} w^k \tag{12.5.4}$$

reduces to the quadratic system

$$\dot{x} = 1 - x + y, \qquad \dot{y} = j + kx - y \tag{12.5.5}$$

via the substitution

$$x = z\frac{w'}{w}, \qquad y = z^j\frac{w^k}{w'}, \qquad t = \log z.$$

Here Emden's equation corresponds to $j = 1 - 2/p$, $k = 1 + 2/p$. The more general Emden-Fowler equation of astrophysics,

$$[z^2 w']' + z^q w^n = 0, \qquad (12.5.6)$$

corresponds to $a_0 = a_1 = a_2 = -1$, $b_0 = q + 1$, $b_1 = n$, $b_2 = 1$. The equation

$$z^2 w'' = w^n, \qquad n \text{ integer} > 1, \qquad (12.5.7)$$

has been encountered by Kurt Mahler (private communication).

The system (12.5.2) has one and only one integral curve passing through any given point $(x_0, y_0)$, with normally four exceptions, namely, the points where $\dot{x}$ and $\dot{y}$ vanish simultaneously. There are four straight lines which determine these points: the coordinate axes $X$ and $Y$ and the lines $L_1 = 0$, $L_2 = 0$. We denote the points by $O$, $U$, $V$, and $W$. See Figure 12.1. In addition, there are normally three infinitary topological singularities, $X_\infty$, $Y_\infty$, and $F_\infty$, which are the end points of the axes and the line

$$F = L_2 - L_1 = 0. \qquad (12.5.8)$$

For our present purposes it is convenient to restrict the discussion to the class $E$, for which the layout is given schematically by the figure; i.e., $U$ lies on the positive $X$-axis, $V$ lies on the positive $Y$-axis, $O$ is the origin, and $W$ lies in the third quadrant. In addition it is required that both $L_1$ and $L_2$ be positive at the origin. The assumptions on the critical points imply that $O$, $U$, $V$, $W$ are the vertices of a nonconvex quadrilateral. For this case Coppel has proved that the critical points are either three saddle points and one nonsaddle point or the other way around. For the class $E$ the origin is always a node, but the other *phase portraits*, in the sense of Poincaré (see S. Lefschetz, 1963), may vary. Analytically the class $E$ is determined by the inequalities

$$a_0 > 0, \qquad a_1 < 0, \qquad a_2 > 0, \qquad b_0 > 0, \qquad b_1 > 0, \qquad b_2 < 0. \qquad (12.5.9)$$

The four lines $X$, $Y$, $L_1$, $L_2$ divide the plane into 11 sectors $S_1$–$S_{11}$, in each of which the four functions

$$x(t), \qquad y(t), \qquad \dot{x}(t), \qquad \dot{y}(t) \qquad (12.5.10)$$

keep a constant sign and at least one signature is changed when a boundary line is crossed. We are chiefly interested in solutions such that $x(t)$ and/or $y(t)$ becomes infinite as $t$ approaches a finite value. Such a solution has a graph which approaches one of the infinitary critical points $X_\infty$, $Y_\infty$, or $F_\infty$. The last case is the most interesting one.

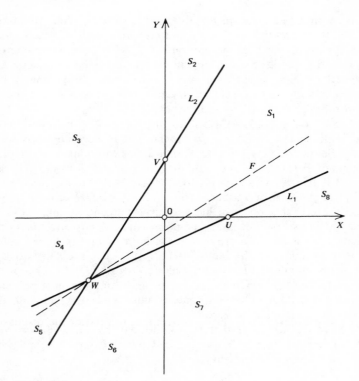

**Figure 12.1**

## THEOREM 12.5.1

*Take a point* $(x_0, y_0)$ *in* $S_1$, *and define a trajectory* $\Gamma = [x(t), y(t)]$ *with* $x(0) = x_0$, $y(0) = y_0$. *As* t *increases from* 0, *the trajectory* $\Gamma$ *stays in* $S_1$ *and tends to infinity as* t *tends to a finite value* $t_0$. *On* $\Gamma$ *the radius vector* $r(t)$ *is monotone increasing, while the argument* $\theta(t)$ *is ultimately monotone and tends to a limit* $\theta_0$ *as* $t \to t_0$ *and*

$$\tan \theta_0 = \frac{b_1 - a_1}{a_2 - b_2}. \tag{12.5.11}$$

*The line* F *or a parallel thereof is an asymptote of* $\Gamma$.

*Proof.* From the system (12.5.2) we obtain

$$r\dot{r} = x^2 L_1(x, y) + y^2 L_2(x, y), \tag{12.5.12}$$

$$\dot{\theta}(t) = \tfrac{1}{2} \sin [2\theta(t)][L_2(x, y) - L_1(x, y)]. \tag{12.5.13}$$

In $S_1$ both $L_1$ and $L_2$ are positive, and so are the four functions of (12.5.10). It follows that $x(t)$, $y(t)$, and $r(t)$ are strictly increasing as long as $\Gamma$ stays in $S_1$. But $\Gamma$ cannot get out of $S_1$ without crossing one of the lines $L_1$ and $L_2$, and the curve tangent would have to be vertical on $L_1$ and horizontal on $L_2$; neither alternative agrees with the geometry of the situation. It follows that $x(t)$, $y(t)$, and $r(t)$ tend to $+\infty$ as $t$ increases toward a limit which may possibly be $+\infty$.

We now look at (12.5.12). Here the sine factor $>0$ in $S_1$ and bounded away from zero. $L_2 - L_1$ is zero on the line $F$, negative above $F$, and positive below. If $(x_0, y_0)$ lies on $F$, then $\Gamma$ has to remain on $F$. Note that the slope of $F$ is $\tan \theta_0$. If $(x_0, y_0)$ lies in $S_1$ below $F$, it has to stay below, and $\theta(t)$ is increasing and tends to a limit $\theta_1 \leq \theta_0$. On the other hand, if $(x_0, y_0)$ lies above $F$ in $S_1$, there is a possibility for $\Gamma$ to cross $F$, provided that at the crossing one can have a curve tangent passing through the origin. At most one crossing is permissible. Thus $\theta(t)$ is certainly ultimately monotone. To prove that the limit of $\theta(t)$ is $\theta_0$, we note that $\dot\theta(t)$ is integrable, and the factor $\sin[2\theta(t)]$ is also integrable so $L_2 - L_1$ must be an integrable function of $t$ along $\Gamma$. Now $L_2 - L_1$ is proportional to the distance of $(x, y)$ from $F$, so the distance function must be bounded; and since it is ultimately monotone, it must tend to a limit $\geq 0$. If $t_0$ is infinite, the limit must be zero.

We have left to prove that $t_0$ is finite. To this end we observe that (12.5.12) implies that $r^{-2}\dot r$ is bounded away from zero and infinity. Integration gives an inequality,

$$mt < [r(0)]^{-1} - [r(t)]^{-1} < Mt.$$

When $t \uparrow t_0$, this yields $mt_0 < [r(0)]^{-1} < Mt_0$, so $t_0$ is finite. Integration of $[r(s)]^{-2}r'(s)$ from $s = t$ to $s = t_0$ gives

$$m(t_0 - t) < [r(t)]^{-1} < M(t_0 - t),$$

suggesting a simple pole of $r(t)$, $x(t)$, and $y(t)$ at $t = t_0$. This proves the theorem. ∎

Similar arguments may be given for solutions which are asymptotic to the $X$-axis or the $Y$-axis.

It is time to start the analytical discussion and consider the various series expansions. Here we must first examine the case where there are solutions with a singularity of $x(t)$ and $y(t)$ which is a simple pole. To assure this a certain parameter $q$ should not be a positive integer. But in addition to the polar solution we obtain also nonlogarithmic psi series. Since the system is autonomous, we can place the singularity anywhere. The choice $t = 0$ simplifies the exposition.

**THEOREM 12.5.2**

*The system will have a solution of the form*

$$x(t) = \alpha t^{-1} + \sum_{n=0}^{\infty} \alpha_n t^n,$$

$$y(t) = \beta t^{-1} + \sum_{n=0}^{\infty} \beta_n t^n, \tag{12.5.14}$$

*iff*

$$q = \frac{(a_1 - b_1)(a_2 - b_2)}{(a_1 b_2 - a_2 b_1)} \tag{12.5.15}$$

*is not a positive integer. Then the coefficients are uniquely determined, and*

$$\alpha = \frac{a_2 - b_2}{a_1 b_2 - a_2 b_1}, \qquad \beta = \frac{b_1 - a_1}{a_1 b_2 - a_2 b_1}. \tag{12.5.16}$$

*In addition, there are solutions of the form*

$$x(t, c) = \sum_{0}^{\infty} \sum_{0}^{\infty} a_{mn} c^n t^{m-1+nq},$$

$$y(t, c) = \sum_{0}^{\infty} \sum_{0}^{\infty} b_{mn} c^n t^{m-1+nq}, \tag{12.5.17}$$

*where c is an arbitrary parameter.*

We omit the proof. The existence of additional solutions in psi series was noted by Russell Smith. Such nonlogarithmic psi series also arise in the discussion of the solutions whose trajectories are associated with $X_\infty$ and $Y_\infty$. Here the exponents are taken from the set

$$M = \left\{ j - \frac{kb_1}{a_1}; j = -1, 0, 1, \ldots, k = 0, 1, 2, \ldots \right\}.$$

If $q$ is a positive integer we have

**THEOREM 12.5.3**

*If q is a positive integer and if*

$$z(t) = \begin{pmatrix} x(t) \\ y(t) \end{pmatrix}$$

*is a solution which becomes infinite when t decreases to zero, then*

$$z(t) = \sum_{n=0}^{\infty} V_n \left[ log \left( \frac{1}{t} \right) \right] t^{n-1}, \tag{12.5.18}$$

*where $V_n(s)$ is a vector-valued polynomial in s of degree $\leq [n/q]$. There* *exists a* $K > 0$ *such that the series*

$$\sum_{n=0}^{\infty} V_n(s) e^{(1-n)s} \tag{12.5.19}$$

*converges in norm for* s *real and greater than the positive root of the equation* $K(s + 2) = e^{qs}$.

Since the Emden equation may be transformed into a quadratic system (12.5.5), the convergence proofs for Theorems 12.5.2 and 12.5.3 will also constitute convergence proofs for the expansions listed in Theorem 12.4.2.

### EXERCISE 12.5

1. Verify the reduction of (12.5.4) to the system (12.5.5).
2. What are the coordinates of $U$, $V$, $W$?
3. Verify that the stated properties of $L_1$ and $L_2$ imply (12.5.9).
4. Discuss the question of $\Gamma$ escaping from $S_1$. Discuss also the geometry of the intersection of $\Gamma$ with $F = 0$.
5. Fill in missing details in the proof of the limit of $\theta(t)$ and for the finiteness of $t_0$.
6. Set up the recurrence relations for the coefficients of (12.5.14). How does the fact that $q$ is not a positive integer come into play? Verify (12.5.17).

### 12.6. OTHER AUTONOMOUS POLYNOMIAL SYSTEMS

This section is devoted to a discussion of higher polynomial systems, in the main reproducing recent results of Russell Smith. In the system of DE's

$$\dot{x} = P(x, y), \qquad \dot{y} = Q(x, y) \tag{12.6.1}$$

$P$ and $Q$ are polynomials in $x, y$ with constant coefficients, and the dot denotes differentiation with respect to a variable $t$. Both $P$ and $Q$ are of degree $n > 2$. The problem is again to study the movable singularities, their nature, and the associated series expansions.

We write $P$ and $Q$ as the sum of homogeneous polynomials

$$P(x, y) = \sum_{\nu=0}^{n} p_\nu(x, y), \qquad Q(x, y) = \sum_{\nu=0}^{n} q_\nu(x, y), \tag{12.6.2}$$

where $p_\nu$ and $q_\nu$ are homogeneous of degree $\nu$. The leading polynomials $p_n$ and $q_n$ contribute most to the behavior of the solutions. Smith defines two accompanying polynomials $\pi(z)$ and $\pi^*(z)$:

$$\pi(z) = zp_n(1, z) - q_n(1, z),$$
$$\pi^*(z) = zq_n(z, 1) - p_n(z, 1). \tag{12.6.3}$$

They are related by the identity

$$\pi^*(z) = -z^{n+1}\pi(z^{-1}).$$ (12.6.4)

Since $n > 2$, there are no longer single-valued solutions having simple poles. We do have algebraic solutions, however, provided that a certain parameter $\lambda$ is not a positive integer. For $n = 2$, $\lambda$ reduces to the parameter $q$ of (12.5.15).
Suppose that $\pi(z)$ has a root $z = \beta$ for which $p_n(1, \beta) \neq 0$. Then set

$$\lambda = \frac{\pi'(\beta)}{p_n(1, \beta)}, \qquad u = [(n-1)p_n(1, \beta)(t_0 - t)]^{1/(n-1)},$$ (12.6.5)

**THEOREM 12.6.1**

*If $\lambda$ is not a positive integer, then* (12.6.1) *has one and only one solution of the form*

$$x = u^{-1}\left(1 + \sum_{j=1}^{\infty} a_j u^j\right), \qquad y = u^{-1}\left(1 + \sum_{j=1}^{\infty} b_j u^j\right),$$ (12.6.6)

*where the power series have positive radii of convergence. If $\lambda > 0$, then* (12.6.1) *has also a family of solutions*

$$x = u^{-1}\{1 + P_1[u, u^\lambda(c + h \log u)]\},$$
$$y = u^{-1}\{1 + P_2[u, u^\lambda(c + h \log u)]\},$$ (12.6.7)

*where c is an arbitrary constant and $P_1(u, v)$, $P_2(u, v)$ are convergent power series without constant terms, whose coefficients are independent of c. Here $h = 0$ unless $\lambda$ is a positive integer, and then h is a rational function of $\lambda$ and the coefficients of P and Q.*

*Proof (sketch).* We use $P_k(u, v)$ as generic notation for a convergent power series in $u, v$ without a constant term. Smith uses several changes of variables and inversions to reduce the system (12.6.1) to a form to which known results on Briot-Bouquet equations will apply (see Section 11.1). The first substitution is $\xi = 1/x, z = y/x$, which leads to

$$\frac{d\xi}{dt} = -\xi^2 P\left(\frac{1}{\xi}, \frac{z}{\xi}\right), \qquad \frac{dz}{dt} = \xi P\left(\frac{1}{\xi}, \frac{z}{\xi}\right) - \xi z Q\left(\frac{1}{\xi}, \frac{z}{\xi}\right).$$ (12.6.8)

Here we take $\xi$ as an independent variable to get

$$\frac{dt}{d\xi} = \frac{-1}{\xi^2 P(1/\xi, z/\xi)}, \qquad \frac{dz}{d\xi} = \frac{zP(1/\xi, z/\xi) - Q(1/\xi, z/\xi)}{\xi P(1/\xi, z/\xi)}.$$ (12.6.9)

The virtue of this system is that the second equation involves only two variables, $z$ and $\xi$. Given any solution of this equation, we can substitute it into the first equation and integrate to obtain a solution $t(\xi), z(\xi)$ of

(12.6.9). By inverting $t = t(\xi)$, we get a solution $\xi(t), z(t)$ of (12.6.8), from which we can obtain a solution $x(t), y(t)$ of (12.6.1) via the relation $x = 1/\xi, y = z/\xi$.

We substitute (12.6.2) into (12.6.9) and use the homogeneity to get

$$\frac{dt}{d\xi} = \frac{-\xi^{n-2}}{p_n + \xi p_{n-1} + \xi^2 p_{n-2} + \cdots + \xi^n p_0}, \tag{12.6.10}$$

$$\xi \frac{dz}{d\xi} = \frac{(zp_n - q_n) + \xi(zp_{n-1} - q_{n-1}) + \cdots + \xi^n(zp_0 - q_0)}{p_n + \xi p_{n-1} + \xi^2 p_{n-2} + \cdots + \xi^n p_0}, \tag{12.6.11}$$

in which $p_j, q_j$ are understood to mean $p_j(1, z), q_j(1, z)$. The polynomial $(zp_n - q_n)$ in (12.6.11) equals $\pi(z)$, whose root is $\beta$. We substitute $z = \beta + \zeta$ in (12.6.11), and expand its right-hand side in powers of $\xi$ and $\zeta$ by Taylor's theorem to get

$$\xi \frac{d\zeta}{d\xi} = \lambda\zeta + \mu\xi + B(\xi, \zeta), \tag{12.6.12}$$

where

$$\lambda = \frac{\pi'(\beta)}{p_n(1, \beta)}, \qquad \mu = \frac{\beta p_{n-1}(1, \beta) - q_{n-1}(1, \beta)}{p_n(1, \beta)},$$

and the power series $B(\xi, \zeta)$ involves all terms of the Taylor expansion of degree $> 1$. The same procedure applied to (12.6.10) gives

$$\frac{dt}{d\xi} = -\frac{\xi^{n-2}}{p_n(1, \beta)} [1 + P_3(\xi, \zeta)] \tag{12.6.13}$$

with the usual convention on the series. Equations (12.6.12) and (12.6.13) form a system equivalent to (12.6.9). Here we use the results on Briot-Bouquet equations as given in Section 11.1. Theorem 11.1.3 applied to (12.6.12) gives

$$\zeta = P_4[\xi, \xi^\lambda(c + h \log \xi)], \tag{12.6.14}$$

provided that $\lambda > 0$. Here $h = 0$ unless $\lambda$ is a positive integer. In the power series $P_4(u, v) = \Sigma_{j,k} a_{jk} u^j v^k$ we have $a_{00} = 0$, $a_{01} = 1$, and the $a_{jk}$ are independent of $c$. When substituted in (12.6.13), this expression for $\zeta$ gives

$$-p_n(1, \beta) \frac{dt}{d\xi} = \xi^{n-2}\{1 + P_5[\xi, \xi^\lambda(c + h \log \xi)]\}. \tag{12.6.15}$$

When this is integrated with respect to $\xi$, the result is of the form

$$(n - 1)p_n(1, \beta)(t_0 - t) = \xi^{n-1}\{1 + P_6[\xi, \xi^\lambda(c + h \log \xi)]\}. \tag{12.6.16}$$

Something will be said later in justification of this step. Here $t_0$ is a constant of integration. Next we take the $(n - 1)$th root on both sides of

(12.6.16) and expand $(1 + P_6)^{1/(n-1)}$ in powers of $P_6$ by the binomial theorem. The powers of $P_6$ are expanded in powers of $\xi$ and of $\xi^\lambda(c + h \log \xi)$; the terms are collected and the result is

$$u = \xi\{1 + P_7[\xi, \xi^\lambda(c + h \log \xi)]\} \tag{12.6.17}$$

with $u$ as defined above. By inversion we express $\xi$ as a function of $u$ to obtain

$$\xi = u\{1 + P_8[u, u^\lambda(c + h \log u)]\}, \tag{12.6.18}$$

and now $x = 1/\xi$ gives the first equation under (12.6.7). Next we note that

$$y = \xi^{-1}z = \xi^{-1}(\beta + \zeta) = \xi^{-1}\{\beta + P_4[\xi, \xi^\lambda(c + \log \xi)]\},$$

where we have now to substitute the value of $\xi$ in terms of $u$ as given by (12.6.18). ∎

This completes the proof of Theorem 12.6.1 except for some comments on the admittedly very heavy analytical machinery. At various stages we are dealing with double series of the form

$$v = \sum_j \sum_k c_{jk} s^{j+p}[s^\lambda(c + h \log s)]^k. \tag{12.6.19}$$

On such series essentially three different operations are performed: (i) substitution of one double series into another of the same type, expansion, rearranging, and collecting terms; (ii) termwise integration and rearrangement and collecting terms; and (iii) inverting (12.6.19), i.e., solving for $s$ in terms of $v$. All these operations preserve the form of the series and preserve convergence for small values of the variables.

Ad (i). This follows from repeated application of the double series theorem of Weierstrass.

Ad (ii). Here the formulas

$$\int_0^x s^i(c + h \log s)^k \, ds = \Gamma x^{j+1}\left[h^k + \sum_{m=1}^k g_m(c + h \log x)^m\right], \tag{12.6.20}$$

$$\Gamma = (-1)^j j!(j+1)^{-k-1}, \qquad g_m = (-1)^m h^{k-m}[(j+1)^m(m-1)!]^{-1}$$

carry the burden of the proof plus again the Weierstrass double series theorem.

Ad (iii). This is too complicated for inclusion here, so the reader is referred to the original paper of Smith.

Smith remarks that one can interchange the roles of the symbols $x$ and $y$ in (12.6.1); this gives a new system but of the same type. In the new system the polynomials playing the roles of $p_n$ and $\pi$ in the old are simply the original $q_n(x, y)$ and $\pi^*(z)$. Here $\beta$ is replaced by $\alpha$, the root of $\pi^*(z) = 0$, now with the condition $q_n(\alpha, 1) \neq 0$. The new parameter $\lambda$ and the independent variable are now

$$\lambda^* = \frac{\pi^{*\prime}(\alpha)}{q_n(\alpha, 1)}, \qquad u^* = [(n-1)q_n(\alpha, 1)]^{1/(n-1)}. \qquad (12.6.21)$$

In terms of these quantities we get expansions for $x$ and $y$ of the types (12.6.6) and (12.6.7) with $\lambda$ and $u$ replaced by $\lambda^*$ and $u^*$, respectively. The reader is encouraged to formulate the result in detail. It is referred to below as THEOREM 12.6.2.

If $\alpha$ is a nonzero root of $\pi^*(z) = 0$, then $\beta = \alpha^{-1}$ is a nonzero root of $\pi(z) = 0$ by virtue of (12.6.4), and calculation shows that the solution given by Theorem 12.6.2 is essentially the same as that of Theorem 12.6.1. It is only when $\alpha = 0$ that something new is obtained.

At this juncture the question arises in (12.6.1) whether the two theorems obtained exhaust the finite singularities. Here Smith proves the following results, which we shall state without proof.

### THEOREM 12.6.3

*Suppose that* (i) $[p_n(x, y)]^2 + [q_n(x, y)]^2 > 0$ *for all real* $(x, y) \neq (0, 0)$, (ii) *at least one of the polynomials* $\pi(z)$, $\pi^*(z)$ *has a real root, and* (iii) *all the real roots of* $\pi(z)$, $\pi^*(z)$ *are simple roots. Then the real solutions of* (12.6.1) *can have singularities at* $t = t_0$ *only of the types given by Theorems 12.6.1 and 12.6.2.*

Here the crux of the argument involves the existence of real roots of $\pi(z)$, $\pi^*(z)$. These are real polynomials of degree $n + 1$, so a real root must exist if $n$ is even. If $n$ is odd, $n + 1$ even, it is possible that there are no real roots. In this case Smith shows that the argument function must tend to $\pm\infty$. This is the result:

### THEOREM 12.6.4

*Suppose that neither of the polynomials* $\pi(z)$, $\pi^*(z)$ *has a real root. If a real solution* $x(t)$, $y(t)$ *has a singularity at* $t = t_0$, *then both the functions* $x(t)$, $y(t)$ *oscillate as* $t \to t_0$ *and their limits of indetermination are* $\pm\infty$.

The possibility is not excluded that the solutions tend to infinity with $t$, even spirally. With this remark we end the discussion of Dr. Smith's basic paper.

## EXERCISE 12.6

1. Verify (12.6.4).
2. Interpret $\lambda$ and the condition on $p_n(1, \beta)$.
3. Show that for $n = 2$ we have $\lambda = q$ as given by (12.5.15), and find $\beta$ and $P_2(1, \beta)$.
4. Verify (12.6.8).
5. Verify (12.6.12).
6. Find the trajectories if $r^2 = x^2 + y^2$ and $P$, $Q$ are given by (i) $yr^2$, $-xr^2$, (ii) $yr^4 + \frac{1}{2}xr^2$, $\frac{1}{2}yr^2 - xr^4$, and (iii) $\frac{1}{4}x + yr^4$, $\frac{1}{4}y - xr^4$.

## LITERATURE

Section 12.4 of the author's LODE has a bearing on the first three sections of this chapter. The higher order elliptic Briot-Bouquet equations are scarcely considered in the literature. Advanced treatises on elliptic functions list the higher derivatives of $\wp(z)$ in terms of $\wp(z)$ and $\wp'(z)$. Since $\wp^{(2k)}(z)$ and $[\wp^{(2k+1)}(z)]^2$ are polynomials in $\wp(z)$, these expressions give examples of higher order elliptic Briot-Bouquet equations. See, for instance, the first collection of formulas in:

Tannery, J. and J. Molk. *Éléments de la théorie des fonctions elliptiques.* Vols. I–IV. Gauthier-Villars, Paris, 1893–1902; 2nd ed., Chelsea, New York, 1972.

For an excellent and detailed presentation of the Painlevé transcendents and the 50-odd equations discussed see Chapter 14 of:

Ince, E. L. *Ordinary Differential Equations.* Dover, New York, 1944.

See also:

Boutroux, P. Recherches sur les transcendents de M. Painlevé et l'étude asymptotique des équations différentielles du second ordre. *Ann. École Norm. Supér.*, (3), **30** (1913), 255–375; **31** (1914), 99–159.

Chazy, J. Sur les équations différentielles du troisième ordre et d'ordre supérieur dont l'intégrale générale a ses points critiques fixes. *Acta Math.*, **34** (1211), 317–385.

Fuchs, R. *Compt. Rend. Acad. Sci. Paris*, **141** (1905), 555.

Gambier, B. Sur les équations différentielles du second ordre et du premier degré dont l'intégrale est à points critiques fixés. *Acta Math.* 33 (1910), 1–55.

Garnier, R. *Ann. École Norm. Sup.*, (3), **27** (1912).

——. *Compt. Rend. Acad. Sci. Paris*, **162** (1916), 937; **163** (1916), 8, 118.

Horn, J. *Gewöhnliche Differentialgleichungen beliebiger Ordnung.* Sammlung Schubert, Vol. 50. Göschen, Leipzig, 1905, Chapter 14.

Painlevé, P., Memoire sur les équations différentielles dont l'intégrale est uniforme, *Bull. Soc. Math. France*, **28** (1900), 201–261.

——. Sur les équations différentielles du second ordre et d'ordre supérieur dont l'intégrale générale est uniforme. *Acta Math.*, **25** (1902), 1–82.

Picard, É. Remarques sur les équations différentielles. *Acta Math.*, **17** (1893), 297–300.

Valiron, G. *Équations Fonctionnelles. Applications.* Masson, Paris, 1950. Chapter 8.

The literature on Emden's and the Thomas-Fermi equations is vast. A few of the papers of more mathematical interest are:

Baker, E. B. Applications of the Thomas-Fermi statistical model to the calculations of potential distributions in positive ions. *Phys. Rev.*, **36** (1930), 630–647.

Coulson, C. A. and N. H. March. Moments in atoms using the Thomas-Fermi method. *Proc. Phys. Soc.*, (A) **63** (1950), 367–374.

Emden, R. *Gaskugeln: Anwendungen der mechanischen Wärmetheorie auf Kosmologie und meteorologischen Probleme.* Teubner, Leipzig, 1907.

Fermi, E. Un metodo statistico par la determinazione di alcune proprietà dell'atome. *Rend. Accad. Naz. Lincei, Cl. Sci. Fis. Mat. Nat.*, (6), **6** (1927), 602–607.

Fowler, R. H. The form near infinity of real continuous solutions of certain differential equations of the second order. *Quart. J. Pure Appl. Math.*, **45** (1914), 289–350.

——. Further studies of Emden's and similar differential equations. *Quart. J. Pure Appl. Math.*, Oxford Ser. 2 (1931) 239–258.

Hille, E. Some aspects of the Thomas-Fermi equation. *J. Analyse Math.*, **23** (1970), 147–170.

——. Aspects of Emden's equation. *J. Fac. Sci. Tokyo*, Sec. 1, **17** (1970), 11–30.

——. On a class of non-linear second order differential equations. *Rev. Unión Mat. Argentina*, **25** (1971), 319–334.

——. On a class of series expansions in the theory of Emden's equation. *Proc. Roy. Soc. Edinburgh*, (A) **71** (1972/73), 95–110.

Karamata, J. Sur l'application des théorèmes de nature tauberienne à l'étude des équations différentielles. *Publ. Inst. Math. Beograd.*, **1** (1947), 93–107.

—— and V. Marić. On some solutions of the differential equation $y''(x) = f(x)y^\lambda(x)$. *Rev. Fac. Arts Nat. Sci., Novi Sad*, **5** (1960), 415–422.

Mambriani, A. Sur un teorema relativo alle equazioni differentiali ordinarie del $2^a$ ordine. *Rend. Accad. Naz. Lincei, Cl. Sci. Fis. Mat. Nat.*, (6) **9** (1929), 142–144, 620–625.

Marić, V. and M. Skendžić. Unbounded solutions of the generalized Thomas-Fermi equation. *Math. Balkanica*, **3** (1973), 312–320.

Rijnierse, P. J. Algebraic solutions to the Thomas-Fermi equation for atoms. Thesis, St. Andrews University, 1968, 127 pp.

Thomas, L. H. The calculation of atomic fields. *Proc. Cambridge Phil. Soc.*, **13** (1927), 542–548.

For quadratic systems we can restrict ourselves to:

Coppel, W. A. A survey of quadratic systems. *J. Diff. Equations*, **2** (1966), 295–304.

Dulac, H. *Points singuliers des équations différentielles.* Mémorial des Sci. Math., No. 61. Gauthier-Villars, Paris, 1934.

Hille, E. A note on quadratic systems. *Proc. Roy. Soc. Edinburgh*, (A) **72** (1972/73), 17–37.

Lefschetz, S. *Differential Equations: Geometric Theory.* 2nd ed. Wiley-Interscience, New York, 1963.

Lotka, A. J. *Elements of Physical Biology.* Williams and Wilkins, Baltimore, 1925.

Volterra, V. *La lutte pour la vie.* Gauthier-Villars, Paris, 1931.

For higher polynomial autonomous systems we refer to Dulac, *loc. cit.*, and also to:

Dulac, H. Solutions d'un système d'équations différentielles dans le voisinage de valeurs singulières. *Bull. Soc. Math. France*, **40** (1912), 324–383.

Ince, E. L. *Loc. cit.*, Chapter 13.

Malmquist, J. Sur les points singuliers des équations différentielles. *Ark. Mat., Astr. Fys.*, **15**, Nos. 3, 27 (1921).

Sansone, G. and R. Conti. *Non-linear Differential Equations* (translated by A. H. Diamond). Pergamon Press. Macmillan, New York, 1964.

Smith, R. A. Singularities of solutions of certain plane autonomous systems. *Proc. Roy. Soc. Edinburgh*, (A) **72**:26 (1973/74), 307–315.

# BIBLIOGRAPHY

Bellman, Richard. *A Survey of the Theory of Boundedness, Stability, and Asymptotic Behavior of Solutions of Linear and Non-linear Differential and Difference Equations.* Office of Naval Research, NAVEXOS P-596. Washington, D.C. (1949).

────. *Stability Theory of Differential Equations.* McGraw-Hill, New York, 1953.

Bieberbach, Ludwig. *Theorie der gewöhnlichen Differentialgleichungen auf funktionen-theoretischen Grundlage dargestellt.* Springer-Verlag, Berlin, 1953.

Bôcher, Maxime. *Leçons sur les méthodes de Sturm dans la théorie des équations différentielles linéaires et leurs dévelopments modernes.* Gauthier-Villars, Paris, 1917.

Borůvka, Otakar. *Lineare Differentialtransformationen 2. Ordnung.* VEB Deutscher Verlag der Wissenschaften. Berlin, 1967.

Bourbaki, N. *Elements de mathématiques.* Vol. IV: *Fonctions d'une variable réelle.* Chapter IV: Équations différentielles. Actualités scientifiques et industrielles, 1132. Hermann, Paris, 1951.

Boutroux, Pierre. *Leçons sur les fonctions définies par les équations différentielles du premier ordre.* Gauthier-Villars, Paris, 1908.

Briot, Charles and Jean Claude Bouquet. *Théorie des fonctions elliptiques.* 2nd ed. Paris, 1875.

Cesari, Lamberto. *Asymptotic Behavior and Stability Problems in Ordinary Differential Equations.* Ergebnisse d. Math., N.S., No. 16, Springer-Verlag, Berlin, 1959.

Coddington, E. A. and Norman Levinson. *Theory of Ordinary Differential Equations.* McGraw-Hill, New York, 1955.

Coppel, W. A. *Stability and Asymptotic Behavior of Differential Equations.* Heath, Boston, 1965.

────. *Disconjugacy.* Lecture Notes in Mathematics, No. 220. Springer-Verlag. Berlin, 1971.

Dulac, Henri. *Points singuliers des équations différentielles.* Mémorial des Sci. Math., No. 61. Gauthier-Villars, Paris, 1934.

Emden, R. *Gaskugeln: Anwendungen der mechanischen Wärmetheorie auf Kosmologie und meteorologischen Probleme.* Teubner, Leipzig, 1907.

Erdelyi, A., W. Magnus, F. Oberhettinger, and F. Tricomi. *Higher Transcendental Functions.* 3 vols. McGraw-Hill, New York, 1953–1955.

Forsyth, A. R. *Theory of Differential Equations.* 6 Vols. Cambridge University Press, London, 1900–1902.

Hartman, Philip. *Ordinary Differential Equations.* Wiley, New York, 1964.

Hayman, W. K. *Meromorphic Functions.* Clarendon Press, Oxford, 1964.

Hilb, Emil. *Lineare Differentialgleichungen im komplexen Gebiet. Encyklopädie der mathemetischen Wissenschaften.* Vol. II:2. Teubner, Leipzig, 1915.

Hille, Einar. *Analytic Function Theory.* 2 vols. Ginn, Boston, 1959, 1962; Chelsea, New York, 1973.

———. *Analysis.* Vol. II. Ginn-Blaisdell, Waltham, Mass., 1966, Chapter 20.

———. *Lectures on Ordinary Differential Equations.* Addison-Wesley, Reading, Mass., 1969.

———. *Methods in Classical and Functional Analysis.* Addison-Wesley. Reading, Mass., 1970.

Horn, Jakob. *Gewöhnliche Differentialgleichungen beliebiger Ordnung.* Sammlung Schubert, Vol. 50. Göschen, Leipzig, 1905.

Hukuhara, M., T. Kimura, and T. Matuda. *Équations différentielles ordinaires du premier ordre dans le champ complexe.* Publications of the Mathematical Society of Japan, No. 7, 1961.

Ince, E. L. *Ordinary Differential Equations.* Dover, New York, 1944.

Kamke, E. *Differentialgleichungen: Lösungen und Lösungsmethoden.* Vol. 1, 3rd ed. Chelsea, New York, 1948.

Klein, Felix. *Vorlesungen über die hypergeometrische Funktion.* Springer-Verlag, Berlin, 1933.

Lappo-Danilevskij, J. A. *Théorie des systèmes des équations différentielles linéaires.* Chelsea, New York, 1953.

Lefschetz, Solomon. *Differential Equations: Geometric Theory.* 2nd ed. Wiley-Interscience, New York, 1963.

Massera, J. L. and J. J. Schaffer. *Linear Differential Equations and Function Spaces.* Academic Press, New York, 1966.

Nevanlinna, Rolf. *Le théorème de Picard-Borel et la théorie des fonctions méromorphes.* Gauthier-Villars, Paris, 1929.

Painlevé, Paul. *Leçons sur la théorie analytique des équations différentielles, profesées à Stockholm, 1895.* Hermann, Paris, 1897.

Poincaré, Henri. *Les méthodes nouvelles de la méchanique céleste.* 3 vols. Gauthier-Villars, Paris, 1892, 1893, 1899.

Sansone, G. and R. Conti. *Non-linear Differential Equations* (translated by A. H. Diamond). Pergamon Press, Macmillan, New York, 1964.

Schlesinger, Ludwig. *Handbuch der Theorie der linearen Differentialgleichungen.* Vols. 1, 2:1, 2:2. Teubner, Leipzig, 1895, 1897, 1898.

———. *Einführung in die Theorie der Differentialgleichungen mit einer unabhängigen Veriablen.* 2nd ed. Sammlung Schubert, Vol. 13. Göschen, Leipzig, 1904.

Titchmarsh, E. C. *Eigenfunction Expansions Associated with Second Order Differential Equations.* Clarendon Press, Oxford, 1947.

Tricomi, Francesco. *Fonctions hypergéométriques confluentes.* Gauthier-Villars, Paris, 1960.

———. *Equazioni differenziale.* Einudi, Turin, 1965.

Trjitzinsky, W. J. *Analytic Theory of Non-linear Differential Equations.* Mémorial des Sci. Math., No. 90. Gauthier-Villars, Paris, 1938.

Valiron, Georges. *Équations fonctionnelles: Applications.* 2nd ed. Masson, Paris, 1950.

Walter, Wolfgang. *Differential- und Integralungleichungen.* Springer-Verlag, Berlin, 1964.

Watson, G. N. *Theory of Bessel Functions.* 2nd ed. Cambridge University Press, London, 1962.

Weiss, Leonard. *Ordinary Differential Equations: 1971 NRL-MRC Conference.* Academic Press, New York, 1972.

Whittaker, E. T. and G. N. Watson. *A Course in Modern Analysis.* 4th ed. Cambridge University Press, London, 1952.

Wittich, Hans. *Neuere Untersuchungen über eindeutige analytische Funktionen.* Ergebnisse d. Math., N.S., No. 8. Springer-Verlag, Berlin, 1955; 2nd ed., 1968.

Yosida, Kôsaku. *Lectures on Differential and Integral Equations.* Wiley-Interscience, New York, 1960.

# INDEX

To facilitate the use of the Index it is partly analytical. Thus, e.g., Bessel's equation is listed under BESSEL and Differential Equations, Linear but not under Equations.

**471**

# A CATALOG OF SELECTED
# **DOVER BOOKS**
## IN SCIENCE AND MATHEMATICS

# A CATALOG OF SELECTED
# DOVER BOOKS
# IN SCIENCE AND MATHEMATICS

QUALITATIVE THEORY OF DIFFERENTIAL EQUATIONS, V.V. Nemytskii and V.V. Stepanov. Classic graduate-level text by two prominent Soviet mathematicians covers classical differential equations as well as topological dynamics and ergodic theory. Bibliographies. 523pp. 5⅜ × 8½. 65954-2 Pa. $14.95

MATRICES AND LINEAR ALGEBRA, Hans Schneider and George Phillip Barker. Basic textbook covers theory of matrices and its applications to systems of linear equations and related topics such as determinants, eigenvalues and differential equations. Numerous exercises. 432pp. 5⅜ × 8½. 66014-1 Pa. $10.95

QUANTUM THEORY, David Bohm. This advanced undergraduate-level text presents the quantum theory in terms of qualitative and imaginative concepts, followed by specific applications worked out in mathematical detail. Preface. Index. 655pp. 5⅜ × 8½. 65969-0 Pa. $14.95

ATOMIC PHYSICS (8th edition), Max Born. Nobel laureate's lucid treatment of kinetic theory of gases, elementary particles, nuclear atom, wave-corpuscles, atomic structure and spectral lines, much more. Over 40 appendices, bibliography. 495pp. 5⅜ × 8½. 65984-4 Pa. $12.95

ELECTRONIC STRUCTURE AND THE PROPERTIES OF SOLIDS: The Physics of the Chemical Bond, Walter A. Harrison. Innovative text offers basic understanding of the electronic structure of covalent and ionic solids, simple metals, transition metals and their compounds. Problems. 1980 edition. 582pp. 6⅛ × 9¼. 66021-4 Pa. $16.95

BOUNDARY VALUE PROBLEMS OF HEAT CONDUCTION, M. Necati Özisik. Systematic, comprehensive treatment of modern mathematical methods of solving problems in heat conduction and diffusion. Numerous examples and problems. Selected references. Appendices. 505pp. 5⅜ × 8½. 65990-9 Pa. $12.95

A SHORT HISTORY OF CHEMISTRY (3rd edition), J.R. Partington. Classic exposition explores origins of chemistry, alchemy, early medical chemistry, nature of atmosphere, theory of valency, laws and structure of atomic theory, much more. 428pp. 5⅜ × 8½. (Available in U.S. only) 65977-1 Pa. $11.95

A HISTORY OF ASTRONOMY, A. Pannekoek. Well-balanced, carefully reasoned study covers such topics as Ptolemaic theory, work of Copernicus, Kepler, Newton, Eddington's work on stars, much more. Illustrated. References. 521pp. 5⅜ × 8½. 65994-1 Pa. $12.95

PRINCIPLES OF METEOROLOGICAL ANALYSIS, Walter J. Saucier. Highly respected, abundantly illustrated classic reviews atmospheric variables, hydrostatics, static stability, various analyses (scalar, cross-section, isobaric, isentropic, more). For intermediate meteorology students. 454pp. 6⅛ × 9¼. 65979-8 Pa. $14.95

RELATIVITY, THERMODYNAMICS AND COSMOLOGY, Richard C. Tolman. Landmark study extends thermodynamics to special, general relativity; also applications of relativistic mechanics, thermodynamics to cosmological models. 501pp. 5⅜ × 8½. 65383-8 Pa. $13.95

APPLIED ANALYSIS, Cornelius Lanczos. Classic work on analysis and design of finite processes for approximating solution of analytical problems. Algebraic equations, matrices, harmonic analysis, quadrature methods, much more. 559pp. 5⅜ × 8½. 65656-X Pa. $13.95

INTRODUCTION TO ANALYSIS, Maxwell Rosenlicht. Unusually clear, accessible coverage of set theory, real number system, metric spaces, continuous functions, Riemann integration, multiple integrals, more. Wide range of problems. Undergraduate level. Bibliography. 254pp. 5⅜ × 8½. 65038-3 Pa. $8.95

INTRODUCTION TO QUANTUM MECHANICS With Applications to Chemistry, Linus Pauling & E. Bright Wilson, Jr. Classic undergraduate text by Nobel Prize winner applies quantum mechanics to chemical and physical problems. Numerous tables and figures enhance the text. Chapter bibliographies. Appendices. Index. 468pp. 5⅜ × 8½. 64871-0 Pa. $12.95

ASYMPTOTIC EXPANSIONS OF INTEGRALS, Norman Bleistein & Richard A. Handelsman. Best introduction to important field with applications in a variety of scientific disciplines. New preface. Problems. Diagrams. Tables. Bibliography. Index. 448pp. 5⅜ × 8½. 65082-0 Pa. $12.95

MATHEMATICS APPLIED TO CONTINUUM MECHANICS, Lee A. Segel. Analyzes models of fluid flow and solid deformation. For upper-level math, science and engineering students. 608pp. 5⅜ × 8½. 65369-2 Pa. $14.95

ELEMENTS OF REAL ANALYSIS, David A. Sprecher. Classic text covers fundamental concepts, real number system, point sets, functions of a real variable, Fourier series, much more. Over 500 exercises. 352pp. 5⅜ × 8½. 65385-4 Pa. $11.95

PHYSICAL PRINCIPLES OF THE QUANTUM THEORY, Werner Heisenberg. Nóbel Laureate discusses quantum theory, uncertainty, wave mechanics, work of Dirac, Schroedinger, Compton, Wilson, Einstein, etc. 184pp. 5⅜ × 8½. 60113-7 Pa. $6.95

INTRODUCTORY REAL ANALYSIS, A.N. Kolmogorov, S.V. Fomin. Translated by Richard A. Silverman. Self-contained, evenly paced introduction to real and functional analysis. Some 350 problems. 403pp. 5⅜ × 8½. 61226-0 Pa. $10.95

PROBLEMS AND SOLUTIONS IN QUANTUM CHEMISTRY AND PHYSICS, Charles S. Johnson, Jr. and Lee G. Pedersen. Unusually varied problems, detailed solutions in coverage of quantum mechanics, wave mechanics, angular momentum, molecular spectroscopy, scattering theory, more. 280 problems plus 139 supplementary exercises. 430pp. 6½ × 9¼. 65236-X Pa. $13.95

ASYMPTOTIC METHODS IN ANALYSIS, N.G. de Bruijn. An inexpensive, comprehensive guide to asymptotic methods—the pioneering work that teaches by explaining worked examples in detail. Index. 224pp. 5⅜ × 8½.     64221-6 Pa. $7.95

OPTICAL RESONANCE AND TWO-LEVEL ATOMS, L. Allen and J.H. Eberly. Clear, comprehensive introduction to basic principles behind all quantum optical resonance phenomena. 53 illustrations. Preface. Index. 256pp. 5⅜ × 8½.
65533-4 Pa. $8.95

COMPLEX VARIABLES, Francis J. Flanigan. Unusual approach, delaying complex algebra till harmonic functions have been analyzed from real variable viewpoint. Includes problems with answers. 364pp. 5⅜ × 8½.    . 61388-7 Pa. $9.95

ATOMIC SPECTRA AND ATOMIC STRUCTURE, Gerhard Herzberg. One of best introductions; especially for specialist in other fields. Treatment is physical rather than mathematical. 80 illustrations. 257pp. 5⅜ × 8½.     60115-3 Pa. $6.95

APPLIED COMPLEX VARIABLES, John W. Dettman. Step-by-step coverage of fundamentals of analytic function theory—plus lucid exposition of five important applications: Potential Theory; Ordinary Differential Equations; Fourier Transforms; Laplace Transforms; Asymptotic Expansions. 66 figures. Exercises at chapter ends. 512pp. 5⅜ × 8½.     64670-X Pa. $12.95

ULTRASONIC ABSORPTION: An Introduction to the Theory of Sound Absorption and Dispersion in Gases, Liquids and Solids, A.B. Bhatia. Standard reference in the field provides a clear, systematically organized introductory review of fundamental concepts for advanced graduate students, research workers. Numerous diagrams. Bibliography. 440pp. 5⅜ × 8½.     64917-2 Pa. $11.95

UNBOUNDED LINEAR OPERATORS: Theory and Applications, Seymour Goldberg. Classic presents systematic treatment of the theory of unbounded linear operators in normed linear spaces with applications to differential equations. Bibliography. 199pp. 5⅜ × 8½.     64830-3 Pa. $7.95

LIGHT SCATTERING BY SMALL PARTICLES, H.C. van de Hulst. Comprehensive treatment including full range of useful approximation methods for researchers in chemistry, meteorology and astronomy. 44 illustrations. 470pp. 5⅜ × 8½.     64228-3 Pa. $11.95

CONFORMAL MAPPING ON RIEMANN SURFACES, Harvey Cohn. Lucid, insightful book presents ideal coverage of subject. 334 exercises make book perfect for self-study. 55 figures. 352pp. 5⅜ × 8¼.     64025-6 Pa. $11.95

OPTICKS, Sir Isaac Newton. Newton's own experiments with spectroscopy, colors, lenses, reflection, refraction, etc., in language the layman can follow. Foreword by Albert Einstein. 532pp. 5⅜ × 8½.     60205-2 Pa. $11.95

GENERALIZED INTEGRAL TRANSFORMATIONS, A.H. Zemanian. Graduate-level study of recent generalizations of the Laplace, Mellin, Hankel, K. Weierstrass, convolution and other simple transformations. Bibliography. 320pp. 5⅜ × 8½.     65375-7 Pa. $8.95

THE ELECTROMAGNETIC FIELD, Albert Shadowitz. Comprehensive undergraduate text covers basics of electric and magnetic fields, builds up to electromagnetic theory. Also related topics, including relativity. Over 900 problems. 768pp. 5⅜ × 8¼.                    65660-8 Pa. $18.95

FOURIER SERIES, Georgi P. Tolstov. Translated by Richard A. Silverman. A valuable addition to the literature on the subject, moving clearly from subject to subject and theorem to theorem. 107 problems, answers. 336pp. 5⅜ × 8½.
63317-9 Pa. $9.95

THEORY OF ELECTROMAGNETIC WAVE PROPAGATION, Charles Herach Papas. Graduate-level study discusses the Maxwell field equations, radiation from wire antennas, the Doppler effect and more. xiii + 244pp. 5⅜ × 8½.
65678-0 Pa. $6.95

DISTRIBUTION THEORY AND TRANSFORM ANALYSIS: An Introduction to Generalized Functions, with Applications, A.H. Zemanian. Provides basics of distribution theory, describes generalized Fourier and Laplace transformations. Numerous problems. 384pp. 5⅜ × 8½.                    65479-6 Pa. $11.95

THE PHYSICS OF WAVES, William C. Elmore and Mark A. Heald. Unique overview of classical wave theory. Acoustics, optics, electromagnetic radiation, more. Ideal as classroom text or for self-study. Problems. 477pp. 5⅜ × 8½.
64926-1 Pa. $12.95

CALCULUS OF VARIATIONS WITH APPLICATIONS, George M. Ewing. Applications-oriented introduction to variational theory develops insight and promotes understanding of specialized books, research papers. Suitable for advanced undergraduate/graduate students as primary, supplementary text. 352pp. 5⅜ × 8½.                    64856-7 Pa. $9.95

A TREATISE ON ELECTRICITY AND MAGNETISM, James Clerk Maxwell. Important foundation work of modern physics. Brings to final form Maxwell's theory of electromagnetism and rigorously derives his general equations of field theory. 1,084pp. 5⅜ × 8½.        60636-8, 60637-6 Pa., Two-vol. set $23.90

AN INTRODUCTION TO THE CALCULUS OF VARIATIONS, Charles Fox. Graduate-level text covers variations of an integral, isoperimetrical problems, least action, special relativity, approximations, more. References. 279pp. 5⅜ × 8½.
65499-0 Pa. $8.95

HYDRODYNAMIC AND HYDROMAGNETIC STABILITY, S. Chandrasekhar. Lucid examination of the Rayleigh-Benard problem; clear coverage of the theory of instabilities causing convection. 704pp. 5⅜ × 8¼.        64071-X Pa. $14.95

CALCULUS OF VARIATIONS, Robert Weinstock. Basic introduction covering isoperimetric problems, theory of elasticity, quantum mechanics, electrostatics, etc. Exercises throughout. 326pp. 5⅜ × 8½.                    63069-2 Pa. $8.95

DYNAMICS OF FLUIDS IN POROUS MEDIA, Jacob Bear. For advanced students of ground water hydrology, soil mechanics and physics, drainage and irrigation engineering and more. 335 illustrations. Exercises, with answers. 784pp. 6⅛ × 9¼.                    65675-6 Pa. $19.95

NUMERICAL METHODS FOR SCIENTISTS AND ENGINEERS, Richard Hamming. Classic text stresses frequency approach in coverage of algorithms, polynomial approximation, Fourier approximation, exponential approximation, other topics. Revised and enlarged 2nd edition. 721pp. 5⅜ × 8½.
65241-6 Pa. $15.95

THEORETICAL SOLID STATE PHYSICS, Vol. I: Perfect Lattices in Equilibrium; Vol. II: Non-Equilibrium and Disorder, William Jones and Norman H. March. Monumental reference work covers fundamental theory of equilibrium properties of perfect crystalline solids, non-equilibrium properties, defects and disordered systems. Appendices. Problems. Preface. Diagrams. Index. Bibliography. Total of 1,301pp. 5⅜ × 8½. Two volumes.               Vol. I 65015-4 Pa. $16.95
Vol. II 65016-2 Pa. $14.95

OPTIMIZATION THEORY WITH APPLICATIONS, Donald A. Pierre. Broad-spectrum approach to important topic. Classical theory of minima and maxima, calculus of variations, simplex technique and linear programming, more. Many problems, examples. 640pp. 5⅜ × 8½.               65205-X Pa. $14.95

THE CONTINUUM: A Critical Examination of the Foundation of Analysis, Hermann Weyl. Classic of 20th-century foundational research deals with the conceptual problem posed by the continuum. 156pp. 5⅜ × 8½.   67982-9 Pa. $6.95

ESSAYS ON THE THEORY OF NUMBERS, Richard Dedekind. Two classic essays by great German mathematician: on the theory of irrational numbers; and on transfinite numbers and properties of natural numbers. 115pp. 5⅜ × 8½.
21010-3 Pa. $5.95

THE FUNCTIONS OF MATHEMATICAL PHYSICS, Harry Hochstadt. Comprehensive treatment of orthogonal polynomials, hypergeometric functions, Hill's equation, much more. Bibliography. Index. 322pp. 5⅜ × 8½.       65214-9 Pa. $9.95

NUMBER THEORY AND ITS HISTORY, Oystein Ore. Unusually clear, accessible introduction covers counting, properties of numbers, prime numbers, much more. Bibliography. 380pp. 5⅜ × 8½.               65620-9 Pa. $9.95

THE VARIATIONAL PRINCIPLES OF MECHANICS, Cornelius Lanczos. Graduate level coverage of calculus of variations, equations of motion, relativistic mechanics, more. First inexpensive paperbound edition of classic treatise. Index. Bibliography. 418pp. 5⅜ × 8½.               65067-7 Pa. $12.95

MATHEMATICAL TABLES AND FORMULAS, Robert D. Carmichael and Edwin R. Smith. Logarithms, sines, tangents, trig functions, powers, roots, reciprocals, exponential and hyperbolic functions, formulas and theorems. 269pp. 5⅜ × 8½.               60111-0 Pa. $6.95

THEORETICAL PHYSICS, Georg Joos, with Ira M. Freeman. Classic overview covers essential math, mechanics, electromagnetic theory, thermodynamics, quantum mechanics, nuclear physics, other topics. First paperback edition. xxiii + 885pp. 5⅜ × 8½.               65227-0 Pa. $21.95

**HANDBOOK OF MATHEMATICAL FUNCTIONS WITH FORMULAS, GRAPHS, AND MATHEMATICAL TABLES,** edited by Milton Abramowitz and Irene A. Stegun. Vast compendium: 29 sets of tables, some to as high as 20 places. 1,046pp. 8 × 10½. 61272-4 Pa. $24.95

**MATHEMATICAL METHODS IN PHYSICS AND ENGINEERING,** John W. Dettman. Algebraically based approach to vectors, mapping, diffraction, other topics in applied math. Also generalized functions, analytic function theory, more. Exercises. 448pp. 5⅜ × 8¼. 65649-7 Pa. $10.95

**A SURVEY OF NUMERICAL MATHEMATICS,** David M. Young and Robert Todd Gregory. Broad self-contained coverage of computer-oriented numerical algorithms for solving various types of mathematical problems in linear algebra, ordinary and partial, differential equations, much more. Exercises. Total of 1,248pp. 5⅜ × 8½. Two volumes. Vol. I 65691-8 Pa. $14.95
Vol. II 65692-6 Pa. $14.95

**TENSOR ANALYSIS FOR PHYSICISTS,** J.A. Schouten. Concise exposition of the mathematical basis of tensor analysis, integrated with well-chosen physical examples of the theory. Exercises. Index. Bibliography. 289pp. 5⅜ × 8½. 65582-2 Pa. $8.95

**INTRODUCTION TO NUMERICAL ANALYSIS (2nd Edition),** F.B. Hildebrand. Classic, fundamental treatment covers computation, approximation, interpolation, numerical differentiation and integration, other topics. 150 new problems. 669pp. 5⅜ × 8½. 65363-3 Pa. $15.95

**INVESTIGATIONS ON THE THEORY OF THE BROWNIAN MOVEMENT,** Albert Einstein. Five papers (1905–8) investigating dynamics of Brownian motion and evolving elementary theory. Notes by R. Fürth. 122pp. 5⅜ × 8½. 60304-0 Pa. $4.95

**CATASTROPHE THEORY FOR SCIENTISTS AND ENGINEERS,** Robert Gilmore. Advanced-level treatment describes mathematics of theory grounded in the work of Poincaré, R. Thom, other mathematicians. Also important applications to problems in mathematics, physics, chemistry and engineering. 1981 edition. References. 28 tables. 397 black-and-white illustrations. xvii + 666pp. 6⅛ × 9¼. 67539-4 Pa. $17.95

**AN INTRODUCTION TO STATISTICAL THERMODYNAMICS,** Terrell L. Hill. Excellent basic text offers wide-ranging coverage of quantum statistical mechanics, systems of interacting molecules, quantum statistics, more. 523pp. 5⅜ × 8½. 65242-4 Pa. $12.95

**STATISTICAL PHYSICS,** Gregory H. Wannier. Classic text combines thermodynamics, statistical mechanics and kinetic theory in one unified presentation of thermal physics. Problems with solutions. Bibliography. 532pp. 5⅜ × 8½. 65401-X Pa. $12.95

**ORDINARY DIFFERENTIAL EQUATIONS**, Morris Tenenbaum and Harry Pollard. Exhaustive survey of ordinary differential equations for undergraduates in mathematics, engineering, science. Thorough analysis of theorems. Diagrams. Bibliography. Index. 818pp. 5⅜ × 8½. 64940-7 Pa. $18.95

**STATISTICAL MECHANICS: Principles and Applications**, Terrell L. Hill. Standard text covers fundamentals of statistical mechanics, applications to fluctuation theory, imperfect gases, distribution functions, more. 448pp. 5⅜ × 8½. 65390-0 Pa. $11.95

**ORDINARY DIFFERENTIAL EQUATIONS AND STABILITY THEORY: An Introduction**, David A. Sánchez. Brief, modern treatment. Linear equation, stability theory for autonomous and nonautonomous systems, etc. 164pp. 5⅜ × 8¼. 63828-6 Pa. $6.95

**THIRTY YEARS THAT SHOOK PHYSICS: The Story of Quantum Theory**, George Gamow. Lucid, accessible introduction to influential theory of energy and matter. Careful explanations of Dirac's anti-particles, Bohr's model of the atom, much more. 12 plates. Numerous drawings. 240pp. 5⅜ × 8½. 24895-X Pa. $6.95

**THEORY OF MATRICES**, Sam Perlis. Outstanding text covering rank, non-singularity and inverses in connection with the development of canonical matrices under the relation of equivalence, and without the intervention of determinants. Includes exercises. 237pp. 5⅜ × 8½. 66810-X Pa. $8.95

**GREAT EXPERIMENTS IN PHYSICS: Firsthand Accounts from Galileo to Einstein**, edited by Morris H. Shamos. 25 crucial discoveries: Newton's laws of motion, Chadwick's study of the neutron, Hertz on electromagnetic waves, more. Original accounts clearly annotated. 370pp. 5⅜ × 8½. 25346-5 Pa. $10.95

**INTRODUCTION TO PARTIAL DIFFERENTIAL EQUATIONS WITH APPLICATIONS**, E.C. Zachmanoglou and Dale W. Thoe. Essentials of partial differential equations applied to common problems in engineering and the physical sciences. Problems and answers. 416pp. 5⅜ × 8½. 65251-3 Pa. $11.95

**BURNHAM'S CELESTIAL HANDBOOK**, Robert Burnham, Jr. Thorough guide to the stars beyond our solar system. Exhaustive treatment. Alphabetical by constellation: Andromeda to Cetus in Vol. 1; Chamaeleon to Orion in Vol. 2; and Pavo to Vulpecula in Vol. 3. Hundreds of illustrations. Index in Vol. 3. 2,000pp. 6⅛ × 9¼. 23567-X, 23568-8, 23673-0 Pa., Three-vol. set $44.85

**CHEMICAL MAGIC**, Leonard A. Ford. Second Edition, Revised by E. Winston Grundmeier. Over 100 unusual stunts demonstrating cold fire, dust explosions, much more. Text explains scientific principles and stresses safety precautions. 128pp. 5⅜ × 8½. 67628-5 Pa. $5.95

**AMATEUR ASTRONOMER'S HANDBOOK**, J.B. Sidgwick. Timeless, comprehensive coverage of telescopes, mirrors, lenses, mountings, telescope drives, micrometers, spectroscopes, more. 189 illustrations. 576pp. 5⅜ × 8¼. (Available in U.S. only) 24034-7 Pa. $11.95

SPECIAL FUNCTIONS, N.N. Lebedev. Translated by Richard Silverman. Famous Russian work treating more important special functions, with applications to specific problems of physics and engineering. 38 figures. 308pp. 5⅜ × 8½.
60624-4 Pa. $9.95

OBSERVATIONAL ASTRONOMY FOR AMATEURS, J.B. Sidgwick. Mine of useful data for observation of sun, moon, planets, asteroids, aurorae, meteors, comets, variables, binaries, etc. 39 illustrations. 384pp. 5⅜ × 8¼. (Available in U.S. only)
24033-9 Pa. $8.95

INTEGRAL EQUATIONS, F.G. Tricomi. Authoritative, well-written treatment of extremely useful mathematical tool with wide applications. Volterra Equations, Fredholm Equations, much more. Advanced undergraduate to graduate level. Exercises. Bibliography. 238pp. 5⅜ × 8½.
64828-1 Pa. $8.95

POPULAR LECTURES ON MATHEMATICAL LOGIC, Hao Wang. Noted logician's lucid treatment of historical developments, set theory, model theory, recursion theory and constructivism, proof theory, more. 3 appendixes. Bibliography. 1981 edition. ix + 283pp. 5⅜ × 8½.
67632-3 Pa. $8.95

MODERN NONLINEAR EQUATIONS, Thomas L. Saaty. Emphasizes practical solution of problems; covers seven types of equations. ". . . a welcome contribution to the existing literature. . . ."—Math Reviews. 490pp. 5⅜ × 8½. 64232-1 Pa. $11.95

FUNDAMENTALS OF ASTRODYNAMICS, Roger Bate et al. Modern approach developed by U.S. Air Force Academy. Designed as a first course. Problems, exercises. Numerous illustrations. 455pp. 5⅜ × 8½.
60061-0 Pa. $9.95

INTRODUCTION TO LINEAR ALGEBRA AND DIFFERENTIAL EQUATIONS, John W. Dettman. Excellent text covers complex numbers, determinants, orthonormal bases, Laplace transforms, much more. Exercises with solutions. Undergraduate level. 416pp. 5⅜ × 8½.
65191-6 Pa. $10.95

INCOMPRESSIBLE AERODYNAMICS, edited by Bryan Thwaites. Covers theoretical and experimental treatment of the uniform flow of air and viscous fluids past two-dimensional aerofoils and three-dimensional wings; many other topics. 654pp. 5⅜ × 8½.
65465-6 Pa. $16.95

INTRODUCTION TO DIFFERENCE EQUATIONS, Samuel Goldberg. Exceptionally clear exposition of important discipline with applications to sociology, psychology, economics. Many illustrative examples; over 250 problems. 260pp. 5⅜ × 8½.
65084-7 Pa. $8.95

LAMINAR BOUNDARY LAYERS, edited by L. Rosenhead. Engineering classic covers steady boundary layers in two- and three-dimensional flow, unsteady boundary layers, stability, observational techniques, much more. 708pp. 5⅜ × 8½.
65646-2 Pa. $18.95

LECTURES ON CLASSICAL DIFFERENTIAL GEOMETRY, Second Edition, Dirk J. Struik. Excellent brief introduction covers curves, theory of surfaces, fundamental equations, geometry on a surface, conformal mapping, other topics. Problems. 240pp. 5⅜ × 8½.
65609-8 Pa. $8.95

ROTARY-WING AERODYNAMICS, W.Z. Stepniewski. Clear, concise text covers aerodynamic phenomena of the rotor and offers guidelines for helicopter performance evaluation. Originally prepared for NASA. 537 figures. 640pp. 6⅛ × 9¼.
64647-5 Pa. $15.95

DIFFERENTIAL GEOMETRY, Heinrich W. Guggenheimer. Local differential geometry as an application of advanced calculus and linear algebra. Curvature, transformation groups, surfaces, more. Exercises. 62 figures. 378pp. 5⅜ × 8½.
63433-7 Pa. $9.95

INTRODUCTION TO SPACE DYNAMICS, William Tyrrell Thomson. Comprehensive, classic introduction to space-flight engineering for advanced undergraduate and graduate students. Includes vector algebra, kinematics, transformation of coordinates. Bibliography. Index. 352pp. 5⅜ × 8½.        65113-4 Pa. $9.95

A SURVEY OF MINIMAL SURFACES, Robert Osserman. Up-to-date, in-depth discussion of the field for advanced students. Corrected and enlarged edition covers new developments. Includes numerous problems. 192pp. 5⅜ × 8½.
64998-9 Pa. $8.95

ANALYTICAL MECHANICS OF GEARS, Earle Buckingham. Indispensable reference for modern gear manufacture covers conjugate gear-tooth action, gear-tooth profiles of various gears, many other topics. 263 figures. 102 tables. 546pp. 5⅜ × 8½.        65712-4 Pa. $14.95

SET THEORY AND LOGIC, Robert R. Stoll. Lucid introduction to unified theory of mathematical concepts. Set theory and logic seen as tools for conceptual understanding of real number system. 496pp. 5⅜ × 8¼.        63829-4 Pa. $12.95

A HISTORY OF MECHANICS, René Dugas. Monumental study of mechanical principles from antiquity to quantum mechanics. Contributions of ancient Greeks, Galileo, Leonardo, Kepler, Lagrange, many others. 671pp. 5⅜ × 8½.
65632-2 Pa. $14.95

FAMOUS PROBLEMS OF GEOMETRY AND HOW TO SOLVE THEM, Benjamin Bold. Squaring the circle, trisecting the angle, duplicating the cube: learn their history, why they are impossible to solve, then solve them yourself. 128pp. 5⅜ × 8½.        24297-8 Pa. $4.95

MECHANICAL VIBRATIONS, J.P. Den Hartog. Classic textbook offers lucid explanations and illustrative models, applying theories of vibrations to a variety of practical industrial engineering problems. Numerous figures. 233 problems, solutions. Appendix. Index. Preface. 436pp. 5⅜ × 8½.        64785-4 Pa. $11.95

CURVATURE AND HOMOLOGY, Samuel I. Goldberg. Thorough treatment of specialized branch of differential geometry. Covers Riemannian manifolds, topology of differentiable manifolds, compact Lie groups, other topics. Exercises. 315pp. 5⅜ × 8½.        64314-X Pa. $9.95

HISTORY OF STRENGTH OF MATERIALS, Stephen P. Timoshenko. Excellent historical survey of the strength of materials with many references to the theories of elasticity and structure. 245 figures. 452pp. 5⅜ × 8½. 61187-6 Pa. $12.95

GEOMETRY OF COMPLEX NUMBERS, Hans Schwerdtfeger. Illuminating, widely praised book on analytic geometry of circles, the Moebius transformation, and two-dimensional non-Euclidean geometries. 200pp. 5⅜ × 8¼.
63830-8 Pa. $8.95

MECHANICS, J.P. Den Hartog. A classic introductory text or refresher. Hundreds of applications and design problems illuminate fundamentals of trusses, loaded beams and cables, etc. 334 answered problems. 462pp. 5⅜ × 8½. 60754-2 Pa. $10.95

TOPOLOGY, John G. Hocking and Gail S. Young. Superb one-year course in classical topology. Topological spaces and functions, point-set topology, much more. Examples and problems. Bibliography. Index. 384pp. 5⅜ × 8¼.
65676-4 Pa. $10.95

STRENGTH OF MATERIALS, J.P. Den Hartog. Full, clear treatment of basic material (tension, torsion, bending, etc.) plus advanced material on engineering methods, applications. 350 answered problems. 323pp. 5⅜ × 8½. 60755-0 Pa. $9.95

ELEMENTARY CONCEPTS OF TOPOLOGY, Paul Alexandroff. Elegant, intuitive approach to topology from set-theoretic topology to Betti groups; how concepts of topology are useful in math and physics. 25 figures. 57pp. 5⅜ × 8½.
60747-X Pa. $3.95

ADVANCED STRENGTH OF MATERIALS, J.P. Den Hartog. Superbly written advanced text covers torsion, rotating disks, membrane stresses in shells, much more. Many problems and answers. 388pp. 5⅜ × 8½. 65407-9 Pa. $10.95

COMPUTABILITY AND UNSOLVABILITY, Martin Davis. Classic graduate-level introduction to theory of computability, usually referred to as theory of recurrent functions. New preface and appendix. 288pp. 5⅜ × 8½. 61471-9 Pa. $8.95

GENERAL CHEMISTRY, Linus Pauling. Revised 3rd edition of classic first-year text by Nobel laureate. Atomic and molecular structure, quantum mechanics, statistical mechanics, thermodynamics correlated with descriptive chemistry. Problems. 992pp. 5⅜ × 8½. 65622-5 Pa. $19.95

AN INTRODUCTION TO MATRICES, SETS AND GROUPS FOR SCIENCE STUDENTS, G. Stephenson. Concise, readable text introduces sets, groups, and most importantly, matrices to undergraduate students of physics, chemistry, and engineering. Problems. 164pp. 5⅜ × 8½. 65077-4 Pa. $7.95

THE HISTORICAL BACKGROUND OF CHEMISTRY, Henry M. Leicester. Evolution of ideas, not individual biography. Concentrates on formulation of a coherent set of chemical laws. 260pp. 5⅜ × 8½. 61053-5 Pa. $7.95

THE PHILOSOPHY OF MATHEMATICS: An Introductory Essay, Stephan Körner. Surveys the views of Plato, Aristotle, Leibniz & Kant concerning propositions and theories of applied and pure mathematics. Introduction. Two appendices. Index. 198pp. 5⅜ × 8½. 25048-2 Pa. $8.95

THE DEVELOPMENT OF MODERN CHEMISTRY, Aaron J. Ihde. Authoritative history of chemistry from ancient Greek theory to 20th-century innovation. Covers major chemists and their discoveries. 209 illustrations. 14 tables. Bibliographies. Indices. Appendices. 851pp. 5⅜ × 8½. 64235-6 Pa. $18.95

DE RE METALLICA, Georgius Agricola. The famous Hoover translation of greatest treatise on technological chemistry, engineering, geology, mining of early modern times (1556). All 289 original woodcuts. 638pp. 6¾ × 11.

60006-8 Pa. $18.95

SOME THEORY OF SAMPLING, William Edwards Deming. Analysis of the problems, theory and design of sampling techniques for social scientists, industrial managers and others who find statistics increasingly important in their work. 61 tables. 90 figures. xvii + 602pp. 5⅜ × 8½.

64684-X Pa. $15.95

THE VARIOUS AND INGENIOUS MACHINES OF AGOSTINO RAMELLI: A Classic Sixteenth-Century Illustrated Treatise on Technology, Agostino Ramelli. One of the most widely known and copied works on machinery in the 16th century. 194 detailed plates of water pumps, grain mills, cranes, more. 608pp. 9 × 12.

28180-9 Pa. $24.95

LINEAR PROGRAMMING AND ECONOMIC ANALYSIS, Robert Dorfman, Paul A. Samuelson and Robert M. Solow. First comprehensive treatment of linear programming in standard economic analysis. Game theory, modern welfare economics, Leontief input-output, more. 525pp. 5⅜ × 8½.

65491-5 Pa. $14.95

ELEMENTARY DECISION THEORY, Herman Chernoff and Lincoln E. Moses. Clear introduction to statistics and statistical theory covers data processing, probability and random variables, testing hypotheses, much more. Exercises. 364pp. 5⅜ × 8½.

65218-1 Pa. $10.95

THE COMPLEAT STRATEGYST: Being a Primer on the Theory of Games of Strategy, J.D. Williams. Highly entertaining classic describes, with many illustrated examples, how to select best strategies in conflict situations. Prefaces. Appendices. 268pp. 5⅜ × 8½.

25101-2 Pa. $7.95

CONSTRUCTIONS AND COMBINATORIAL PROBLEMS IN DESIGN OF EXPERIMENTS, Damaraju Raghavarao. In-depth reference work examines orthogonal Latin squares, incomplete block designs, tactical configuration, partial geometry, much more. Abundant explanations, examples. 416pp. 5⅜ × 8¼.

65685-3 Pa. $10.95

THE ABSOLUTE DIFFERENTIAL CALCULUS (CALCULUS OF TENSORS), Tullio Levi-Civita. Great 20th-century mathematician's classic work on material necessary for mathematical grasp of theory of relativity. 452pp. 5⅜ × 8½.

63401-9 Pa. $11.95

VECTOR AND TENSOR ANALYSIS WITH APPLICATIONS, A.I. Borisenko and I.E. Tarapov. Concise introduction. Worked-out problems, solutions, exercises. 257pp. 5⅜ × 8¼.

63833-2 Pa. $8.95

THE FOUR-COLOR PROBLEM: Assaults and Conquest, Thomas L. Saaty and Paul G. Kainen. Engrossing, comprehensive account of the century-old combinatorial topological problem, its history and solution. Bibliographies. Index. 110 figures. 228pp. 5⅜ × 8½. 65092-8 Pa. $6.95

CATALYSIS IN CHEMISTRY AND ENZYMOLOGY, William P. Jencks. Exceptionally clear coverage of mechanisms for catalysis, forces in aqueous solution, carbonyl- and acyl-group reactions, practical kinetics, more. 864pp. 5⅜ × 8½. 65460-5 Pa. $19.95

PROBABILITY: An Introduction, Samuel Goldberg. Excellent basic text covers set theory, probability theory for finite sample spaces, binomial theorem, much more. 360 problems. Bibliographies. 322pp. 5⅜ × 8½. 65252-1 Pa. $9.95

LIGHTNING, Martin A. Uman. Revised, updated edition of classic work on the physics of lightning. Phenomena, terminology, measurement, photography, spectroscopy, thunder, more. Reviews recent research. Bibliography. Indices. 320pp. 5⅜ × 8¼. 64575-4 Pa. $8.95

PROBABILITY THEORY: A Concise Course, Y.A. Rozanov. Highly readable, self-contained introduction covers combination of events, dependent events, Bernoulli trials, etc. Translation by Richard Silverman. 148pp. 5⅜ × 8¼. 63544-9 Pa. $6.95

AN INTRODUCTION TO HAMILTONIAN OPTICS, H. A. Buchdahl. Detailed account of the Hamiltonian treatment of aberration theory in geometrical optics. Many classes of optical systems defined in terms of the symmetries they possess. Problems with detailed solutions. 1970 edition. xv + 360pp. 5⅜ × 8½. 67597-1 Pa. $10.95

STATISTICS MANUAL, Edwin L. Crow, et al. Comprehensive, practical collection of classical and modern methods prepared by U.S. Naval Ordnance Test Station. Stress on use. Basics of statistics assumed. 288pp. 5⅜ × 8½. 60599-X Pa. $7.95

DICTIONARY/OUTLINE OF BASIC STATISTICS, John E. Freund and Frank J. Williams. A clear concise dictionary of over 1,000 statistical terms and an outline of statistical formulas covering probability, nonparametric tests, much more. 208pp. 5⅜ × 8½. 66796-0 Pa. $7.95

STATISTICAL METHOD FROM THE VIEWPOINT OF QUALITY CONTROL, Walter A. Shewhart. Important text explains regulation of variables, uses of statistical control to achieve quality control in industry, agriculture, other areas. 192pp. 5⅜ × 8½. 65232-7 Pa. $7.95

THE INTERPRETATION OF GEOLOGICAL PHASE DIAGRAMS, Ernest G. Ehlers. Clear, concise text emphasizes diagrams of systems under fluid or containing pressure; also coverage of complex binary systems, hydrothermal melting, more. 288pp. 6½ × 9¼. 65389-7 Pa. $10.95

STATISTICAL ADJUSTMENT OF DATA, W. Edwards Deming. Introduction to basic concepts of statistics, curve fitting, least squares solution, conditions without parameter, conditions containing parameters. 26 exercises worked out. 271pp. 5⅜ × 8½. 64685-8 Pa. $9.95

TENSOR CALCULUS, J.L. Synge and A. Schild. Widely used introductory text covers spaces and tensors, basic operations in Riemannian space, non-Riemannian spaces, etc. 324pp. 5⅜ × 8¼. 63612-7 Pa. $9.95

A CONCISE HISTORY OF MATHEMATICS, Dirk J. Struik. The best brief history of mathematics. Stresses origins and covers every major figure from ancient Near East to 19th century. 41 illustrations. 195pp. 5⅜ × 8½. 60255-9 Pa. $7.95

A SHORT ACCOUNT OF THE HISTORY OF MATHEMATICS, W.W. Rouse Ball. One of clearest, most authoritative surveys from the Egyptians and Phoenicians through 19th-century figures such as Grassman, Galois, Riemann. Fourth edition. 522pp. 5⅜ × 8½. 20630-0 Pa. $11.95

HISTORY OF MATHEMATICS, David E. Smith. Nontechnical survey from ancient Greece and Orient to late 19th century; evolution of arithmetic, geometry, trigonometry, calculating devices, algebra, the calculus. 362 illustrations. 1,355pp. 5⅜ × 8½. 20429-4, 20430-8 Pa., Two-vol. set $26.90

THE GEOMETRY OF RENÉ DESCARTES, René Descartes. The great work founded analytical geometry. Original French text, Descartes' own diagrams, together with definitive Smith-Latham translation. 244pp. 5⅜ × 8½. 60068-8 Pa. $7.95

THE ORIGINS OF THE INFINITESIMAL CALCULUS, Margaret E. Baron. Only fully detailed and documented account of crucial discipline: origins; development by Galileo, Kepler, Cavalieri; contributions of Newton, Leibniz, more. 304pp. 5⅜ × 8½. (Available in U.S. and Canada only) 65371-4 Pa. $9.95

THE HISTORY OF THE CALCULUS AND ITS CONCEPTUAL DEVELOP-MENT, Carl B. Boyer. Origins in antiquity, medieval contributions, work of Newton, Leibniz, rigorous formulation. Treatment is verbal. 346pp. 5⅜ × 8½. 60509-4 Pa. $9.95

THE THIRTEEN BOOKS OF EUCLID'S ELEMENTS, translated with introduction and commentary by Sir Thomas L. Heath. Definitive edition. Textual and linguistic notes, mathematical analysis. 2,500 years of critical commentary. Not abridged. 1,414pp. 5⅜ × 8½. 60088-2, 60089-0, 60090-4 Pa., Three-vol. set $31.85

GAMES AND DECISIONS: Introduction and Critical Survey, R. Duncan Luce and Howard Raiffa. Superb nontechnical introduction to game theory, primarily applied to social sciences. Utility theory, zero-sum games, n-person games, decision-making, much more. Bibliography. 509pp. 5⅜ × 8½. 65943-7 Pa. $12.95

THE HISTORICAL ROOTS OF ELEMENTARY MATHEMATICS, Lucas N.H. Bunt, Phillip S. Jones, and Jack D. Bedient. Fundamental underpinnings of modern arithmetic, algebra, geometry and number systems derived from ancient civilizations. 320pp. 5⅜ × 8½. 25563-8 Pa. $8.95

CALCULUS REFRESHER FOR TECHNICAL PEOPLE, A. Albert Klaf. Covers important aspects of integral and differential calculus via 756 questions. 566 problems, most answered. 431pp. 5⅜ × 8½. 20370-0 Pa. $8.95

CHALLENGING MATHEMATICAL PROBLEMS WITH ELEMENTARY SOLUTIONS, A.M. Yaglom and I.M. Yaglom. Over 170 challenging problems on probability theory, combinatorial analysis, points and lines, topology, convex polygons, many other topics. Solutions. Total of 445pp. 5⅜ × 8½. Two-vol. set.
Vol. I 65536-9 Pa. $7.95
Vol. II 65537-7 Pa. $7.95

FIFTY CHALLENGING PROBLEMS IN PROBABILITY WITH SOLUTIONS, Frederick Mosteller. Remarkable puzzlers, graded in difficulty, illustrate elementary and advanced aspects of probability. Detailed solutions. 88pp. 5⅜ × 8½.
65355-2 Pa. $4.95

EXPERIMENTS IN TOPOLOGY, Stephen Barr. Classic, lively explanation of one of the byways of mathematics. Klein bottles, Moebius strips, projective planes, map coloring, problem of the Koenigsberg bridges, much more, described with clarity and wit. 43 figures. 210pp. 5⅜ × 8½. 25933-1 Pa. $6.95

RELATIVITY IN ILLUSTRATIONS, Jacob T. Schwartz. Clear nontechnical treatment makes relativity more accessible than ever before. Over 60 drawings illustrate concepts more clearly than text alone. Only high school geometry needed. Bibliography. 128pp. 6⅛ × 9¼. 25965-X Pa. $7.95

AN INTRODUCTION TO ORDINARY DIFFERENTIAL EQUATIONS, Earl A. Coddington. A thorough and systematic first course in elementary differential equations for undergraduates in mathematics and science, with many exercises and problems (with answers). Index. 304pp. 5⅜ × 8½. 65942-9 Pa. $8.95

FOURIER SERIES AND ORTHOGONAL FUNCTIONS, Harry F. Davis. An incisive text combining theory and practical example to introduce Fourier series, orthogonal functions and applications of the Fourier method to boundary-value problems. 570 exercises. Answers and notes. 416pp. 5⅜ × 8½. 65973-9 Pa. $11.95

AN INTRODUCTION TO ALGEBRAIC STRUCTURES, Joseph Landin. Superb self-contained text covers "abstract algebra": sets and numbers, theory of groups, theory of rings, much more. Numerous well-chosen examples, exercises. 247pp. 5⅜ × 8½. 65940-2 Pa. $8.95

*Prices subject to change without notice.*
Available at your book dealer or write for free Mathematics and Science Catalog to Dept. GI, Dover Publications, Inc., 31 East 2nd St., Mineola, N.Y. 11501. Dover publishes more than 175 books each year on science, elementary and advanced mathematics, biology, music, art, literature, history, social sciences and other areas.